Tradeoffs or Synergies? Agricultural Intensification, Economic Development and the Environment

Tradeoffs or Synergies? Agricultural Intensification, Economic Development and the Environment

Edited by

D.R. Lee and C.B. Barrett

Department of Applied Economics and Management
Cornell University
Ithaca
New York
USA

CABI *Publishing*

CABI *Publishing* is a division of CAB *International*

CABI Publishing
CAB International
Wallingford
Oxon OX10 8DE
UK

Tel: +44 (0)1491 832111
Fax: +44 (0)1491 833508
Email: cabi@cabi.org
Web site: http://www.cabi.org

CABI Publishing
10 E 40th Street
Suite 3203
New York, NY 10016
USA

Tel: +1 212 481 7018
Fax: +1 212 686 7993
Email: cabi-nao@cabi.org

A catalogue record for this book is available from the British Library, London, UK.

Library of Congress Cataloging-in-Publication Data

Tradeoffs or synergies? : agricultural intensification, economic development, and the environment / edited by D.R. Lee and C.B. Barrett.

p. cm.

Based on papers presented at an international conference held in Salt Lake City, Utah, late July-early Aug. 1998.

Includes bibliographical references.

ISBN 0-85199-435-0 (alk.)

1. Agricultural intensification--Developing countries. 2. Agricultural ecology--Developing countries. 3. Rural development--Developing countries I. Lee, David R. (David Robinson), 1950- II. Barrett, Christopher B. (Christopher Brendan)

S482.T72 2000
338.1'6'091724--dc21

00-030396

ISBN 0 85199 435 0

Typeset by AMA DataSet Ltd, UK
Printed and bound in the UK by Biddles Ltd, Guildford and King's Lynn

Contents

Contributors		ix
Preface		xv
Foreword		xix
1	**Introduction: Changing Perspectives on Agricultural Intensification, Economic Development and the Environment** *David R. Lee, Paul J. Ferraro and Christopher B. Barrett*	1
Part I:	**Balancing Food Production, Economic and Environmental Goals: Concepts and Methods**	
2	**The Doubly Green Revolution: Balancing Food, Poverty and Environmental Needs in the 21st Century** *Gordon Conway*	17
3	**Population, Agricultural Land Use and the Environment in Developing Countries** *Richard E. Bilsborrow and David L. Carr*	35
4	**The Economics of Biodiversity Loss and Agricultural Development in Low-income Countries** *Charles Perrings*	57
5	**Farm Household Intensification Decisions and the Environment** *Stefano Pagiola and Stein Holden*	73

6 **When Does Technological Change in Agriculture Promote Deforestation?** 89
Arild Angelsen and David Kaimowitz

7 **Bioeconomic Models and Ecoregional Development: Policy Instruments for Sustainable Intensification** 115
Ruerd Ruben, Arie Kuyvenhoven and Gideon Kruseman

8 **Tradeoffs in Agriculture, the Environment and Human Health: Decision Support for Policy and Technology Managers** 135
Charles C. Crissman, John M. Antle and Jetse J. Stoorvogel

Part II: Empirical Studies of Tradeoffs and Synergies in Agricultural Intensification, Economic Development and the Environment

9 **Balancing Regional Development Priorities to Achieve Sustainable and Equitable Agricultural Growth** 151
Peter Hazell and Shenggen Fan

10 **Pathways of Development in the Hillside Areas of Honduras: Causes and Implications for Agricultural Production, Poverty and Sustainable Resource Use** 171
John Pender, Sara J. Scherr and Guadalupe Durón

11 **Implications of Resource-use Intensification for the Environment and Sustainable Technology Systems in the Central African Rainforest** 197
James Gockowski, G. Blaise Nkamleu and John Wendt

12 **Agricultural Intensification, Deforestation and the Environment: Assessing Tradeoffs in Sumatra, Indonesia** 221
Thomas P. Tomich, Meine van Noordwijk, Suseno Budidarsono, Andy Gillison, Trikurniati Kusumanto, Daniel Murdiyarso, Fred Stolle and Achmad M. Fagi

13 **Intensifying Small-scale Agriculture in the Western Brazilian Amazon: Issues, Implications and Implementation** 245
Stephen A. Vosti, Julie Witcover, Chantal Line Carpentier, Samuel José Magalhães de Oliveira and Jair Carvalho dos Santos (with collaborators)

14 **Integrated Bioeconomic Land-use Models: an Analysis of Policy Issues in the Atlantic Zone of Costa Rica** 267
Robert A. Schipper, Hans G.P. Jansen, Bas A.M. Bouman, Huib Hengsdijk, André Nieuwenhuyse and Fernando Sáenz

15 **Land Tenure and the Management of Land and Trees: a Comparative Study of Asia and Africa** 285
Keijiro Otsuka and Frank Place

16 **Sustainable Agriculture and Natural Resource Management in India's Semi-arid Tropics** 303
John Kerr, Ganesh Pangare, Vasudha Lokur Pangare and P.J. George

Part III: Evaluating Tradeoffs and Synergies: Technology, Institutions and Policies

17 **Soil Fertility, Small-farm Intensification and the Environment in Africa** 325
Pedro A. Sanchez, Bashir Jama, Amadou I. Niang and Cheryl A. Palm

18 **Livestock–Environment Interactions under Intensifying Production** 345
Steven J. Staal, Simeon Ehui and Jon Tanner

19 **Sustainable versus Unsustainable Agricultural Intensification in Africa: Focus on Policy Reforms and Market Conditions** 365
Thomas Reardon, Christopher B. Barrett, Valerie Kelly and Kimseyinga Savadogo

20 **Intensive Food Systems in Asia: Can the Degradation Problems be Reversed?** 383
Prabhu L. Pingali and Mark W. Rosegrant

21 **Biodiversity and Agricultural Development: the Crucial Institutional Issues** 399
Jeffrey A. McNeely

22 **Moving Beyond Integrated Conservation and Development Projects (ICDPs) to Achieve Biodiversity Conservation** 417
Katrina Brandon

23 **Balancing Development and Environmental Goals through Community-based Natural Resource Management** 433
Norman Uphoff

24 **Assessing Tradeoffs and Synergies among Agricultural Intensification, Economic Development and Environmental Goals: Conclusions and Implications for Policy** 451
David R. Lee, Christopher B. Barrett, Peter Hazell and Douglas Southgate

References 465

Index 521

Contributors

Arild Angelsen *is Associate Professor in the Department of Economics and Social Sciences, Agricultural University of Norway, PO Box 5033, N-1432, Ås, Norway.*

John M. Antle *is Professor and Director of the Trade Research Center at Montana State University, 312 Lindfield Hall, Bozeman, MT 59717, USA.*

Christopher B. Barrett *is Associate Professor in the Department of Applied Economics and Management, 351 Warren Hall, Cornell University, Ithaca, NY 14853, USA.*

Richard E. Bilsborrow *is Research Professor in the Carolina Population Center, University of North Carolina at Chapel Hill, Campus Box 8120, University Square, 123 West Franklin Street, Chapel Hill, NC 27516-3997, USA.*

Bas A.M. Bouman *is a Water Scientist at the International Rice Research Institute (IRRI), PO Box 3127, Makati Central Post Office, 1271 Makati City, Philippines.*

Katrina Brandon *is a Senior Research Fellow at the Center for Applied Biodiversity Science, Conservation International, 2501 M Street, NW, Suite 200, Washington, DC 20037, USA.*

Suseno Budidarsono *is Associate Research Officer with the International Centre for Research in Agroforestry (ICRAF – SE Asia), PO Box 161, Bogor 16001, Indonesia.*

Chantal Line Carpentier *is Program Manager in the Environment, Economy and Trade division at the Commission for Environmental Cooperation (CEC), 393 rue St Jacques O., Bureau 200, Montréal, Quebec, H2Y 1N9, Canada.*

David L. Carr *is a PhD candidate in the Department of Geography at the University of North Carolina at Chapel Hill, Campus Box 8120,*

University Square, 123 West Franklin Street, Chapel Hill, NC 27516-3997, USA.

Gordon Conway *is President of the Rockefeller Foundation, 420 Fifth Avenue, New York, NY 10018-2702, USA.*

Charles C. Crissman *is Economist and Country Representative for the International Potato Center (CIP), Box 17-21-1977, Quito, Ecuador.*

Samuel J.M. de Oliveira *is Director of Research, Empresa Brasileira de Pesquisa Agropecuária (Embrapa), Rondônia, Brazil.*

Jair Carvalho dos Santos *is a researcher with the Empresa Brasileira de Pesquisa Agropecuária (Embrapa), Acre, Brazil.*

Guadalupe Durón *is a Research Assistant at the International Center for Women, 1717 Massachusetts Ave., NW, Suite 302, Washington, DC 20036, USA.*

Simeon Ehui *is Programme Coordinator of the Livestock Policy Analysis Programme, International Livestock Research Institute (ILRI), PO Box 5689, Addis Ababa, Ethiopia.*

Achmad M. Fagi *is South-east Asian Regional Coordinator for the Alternatives to Slash-and-Burn Programme, Agency for Agricultural Research and Development, Indonesia.*

Shenggen Fan *is a Research Fellow in the Environment and Production Technology Division, International Food Policy Research Institute (IFPRI), 2033 K Street NW, Washington, DC 20006, USA.*

Paul J. Ferraro *is a PhD candidate in resource economics in the Department of Applied Economics and Management, 248 Warren Hall, Cornell University, Ithaca, NY 14853, USA.*

P.J. George *is a Consultant with Farming Systems International, Hill View Apartments, Near NGO Quarters, Trikakkara, Cochin-21, Kerala, India.*

Andy Gillison *is Senior Associate Scientist with the Centre for International Forestry Research (CIFOR), Bogor, Indonesia.*

James Gockowski *is Scientist (Agricultural Economics) at the Humid Forest Ecoregional Centre of the International Institute of Tropical Agriculture (IITA), BP 2008 (Messa), Yaoundé, Cameroon.*

Peter Hazell *is Director of the Environment and Production Technology Division, International Food Policy Research Institute (IFPRI), 2033 K Street NW, Washington, DC 20006, USA.*

Huib Hengsdijk *is a Production Ecology Scientist in the Laboratory of Theoretical Production Ecology, Department of Crop Science, Wageningen University, PO Box 430, 6700 AK Wageningen, The Netherlands.*

Stein Holden *is Associate Professor in the Department of Economics and Social Sciences, Agricultural University of Norway, PO Box 5033, N-1432 Ås, Norway.*

Bashir Jama *is Senior Scientist at the International Centre for Research in Agroforestry (ICRAF), PO Box 30677, Nairobi, Kenya.*

Hans G.P. Jansen *is Senior Agricultural Development Economist at the Agricultural Economics Research Institute (LEI – Wageningen UR), PO Box 29703, 2502 LS The Hague, The Netherlands.*

David Kaimowitz *is Principal Economist at the Centre for International Forestry Research (CIFOR), PO Box 6596 JKPWB, Jakarta 10065, Indonesia.*

Valerie Kelly *is Visiting Associate Professor in the Department of Agricultural Economics, Michigan State University, Agricultural Hall, East Lansing, MI 48824, USA.*

John Kerr *is Assistant Professor in the Department of Resource Development at Michigan State University, 323 Natural Resources, East Lansing, MI 48824, USA.*

Gideon Kruseman *is a PhD researcher in the Development Economics Group, Department of Social Sciences, Wageningen University, Hollandseweg 1, 6706 KN, Wageningen, The Netherlands.*

Trikurniati Kusumanto *is Research Officer with the International Centre for Research in Agroforestry (ICRAF – SE Asia), PO Box 161, Bogor 16001, Indonesia.*

Arie Kuyvenhoven *is Professor in the Development Economics Group, Department of Economics and Management, Wageningen University, Hollandseweg 1, 6706 KN, Wageningen, The Netherlands.*

David R. Lee *is Professor of Agricultural Economics in the Department of Applied Economics and Management, 248 Warren Hall, Cornell University, Ithaca, NY 14853, USA.*

Jeffrey A. McNeely *is Chief Scientist at IUCN – The World Conservation Union, 1196 Gland, Switzerland.*

Daniel Murdiyarso *is Head of the BIOTROP-GCTE Impacts Centre for South-east Asia (IC-SEA), Bogor, Indonesia.*

Amadou I. Niang *is Principal Scientist at the International Centre for Research in Agroforestry (ICRAF), PO Box 30677, Nairobi, Kenya.*

André Nieuwenhuyse *is a Soil Scientist and GIS specialist in Proyecto ZONISIG, Casilla 14533, La Paz, Bolivia.*

Blaise Nkamleu *is a Research Associate in agricultural economics at the Humid Forest Ecoregional Centre of the International Institute of Tropical Agriculture (IITA), BP 2008 (Messa), Yaoundé, Cameroon.*

Keijiro Otsuka *is Professor in the Economics Department at Tokyo Metropolitan University (1-1 Minami-Ohsawa, Hachiohji, Tokyo 192-0397, Japan) and a Visiting Research Fellow at the International Food Policy Research Institute (IFPRI; 2033 K Street NW, Washington, DC 20006, USA).*

Stefano Pagiola *is an Economist in the Environment Department of the World Bank, 1818 H Street NW, Washington, DC 20433, USA.*

Cheryl A. Palm *is Senior Scientific Officer in the Tropical Soil Biology and Fertility Programme, PO Box 30592, Nairobi, Kenya.*

Ganesh Pangare *is Executive Director of the Indian Network for Participatory Irrigation Management, Room No. 218 (Old Building), Central Soil and*

Materials Research Station, Olafe Palme Marg, Hauz Khas, New Delhi 110 016, India.

Vasudha Lokur Pangare *is a Consultant with OIKOS, 1st Floor M-60, Saket, New Delhi 110 017, India.*

John Pender *is a Research Fellow in the Environment and Production Technology Division, International Food Policy Research Institute (IFPRI), 2033 K Street NW, Washington, DC 20006, USA.*

Charles Perrings *is Professor and Head of the Environment Department, University of York, Heslington, York YO10 5DD, UK.*

Prabhu L. Pingali *is Director of the Economics Program at the International Maize and Wheat Improvement Center (CIMMYT), Lisboa 27, Apdo. Postal 6-641, 06600 Mexico DF, Mexico.*

Frank Place *is an Agricultural Economist at the International Centre for Research in Agroforestry (ICRAF), PO Box 30677, Nairobi, Kenya.*

Thomas Reardon *is Associate Professor in the Department of Agricultural Economics, 211F Agriculture Hall, Michigan State University, East Lansing, MI 48824-1039, USA.*

Mark W. Rosegrant *is Senior Research Fellow in the Environment and Production Technology Division, International Food Policy Research Institute (IFPRI), 2033 K Street NW, Washington, DC 20006, USA.*

Ruerd Ruben *is Associate Professor in the Development Economics Group, Department of Social Sciences, Wageningen University, Hollandseweg 1, 6706 KN, Wageningen, The Netherlands.*

Fernando Sáenz *is an Economist at the Centro Internacional de Política Económica, Universidad Nacional Autónoma, Apartado 555-3000, Heredia, Costa Rica.*

Pedro A. Sanchez *is Director General of the International Centre for Research in Agroforestry (ICRAF), PO Box 30677, Nairobi, Kenya.*

Kimseyinga Savodogo *is Professor Aggrège and Dean of the School of Economics, University of Ouagadougou, 03 BP 7021, Ouagadougou 03, Burkina Faso.*

Sara J. Scherr *is a Visiting Fellow, Department of Agricultural and Resource Economics, Symons Hall, University of Maryland, College Park, MD 20742-5500, USA.*

Robert A. Schipper *is an Agricultural Economist and Lecturer in the Development Economics Group, Department of Social Sciences, Wageningen University, PO Box 8130, 6700 EW Wageningen, The Netherlands.*

Douglas Southgate *is Professor in the Department of Agricultural Economics at Ohio State University, 2120 Fyffe Road, Columbus, OH 43210, USA.*

Steven J. Staal *is an Agricultural Economist with the International Livestock Research Institute (ILRI), PO Box 30709, Nairobi, Kenya.*

Fred Stolle *is Associate Geographer with the International Centre for Research in Agroforestry (ICRAF – SE Asia), PO Box 161, Bogor 16001, Indonesia.*

Jetse J. Stoorvogel *is an Instructor at the Laboratory of Soil Science and Geology, Wageningen University, PO Box 37, 6700 AA Wageningen, The Netherlands.*

Jon Tanner *is an Animal Nutritionist with the International Livestock Research Institute (ILRI), PO Box 30709, Nairobi, Kenya.*

Thomas P. Tomich *is Principal Economist and Global Coordinator, Alternatives to Slash-and-Burn Programme (ASB), International Centre for Research in Agroforestry (ICRAF), United Nations Avenue, Gigiri, PO Box 30677, Nairobi, Kenya.*

Norman Uphoff *is Professor and Director of the Cornell International Institute for Food, Agriculture and Development, Box 14, Kennedy Hall, Cornell University, Ithaca, NY 14853, USA.*

Meine van Noordwijk *is Principal Soil Scientist with the International Centre for Research in Agroforestry (ICRAF – SE Asia), PO Box 161, Bogor 16001, Indonesia.*

Stephen A. Vosti *is Visiting Assistant Professor in the Department of Agricultural and Resource Economics, University of California, One Shields Avenue, Davis, CA 95616-5270, USA.*

John Wendt *is Scientist (Soil Chemistry) at the Humid Forest Ecoregional Centre of the International Institute of Tropical Agriculture (IITA), BP 2008 (Messa), Yaoundé, Cameroon.*

Julie Witcover *is a PhD candidate in the Department of Agricultural and Resource Economics, University of California, One Shields Avenue, Davis, CA 95616-5270, USA.*

Preface

In recent years, issues associated with the linkages and interdependence among the multiple goals of agricultural and rural development have achieved a high degree of prominence among development researchers, practitioners and policy-makers. Among the multiple objectives facing developing nations are longstanding concerns about generating adequate food production and income growth to feed the millions of poor and impoverished people who continue to be hungry and malnourished. In addition, though, growing concerns about other human needs and environmental sustainability have broadened the set of development goals which it is expected that technologies, policies and institutions will address. These include the reduction of poverty, income and employment growth, the long-term sustainability of farming systems and the soils on which they are based, and improvement across a broad set of environmental indicators ranging from reducing pollution, runoff and resource depletion resulting from agriculture to maintaining on-farm and local biodiversity and successfully addressing global concerns such as greenhouse-gas emissions and global warming.

Broad-based recognition of the severity of and linkages among these problems as well as the challenges they present for researchers and policy-makers has been heightened by a series of high-profile international conferences, policy initiatives and reports that have emphasized environment and development interactions. These have included the 1972 United Nations Conference on the Human Environment, the United Nations Educational, Scientific and Cultural Organization's (UNESCO) Man and the Biosphere Programme, the 1987 Report of the World Commission on Environment and Development (the 'Brundtland Report'), the 1992 United Nations Conference on Environment and Development (the 'Rio Earth Summit'), policy documents written by major international conservation organizations (such as the International Union for the Conservation of Nature–United Nations Environment Programme–World Wildlife Fund (IUCN–UNEP–WWF) *Caring for the Earth* report in 1991) and many others. While the problem diagnoses and policy prescriptions issuing from these initiatives differ, many have echoed similar themes regarding the critical linkages between economic development and environmental outcomes. Some have gone even further, in assuming or asserting (often based on little evidence) that synergistic relations between development and environmental outcomes commonly characterize the development process and thus that achieving 'win–win' outcomes is straightforward.

Simultaneous with these developments – and impelled and influenced by them – have been numerous developments in research and applied development, often at the farm or household level, which have also recognized the importance of agronomic–economic–environmental linkages and have attempted to address them in more formal ways. Developments in applied research have included the emergence of bioeconomic and ecoregional modelling, varying systems approaches in the biophysical sciences, such as soil ecology and systems ecology, and the emergence of agroecological analysis. Within the community of development and conservation practitioners, there has been increasing mutual recognition of the fact that human needs and environmental conservation goals in developing countries are inextricably interlinked and that both must be addressed, often simultaneously. This has led to the broadening of earlier efforts, which had focused mostly on biodiversity conservation, to incorporate economic development goals through such strategies as integrated development and conservation projects and, more recently, the emergence of 'ecoregional' or 'landscape' approaches to conservation and development.

Growing recognition of the importance of the 'tradeoffs versus synergies' theme – along with the emergence of analytical methods and approaches which have been applied to a variety of specific problems and regions – stimulated the plans for this volume. Most of this book's chapters are drawn from papers presented at an international conference on agricultural intensification, economic development and the environment held in Salt Lake City, Utah, in late July–early August 1998. Conference participants included a broad range of social scientists, biophysical scientists, development practitioners and representatives of international agencies, private foundations, and conservation and development organizations. In addition to discussion of earlier versions of many of these chapters, the conference focused on synthesizing their results and on the implications for policy and research. The editors would like to thank all of the conference participants, and especially those whose work is incorporated in this volume, for their contributions to both the conference and this book.

We are particularly grateful to the sponsors of the earlier conference, whose generous assistance and support were responsible for both its success and the subsequent publication of this volume. In this connection, we especially wish to thank Per Pinstrup-Andersen and Peter Hazell of the International Food Policy Research Institute, Walter Armbruster of the Farm Foundation, Norman Uphoff of the Cornell International Institute for Food, Agriculture and Development, Wallace Tyner of Purdue University and former Chair of the International Committee of the American Agricultural Economics Association (AAEA), the AAEA Executive Board and the AAEA Foundation Board. For unstinting assistance in conference planning and support, we wish to express sincere thanks to AAEA staff members Lona Christoffers and Nancy Herselius and to student assistants Travis Lybbert and Shane Sherlund.

In addition to the authors and co-authors of the chapters included in this volume, we wish to thank many others who provided input in the planning and completion of the earlier conference and who, directly or indirectly, contributed in important ways to this volume. These include Jock Anderson, Louise Buck, Derek Byerlee, Chris Delgado, John Dixon, Erick Fernandes, Mark Freudenberger, Robert Herdt, Randall Kramer, Alex McCalla, John Schelhas and Ann Thrupp.

We also wish to express our gratitude to those whose dedicated efforts contributed substantially to the production of this book: Jonell Blakeley, Dan Chapman, Paul Ferraro, Nelson Villoria, Joyce Knuutila and Jim Houghton. In particular, we are greatly indebted to Maria Burdett, whose detailed and painstaking technical editing greatly improved the quality of this volume. Sincere thanks also go to Tim Hardwick and Rachel Robinson, our editors at CABI *Publishing*, for their assistance and patience in the course of completing this book.

Finally, the senior editor would like to thank Cornell University for the opportunity to spend a sabbatical leave at Wageningen University in the Netherlands, during which much of the work involved in completing this book was accomplished. Special thanks also go to members of the Development Economics Group at Wageningen – in particular, Arie Kuyvenhoven, Ruerd Ruben and Rob Schipper – for helping to provide a most stimulating and hospitable environment.

Foreword

Developing countries must produce more food in the future. According to projections of the International Food Policy Research Institute (IFPRI), the demand for cereals, roots and tubers will increase by more than 50% by 2020, and developing countries will double their demand for livestock commodities. IFPRI projections further show that developing country production will not keep pace with demand and these countries are likely to almost double their net imports of food by the year 2020. Although self-sufficiency in food is not usually a sound goal, it is clear that most low-income developing countries would benefit from producing greater quantities of food and other agricultural commodities in the future.

One of the critical questions that these countries face is whether agricultural production can be intensified without doing harm to the environment. Are there conditions under which intensification could actually lead to improved environmental outcomes or will countries always be faced with tradeoffs between the intensification of agriculture and achieving environmental goals?

Poverty can also be a major source of natural resource degradation. Thus, if agricultural intensification reduces poverty, one might expect that the environment would benefit as well. Of course, the real world is not quite so simple. Agricultural intensification can benefit natural resource utilization in a 'win–win' outcome or it can do great harm.

This book provides much needed knowledge about how agricultural intensification, economic development and natural resource management interact under various circumstances. It improves our understanding of how policies and institutions can best be designed and implemented to achieve the triple goal of agricultural intensification, agricultural development and the sustainable management of natural resources. It deserves close scrutiny by those who are responsible for designing and implementing such policies and institutions and those who wish to understand this relationship better.

Per Pinstrup-Andersen, *Director-General, International Food Policy Research Institute (IFPRI), Washington DC, USA*

1 Introduction: Changing Perspectives on Agricultural Intensification, Economic Development and the Environment

DAVID R. LEE, PAUL J. FERRARO AND CHRISTOPHER B. BARRETT

Department of Applied Economics and Management, Cornell University, Ithaca, New York, USA

Introduction

Growing concerns over the sustainability of agriculture and natural resource management in the developing world have, in recent years, focused attention on the multiple challenges confronting growth-promoting development strategies and have broadened the expectations of those strategies. The Green Revolution of the 1960s and 1970s was considered a success by most observers, in that it achieved major gains in food production and food security in many low-income countries where population growth had previously outstripped the growth in crop yields. Concerns about equity and environmental impacts were raised by some, but the consensus view was that these were secondary to alleviating hunger and malnutrition through increasing food production and agricultural productivity.

Over the past two decades, however, this production-centred view of agricultural development objectives has been challenged from a variety of sources. Agricultural and rural development strategies are still widely expected to intensify agricultural production and enhance rural food security through food production and income growth. In addition, however, development strategies are now commonly expected to address a broader range of concerns, such as poverty reduction and employment generation, and to be environmentally sustainable – that is, not unduly compromising the natural resource base of future generations of users. One prominent contribution to this debate has, for example, posited a 'critical triangle' of development goals: agricultural growth, poverty alleviation and sustainable resource use (Vosti and Reardon, 1997). The continuing debate over achieving sustainable growth has at times emphasized one objective over another, but a widely held view is that agricultural and rural development strategies should contribute, at one level or another, to the enhancement of these multiple objectives.

The broadened agenda of economic and environmental sustainability has had many sources and has been evidenced in a variety of ways. A series of major global policy initiatives in the 1980s and 1990s have highlighted the broad set of concerns facing rural development in both developing and industrialized countries. A range of approaches taken in different academic fields – ecoregional analysis, farming systems analysis, bioeconomic modelling, systems ecology – have contributed to basic and applied research, emphasizing systemic approaches to solving agricultural and environmental problems and the linkages tying agricultural production to economic and ecosystem outcomes. In applied development circles, heightened concerns that development strategies, in order to be successful, must be relevant to the circumstances faced by local people have intensified interest in decentralized strategies that are accountable to households and communities. This has stimulated interest in participatory development, local institutional development and community-based natural resource management, and is one of the factors contributing to the paradigm represented by integrated conservation and development projects (ICDPs).

As we review in greater detail below, the notion that there exist important synergies, rather than tradeoffs, in efforts to advance food production, economic growth and environmental sustainability has been central to much recent thinking about the mechanisms used to address these goals. In a wide-ranging set of high-profile international conferences, international policy and planning documents and institutional mission statements, complementarity among these multiple goals is widely presumed. Often, however, the existence of synergies appears to be accepted on faith, rather than concluded as a result of careful analysis, research and observation. Yet economic intuition and an increasing base of research and applied work in developing countries suggest that – at least in the short to medium run – tradeoffs often, although not always, characterize the simultaneous pursuit of development goals. 'Win–win–win' opportunities exist and must be pursued, to be sure, but they are less ubiquitous than is often assumed, and, in the shorter term, hard choices must typically be made in the allocation of resources among multiple desired objectives. A central theme of this volume is that agricultural intensification is a necessary condition for satisfying economic growth, environmental sustainability and poverty alleviation goals in most developing countries, but it is by no means sufficient. A host of factors condition the complementarity or competition between these objectives. The chapters that follow describe and explore these linkages in considerable detail. But, first, it is useful to briefly consider the origins of current thinking on the 'tradeoffs versus synergies' debate.

The 'Tradeoffs versus Synergies' Debate: a Brief Review

Policies for sustainable development

In the 1960s and early 1970s, policy-makers, development scholars and practitioners grew increasingly aware of the interactions between economic development and environmental quality. This interaction, however, was widely deemed to be rife with tradeoffs – development practitioners believed that environmental protection would lead to reduced growth, while conservationists generally opposed economic development initiatives and were primarily concerned with alleviating the pollution and environmental degradation associated with economic growth. The impetus for the 1972 United Nations Conference on the Human Environment, for example, came from representatives of high-income nations concerned with environmental quality, yet representatives from low-income nations made a point of emphasizing that environmental protection was a luxury they could not afford (Sandbrook, 1992).

Starting in the mid-1970s, however, policy-makers and practitioners began to question the existence of such tradeoffs. An increasingly common viewpoint was that environmental conservation objectives could only succeed if human needs were attended to, and human needs could not be met in the long run with continuous degradation of the environment. To cite one prominent example, after originating in the 1970s as a programme focused chiefly on biodiversity conservation and protection, the United Nations Educational, Scientific and Cultural Organization's (UNESCO) Man and the Biosphere Programme increasingly encompassed human development priorities in buffer zones, where extractive and other economic activities were permitted (UNESCO, 1987). Efforts were begun to identify 'win–win' scenarios that could promote both environmental protection and economic development.

Throughout the 1980s and early 1990s, international institutions and coalitions of development and conservation practitioners published high-profile policy and planning documents that advocated synergies between environmental protection and economic development (IUCN, UNEP and WWF, 1980, 1991; WCED, 1987a,b; World Bank, 1992). Not surprisingly, for rural areas of the developing world heavily dependent on agriculture for food production and employment, agricultural intensification played an important role in the 'win–win' scenarios outlined in these documents. Perhaps the best-known example is the 1987 Report of the World Commission on Environment and Development (WCED, 1987b) – commonly known as the Brundtland Report – in which the authors emphasized that '[n]ew technologies provide opportunities for increasing productivity while reducing pressures on resources' (p. 144). In its report to the United Nations, the WCED identified 'increasing yields and productivity' (WCED, 1987a, pp. 26–27) as one of three essential strategies for achieving a sustainable world agricultural system. The report noted that increasing productivity was critical, among other things, to 'halting indiscriminate deforestation'. The World Bank's 1992 review of the status of the global economy and environment echoed a similar theme: 'If more food can be grown on the same land, that will ease the pressure to cultivate new land and will permit the preservation of natural intact areas' (p. 134).

Among groups with more explicit conservation agendas, similar themes were emphasized. In *Caring for the Earth* (IUCN, UNEP and WWF, 1991), a policy document published by three prominent international conservation organizations, the authors wrote: 'There are no great unused resources of cultivable land that can safely be taken from nature. Consequently, the land now used for agriculture will need to be cropped more intensively' (p. 33). In reference to land degradation, the same document stated: 'Today's losses of soil and productivity through erosion, salinization, desertification and misuse are intolerable [in poor nations]. The development of techniques for more intensive, more sustainable agriculture applicable at the local level in the lower-income countries has highest priority' (p. 35). At the field level, these themes are echoed in myriad project work plans and similar documents written by project managers and conservation and development practitioners. In the African rainforest, for example, a conservation project listed agricultural intensification as one of its priority tasks, noting that intensification would 'decrease the pressures on the protected area by reducing the need for upland agricultural surface area' (Conservation International, 1993, p. 77)

While policy-makers and practitioners over the past two decades have widely asserted that environmental and development goals are complementary, it is useful to consider the evidence upon which these assertions are based. In particular, how has the perceived relationship between agricultural intensification and associated economic development goals and environmental objectives been transformed from what was often an antagonistic relationship to a more cooperative one? The literature from several different traditions over the past three decades is useful for understanding this process and is briefly reviewed here.

Endogenous intensification and policy-led intensification strategies

Boserup (1965) began an important path of enquiry in the study of agricultural intensification by arguing that population growth was a major determinant of technological change in agriculture. In the Boserupian framework, population growth in a given area leads to growing demand for agricultural products, land degradation and the disappearance of the frontier. These conditions exert pressures on farmers to intensify (reduce fallow length, double-crop, etc.) by using more labour or capital per unit of land. Until some constraint to land extensification is reached, it may not be profitable for farmers in an area to switch from extensive practices to more intensive practices (Boserup, 1965; Holden, 1993b).

The Boserupian perspective suggests that intensification can be viewed as an endogenous outcome – e.g. that soil fertility is a 'dependent variable' – in a development process stemming from forces affecting agricultural communities. Such intensification, while consistent with high external input use, does not necessarily help to stem the disappearance of natural ecosystems, since the intensification process does not begin until land and ecosystem services are limited in supply.[1] The Boserupian perspective is also generally silent on issues related to off-site agricultural externalities (e.g. nutrient runoff), although some authors (Edwards and Wali, 1993; Tiffen *et al.*, 1994) have attempted to connect population growth to the protection of productive agricultural resources (as opposed to environmental amenities).

Most proponents of the positive links between agricultural intensification and the environment, however, do not advocate that less developed nations simply wait for a gradual change from extensive to intensive systems as population grows and agriculture expands at the extensive margin. Boserup herself recognized that this shift may, in fact, not occur in those densely populated communities characterized by a very high rate of population growth and limited ability to change tenure regimes and make investments in intensification (Boserup, 1965, pp. 77–111, 118). As emphasized in the WCED's United Nations report (1987a), the linkages between agricultural intensification and the environment suggest that '[n]ational and . . . provincial governments will have to develop a package of incentives and disincentives to promote conservation-based development' (p. x).

In order to avoid serious environmental degradation and habitat loss, the logical and commonly voiced policy prescription is, then, to promote a process of intensification before these losses are realized through Boserupian intensification and induced technological innovation (which may lag behind changes in relative factor prices that do not fully reflect social costs). Such 'policy-led intensification strategies' can play a role in 'speeding up the natural evolution of intensification' (Lele and Stone, 1989) and may forestall resource degradation stemming not only from rapidly rising agricultural populations but from rising food demands from the non-agricultural population (Pingali and Binswanger, 1984). This process typically implies interventions by outside agents, including policy changes (WCED, 1987a), investments in research and extension (Aldy *et al.*, 1998) and local investments (Wells and Brandon, 1992). These interventions aim to increase agricultural productivity by introducing, or facilitating access to, external inputs and improved agricultural practices – improved crop varieties, inorganic and organic fertilizers, pesticides, credit, conservation agriculture practices, irrigation, crop diversification, improved market access and a variety of other inputs and investments.

To be successful, however, such interventions typically require money and the ability to coordinate resources at large scales. Thus, it should not be surprising that, by the early 1990s, the largest global development aid organizations had also become the largest global environmental organizations. Between 1988 and mid-1995, for example, the World Bank committed US$1.25 billion in loans, credits and grants for projects with explicit objectives of conserving biodiversity. This money leveraged an additional US$0.5 billion (Jana and Cooke, 1996). Moreover, much of the Bank's broader agricultural development portfolio has also explicitly involved conservation and natural resource management objectives (e.g. 10% of projects have included biodiversity conservation objectives). Among bilateral donors, the US Agency for International Development (USAID), in particular, has also become an important financial and intellectual leader in international conservation, spending US$650 million each year on its environmental portfolio during the early 1990s (USAID, 1994).

The environmental Kuznets curve

Another tradition, that associated with the 'environmental Kuznets curve' (EKC) hypothesis, has also played an important role in research and policy debates regarding development and the environment. The EKC hypothesis posits that there is an inverted U-shape relationship between environmental degradation and per capita income similar to the original Kuznets relation applied to income inequality. This suggests that, as economic development proceeds from very low income levels, pollution, resource use and waste generation per capita increase rapidly. Then, '[a]t higher levels of development, structural change towards information-intensive industries and services, coupled with increased environmental awareness, enforcement of environmental regulations, better technology and higher environmental expenditures, result in leveling off and gradual decline of environmental degradation' (Panayotou, 1995, p. 13).

The EKC hypothesis implies that economic growth will eventually redress the negative environmental impacts associated with the early stages of economic development. Rather than being a threat to the environment, economic development is complementary to – or necessary for (Beckerman, 1992) – the maintenance or improvement of environmental quality. The logic behind the EKC hypothesis had an important influence on the sustainable development arguments of the WCED's *Our Common Future* (1987b) and the World Bank's *World Development Report 1992*. The latter report noted that:

> The view that greater economic activity inevitably hurts the environment is based on static assumptions about technology, tastes and environmental investments. . . As incomes rise, the demand for improvements in environmental quality will increase, as will the resources available for investment.
>
> (World Bank, 1992, pp. 38–39)

After a decade of debate, however, the empirical evidence for the EKC hypothesis is decidedly mixed. First, the inverted U-shape relation appears to apply only to a subset of impacts, mainly airborne pollutants (Arrow *et al.*, 1995). Secondly, the assertion that environmental quality is a 'luxury good' has not been conclusively demonstrated (Barbier, 1997). Thirdly, the relationship between income and the environment is not static but can be influenced by policy changes (Panayotou, 1995; Barbier, 1997). Finally, it is not at all clear why the posited relation should necessarily hold to begin with (Rothman, 1998; Suri and Chapman, 1998). The EKC hypothesis makes no explicit arguments about the role that growth in agricultural productivity can play in environmental protection. Proponents of a positive synergy between intensification and the environment infer a synergistic relationship based on known correlations between increases in productivity and increases in per capita income (e.g. WCED, 1987b). To date, there have been relatively few studies on the EKC hypothesis in the specific context of agriculture and environmental degradation. Studies that could be considered relevant tend to focus on deforestation or water quality and their results do not consistently confirm or disconfirm the EKC hypothesis. Panayotou (1995), for example, finds support for the notion that industrialization is linked to declines in deforestation, while Koop and Tole (1999), using a comprehensive data set on more than 70 developing nations, find no support for an inverted U-shaped relationship.

Technology-driven intensification and environmental effects

Despite the ambiguous results derived from empirical tests of the EKC hypothesis, there are other studies in the literature that attempt to more explicitly link agricultural intensification with improvements in environmental quality. These studies have often turned to the experiences of industrialized countries, particularly the USA, and the Green Revolution in Asia, both of which suggest that increases in agricultural productivity can relieve pressure on natural ecosystems.

In the US context, it has been argued that technological progress has made it possible to limit US land conversion to agricultural uses, thereby protecting natural habitats and biological diversity (Goklany and Sprague, 1992; Waggoner *et al.*, 1996). The USA has doubled its harvests over the last 50 years while keeping the area under cultivation steady (Raloff, 1997). Waggoner *et al.* (1996), for example, note that the clearing of forest for agriculture had largely stopped in the USA by 1920 and state that intensification over the last 50 years has 'spared' 90 million ha for nature. Goklany and Sprague (1992) use observed correlations between agricultural productivity and farmland area to make a similar case with respect to agriculture in New York and the north-east. In the same vein, attention has also been given to increases in the area of forests (MacCleery, 1993) and wildlife (Kolbert, 1986) in the USA. Sedjo (1995), in examining forest cover increases in the north-eastern USA in the period 1850–1990 and deforestation and reforestation patterns across the world, concludes that 'economic development promotes forest stability through well-defined and recognized property rights, the enforcement of property rights, the absence of government subsidies to encourage land clearing, and high levels and growth rates of agricultural productivity' (p. 205).

It is with respect to the Green Revolution, particularly in the Asian context, that the 'land-sparing' argument is most frequently made. For example, in extending earlier estimates by Borlaug (1987), Waggoner *et al.* (1996) maintain that Green Revolution technologies increased wheat production in India five-fold between 1961–1966 and 1991, while acreage only expanded by about three-quarters. Had traditional low-input technologies not been supplanted by modern technologies, 42 million additional hectares would thus have been required to generate the same levels of production as occurred in 1991. It is argued, then, that the exploitation of many millions of hectares of land was spared through technological improvements and crop intensification.

Elsewhere in the developing world, complementarities between intensified agriculture and environmental indicators have been reported in many places. Studies from Africa have reported that, when fertilizer prices increased or fertilizers became more scarce, farmers changed from sedentary farming to shifting cultivation, leading to more land degradation and more deforestation (Ferraro *et al.*, 1997; Holden, 1997). In Honduras, Bunch (1988) reports that agricultural intensification has led to positive environmental effects, such as air and water quality improvements through reductions in biomass burning and increases in the farm demand for manure formerly dumped into rivers. Forest area and quality increased through reductions in erosion, migration, forest fires and turpentine collection. Similarly, regression analysis using household data from Honduras (Godoy *et al.*, 1997) indicates a positive association between increased productivity per hectare and low levels of agricultural expansion. Other empirical models have found similar associations in other areas of the world and at larger scales (see Kaimowitz and Angelsen (1998) for an excellent review).

Inspired by historical data and field experimentation, numerous scientists have emphasized the potentially positive impacts of using existing, but thus far non-adopted, agricultural technologies in developing nations. Based largely on experimental work in Latin America (Sanchez *et al.*, 1982), soil scientists have argued that, for every hectare converted to more productive, sustainable technologies, 5–10 ha of tropical rainforest are conserved (Sanchez *et al.*, 1990). A similar argument has been made for the intensification of Amazonian pastures (Serrao and Toledo, 1990).

Despite the many empirical examples suggesting a positive relationship between agricultural intensification and environmental quality, it is difficult to argue *a priori* that one should expect the relationship to be positive. Where agricultural intensification has enjoyed widespread success, this has not occurred without environmental degradation, ranging from a loss of native ecosystems to waterlogging and salinization due to improper irrigation and water pollution resulting from the use of pesticides, fertilizers and animal wastes. In the case of the Green Revolution in South Asia, for example, Pingali *et al.* (1997) and other scientists have documented widespread environmental problems which threaten future production (see also Pingali and Rosegrant, Chapter 20 of this volume). Even the use of basic intensification technologies has been demonstrated to have mixed effects. Reductions in (inorganic) fertilizer prices, for example, can have both substitution effects (as farmers substitute fertilizer for land) and output effects (as all input use increases because farming is more profitable). The total use of land may rise, even though land use at the intensive margin falls. In a recent empirical study of agricultural assistance in low-income countries, Lewandrowski *et al.* (1997) found that the coefficient of the fertilizer price variable in an equation explaining arable land use in eight nations was negative and significant, suggesting that the output effect may dominate the positive substitution effect. A study in Brazil (Ozorio de Almeida and Campari, 1995) found that, when the prices of inputs other than fertilizer increased, there was a decrease in land clearing.

The linkages between intensification and tropical deforestation have been the focus of perhaps the greatest body of work in this area. The 'land-saving' arguments associated with the Green Revolution in Asia are commonly cited (Waggoner, 1994), but recent empirical analysis of deforestation in India, rather than confirming a positive relationship between intensification and reductions in deforestation, suggests that agricultural technology improvements have promoted deforestation in India by pushing up the value of land for growing crops (Foster *et al.*, 1999). In an evaluation of four case studies of agricultural intensification in Africa, Wiersum (1986) concluded that the adoption of intensified cash-cropping systems did not necessarily lead to forest conservation due to increases in the demand for land, widespread land speculation and cultivated area. In another case study review, the adoption of more productive technologies was found to be correlated with the expansion of market opportunities, leading to an increase in the use of all factors of production, including land (Barraclough and Ghimire, 1995). Kaimowitz and Angelsen's (1998) extensive assessment of 148 economic models and empirical studies of tropical deforestation finds contradictory conclusions regarding the effects of higher agricultural productivity on deforestation. Conclusions and results vary with model specification, assumptions and the nature and source of the data employed. The principal general conclusion reached is that the relationship between intensification and deforestation is indeterminate.

The argument that land-sparing technological change is unambiguously beneficial for the environment is not clear-cut even in the context of high-income nations. In the USA, although some eastern states have experienced a return of their native ecosystems, the prairie ecosystem of the Midwest has been largely lost (Bultena *et al.* (1996) document this for Iowa, for example). In Australia, a similar evolution has taken place in the central-western region of New South Wales, one of the most productive farming areas in the world, where the only remaining woodland is confined mainly to areas unsuitable for agriculture (Goldney and Bauer, 1998).

Even if developed countries do offer some clear examples of the long-run synergies between agricultural intensification and environmental quality, it is not clear that these cases are necessarily transferable to low-income countries. The USA and other developed countries have increased agricultural productivity, but they have also increased their land–labour ratios (Hayami and Ruttan, 1985; Thirtle, 1985). Although the area under cultivation in the USA has remained steady over the last 50 years, the number of farmers has shrunk by more than two-thirds. While excess rural labour has been successfully absorbed into growing manufacturing and service sectors (Sedjo, 1995), it is not clear whether these sectors in low-income countries can fully accommodate the large rural populations looking for agroindustrial and urban employment opportunities.

The point of this brief review is not to take issue with the important role of technological change in agriculture or with the enormously beneficial effects of agricultural intensification in increasing food production and farmers' incomes around the world. Indeed, the record of technology change in agriculture has been highly impressive on these (and other) scores; consequently, hunger and malnutrition among millions of rural and urban poor have been reduced and food security enhanced. Rather, the point is simply to indicate that the empirical evidence does not support the argument – sometimes made explicitly, sometimes implicitly – that agricultural intensification and the economic growth associated with it are *necessarily* beneficial for the environment, particularly in the short term. As *Caring for the Earth* (IUCN, UNEP and WWF, 1991) concludes, 'The pressures on agricultural land . . . can be partly relieved by increasing productivity. But short-sighted, short-term improvements in productivity can create different forms of ecological stress' (p. 57).

Integrated conservation and development projects (ICDPs)

During the 1980s and early 1990s, conservation practitioners also began to embrace the 'synergy' perspective on agricultural intensification and the environment. They added large community development initiatives to their portfolios, initiated partnerships with development-oriented organizations and hired more social scientists to facilitate their work with communities (Wells and Brandon, 1992; Western and Wright, 1994). A new approach called the 'integrated conservation and development project' (ICDP) became a popular vehicle for channelling conservation funds into community development initiatives.

Unlike conservation projects in the 1960s and 1970s, ICDPs tend to focus their efforts on lands outside, rather than within, protected areas. More importantly, 'people' were explicitly identified as part of the solution. *Caring for the Earth* (IUCN, UNEP and WWF, 1991), a revision of *World Conservation Strategy* (IUCN, UNEP and WWF, 1980), emphasized that economic development was not necessarily antithetical to nature conservation and that local communities should be given a significant role in the design and management of integrated conservation and development activities.

An important element of many ICDPs was the introduction of technologies that offered substitution possibilities for the existing production methods that were leading to deforestation. Proponents argued that farm-level interventions in the 'buffer zones' around protected areas could improve the productivity of agriculture and thereby reduce the incentives for local residents to expand cultivation and animal husbandry into the protected areas or to engage in resource extraction through activities such as fuelwood collection or wildlife poaching (see McNeely and Brandon, Chapters 21 and 22, respectively, of this volume).

However, it is well known that the introduction of new technologies and practices in rural environments can be a challenge. Rotational crop–fallow land use with slash-and-burn methods remains widespread in the tropics, despite decades of agricultural research and technical assistance (Weischet and Caviedes, 1993). A review of over 450 articles on agroforestry and other intensified cropping systems published between 1972 and 1989 found some 'promising technologies', but few clear successes (Robison and McKean, 1992).[2] The same source reviewed more than 85 soil conservation technology studies, and concluded that promoting the adoption of these technologies is difficult and, even when they are adopted, they do not always have the desired effects.

It is perhaps not surprising, then, that several recent comprehensive surveys of the experiences of conservation projects with explicit development objectives have suggested that ICDPs are characterized by numerous problems in design and execution, are complicated to manage effectively and have typically not fared any better than projects with a strict development focus (Wells and Brandon, 1992; Brandon *et al.*, 1998; Larson *et al.*, 1998; Oates, 1999; see also Brandon, Chapter 22 of this volume). Although substantial resources continue to be allocated to ICDPs, the role that agricultural intensification can play in helping to achieve their conservation objectives is still inconclusive.

Conclusions

The relationship between agricultural intensification and the environment is an important one, and not simply because 70% of the total land surface of the earth is in agriculture or managed forests (Pimentel *et al.*, 1992) and the impact of agriculture on the environment is expected to grow in the future (FAO, 1993a). It is also important because many policy-makers, international organizations and influential policy and planning efforts have appeared to accept on faith the assumption that agricultural intensification and accompanying economic growth in rural areas will necessarily lead to improvements in environmental quality. Guided by this critical assumption, international donors and development and conservation organizations have allocated, and continue to allocate, significant resources to encouraging agricultural intensification strategies, with the expectation of achieving progress not only in food production and income generation, but also in poverty reduction and environmental goals.

Our brief survey of the past literature on agricultural intensification and the environment, however, suggests that the question of 'tradeoffs versus synergies' is a complex one with few easy answers or unambiguous conclusions. Simple assertions of complementarities in the realization of multiple goals have, in many instances, been shown to be unrealistic and overly simplistic or, at best, to pertain to mostly long-run and aggregate-level relationships. This is demonstrated time and again in the rethinking that has occurred in areas as diverse as assessing the impacts of green-revolution technologies, the environmental Kuznets curve, agriculture and deforestation linkages and ICDPs. The conclusion one draws from the mass of theoretical and empirical evidence is that, in most developing-country circumstances, while agricultural intensification is generally necessary for achieving conservation objectives in concert with rural economic growth and poverty alleviation goals, it is by no means sufficient. Further critical consideration regarding the relationship between agricultural intensification, economic development and the environment in developing countries is clearly warranted. The chapters that follow offer important contributions to the extension of that discourse.

Outline of this Volume

The central objective of this volume is to review and consider a range of evidence on the 'tradeoffs versus synergies' theme as it pertains to the multiple objectives of agricultural intensification in low-income countries. This evidence is wide-ranging and includes theoretical and conceptual analysis, numerous empirical examples and case studies from Asia, Africa and Latin America and synthetic analyses of selected technology, policy and institutional issues. The intention is to directly address the 'tradeoffs versus synergies' question and related issues this question engenders.

Part I considers a number of cross-cutting themes and background issues relevant to the empirical studies discussed later. In Chapter 2, Conway addresses the enormity of the global challenge of generating adequate food supplies for the estimated 800 million people around the world who are hungry or malnourished, while at the same time ensuring equitable access to food and mitigating the many negative environmental externalities associated with agricultural production. He suggests that a 'Doubly Green Revolution' – exploiting scientific developments in biotechnology and ecology as well as more participatory development strategies – is required, which will repeat the successes of the Green Revolution and do so in a manner that is equitable, sustainable and more conserving of natural resources.

Bilsborrow and Carr, in Chapter 3, consider the dynamics of population size, movement and density in relation to changes in forest cover and trends in agricultural intensification and extensification, principally in Latin America. While population pressures (mainly rural–rural migration) are found to result in some of the expected consequences for land use consistent with Boserup, they find that population pressure is generally a secondary factor in driving land clearing, with road construction and economic, market and political factors dominant. They also find substantial variations in the nature of the driving forces behind land-use trends in different countries and different environments.

In Chapter 4, Perrings addresses the biodiversity implications of agricultural development in developing countries, and identifies the major economic factors behind biodiversity loss in agriculture, forestry and fisheries. He contrasts the benefits realized in these sectors with the costs of biodiversity loss, principal among them being the reduced resilience of agroecosystems in confronting environmental and market shocks. He posits an 'optimal level' of biodiversity that mediates benefits and costs, and highlights the role of poverty and market failures in reducing the effectiveness of economic instruments in stemming biodiversity loss.

In the following two chapters, Pagiola and Holden (Chapter 5) and Angelsen and Kaimowitz (Chapter 6) develop the theoretical basis for farm households' decisions regarding intensification and extensification of agricultural production. Pagiola and Holden develop a stylized model of farm household decision-making that illustrates the roles played by clearing efficiency and the productivity of newly cleared lands, output prices, labour costs and the household's rate of time preference in determining household land-use strategies. The analytical model developed by Angelsen and Kaimowitz highlights households' subsistence versus profit-maximizing behaviour, capital- versus labour-intensive technological change and product market characteristics in accounting for deforestation outcomes. Both chapters conclude that *a priori* theoretical results are ambiguous and that intensification–income–environment outcomes are essentially empirical questions.

Chapters 7 and 8 together illustrate the benefits of using ecoregional and bioeconomic models jointly to address economic and environmental objectives. Ruben *et al.* (Chapter 7) discuss the advantages and disadvantages of different types of bioeconomic models that can be employed in joint economic and biophysical analysis. They then employ three variants of a bioeconomic model developed for southern Mali to address the tradeoffs and complementarities among production, economic welfare and sustainability indicators that result from a simulated fertilizer subsidy. Crissman and co-authors (Chapter 8) present the conceptual framework and structure of a model that explicitly addresses tradeoffs among economic, biophysical and health outcomes in an application to the potato–dairy system of the Andean highlands. They highlight the economic–environmental and economic–health tradeoffs that stem from policy alternatives and stress the importance of making the results of research on tradeoffs understandable by and accessible to policy-makers.

Part II of this volume encompasses a series of empirical studies which, using different analytical approaches and models, address issues of tradeoffs and synergies among multiple development goals under different developing country circumstances. In Chapter 9, Hazell and Fan analyse contributions to agricultural productivity growth in high- versus low-potential areas in India. They find that investments in rural infrastructure, agricultural technology and human capital are at least as productive (and typically more so) in many rain-fed areas as in irrigated areas, and reach the important conclusion that 'win–win' strategies for addressing productivity and poverty problems may thus be possible in the case of India.

Pender and co-authors (Chapter 10) analyse 'pathways of development' in a set of diverse communities in central Honduras, where economic restructuring and land-use changes have been particularly great since the 1970s. They identify the characteristics and determinants of six distinct pathways, which differ substantially in cropping practices and resource management strategies, and suggest that a 'one-size-fits-all' approach to technical assistance is unlikely to be successful. Poverty reduction in these communities, rather than being dependent on natural resource management strategies, is found to be highly dependent on the provision of public services.

Chapters 11–13 report empirical results from the three principal benchmark sites of the Global Initiative for Alternatives to Slash-and-Burn (ASB) of the Consultative Group on International Agricultural Research (CGIAR). Coordinated by the International Centre for Research in Agroforestry (ICRAF), the ASB programme is using common research strategies and methodologies to: (i) develop and test alternative technologies for smallholder farms; (ii) examine policies that may create disincentives for deforestation; and (iii) promote sustainable alternatives to slash-and-burn agriculture. Specific cropping-system alternatives are evaluated, using the 'ASB matrix' approach, according to a wide set of indicators of agronomic and environmental sustainability, economic viability and indicators of interest to policy-makers (e.g. food security). 'Best bet' alternatives are identified which merit further attention by researchers and development practitioners.

In the first of the ASB studies, Gockowski and co-authors (Chapter 11) analyse economic and environmental tradeoffs characterizing resource use alternatives in the rainforest of the Congo River basin in Central Africa. In Chapter 12, Tomich and colleagues apply the 'ASB matrix' of indicators to a series of crop monoculture, agroforestry and forest management land uses on the island of Sumatra, Indonesia. Vosti and colleagues (in Chapter 13) also employ the ASB matrix approach in their evaluation of alternative cropping systems facing small-scale agriculturalists in the western Brazilian Amazon. In each case, 'best bet' farming systems are identified, often based on diversified agroforestry or perennial cropping systems. These systems appear to offer considerable scope for the joint realization of farmers' economic and agronomic goals, as well as addressing national- and international-level environmental and food-security goals.

Using a bioeconomic modelling approach, Schipper and colleagues, in Chapter 14, analyse land-use alternatives in northern Costa Rica, giving explicit attention to the incorporation of local labour markets, product markets, crop and animal production and environmental indicators. Their model is used to analyse the effects of technology and policy alternatives, including a pesticide tax and forestry subsidies. Labour and product market conditions are found to substantially influence the outcomes.

In Chapter 15, Otsuka and Place summarize the results of an extensive comparative study of land tenure and land and forest management in seven Asian and African countries. They identify a number of key factors that account for deforestation and poor forest management, including farming on marginal lands, weak individual land rights and major deficiencies in state ownership and communal tenure regimes. They discuss several policy and institutional changes which may enhance the incentives for tree planting and the adoption of agroforestry systems.

In Chapter 16, Kerr and co-authors examine factors contributing to incentives for improved agricultural productivity and natural resource management across a broad sample of watershed management projects in India's semi-arid tropics. A variety of factors are found to affect these incentives, including population density, infrastructure, social organization and agroclimatic conditions. Importantly, participatory projects that focus as much on social organization as on technology transfer are shown to be generally the most successful.

Part III addresses a series of technology, policy and institutional concerns that affect the joint realization of intensification, environmental and development objectives. In Chapter 17, Sanchez and colleagues discuss factors contributing to what they argue to be the critical factor limiting sustainable agricultural intensification in Africa – declining soil fertility. They outline a detailed strategy encompassing specific technological and policy measures to address problems of soil fertility and agricultural intensification, focusing largely on soil fertility replenishment and diversification into high-value crops.

In Chapter 18, Staal and co-authors address the problems associated with livestock intensification in developing countries, particularly in Africa. They identify the significant potential that successfully intensified livestock systems can play in contributing to smallholders' goals and discuss a series of factors that have thus far limited the success of these systems. Strategies are outlined to redress these problems, including better integrated crop–livestock systems, improved manure management and improved soil fertility management.

Still with a focus on Africa, Chapter 19 turns to the realm of policy. Reardon and colleagues review a wide set of factors that have limited sustainable agricultural intensification in African agriculture. The limitations of past macroeconomic, sectoral, marketing and credit policies are identified and applications to a number of diverse settings in sub-Saharan Africa are reviewed. A strategy of capital-led intensification and selected public investments, accompanied by continued policy and market reforms, is proposed to generate successful intensification in African agriculture.

Chapter 20 treats a dramatically different situation – the intensified agricultural systems of Asia and the factors contributing to recent declines in productivity growth in the important rice and rice–wheat systems of the Indo-Gangetic plains of South Asia. Pingali and Rosegrant identify a host of agronomic and soil-related constraints (salinity, waterlogging, soil nutrient deficiencies, etc.) which have contributed to these productivity problems. These are accompanied by policy and institutional constraints that have distorted input markets (notably groundwater for irrigation) and exacerbated intensification-related problems. A range of technology and policy solutions are suggested for improving environmental outcomes while increasing the likelihood of continued growth in food production.

Chapters 21–23 shift the focus of the book to institutional strategies for enhancing the joint realization of intensification and economic objectives. In Chapter 21, McNeely outlines the critical institutional issues faced by those seeking to jointly promote agricultural development and biodiversity conservation. He emphasizes the contributions provided by ecosystem services and protected areas for agricultural development and rural communities. Alternatives, including economic alternatives, for involving local communities in protected area and buffer-zone management are stressed as mechanisms to make the conservation of natural resources attractive to local people and thus ensure the availability of natural resources for the future.

In Chapter 22, Brandon provides an overview of goals, underlying assumptions and operational issues characterizing ICDPs, which, as we discussed earlier in this chapter, have been a popular mechanism in seeking to jointly accomplish economic development (including agricultural intensification) and conservation goals. She finds that successes have been rare, due to a host of problems related to project design and execution, and suggests that the current movement toward ‘ecoregional’ or ‘landscape’ approaches to integrated development and conservation projects may address some of the fundamental limitations encountered in ICDPs to date.

Highlighting the theme of community-based approaches identified in earlier chapters, Uphoff (Chapter 23) emphasizes the fundamental need to involve local people and communities in the design and execution of projects with development and environmental goals. He describes a number of developing-country case studies where community-based natural resource management strategies have been a cornerstone of local development efforts. He further argues that addressing critical social and community-related aspects of development – that is, those beyond purely individual economic motivations – is essential to making local development efforts successful. The volume then concludes with a summary chapter (Chapter 24), which synthesizes some of the important results of the chapters of this book.

Notes

1 Boserup highlights the island of Java as an important example of endogenous intensification. The Javanese, however, largely removed the entire extent of natural forest cover on the island; most modern forest cover on Java is in the form of plantations and secondary forest (Gillis, 1988).

2 The vast majority of the successes were systems developed endogenously by local populations, with little or no outside technical assistance. Their development was more akin to a Boserupian notion of intensification.

2 The Doubly Green Revolution: Balancing Food, Poverty and Environmental Needs in the 21st Century[1]

GORDON CONWAY

Rockefeller Foundation, New York City, New York, USA

To die of hunger is the bitterest of fates.

(Homer, *The Odyssey*, 12,342 (Ash, 1941))

Homer, in his epic poem, *The Odyssey*, recounts how Odysseus and his companions have resisted the lure of the Sirens, sailed safely between Scylla and Charybdis and come to the island of Thrinacie, where the 'Sun-god's cattle and plump sheep graze'. Odysseus has been warned they are not to be harmed, but his companions succumb to the temptation. 'To die of hunger', declares Eurylochus, 'is the bitterest of fates.' They kill the cattle and feast. No sooner have they set sail again than Zeus sends a hurricane as a punishment. All perish except for Odysseus.

Today there are more than three-quarters of a billion people who, like Odysseus's companions, live in a world where food is plentiful and yet it is denied to them. If we were to add up all the world's production of food and then divide it equally among the world's population, each man, woman and child would receive a daily average of over 2700 calories of energy (Alexandratos, 1995). This is just about enough to prevent hunger and probably sufficient for everyone to lead active, healthy lives.

Yet the harsh reality is great inequality. While in Western Europe and North America average supplies exceed 3500 calories a day, caloric availability is less than two-thirds this amount in sub-Saharan Africa and South Asia (Fig. 2.1). Thirty-five developing countries, including nearly half the countries of Africa, have average supplies of less than 2200 cal day^{-1}. According to recent estimates, over 800 million people, equivalent to 15% of the world's population, get less than 2000 cal day^{-1}, live a life of permanent or intermittent hunger and are chronically undernourished (FAO, 1992).

Unlike Odysseus's companions, many of the hungry are women and children. More than 180 million children under 5 years of age are underweight, that is, they are well below the standard weight for their age.[2] This represents

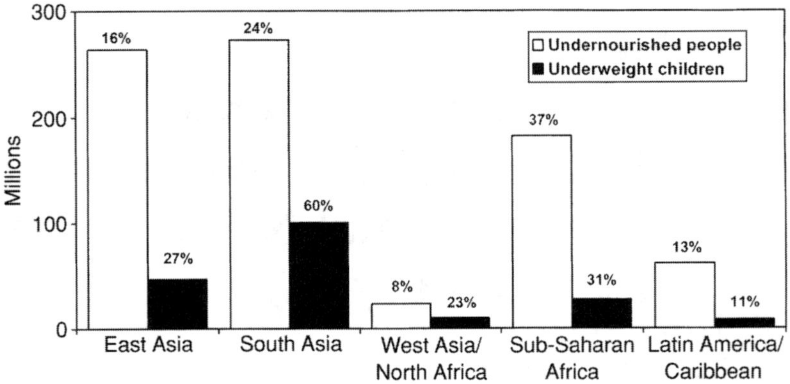

Fig. 2.1. Hunger by region in the developing world (from FAO, 1992; United Nations Administrative Committee on Coordination/Sub-Committee on Nutrition, 1992; Grigg, 1993).

one-third of those under 5 years of age in the developing countries. Young children crucially need food because they are growing fast and, once weaned, are liable to succumb to infections. And women, in addition to their own needs, require an adequate diet if they are to give birth to and raise healthy children.

For many, undernutrition and malnutrition lead to death, not necessarily through starvation, although this may happen in famine situations, but because a poor diet reduces the capacity to fight disease (WRI, 1992). Diarrhoea, measles, respiratory infections and malaria are common in many parts of the developing world. Well-fed people can fight them off; the malnourished, especially children, will succumb. Seventeen million children under 5 die each year and malnourishment contributes to at least one-third of these deaths (UNICEF, 1990; United Nations Administrative Committee on Coordination/ Sub-Committee on Nutrition, 1992). Over 100 million children suffer from vitamin A deficiency. Lack of minerals in the diet can have equally severe effects. Iron deficiency is common in the developing world, affecting as many as a billion people. Anaemia caused by iron deficiency afflicts nearly 400 million women of childbearing age (15–49 years old). Anaemic women tend to produce stillborn or underweight children and are more likely to die in childbirth.

Paradoxically, hunger is common despite 20 years of rapidly declining world food prices. In many developing countries, there is enough food to meet demand, and yet large numbers of people still go hungry. Although food prices are low, they remain high relative to the earning capacity of the poor. Market demand is satisfied, but there are many whose low incomes leave them unable to purchase the food they need and, hence, the market is oblivious to them.

As Amartya Sen (1982) points out, hunger occurs because, in one way or another, people are not entitled to the wherewithal to obtain food. They may be unable to:

- grow enough food on the land they own or rent or are otherwise entitled to cultivate;

- buy enough food because their income is too low or they are unable to borrow, beg or steal enough money;
- acquire enough food as a gift or loan from relatives or neighbours or through entitlement to government rations or aid donations.

Not surprisingly, hunger is closely related to poverty. Poor people have few or no assets, are unemployed or earn less than a living wage and thus cannot produce or buy the food they need. According to World Bank estimates, over one billion people – that is, a quarter of the developing world's population – are in poverty, defined as living on less than US$1 day^{-1}.

To the casual observer, poverty seems to be worse in the cities, but, in reality, the urban poor fare better. Although the cost of living may be low in rural areas, there are fewer opportunities to make a living. In the extreme, the urban poor can at least beg or steal. About 130 million of the poorest 20% of developing country populations live in urban settlements, most of them in slums and squatter settlements, while 650 million of the poorest live in rural areas. In sub-Saharan Africa and Asia, most of the poor are rural poor (Leonard, 1989). Some live in rural areas with high agricultural potential and high population densities – the Gangetic plain of India and the island of Java. But the majority, about 370 million, live where the agricultural potential is low and natural resources are poor, such as the Andean highlands and the Sahel.

The first question we ought to ask ourselves is: why should we be concerned? Probably everyone who reads this volume is getting an adequate diet. Does it matter to us that others are not so fortunate? Does it matter to the industrialized countries that many people in the developing countries are malnourished? Part of the answer to these questions is political. The end of the Cold War has not brought about an increase in global stability. While conflict between East and West has declined, there is a fast-growing divide between the world of the peoples, countries and regions who 'belong' in terms of global power and those who are excluded. Over 2 billion people in the world regularly watch television. For the rich, the images on their screens provide a constant reminder of the horrors of natural disasters, civil war and famine. For the poor, the screens portray the everyday luxuries of the affluent and well-fed. Globally, the consequence is a potentially explosive mix of fears, threats and unsatisfied hopes.

Yet this growing conflict receives relatively little attention in the industrialized countries. The volume of aid going to developing countries is stagnating in real terms. However, unless the developing countries are helped to realize sufficient food, employment and shelter for their growing populations or to gain the means to purchase the food internationally, the political stability of the world will be further undermined. In today's world, poverty and hunger, however remote, affect us all.

At the same time, the growing interconnectedness of the world – the process commonly referred to as globalization – holds the promise of alleviating, if not eliminating, poverty and hunger. Globalization, while threatening, on the one hand, to concentrate power and increase division, on the other, contains the economic and technological potential to transform the lives of rich and poor alike. Much depends on where our priorities lie and, in particular, whether there is sufficient access by the poor to the economic opportunities created by the products of the new technologies.

Prospects for the Year 2020

We have little time. If nothing new is done, the numbers of poor and hungry will grow. Partly this is because most populations in the developing world are still growing rapidly. By the year 2020, only two decades from now, there will be another 1.5 billion people in the developing world who will require food. This is in addition to the 0.80 billion people who are chronically undernourished today.

While the growth rate of the world's population has declined from a high of about 2.0% $year^{-1}$ during the late 1960s to 1.4% in the early 1990s, the size of the current annual increment is unprecedented (UN, 1998). Nearly 900 million people were added to the world's population in the 1990s, the largest increase for any decade in history. Until well into the 21st century, a further 80 million people will be added each year, close to 0.25 million people per day. If the proportion of the population of the developing countries deprived of an adequate diet remains constant, the number who are undernourished by the year 2020 could be greater than 1.4 billion.

What is the prognosis for feeding the world's population in the 21st century? It is not possible to foresee, with any accuracy, the situation in the latter half of this century. Predicting the next two decades is more feasible, and this will be the most critical period; after 2020, the annual increments in world population will begin to decrease significantly. If we can achieve a well-fed world by then, it should be possible to meet future demands, providing the resource base has been adequately protected.

Producing forecasts of world food production is complicated. There are numerous variables – and many unknowns – that have to be considered, and they interrelate with one another in complex and often circuitous ways. At the time of this writing, most attention is being paid to three models. These differ in their scope and level of detail (Alexandratos, 1995; Mitchell and Ingco, 1995; Rosegrant *et al.*, 1995; Queen Elizabeth House, 1996). As with all models, their outcomes depend on the assumptions built into them and the validity of the baseline data. They are essentially econometric models: they assume that supply meets demand in a market economy and they are constructed to ensure this occurs. In addition:

- regional or country food demand is calculated by multiplying population size by per capita demand. Population estimates are based on the United Nations (UN) medium variant, and per capita demand is a function of per capita income, which, in turn, is derived (with some modifications) from the World Bank projections of gross domestic product (GDP).
- Food production is determined by multiplying the amount of cropped land by yield, which in the In ernational Food Policy Research Institute (IFPRI)

model, is predicted to increase annually by 1.5–1.8% year^{-1}, depending on the specific cereal product.

For each of the three models, the forecast is reasonably optimistic. The world population growth rate is matched by a similar growth in food production. World food prices continue to decline. However, the developing countries as a whole will not be able to meet their market demand. In the IFPRI model, the shortfall is some 190 million tonnes (Mt); however, the model predicts that this can be met by imports from the developed countries (Fig. 2.2).

In 1990, the developing countries imported about 9% of the total grain they consumed, or roughly 91 Mt. The predicted imports for 2020 are double this figure and comprise 11% of consumption. Of the 190 Mt, over 140 Mt will go to satisfying demand in East Asia, West Asia/North Africa and Latin America. This will be primarily wheat for human consumption and maize and other coarse grains for pig and poultry feed. Especially in East Asia, a rising demand for wheat for bread and pasta will be difficult to meet from within the region since much of the land is more suitable for rice production. Increasing incomes in these regions will result in dietary changes, in particular, growth in livestock consumption, as well as the capacity to pay for the necessary imports. However, in sub-Saharan Africa and South Asia, the proportion of grain going to livestock will remain very small and imports will be required to meet human food needs.

Although overall the models are optimistic, there are pessimistic scenarios for significant regions of the developing world. Food production in sub-Saharan Africa will be hard pressed to keep up with population increases for a long time to come. According to the IFPRI model, by the year 2020, the excess of market demand for grain over production will be nearly 26 Mt; this compares with current net imports of 9 Mt. And South Asia will require more than 22 Mt, compared with 1 Mt today.

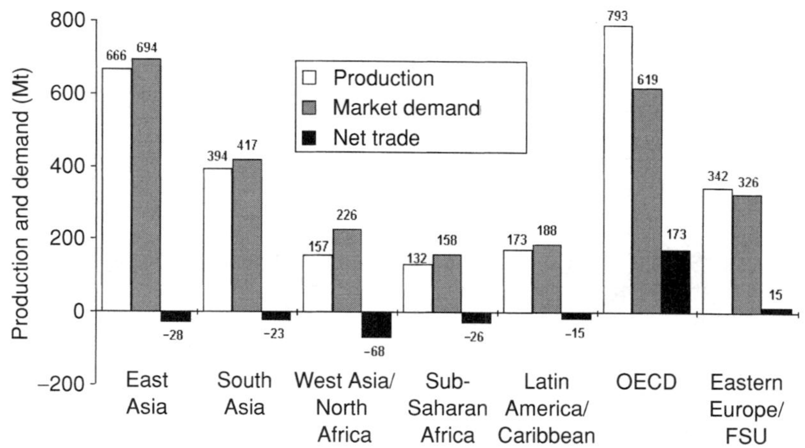

Fig. 2.2. Grain production and market demand (for human and livestock feed) from the IFPRI model for the year 2020 (from Rosegrant *et al.*, 1995). FSU; former Soviet Union; OECD, Organization for Economic Cooperation and Development.

Inevitably, models of this kind raise more questions than they answer. The most important omission from the calculations is the food needs of the poor and hungry. As in the real world, they are simply priced out of the market and their needs are 'hidden'. The gap between demand and supply that the model closes is the market gap. Then, as now, there will be a substantial hidden food gap, particularly in South Asia and sub-Saharan Africa, where the average calorie availability will remain below 2200 cal $person^{-1}$ day^{-1}.

This hidden food gap is the cereal requirement to meet the energy needs of the population, less the sum of domestic production and imports. Hazell and Rosegrant of IFPRI have calculated the total cereal requirement by assuming a minimum need per person of 3000 cereal cal day^{-1} – which covers food, livestock feed, seed, storage losses and waste during processing (P. Hazell, personal communication). Using a conversion rate of 1 kg of cereals equivalent to 3600 cal, each person then requires a little over 300 kg cereals $year^{-1}$. The difference between the total need for the whole population and the market demand in the IFPRI model is, then, the 'hidden gap' (Table 2.1).

The assumption of 3000 cal $person^{-1}$ day^{-1} is hardly generous, but it leads to total food gaps, in terms of cereals, of 214 Mt for sub-Saharan Africa and 183 Mt for South Asia in 2020. If all this food were to be supplied by the developed countries, it would require nearly 550 Mt, nearly three times that predicted by the market model.

In human terms, the hidden gap can be translated into a persistence of large numbers of malnourished children. By 2020, the total numbers will have declined slightly from the current 180 million to 155 million, but in sub-Saharan Africa, they will have increased by nearly 50%. And there will probably still be close to 0.75 billion people chronically undernourished (the Food and Agriculture Organization (FAO) model predicts over 600 million) (Alexandratos, 1995).

Table 2.1. Hidden and total food needs by the year 2020 (from Peter Hazell, personal communication).

	Food needs (Mt)		
	Hidden food gap (need less production plus imports)	Imports	Total food gap
East Asia	–	55.8	55.8
South Asia	160.0	2.7	182.7
West Asia/North Africa	–	68.5	68.5
Sub-Saharan Africa	187.5	26.1	213.6
Latin America/Caribbean	11.6	15.0	26.6
Total developing countries	359.1	188.1	547.2

Note: Assumes cereal needs equivalent to 3000 cal $person^{-1}$ day^{-1}.

Yield Trends

These models also predict continuing increases in crop yields and production in line with recent trends. There are a number of grounds for questioning this assumption. While production per capita continues to grow in East Asia and, in recent years, in Latin America, there is a measurable slowing of growth in South Asia. In the Near East/North-east Africa and sub-Saharan Africa regions, per capita production is stagnant or declining and is coupled with a high degree of variability in production (Fig. 2.3).

Recent data confirming declines in the rates of yield growth are worrying (Conway, 1997). In South Asia, the growth of Pakistan's wheat yields slowed in the 1980s, although in India there has been no slackening. Rice yields continued to grow in the 1980s in India and Bangladesh, and at a somewhat higher rate in India, but there are signs of a plateau in both countries in recent years. In East Asia, the slowing down of yield growth is more apparent. In both the Philippines and Indonesia, rice yields have levelled off in the 1990s and, in

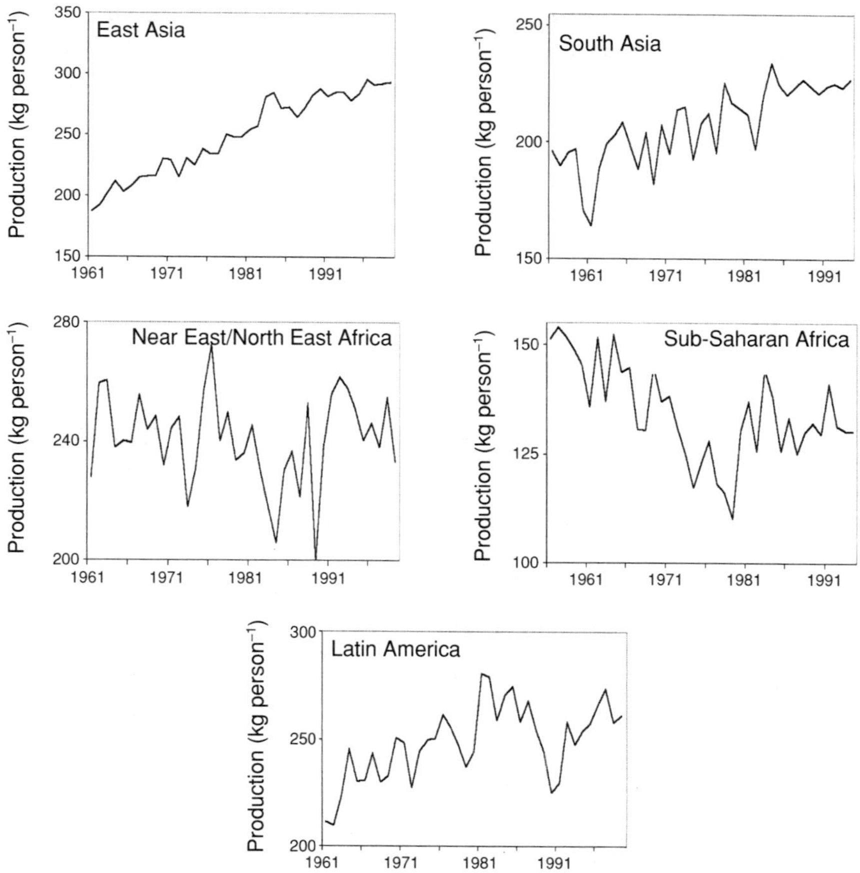

Fig. 2.3. Cereal production per capita in the developing world (from FAOSTAT TS: SOFA, 1999).

China, rice and wheat yields grew more slowly over the past decade than in the 1970s and 1980s (Fig. 2.4).

Similar slow-downs are apparent in wheat yields in Latin America and North Africa. The growth of maize yields has remained low, although in Mexico and other countries of Latin America there has been a sudden increase in yields, after a plateau in the 1980s. Most of the yield gains in maize in sub-Saharan Africa were in the 1950s and early 1960s. Subsequent yield increases have been very erratic. Yields of millet and sorghum have remained more or less constant, partly because of their replacement on the better lands by maize.

There is also evidence, albeit largely anecdotal, of increasing production problems in those places where yield growth has been most marked. For example, in the Punjab, although wheat yields are still growing, the record of previous high yield growth is now being seriously threatened (Randhawa, n.d.). Of greatest concern is the growing scarcity of water. According to several estimates, good quality water availability in the state is about 25 million acre feet, but the demand from existing cropping systems is about 37 million acre feet. There are some three-quarters of a million tube-wells drawing water at greater than the recharge rate. In the most intensively cultivated districts of Patiala and Ludhiana, the ground water table has fallen to a depth of 9–15 m and is falling at about 0.5 m $year^{-1}$. Salinization is also serious, affecting 9% of the total cropped area, as is waterlogging. All this and other, albeit largely anecdotal, evidence from Luzon, Java and Sonora suggest there are serious and

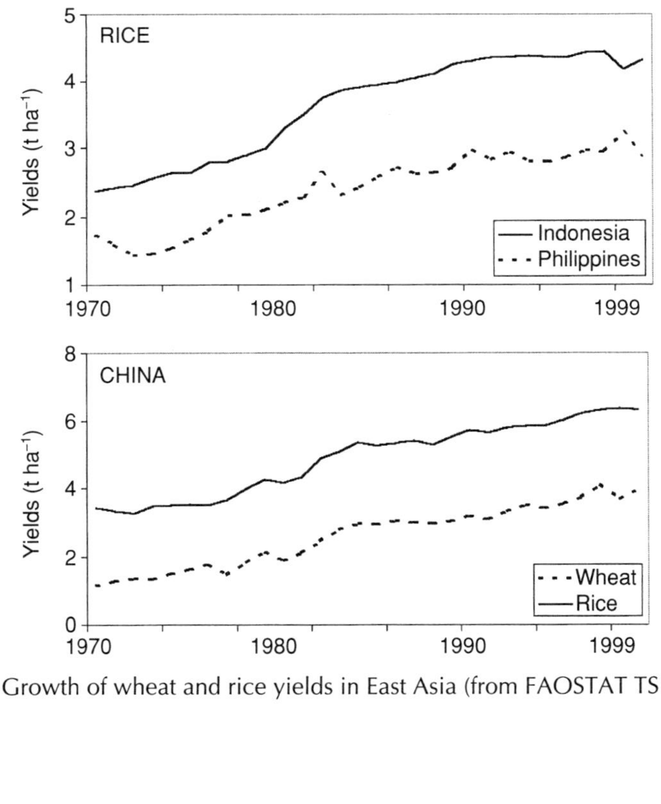

Fig. 2.4. Growth of wheat and rice yields in East Asia (from FAOSTAT TS : SOFA, 1999).

growing threats to the sustainability of the yields and production of the Green-Revolution lands (see Pingali and Rosegrant, Chapter 20 of this volume).

Agriculture and the Environment

The causes of this slowing in yield growth are not clear, although one factor is likely to be the cumulative effect of environmental degradation, partly caused by agriculture itself. The litany of loss is familiar. Soils are eroding and losing their fertility, precious water supplies are being squandered, rangeland overgrazed, forests destroyed and fisheries overexploited.

In the 1960s, when the Green Revolution was beginning to make its impact, little thought was given to environmental consequences. They were deemed either insignificant or, at least, capable of being easily redressed at a future date, once the main task of feeding the world was accomplished. There was also a strongly held view, one still commonly voiced, that a healthy, productive agriculture would necessarily benefit the environment. Good agronomy was good environmental management. It is a point with some force. Traditional agriculture is usually informed by ecological wisdom. Modern technologies can be as environmentally sensitive, but only if they are designed and used with the benefit of modern ecological knowledge.

Too often, over the past 30 years, the technologies accompanying the Green Revolution have turned out to have adverse environmental effects (Conway and Pretty, 1991). The heavy use of pesticides has caused severe problems. There is growing human morbidity and mortality while, at the same time, pest populations are becoming resistant and escaping from natural control. In the intensively farmed lands of both the developed and developing countries, heavy fertilizer applications are producing nitrate levels in drinking-water that approach or exceed permitted levels, increasing the likelihood of government restrictions on fertilizer use.

Increased and inefficient use of pesticides and nitrogen fertilizers produces severe pollution, but it is mostly local in its effect. Other agricultural pollutants have the potential for damage on a much larger scale. While industry is often to blame, agriculture is becoming a major contributor to regional and global pollution, producing significant levels of methane, carbon dioxide (CO_2) and nitrous oxide (Fig. 2.5). Natural processes generate these gases, but the intensification of agriculture in both the developed and developing countries has increased the rates of emission. Irrigated rice-fields in Asia have grown 40% in area since 1970, contributing to increased production of methane and ammonia (Conway and Pretty, 1991). Nitrous oxide emissions have grown in parallel with the use of nitrogen fertilizers. Ammonia and methane emissions have increased as a result of the intensification of livestock husbandry, and the clearance of forests and grasslands for arable land has raised the production of carbon and nitrogen oxides.

Agriculture is a highly significant and growing contributor to the total production of globally important gas emissions. Individually or in combination, these gases are contributing to acid deposition, the depletion of the stratospheric

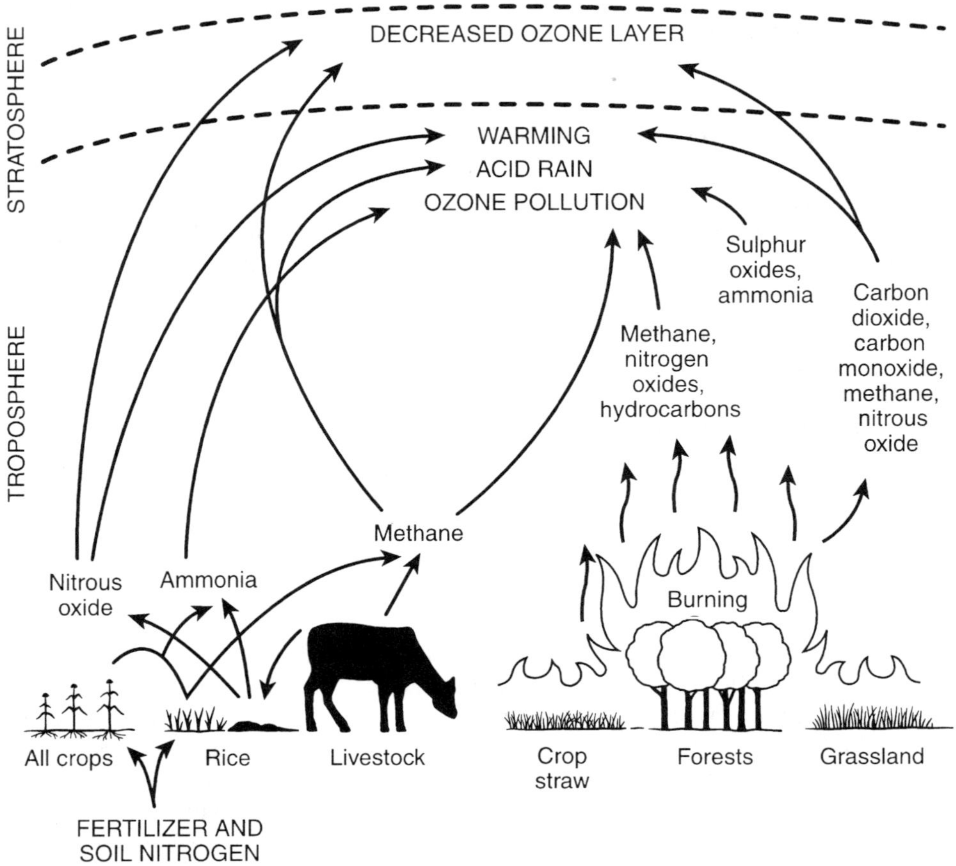

Fig. 2.5. Global pollution caused by agriculture (from Conway and Pretty, 1991).

zone, the build-up of ozone in the lower atmosphere and global warming. The effects of gas emissions on the natural environment and on human well-being are well known and do not need to be repeated here, but in each case there are significant adverse effects on agriculture.

The existence of the ozone layer is important for life on the planet since it serves to screen out the incoming ultraviolet (UV) radiation. High levels of UV light cause skin cancers in human beings and damage to plants (Worrest and Grant, 1989). Cereals are relatively tolerant, but many other crops, including legumes, squashes and cabbages, are easily damaged. Yields are reduced and, in soybeans and potatoes, the oil and protein contents are lowered.

Assessing the effects of global warming on agriculture is difficult because the temperature changes and their effects will vary from place to place and in ways that are not yet fully predictable. The greatest temperature changes will be at high latitudes, but water availability may worsen at lower latitudes. Heat and water stress may result in yield reductions, especially in the low latitudes where most of the developing countries are situated (Parry, 1990; Rosenzweig and

Parry, 1994). In contrast, in the middle and high latitudes, the increased CO_2 will have a physiological effect encouraging crop growth, particularly of so-called C_3 crops, such as wheat, barley, rice and potatoes.[3] On average, a doubling of CO_2 produces a 30% increase in yields in these crops (Reilly, 1996). Combined with higher average temperatures, this may increase production of grain and other crops in the developed countries.

Sea levels are also expected to rise, initially from the thermal expansion of the oceans and perhaps, eventually, as a result of the melting of the polar ice-caps. The best estimate, at current rates of global warming, is that the average sea level will have risen by up to 15 cm by the year 2020 (Houghton *et al.*, 1996). This is not a great amount, but it could lead to a greater risk of flooding in countries such as Bangladesh, where much of the cultivation is precariously sited in the delta of the Brahmaputra.

The most serious consequences are some time in the future, toward the end of the next century. Many doubt whether a rise of a further 0.4°C by the year 2020 is likely to have a major effect on agricultural production (Dyson, 1996). But there may be grounds for concern. There has already been a significant fall in precipitation in the subtropics and tropics since 1960. If this continues, it could soon have severe effects on agricultural production, particularly in the semi-arid tropics. The latest projections by the International Panel on Climate Change indicate a reduction in the rainfall produced by South Asian monsoons as a result of global warming (Houghton *et al.*, 1996).

Another consequence of global warming may be a greater variability in the weather and a higher incidence of extreme weather conditions with unpredictable effects (Katz and Brown, 1992; Reilly, 1996). Floods, droughts, hurricanes, extremely high temperatures and severe frosts may become more common. In the developing countries, rainfall may become more variable, possibly with a greater frequency of heavy rainstorms creating flooding and exacerbating soil erosion (Parry, 1990). The rainy season may also shorten, reducing the pre-monsoon rains, which are crucial for crop germination. There are some recent signs of a greater frequency of such extreme events. In both North America and sub-Saharan Africa, there were severe droughts in the 1980s and they may be responsible for the increased variability in cereal production over the same period in these regions.

The Doubly Green Revolution

These concerns, I believe, add up to a formidable challenge. If, over the next two to three decades, we are to provide enough food for everyone, we shall have to: increase food production at a greater rate than in recent years; do so in a sustainable manner, without significantly damaging the environment; and ensure that food is accessible to all. It is a daunting prospect, the magnitude of which becomes clear when we examine two contrasting scenarios of how this goal may be achieved.

Under the first scenario, the developed countries continue to produce food well in excess of their requirements and export this excess to meet the demand of the developing countries. If, as the aforementioned models assume, environmental constraints to increased food production can be overcome and if the existing food deficiencies of the poor are ignored, then there is little cause for concern. In the IFPRI model forecast, this would entail some 190 Mt of cereals being sold to the developing countries by the developed world in 2020.

However, if the food needs of the poor are not ignored, then under this scenario a further 300 Mt would be required in 2020 as subsidized or free food aid (assuming a global requirement for the undernourished of 3000 cal person^{-1} day^{-1}). This is equivalent to 30 times the current supply of direct food aid and would be extremely costly. Such massive food aid would place heavy burdens on both the donors and the recipients. For example, the environmental costs of such a scenario for the developed countries would be high. Most importantly, the availability of free or subsidized aid in such large quantities will depress local prices and add to existing disincentives for local food production.

These issues raise doubts about the viability of such a scenario, but a more fundamental objection is to an assumption implicit in the scenario – that a large proportion of the population in the developing world would fail to participate in global economic growth. An alternative scenario, which explicitly addresses this objection, is for the developing countries to undertake an accelerated, broad-based growth, not only in food production, but also in agricultural and natural resource development, as part of a larger development process aimed at meeting most of their own food production needs, including the needs of the poor.

Implicitly, this scenario recognizes that food security is not a matter solely of producing sufficient food. It is too simplistic to add up a nation's food production and divide it by the size of population. Nor is it enough to point to declining food prices. A nation is food-secure only if each and every one of its inhabitants is food-secure, that is, has access at all times to the food required to lead a healthy and productive life. To achieve this, each individual or, in practice, each household must grow sufficient food or be able to purchase the food from income earned either through selling agricultural products or by engaging in agricultural or non-agricultural employment. For urban dwellers, the only option is to engage in non-agricultural employment, but, for the vast numbers of rural poor, if they are not growing enough food to meet their needs, they must have the means to purchase the food they require. For them, food security depends as much on employment and income as it does on food production, and agricultural and natural resource development is crucial in both respects.

Food security, so defined, is also a key determinant of family size. The more confident women are about the immediate and longer-term future, the more likely they are to produce fewer children. Enhanced earning opportunities for women, as provided by the production, processing and trading activities generated by broad-based agricultural and natural resource development, can contribute to lower fertility rates. The greater the degree of security and the higher the level of their education, the more will women take advantage of new opportunities and plan ahead for themselves and their families.

In addition, appropriate agricultural and natural resource development can significantly contribute to greater environmental protection and conservation.

Properly designed, sustainable approaches to food production and to forestry and fishery management can reverse land degradation, reduce pollution from agrochemicals, remove pressure on national parks and reserves, conserve biodiversity and, at the same time, increase food security. Finally, vigorous agricultural and economic growth can stimulate world trade, providing significant benefits for all countries, developed and developing.

These arguments, when taken together, point to the need for a second Green Revolution, but a revolution that does not simply reflect the successes of the first. The technologies of the first Green Revolution were developed on experiment stations that were favoured with fertile soils, well-controlled water sources and other factors suitable for high production. There was little perception of the complexity and diversity of farmers' physical environments, let alone the diversity of the economic and social environment. The new Green Revolution must not only benefit the poor more directly, but must also be applicable under highly diverse conditions and be environmentally sustainable. By implication, it must make greater use of indigenous resources, complemented by a far more judicious use of external inputs.

In effect, we require a 'Doubly Green Revolution', a revolution that is even more productive than the first Green Revolution and even more 'green' in terms of conserving natural resources and the environment (Conway *et al.*, 1994). Over the next three decades, the Doubly Green Revolution must aim to repeat the successes of the Green Revolution, on a global scale and in many diverse localities, and be equitable, sustainable and environmentally friendly.

The complexity of these challenges is daunting, in many respects because of a greater order of sophistication than has gone before. There is certainly no case for abandoning technology. Indeed, at the outset we have to recognize there is much technology that has yet to be fully applied. In many regions, average farm yields are below those possible with only a modest increase in inputs and well below those achievable under experiment station conditions.

Yet, despite what can be done with these well-tried technologies, I believe the challenge of the Doubly Green Revolution is only likely to be met by exploiting two key, recent developments in modern science. The first is the emergence of molecular and cellular biology, a discipline, with its associated technologies, which is having far-reaching consequences in our ability to understand and manipulate living organisms. The second is the application of modern principles of ecology.

Biotechnology

Hitherto, the success of the Green Revolution has depended on working on blueprints of desirable new plant and animal types through painstaking conventional plant breeding. Biotechnology and genetic engineering offer a faster route and also the means of tackling the particularly intractable problems of drought, salinity and toxicity that typically face the poorest farmers on marginal lands.

A good start has been made in improving rice varieties using these technologies. In 1984, the Rockefeller Foundation launched its International Program on

Rice Biotechnology with the aim of facilitating the creation of a number of Asian centres of excellence in biotechnology. To date, over US$80 million has been spent on collaborative programmes with laboratories in the industrialized world, involving a network of some 700 researchers, fellows and advisers.

Practical results include the development, through tissue culture, of a new rice variety in China, named La Fen Rockefeller, now widely grown in the Shanghai area and producing yields 25% above previous varieties. Genetically engineered rices are now being developed that incorporate the *Bacillus thuringiensis* (*Bt*) gene, which confers resistance to insect pests. Other genetically engineered varieties confer resistance to bacterial blight, rice stripe virus and *hoja blance* virus. Molecular markers have been used successfully to incorporate multiple resistance in rice in India, China and Colombia. Of even greater significance in the long term is that rice is proving to be a model plant for cereal biotechnology. The techniques that have been acquired, and many of the genes themselves, have the capability of being utilized to improve varieties of other cereal crops.

The potentials for genetic engineering are almost endless. But there are serious hazards, some easily perceived, others yet to become apparent (Mooney and Bernardi, 1990; Casper and Landsmann, 1992). Perhaps the most obvious hazard is the possibility of a transferred gene being further passed through natural processes to another organism, with detrimental effects (Rissler and Mellon, 1996). Many crops have wild relatives and hybrids may occur naturally. For example, rice plants are self-pollinating, but not exclusively so. A degree of cross-pollination does occur under natural conditions, both among cultivated rices and between cultivated and wild rices. There is thus a possibility of the *Bt* gene transferring to wild relatives, particularly *Oryza nivara* and *Oryza rufipogon* (Clegg *et al*., 1993), which, by escaping natural insect control, could then become serious weeds. Weediness is a complex phenomenon resulting from a combination of many characteristics. The transfer of genes that increase a plant's competitiveness or resistance to stress will make it more likely to become a weed. As such examples show, we are increasingly dealing with risks that are unknown or overlooked because of the limited understanding of disciplines such as physiology, genetics and evolution (Rissler and Mellon, 1996).

More important than the potential hazards of biotechnology may be the question of who benefits from genetic engineering, and indeed from conventional breeding processes. Genetic engineering is a highly competitive business and, inevitably, the focus of biotechnology companies has been on developed-country markets, where potential sales are large, patents are well protected and the risks are lower. In this situation, public–private partnerships are going to be essential if developing countries are to benefit (Greeley, 1992).

The Application of Ecology

The second development important to meeting the challenges of the Doubly Green Revolution is the emergence of modern ecology, an equally powerful discipline, which is rapidly increasing our understanding of the structure and dynamics of agricultural and natural resource ecosystems and providing clues to their productive and sustainable management. Specifically, modern population, community and ecosystem ecology has direct relevance for pest, disease and weed control and for improvements to cropping systems and rangeland management.

The widely successful application of integrated pest management (IPM) to control rice pests in South-east Asia is proof of what can be achieved. IPM looks at each crop and pest situation as a whole and then devises a programme that integrates the various control methods in the light of all the factors present. As practised today, it combines modern technology – including the application of synthetic, yet selective, pesticides and the engineering of pest resistance – with natural methods of control, including agronomic practices and the use of natural predators and parasites. The outcome is sustainable, efficient pest control that is often cheaper than the conventional use of pesticides.

A recent, highly successful example is IPM developed for the brown planthopper and other rice pests in Indonesia (see descriptions in Kenmore, 1991a,b; Stone, 1992; Gallagher *et al.*, 1994). Under the programme, farmers are trained to recognize and regularly monitor the pests and their natural enemies. They then use simple, but effective, rules to determine the minimum necessary use of pesticides. The outcome is a reduction in the average number of sprayings from over four to less than one per season, while yields have grown from 6 to nearly 7.5 t ha^{-1}. Since 1986, rice production has increased 15% while pesticide use has declined 60%, saving US$120 million $year^{-1}$ in subsidies. The total economic benefit for 1990 was estimated to be over US$1 billion. Farmers' health has improved and an additional benefit has been the return of fish to the rice-fields.

The next challenge is to extend the principles of integration established in IPM to other subsystems of agriculture, for example, to nutrient conservation and to the management of soil, water and other natural resources.

Participation

However, a successful Doubly Green Revolution will not come from the application of biology alone. If the first Green Revolution started with the biological challenge inherent in producing new high-yielding food crops and then looked to determine how the benefits could reach the poor, this new revolution has to reverse the chain of logic, starting with the socio-economic demands of poor households and then seeking to identify the appropriate research priorities.

Biologists will have to listen as well as instruct. There will be no easy solutions and few, if any, miracles in the new revolution. Greater food production will come from targeting local agroecosystems, making the most of indigenous resources, knowledge and analysis. More than ever before, we shall have to forge genuine partnerships between biologists and farmers. It will not be enough simply to test new varieties on farmers' fields at the end of the breeding process. Experiments in many parts of the developing world are showing very effective ways of involving farmers right at the beginning, in the design of new varieties and in the breeding process itself (Sperling and Scheidegger, 1995).

Participation has long been a slogan of development. For the first time we now have effective techniques to make it a reality. Under the heading of participatory rural appraisal (PRA), there is a formidable array of methods which permit farmers to analyse their own situations and, most importantly, to engage in productive dialogue with research scientists and extension workers. PRA arose in the late 1980s out of earlier participatory approaches by combining semi-structured interviewing and diagram-making of farm production and resource utilization. It enables rural people to take the lead, producing their own diagrams, undertaking their own analyses and developing solutions to problems and recommendations for change and innovation. Maps are readily created by simply providing villagers with chalk and coloured powder and no further instruction other than the request to produce a map of the village, the watershed or a farm. A threshing floor or a cleared space in the village square is all that is needed to produce such a map, often of considerable complexity.

The approach has been rapidly taken up with enthusiasm, particularly by leaders of non-governmental organizations (NGOs) eager to find ways of creating greater levels of participation. The range of diagrams has quickly expanded: people who are illiterate and barely numerate can construct seasonal calendars using pebbles or seeds. Pie diagrams – pieces of straw and coloured powder laid out on an earthen floor – are used to indicate relative sources of income. Although this is, in itself, encouraging, it is the actual use of the diagrams that is important. Maps and seasonal and pie diagrams not only reveal existing patterns but point to existing problems and opportunities, and are often seized upon by rural people to make their needs felt. The diagrams have become a basis for collective planning, and the approach has begun to change the relationship between 'expert outsiders' and village people. The traditional position of rural people being passive recipients of knowledge and instruction has been replaced by the creation of productive dialogues.

PRA has now spread to most countries of the developing world and has been adopted by government agencies, research centres and university workers, as well as by NGOs. As a deliberate policy, no central guidebook has been produced, although much has been written and there is an extensive network of practitioners. The methodologies, which are described by a bewildering variety of names, have evolved according to local needs and customs, and reflect local ingenuity (Cornwall *et al.*, 1994; Chambers, 1997). *PLA Notes*, produced by the International Institute for Environment and Development in London and distributed to several thousand individuals worldwide, disseminates good practice and new ideas, so that innovations in an approach reported from an African village, for example, are being tried out in an Asian village only a few weeks later.

In some ways it has been a revolution, a set of methodologies, an attitude and a way of working that have finally challenged the traditional top-down process that has characterized so much development work. Participants from NGOs, government agencies and the research centres rapidly find themselves, usually unexpectedly, listening as much as talking, experiencing close to firsthand the conditions of life in poor households and changing their perceptions about the kinds of interventions and the research needs that are required.

Conclusion

The Doubly Green Revolution, as its name is meant to imply, is not simply about producing more food. That is a relatively easy task. If everyone is to be fed in this century, we need to pay attention not only to productivity, but also to sustainability – the preservation of the natural resource base and the avoidance of unacceptable levels of pollution – and to equitability – the access by the poor to the benefits of new productive technologies. But there are significant tradeoffs between these desirable outcomes. High productivity (entailing natural resource damage) may be at the expense of sustainability. Too much emphasis on sustainability – for instance, tough environmental protection – may inhibit the search for major gains in productivity, and both may, by concentrating technological access or by denying essential inputs, reinforce inequity. These are often difficult tradeoffs, involving tough choices. But there are ways, as this volume demonstrates, of seeking out situations involving – in various combinations – new technologies, the application of ecological knowledge and farmer participation, in which the tradeoffs can be minimized and 'win–win' solutions achieved.

Notes

1 This chapter is largely based on my recent book (Conway, 1997).

2 Underweight children (under 5 years old) are more than two standard deviations below the standard weight for their height. The undernourished are those individuals whose average daily intake of calories is not sufficient to 'maintain body weight and support light activity' (set at 1.54 times the basal metabolic rate) (United Nations Administrative Committee on Coordination/Sub-Committee on Nutrition, 1992). The threshold varies from 1760 cal day^{-1} in Asia to 1985 cal day^{-1} in Latin America.

3 C_3 and C_4 refer to different photosynthetic mechanisms. Most crops, especially in cooler and wetter habitats, are C_3. Plants such as tropical grasses, maize, sugar cane and sorghum are C_4.

3 Population, Agricultural Land Use and the Environment in Developing Countries

RICHARD E. BILSBORROW[1] AND DAVID L. CARR[2]

[1]*Carolina Population Center and* [2]*Department of Geography, University of North Carolina at Chapel Hill, Chapel Hill, North Carolina, USA*

Introduction

The earth's human population exceeded 1 billion for the first time in 1850. By 1999, the world's population reached 6 billion. The primary reason for such rapid growth during the past century and a half is that, despite the ongoing declines in mortality due to improved health conditions, fertility remained high for another generation or two in most countries (Teitelbaum, 1975). It is only since the 1970s that fertility has declined in most parts of the developing world.

In fact, the developing nations will be responsible for the entire increase of 2–4 billion persons that will be added to the global population during the next 50 years (based upon the United Nations (UN) medium fertility projections). This will increase their share of the global population from three-quarters to eight-ninths. Concomitantly, the loss of global forest cover is linked solely to deforestation in the developing world.[1] Once covering most of the earth, global forests have fallen to approximately a quarter of the terrestrial surface (FAO, 1995). If rates continue to accelerate as they did from 1984 to 1994 (by 40%), all primary tropical forests will be destroyed in another 50 years (Houghton, 1994).

Since sedentary agriculture dawned on the fertile banks of the Ganges, Tigris and Nile Rivers some 10,000 years ago (although recent evidence indicates that the first cultivated maize in central Mexico goes back nearly as far), growing populations have been fed by the expansion of agricultural land, mostly at the expense of forested areas. Despite technological advances and an increase in crop intensification, agricultural conversion was most dramatic during the 20th century. In 1900, 40 Mha of land were in crops around the world; by 1993, cropland had risen to 248 Mha (Brown, 1997). Resource degradation linked to rapid deforestation has led to chronic underdevelopment in rural areas where primary resource extraction is necessary for subsistence (Stonich, 1989; Dasgupta, 1995b).

Since the earth's land is finite, continued high rates of population growth in developing countries are thought by many to present urgent challenges to the sustainability of systems of food production and consumption (e.g. Meadows *et al.*, 1992; Pimentel, 1995; Brown, 1997). So far, with the exception of sub-Saharan Africa, agricultural production has outpaced population growth. However, global production has recently grown at a slower rate, the annual average rates of growth being 3.0, 2.3 and 2.0 in the 1960s, 1970s and 1980s, respectively (WRI, 1996). Further declines may be occurring in the 1990s (Brown, 1997).

The next section considers theoretical aspects of the relationship between trends in population, land use and agriculture; this is followed by an overview of the major world regions. Recent trends in the extensification and intensification of agriculture and population change will then be examined at the cross-country level for Latin America. This region is of particular interest since, during the first half of the 1990s, twice as much forestland was converted to agriculture there as in any other world region (World Bank, 1998b). A summary of some case studies illustrates the wide variability of relationships between population change and land-use change and the key roles played by contextual factors. Finally, we summarize and note gaps in knowledge and the need for more research at the micro or farm level.

Land-use Responses to Population: Some Theoretical Issues

Musing on human responses to population growth and land use has a long and diverse ancestry dating back to the Zoroastrians around 325 BC, the Indian sage Kautilya in 300 BC, the Bible, and Aristotle in his *Politics* (Petersen, 1972). However, the parson Thomas Malthus is credited with developing the first comprehensive theory of population–land use relationships. Malthus is most commonly attributed with having predicted that population growth would lead to famine and an eventual population crash, since, he noted, whereas human populations grow geometrically, food production tends to increase only arithmetically (Malthus, 1803).

But, in other passages rarely cited, Malthus also said that, since the most productive land tends to be used first, as population grows and the area used for agriculture expands with it, the average quality of new agricultural land brought into production declines, and thus mean land productivity also declines (Malthus, 1803). In addition, where land is fixed, classical economists noted that increased applications of labour lead to a fall in mean output per worker through the law of variable proportions, more commonly referred to as the law of diminishing returns (e.g. Ricardo, 1887).

Following more than a century of technological advances, including advances in agriculture that Malthus could not have foreseen, Boserup (1965) introduced the notion that technological change could mitigate the effect of population growth on food supply by facilitating increases in food production;

that is, as available arable land becomes scarce relative to labour, societies adopt more labour-intensive techniques, which take advantage of increased labour–land factor ratios. Boserup enumerated five stages of land intensification through which agrarian societies evolve over time: (i) forest or long fallow (20–25 years between crops); (ii) bush fallow (6–10 years); (iii) short fallow (1–2 years); (iv) annual cropping; and (v) multiple cropping (more than one crop per year on the same land). These forms of increasingly intensive use of land embody increasing inputs of labour per unit of land per unit of time.[2]

Based on a review of historical processes in Japan and Europe, Davis (1963) proposed that demographic responses to population pressures on the land may occur: that is, rural populations may respond by reducing fertility through postponement of marriage and increased celibacy, increased regulation of fertility within marriage and/or increasing abortions. Davis considered out-migration as a last resort if these responses proved inadequate. He also proposed that the various responses may occur simultaneously (or 'multiphasically'), and that the more one response occurs – and the more the effects of population pressures on the land are thereby relieved – the less likely other responses are to occur.

In support of Davis's hypothesis, Friedlander (1969) found that fertility reduction preceded out-migration in preindustrial France. Scarce employment opportunities in French urban centres deterred rural–urban migration. With shrinking farmland per capita due to high fertility and the absence of out-migration possibilities, France thus became the first European nation to experience declining marital fertility, even prior to industrialization. According to Friedlander, fertility will decline to the extent that communities are constrained in relieving population pressures through out-migration.

However, studies of demographic responses (e.g. Davis, 1963; Friedlander, 1969; Mosher, 1980) neglect economic and land-use responses and vice versa. An attempt at synthesizing theoretical views is found in Bilsborrow (1987), who classified responses as economic (land intensification and extensification), demographic (fertility responses) and economic–demographic (out-migration). He hypothesizes that households traditionally exhaust economic options first, beginning with land expansion. If that is insufficient, then available land intensification technologies may be adopted, involving a decline in leisure time. If such adjustments together still prove inadequate, the next reaction is likely to be out-migration, temporary or seasonal migration at first, since that does not require giving up one's home and community. Permanent out-migration is likely to follow if prior adjustments, along with the remittances and savings brought by the temporary migrant, prove inadequate. This may take the form of either migration to urban areas, especially if wage levels are much higher there, or migration to other rural areas. The latter is more likely in countries with lands that have potential value for agriculture but are still unoccupied. This is particularly true of countries with significant endowments of tropical forests – for example, in South America and Central Africa. Finally, since customs producing high fertility are so deeply rooted in culture, fertility reduction is postulated to occur in traditional societies only as a *last* resort, when other adaptations have proved insufficient (see Bilsborrow and Geores (1992) for a more detailed discussion).

As illustrated in Fig. 3.1, these responses are mediated by factors operating at multiple scales within the physical and human environments (Bilsborrow, 1997). The following are some of the factors shown to influence household responses: (i) quantity and propinquity of potentially arable land (Boserup, 1965, 1981; Bilsborrow and Geores, 1994; Sambrook *et al.*, 1999); (ii) quality of natural resources, as manifest in climate, rainfall, topography and soil quality (Sanchez and Cochrane, 1980; Blaikie, 1987; Barbier, 1990; Buol, 1995); (iii) type of land tenure regime (Forster and Stanfield, 1993; Utting, 1993; Schmink, 1994); (iv) land distribution and the extent of landlessness (Stonich, 1989; World Bank, 1995a; Valenzuela de Pisano, 1996); (v) urban employment opportunities that may attract migrants (Stern, 1976; Alberts, 1977; Gilbert, 1994; Altamirano *et al.*, 1997); (vi) proximity and accessibility to product and labour markets (Rudel, 1983; Rudel and Richards, 1990); (vii) knowledge of and access to alternative forms of agricultural technology (Boserup, 1965; Zimmerer, 1993); and (viii) government policies, including those related to land distribution and tenure; tax and expenditure policies (including subsidies), especially those specifically encouraging extensification and intensification; transportation infrastructure (Bilsborrow, 1987; Hecht and Cockburn, 1990; Rudel and Richards, 1990); the fixing of prices for agricultural products (Garland Bedoya, 1991; Stewart, 1994); and family planning, education and health policies which affect fertility and mortality (Bilsborrow and Geores, 1992). With these

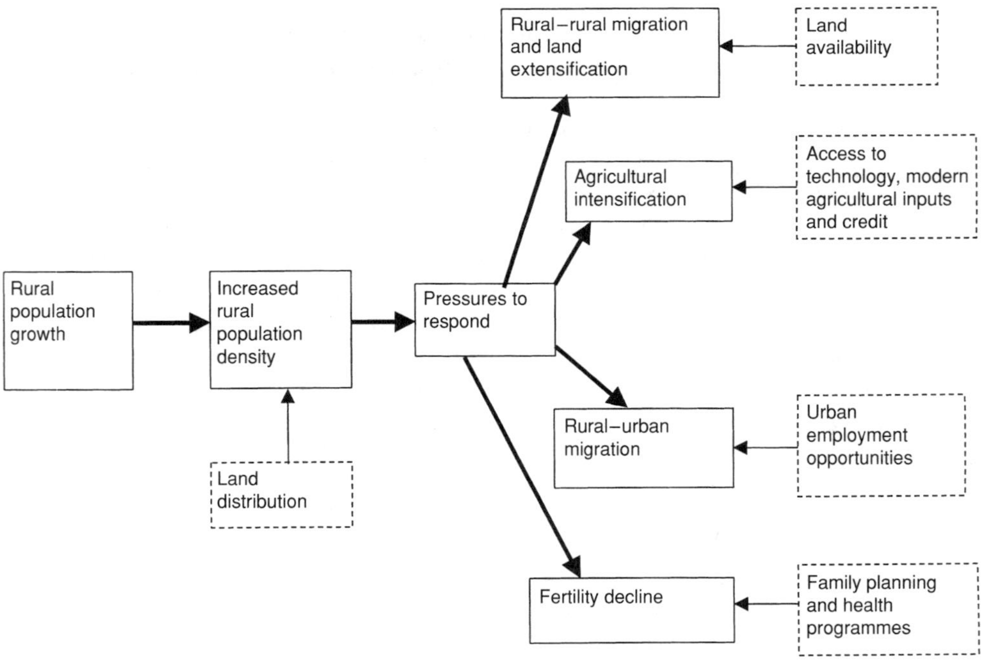

Fig. 3.1. Economic, demographic and economic–demographic responses to rural population growth.

conditioning mechanisms in mind, what does empirical evidence tell us about responses to population change among the major world regions?

A Glance at Global Relations

In an earlier paper, the first author looked at continent-wide relationships from a broad multiphasic perspective, based on estimates of trends over time (Bilsborrow, 1987). Data on demographic and economic changes in five developing country regions from the 1960s to 1980 were presented. Demographic change was measured by the change in the total fertility rate and economic–demographic change by the annual rate of out-migration from rural areas. Economic (or agricultural) changes comprised land extensification, measured by the percentage change in the share of arable and cropped land in total land, and land intensification, measured by the percentage change in rural labour productivity. The developing-country regions with the largest responses of one type did indeed tend to have smaller responses of the other types, as predicted by the multiphasic approach, and major deviations from the expected tradeoffs were generally explainable by exogenous factors, such as differences across regions in natural resource endowments or government policies.

Now that over a decade has passed, it is desirable to update that overview. We have prepared estimates of changes in the same four factors, up to and including 1995 (Table 3.1), revising the earlier period estimates using data sources appropriate for the latter period. The measures of all variables are the same as earlier, except that better estimates of rural out-migration have become available recently from the UN (1997). The caveats pertaining to the earlier study – especially the riskiness of cross-country comparisons, due to inconsistencies across countries and deficiencies in the underlying data – should still be kept in mind. We have presented data here for medians as well as for means, and we note where the means would be considerably different with the exclusion of outliers.[3]

Note that what we are concerned with here is the association between the change in rural population density and the associated demographic, economic–demographic and agricultural changes over two time periods. We observe at the outset that, at a continent-wide level, there was a slight increase in rural population density in Latin America in the first period (1965–1980) but none in the second (1981–1995), as fertility and natural population growth declined to equal out-migration from rural areas. In all other regions, however, rural population density rose substantially, most notably in China. In Africa, the Near East, China and especially 'Other Asia', the increase in rural density was much greater in the second period than in the first, implying that there should be greater responses.

Regarding economic or land-use responses, we expect agricultural land to increase where there is untapped land, which occurred in Latin America and 'Other Asia', as expected. Secondly, agricultural labour productivity rose substantially, except in sub-Saharan Africa, but this occurred more in the first time period (except for the Near East), probably due to the earlier exploitation

Table 3.1. Population density and economic, economic–demographic and demographic responses.

	Economically active rural population/A&P land[a]			Percentage change in arable and permanently cropped land		Percentage change in total fertility rate		Percentage change in rural labour productivity[b]		Rate of rural out-migration	
Regions	1965–1966	1979–1980	1994–1995	1966–1978	1979–1994	1960–1965 to 1975–1980	1975–1980 to 1990–1995	1966–1980	1980–1995	1960s–1970s	1980s–1990s
Latin America	0.5	0.5	0.5	19	9	–22	–26	29	11	–1.7	–1.8
Other Asia	1.1	1.2	1.7	6	19	–12	–26	27	5	–0.7	–2.0
China	2.0	2.8	5.4	–18	–5	–42	–42	22	0	0.0	–1.5
Near East	0.5	0.7	1.0	–10	11	–11	–24	2	71	–1.7	–1.3
Africa	0.7	1.0	1.6	–7	9	2	–8	1	–13	–0.8	–1.3
Developing world means	0.9	1.2	2.0	–2	9	–17	–25	25	13	–1.4	–1.6
Latin America	0.5	0.5	0.4	8	6	–21	–25	22	12	–1.7	–1.7
Other Asia	0.9	0.9	1.5	2	7	–13	–24	33	–7	–0.5	–1.6
China	2.0	2.8	5.4	–18	–5	–42	–42	22	0	0.0	–1.5
Near East	0.2	0.7	1.0	1	6	–10	–26	–5	60	–1.5	–1.7
Africa	0.6	0.9	1.4	0	5	1	–9	–3	–14	–0.8	–0.7
Developing world medians	0.6	0.9	1.4	1.4	6	–13	–25	23	11	–1.2	–1.3

[a]A&P or 'arable and permanently cropped land' is land that is under temporary crops, pasture, gardens and land that is in fallow under 5 years. 'Permanent crops' include land cultivated with crops for long periods of time and which do not need to be replanted after each harvest (e.g. coffee, rubber and fruit trees and vines).

[b]Calculated as follows for 1966–1980: (1980 agricultural production index per 1980 economically active population in agriculture)–(1966 agricultural production index per 1965 economically active population in agriculture)/(1966 agricultural production index per 1965 economically active population in agriculture). The procedure for 1980–1995 is similar.

All land use and production data calculated from FAO (1977a, 1984, 1993c, 1995).

Fertility data calculated from UN (1999).

Migration data are taken from UN (1997).

of Green Revolution technologies and the associated use of inputs. Fertility changes mainly occurred starting in the latter part of the first period – most of the decline in Latin America before 1980 was during 1975–1979 – and accelerated during the second period, as family planning spread through most of the developing world. Meanwhile, out-migration from rural areas was substantial in most regions in both periods, though again slightly higher in the second period.

We now examine the evidence on economic, economic–demographic and demographic changes and tradeoffs in each of the developing-country regions, and compare and contrast them for the two time periods. First, Africa, with a *prima facie* low rural population density (but not low when water supply limitations are taken into account), experienced a modest decline in arable land and in per capita agricultural production during the 1960s–1970s. During the later 1980s–1990s period, rural labour density increased rapidly, despite notable expansion of agricultural land and some out-migration because of continuing high fertility. The economic crisis pervading the region through both time periods, combined with low levels of education, undoubtedly contributed to the lack of increase in agricultural productivity. In the first period, the only response to rising population density was out-migration, but, as pressures on the land increased considerably, in the second period multiple responses are evident, with land extensification, higher rural out-migration and the onset of a fertility decline. Since we expect extensification to precede intensification, in the case of Africa in the latter period, the lack of intensification and the increasing desperation of rural populations may have led to pushing out the agricultural frontier as populations sought land to feed their families.

The Near East has little arable land that is not already used, so increased production has generally been possible only through increased irrigation. However, during the first time period, rural population density remained low and the only response observed was out-migration. But, as the rural population became denser during the second period, the full range of responses occurred, including a large increase in agricultural output per rural worker (presumably as the Green Revolution technology made its way into the region) and an acceleration in the pace of fertility decline, which began to approximate that of other regions. Looking at each period in terms of tradeoffs, in the first period the lack of significant other responses put more pressure on out-migration; in the later period, continuing increases in rural density are associated with multiple responses.

China experienced rapid increases in rural population density in both periods, especially in the second. Agricultural land actually declined significantly in the first period and slightly in the second (the latter confirmed in Heilig, 1999). China rapidly adopted new agricultural technology in the earlier period, but the Food and Agriculture Organization of the UN (FAO) data suggest no further intensification in the latter period (which is inconsistent with recently reported macroeconomic indicators of the agricultural sector). Also, in the first period it is known that China restricted migration to urban areas and imposed a one-child policy, which forced a substantial fertility decline. Whether such policies should be considered completely exogenous or as being themselves induced by population pressures is a moot point, but the changes did occur. In the more recent period, with economic liberalization, changes appear to be primarily demographic, with a continuing rapid decline in fertility and accelerating rural–urban migration. China has evidently experienced the swiftest fertility decline of all regions, in both time periods. In terms of tradeoffs, in the absence of a land response, in the first period, China responded to an increase in rural population density by land intensification and a draconian fertility control policy. In the second period, both fertility decline and out-migration occurred.

In the 'Other Asia' region, rural density increased only slightly in the first period but rapidly in the second, so we would expect much more in the way of responses in the second period. This is indeed the case, with little change in agricultural land or fertility in the first period, and only modest rural–urban migration. The only major change was agricultural intensification, associated with the Green Revolution in India and other countries in the region. In the second period, the changes that occurred were completely the inverse, with little intensification (this having seemingly run its course) but a large increase in cropped land, a decrease in fertility and out-migration from rural areas. In the first time period, the lack of increase in density and the success of intensification deferred pressures to respond in other ways, while in the later period rapidly rising density tended to stimulate multiple responses. Government policies also played a role in encouraging fertility declines in some countries such as India, Indonesia and Thailand. Furthermore, land availability in many countries was associated – as expected – with continued land extensification in the second period.

Finally, in Latin America, rural population density rose slightly in the first period and not at all in the second. Thus, overall, one would not expect many of the changes observed in the suite of economic, economic–demographic and demographic responses to be strongly linked to population pressures in general. In the first time period, substantial extensification occurred, befitting the region's plentiful land endowments. At the same time, agricultural productivity rose and rapid rural out-migration occurred, associated with urbanization, which had begun earlier. Fertility started to drop only in the 1970s in most countries, associated with public- and private-sector family-planning provision and the demand to reduce actual fertility to desired levels, impelled by high levels of urbanization and improved women's education, following cumulative trends over previous decades. In the second period, associated with no increase in density, there was much less agricultural extensification and intensification, but rural out-migration and fertility decline continued. Scant economic opportunities in rural areas, land inequality, considerably higher wages in cities, and the urban bias in development policies contributed to consistently high rural–urban migration: from the 1960s to the 1980s, nearly 2% of rural Latin Americans out-migrated annually (Chen *et al*., 1998, p. 82). While the changes observed in Table 3.1 in Latin America would appear exogenous relative to changes in population density, substantial differences exist from country to country which may be related to changes in rural population growth and density at the country level (see next section).

Although the data here illustrate broad differences across regions, more empirical work is needed to understand the interrelationships between tradeoffs. In the next section, we examine the agricultural sector in Latin America at the country level, looking at trends in agricultural extensification and intensification and their possible relationships to population change.

Population Density, Land Use and Deforestation in Latin America

Central and South America are different from other developing regions in having much more potential arable land available, more current agricultural land in pasture, higher levels of forests being converted to agricultural land and in being far more urbanized. Amounts of land being deforested were more than double those of other continents during the first half of the 1990s (World Bank, 1998b). But differences are substantial within the region. For the present analysis, it is important to distinguish between Central and South America. First, resource endowments are very different, largely due to the presence of the world's largest tropical forest, the Amazon, which covers parts of eight South American nations. Secondly, fertility has declined far more broadly in South America than in Central America (taken here to refer to Mexico and the Caribbean as well). Taking into account both lower rural fertility and significant net rural–urban migration, the rural population in South America as a whole actually declined slightly from 1984 to 1994. As a result, while mean rural population density in South America was less than half that of Central America in 1966, it has since fallen to about a third.

That there is a strong relationship between population density and the percentage of land area covered by forests is hardly surprising, and is evident at a global level in the differences in the percentage of land area covered by forests in Asia and Europe compared with sub-Saharan Africa and the western hemisphere. A much more sensitive – as well as contemporary and dynamic – test of the relationship is that of first differences – that is, of the relationship between population change and reduction in forest area.

The difference in the two regions of Latin America is significant not only because greater population densities can be expected to be associated with greater cumulative levels of deforestation, but also because countries with small areas remaining in forests can have very high rates of deforestation even when the absolute areas of forests being lost are small (e.g. countries in Central America). Other countries with large forest stocks may experience very low annual rates of deforestation even when the absolute area being cleared is huge (e.g. Brazil).

The usual simple measure of national population density (total population/total land area) ignores the fact that countries have large areas unusable for agriculture (e.g. mountain ranges, deserts, bodies of water and urban areas), and also confuses the issue by combining urban and rural populations. We therefore measure population density by rural population density (rural population per unit of arable land plus land under permanent crops), and use this for analysing population impacts on land use and land-cover change. The numerator is modified through population change (fertility, mortality and net migration), while the denominator is affected by changes in the amount of arable and permanently cropped land. We exclude pastureland from the denominator since it has little direct relationship to population density.

Rural population dynamics, forest-cover change and agricultural extensification

Urbanization and rural fertility decline occurred earlier in South America than in Central America. In fact, during the last three decades, Central America gained approximately 8 million rural residents, while South America lost 3 million. With the exception of pioneering countries, such as Costa Rica and Cuba, where fertility decline began in earnest in the 1960s, it was not until the 1970s that fertility began to fall generally in Latin America. None the less, fertility decline has been delayed in a number of Central American countries (Guatemala, Honduras, Nicaragua and Haiti).

Table 3.2 shows that in Central America, from the 1960s to the mid-1990s, rural population density grew (from 2.2 to 2.4 persons ha^{-1} arable and cropped land), as modest increases in the area of land in crops (29%) were still outpaced by rural population growth. Only in the Dominican Republic and Honduras did growth in agricultural land exceed population growth. In South America, in contrast, greater agricultural expansion, earlier and greater rates of rural–urban migration and earlier fertility decline together led to a slight decrease in overall rural population density, from 1.0 to 0.9 persons per hectare.

Latin America harbours the greatest area of closed tropical forests in the world, with Brazil alone containing a third of the world's tropical forests. The Amazon basin also contains 45% of all the fresh water on the earth, and is the planet's largest carbon sink. Its preservation is thought to be crucial for moderating global warming and regulating global weather systems (e.g. Adger and Brown, 1994; Tinker *et al.*, 1996). Yet Latin Americans have been clearing much more forest than those elsewhere: during the first half of the 1990s, Latin Americans deforested five times more forest per rural person than Africans and 40 times more than Asians (derived from data in FAO, 1997b). Furthermore, reforestation has only minimally compensated for forest losses in Latin America, and far less than in other regions. Thus, in the Asia–Pacific region, reforestation compensated for over 50% of the forest loss from 1981 to 1992, while in Latin America it was around 6% (FAO, 1997b).

It is instructive to examine more closely trends in population and forest cover in the two regions by decade. As is evident from Table 3.3, from 1966 to 1976, Latin America's forest cover decreased only slightly (by 3%), accompanied by a 5% increase in rural population density. But in the ensuing decade or so, from 1976 to 1984, deforestation accelerated sharply in both Central America and South America. High rates of deforestation were accompanied by increases in rural population density of 40% and 8%, respectively. Finally, deforestation in the latest decade (1984 to 1994) was much lower in both regions compared with the previous decade, and was accompanied by increased agricultural intensification, particularly in Central America.

We expect deforestation to be linked to the expansion of cropland and that, at higher rural population densities, a greater proportion of land is likely to be in

Table 3.2. Latin America: rural population density and percentage of land forested.

	Rural population per hectare of agricultural land				Forest land/total land (%)[a]			
Country	1966	1976	1984	1994	1966	1976	1984	1994
Central America								
Costa Rica	1.9	2.4	5.0	3.2	54	50	32	24
Cuba	1.9	1.1	1.1	0.8	14	11	24	22
Dominican Republic	2.8	2.8	2.6	1.9	23	23	13	10
El Salvador	2.9	3.7	4.8	4.3	11	13	6	4
Guatemala	2.1	2.3	3.7	3.2	56	54	42	34
Haiti	4.6	4.5	7.8	5.3	11	7	5	3
Honduras	2.1	2.4	1.6	1.5	63	63	54	43
Mexico	0.8	0.8	1.0	0.9	40	36	25	22
Nicaragua	0.7	0.8	1.3	1.3	49	53	33	26
Panama	1.3	1.6	2.3	1.8	49	55	51	41
C. Am. total	2.1	2.2	3.1	2.4	37	36	28	23
C. Am total[b]	1.1	1.1	1.4	1.3	40	38	27	24
South America								
Argentina	0.2	0.1	0.2	0.2	22	22	19	18[c]
Bolivia	1.3	0.8	1.3	1.2	55	52	53	47
Brazil	1.3	1.1	0.8	0.7	62	60	60	57
Chile	0.5	0.4	0.5	0.5	28	28	22	21[c]
Colombia	1.7	1.8	1.9	1.8	74	74	50	47
Ecuador	1.0	0.8	1.8	1.6	66	65	56	47
Paraguay	1.4	1.6	1.0	1.0	52	51	46	35
Peru	2.1	1.8	1.7	1.6	58	58	53	53[c]
Uruguay	0.3	0.2	0.3	0.2	3	3	5	5[c]
Venezuela	0.5	0.5	0.6	0.4	54	54	36	32
S. Am. total	1.0	0.9	1.0	0.9	47	47	41	37
S. Am. total[b]	0.8	0.7	0.7	0.7	53	52	48	45

[a]Unless otherwise mentioned, all 1994 data on forested land were calculated averaging deforestation rates from 1981 to 1990 from FAO.
[b]Weighted totals.
[c]Estimated from deforestation rates from 1990 to 1995 in FAO (1997a).

crops for human consumption. A correlation between land in crop production and population density is strikingly evident at the regional level. Thus, when rural population densities are computed with the total land area in the denominator, rural Central America was seven times more densely populated than rural South America in 1966 and 11 times more in 1994. When rural population density was measured using only arable and permanently cropped land in the denominator, it was never more than three times as high in Central America during the period. As is evident from Table 3.3, the difference is because Central America had much more arable and permanently cropped land (15% compared with 6% in 1966, and 21% compared with 6% in 1994).

Table 3.3. Extensification: percentage of land in pasture and arable and cropped land.

	Pastureland as % of total land		A&P land and pastureland as % of total land		A&P land as % of total land	
Country	1966	1994	1966	1994	1966	1994
Central America						
Costa Rica	23	46	32	56	10	10
Cuba	24	27	39	58	15	31
Dominican Republic	24	43	43	74	18	31
El Salvador	29	29	59	65	30	35
Guatemala	9	24	23	42	14	18
Haiti	22	18	49	51	27	33
Honduras	18	14	25	32	7	18
Mexico	37	39	50	52	13	13
Nicaragua	28	45	40	56	12	10
Panama	13	20	21	29	7	9
Total (weighted)	32	36	46	51	13	15
Total (unweighted)	23	31	38	51	15	21
South America						
Argentina	53	52	64	62	11	10
Bolivia	26	24	27	27	2	2
Brazil	17	22	20	28	4	6
Chile	14	18	20	24	6	6
Colombia	16	39	21	44	5	5
Ecuador	8	18	20	29	12	11
Paraguay	35	55	38	60	2	6
Peru	21	21	23	24	2	3
Uruguay	78	77	89	85	10	7
Venezuela	17	20	23	25	6	4
Total (weighted)	24	29	29	35	5	6
Total (unweighted)	29	35	35	41	6	6

A&P, arable and permanently cropped.

As illustrated in Fig. 3.2, most agricultural land in Latin America is in pasture. The conversion of forest to pasture cannot be directly linked to rural population growth. Thus, the more densely populated an area, the more likely it is to have land in crops rather than pasture. Indeed, as the rural population grew by 47% from 1966 to 1994 in the already densely populated region of Central America, cropland expansion kept pace with growth in pastureland. Conversely, in less densely populated South America, the rural population increased by only 13% per decade on average and the percentage of cropland remained fixed at only 6% of the continent's land area. Thus, all of South America's 21% net growth in cleared land over the 30 years was due to the expansion of pasture or other land uses. In 1994, South America had five times more pastureland than cropland. Even in Central America the ratio was 3 : 1.

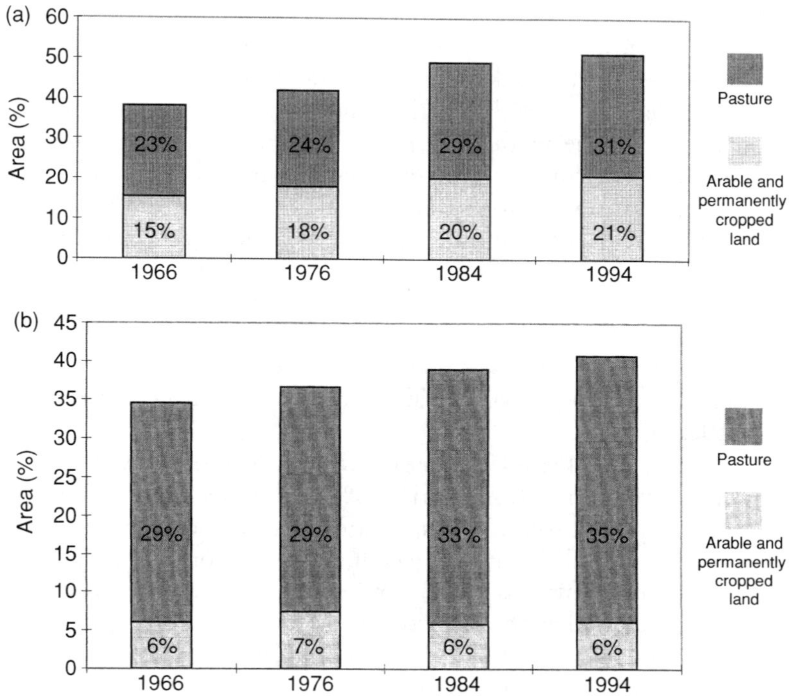

Fig. 3.2. Pasture and arable and permanently cropped land as a percentage of total land (a) in Central America and (b) in South America.

However, it is important to note that much pastureland in Latin America is converted from cropland and not from forests directly (see, for example, Hecht and Cockburn, 1990; Nations, 1992). This is most clear in the case of Brazil, where the total growth in pastureland from 1966 to 1994 equalled the total area of forest eliminated. Below we review land-use changes in more detail, first in Central America and secondly in South America.

Central America

The percentage of forests cleared in Central America in recent decades is astounding. The area in forest extant in 1966 that was cleared by 1994 surpassed the area of forestland remaining in 1994 in all but three countries in the region. Still, the percentage of forests lost was less than the percentage growth in the rural population in every country. For example, Guatemala cleared 38% of its forests, and was left with but 35% of its initial level, while the rural population grew by 97%. Between 1966 and 1994, more than 20% of the total national territory was deforested in Guatemala, Nicaragua and Costa Rica; at over 7% per annum, the deforestation rate in Costa Rica during the 1970s was probably the highest in the world, reducing its stock of forests by over a half in a single decade. Three nations in the region cleared more forests in the 28 years than they had remaining in 1994: El Salvador, the Dominican Republic and Haiti. In these nations, deforestation rates must fall in the future, since there is little left to clear.

South America

As in Central America, forest loss in South America has proceeded at a rapid pace in recent decades. But rural population density is generally much lower than in Central America, and many of the South American countries had no increase in the size of their rural populations over the observation period. As a consequence, changes in rural population density due to net rural population growth cannot explain much of the deforestation in South America. Government policies favouring cattle, especially including subsidies for ranching in Brazil in the 1980s, strong export markets and rising domestic demand from growing and increasingly prosperous urban populations help to explain the large growth of pastureland in South America. In addition, policies to keep the prices of basic grains low to benefit urban consumers at the expense of rural producers have provided incentives for farmers to switch from crops to cattle.

Over the 1966–1994 period, Brazil had the lowest percentage rate of deforestation in Latin America (8%), while its rural population declined by 16%. But, in absolute terms, its area of forest loss, primarily in the Amazon basin, far exceeded that of any country in the world. In industrialized countries such as Argentina and Chile, rural populations also declined but deforestation rose (associated with an increase in timber harvesting and an expansion of export agriculture, rather than with population growth). On the other hand, in a few South American nations – for example, Ecuador and Paraguay – rural population growth may well have contributed to deforestation. In both, forest clearing facilitated an increase in arable and permanently cropped land by nearly a half and a more than doubling of pastureland. By 1994, Ecuador's share of total land in cropland (11%) was the highest in South America, while its share in pastureland remained the lowest, albeit rapidly growing.

Even more than in Central America, however, it is the conversion of cropland and forest to pasture that has been the most dramatic land-cover change. Thus, while Colombia's rural population grew slowly, by only 10% over three decades, pasture grew by 144%, while its crop area remained constant at only 5% of its territory – barely one-ninth the area in pasture in 1994. In Venezuela, the rural population declined 44%, and yet 41% of its forested land in 1966 was still cut by 1994. Evidently, even where rural populations are declining, agricultural frontiers have still expanded.

Although rural–urban migration continues to be significant and to contribute to urban growth,[4] rural–rural migration is of growing importance, with people migrating from rural areas characterized by high poverty and unequal land distributions to other, more remote, rural areas in search of land to continue agricultural livelihoods. Thus, contrary to what some scholars expect, even in countries and areas of overall rural population decline, substantial deforestation can still occur. On the remote frontiers, colonists are migrating along newly established roads, converting forests to cropland and pasture (e.g. Rudel, 1983; Rudel and Richards, 1990; Pichón, 1997a,b; Sader *et al.*, 1997).

A comparison

Despite considerable country-to-country variation, some patterns do emerge. First, where population density is lower, the level of deforestation is lower, but the rate of deforestation per capita is higher (e.g. South American nations compared with Central America; on a global scale, the same comparisons hold for Latin America relative to Asia). Secondly, countries with the highest rural population densities (e.g. Haiti and El Salvador) deforested their land earlier. Deforestation rates remain high because of a shrinking base, but these countries have increasingly lower deforestation rates per rural dweller as forest reserves approach extinction. Thirdly, pastureland accounts for a greater proportion of agricultural land than cropland in Latin America, and its share has risen in recent decades.

To maintain current per capita food consumption levels in coming decades, Latin America will need to greatly increase food imports or increase domestic production through intensification of current agricultural land and/or expansion of the land area in food production. In countries where little forest remains, as in many Central American nations, intensification will be the main option for increasing food production. On the other hand, among South American nations with large endowments of Amazon jungle, extensification is still occurring on a large scale, and could continue. But the sustainability of Amazonian agriculture is dubious as most of the region's biomass is rooted in nutrient-impoverished soils (Buol, 1995).

Agricultural intensification

Starting in the 1950s, international agencies promoted technological advances in agriculture in developing countries in what became known as the Green Revolution. During these years, new seeds were developed for basic grains, and irrigation, pesticide and fertilizer use all grew dramatically. Most of the growth in agricultural production of the past 30 years is attributable to increases in yields per hectare: from 1961 to 1996, global cereal yields increased 107%, while the area harvested increased by only 10% (Bender and Smith, 1997). Only in Latin America has the expansion of the agricultural land area been an important contributor to the expansion of agricultural output, but this share has been declining and is now less than that of increases in yields.

In developing countries, agriculture accounts for 80% of water use (IFPRI, 1996). The growing scarcity of fresh water limits its contribution to increasing agricultural yields in much of the world, though this is not a pervasive problem in Latin America. The percentage of agricultural land irrigated rose from 9% to 14% in Central America and from 11% to 14% in South America from 1966 to 1994. As is evident from Table 3.4, the countries that increased irrigation most were not those with the highest rural population densities or the most rapidly growing populations, but, rather, those with large export operations based on semi-arid and arid lands (Mexico and Peru).

Table 3.4. Indicators of agricultural intensification in Latin America.

Country	Percentage of A&P land irrigated				Fertilizer use (kg ha^{-1} of cropland)			
	1966	1976	1984	1994	1964–1966	1974–1976	1983	1993
Central America								
Costa Rica	5	5	21	24	59	137	171	208
Cuba	29	23	25	27	134	103	164	52
Dominican Republic	13	14	13	17	12	68	39	61
El Salvador	3	5	15	16	71	147	113	106
Guatemala	3	4	6	7	14	36	38	87
Haiti	6	8	8	8	0	2	4	5
Honduras	8	9	4	4	6	12	16	32
Mexico	14	17	20	25	15	42	60	71
Nicaragua	1	5	7	7	17	22	56	21
Panama	3	4	5	5	17	47	40	48
Average[a]	9	9	12	14	34.5	61.6	70.1	69.1
South America								
Argentina	5	5	6	6	1	2	5	11
Bolivia	4	4	6	4	1	1	4	6
Brazil	2	3	4	6	5	31	45	85
Chile	24	22	29	30	26	25	18	58
Colombia	5	6	9	14	28	44	61	94
Ecuador	14	10	21	18	6	14	30	31
Paraguay	4	5	3	3	2	1	5	14
Peru	41	35	33	41	29	38	22	44
Uruguay	2	3	7	11	29	40	31	72
Venezuela	5	6	5	5	10	38	41	65
Average[a]	11	10	12	14	13.7	23.4	26.2	48

[a]Unweighted.
A&P, arable and permanently cropped.

A Sampling of Country Case Studies

The studies summarized below are selected to represent the diversity of experiences across countries and the fundamental effects of contextual factors, such as natural resource endowments and government policy. This is by no means intended to indicate that all cases are equally likely.

Population growth and agricultural extensification

In Ecuador (as well as many other countries with tropical forests), population growth in the Amazon region has led to a widespread process of land clearing and deforestation by agricultural colonists in recent decades. With the discovery of significant petroleum deposits in 1967, oil companies built roads to extract and transport the oil from the northern Amazon starting in the 1970s. This facilitated an influx of migrant colonists, coming largely from the highlands (Pichón, 1997a; Pichón and Bilsborrow, 1999). Population grew at annual rates of 8% in 1974–1982 and 6% in 1982–1990 – in both cases more than double the national average – while deforestation was widespread in the region and led to Ecuador's overall annual deforestation rate of 1.8%, the highest of any Amazon basin country. By the time of a detailed household survey of migrant colonists in 1990, 44% of the colonists' plots (most 40–50 ha) had been deforested. Ongoing work in the region by Bilsborrow and others indicates continuing rapid population growth and agricultural extensification, due as much to high fertility in the region as to new in-migration.

Population decline and agricultural extensification

In apparent contradiction to theory, in the Brazilian Amazon during recent decades, deforestation has accompanied a declining rural population; the rural population decreased by 16% from 1966 to 1994 and, since the early 1980s, even the rural population in the Amazon has not grown. Poor tropical soils, underdeveloped transportation infrastructure, titling delays, tenure insecurity and lack of credit severely constrained the agricultural development prospects of many migrant farmers within a few years of their arrival. Many migrants abandoned their farms to work in mining or timber harvesting or to move to cities. Large-scale ranchers engaged in speculation, buying up land and sometimes forcefully removing farmers (Hecht and Cockburn, 1990; Stewart, 1994). Much of the abandoned land was converted to pastureland, which is a far more extensive use of land than growing crops, contributing to ongoing deforestation, despite a decreasing rural population.

While demographic factors appear absent as proximate causes of this process, both high (earlier) fertility and migration are among the underlying factors. Many of the migrant settlers initially came from north-eastern Brazil, where high fertility, extreme land inequality and recurrent droughts led to population pressures on the land and poverty, which impelled the population to migrate in search of solutions, mostly to cities but many to the Amazon frontier as well. It is striking that, in both Brazil and Ecuador, a decline in overall rural population density is associated with rapid deforestation at the national level, due, in large part, to rural–rural migration from densely populated areas to the Amazon.

Population growth and agricultural intensification

A direct link between rural population change and land-use change can be ascertained better when confounding economic and political variables are absent. Such is the case in traditional (non-monetized) agrarian societies, which are rare today. However, some scattered evidence providing a quantitative test of Boserup's intensification hypothesis has been brought together based on data from studies of 29 tropical subsistence agricultural societies around the world

(Turner *et al.*, 1977). Turner and colleagues observed a significant positive cross-sectional relationship between population density and agricultural intensity, with the latter measured as the proportion of time the agricultural land was in use.[5] The best fit was from an exponential model, suggesting that agricultural intensity is much higher as population density rises – evidence in support of diminishing returns.

A later study by Turner and Ali (1996) examined the link between growing population density in six villages (265 households) in Bangladesh from 1950 to 1986. They noted that the lack of available unused land dictated intensification as the only agricultural response (91% of the land was cultivated at least once per year). Yet population density differed greatly across the six villages, with three much more densely populated than the others (means of about 1200 persons per square kilometre in 1985/86 versus about 500 in 1950); land productivity was over twice as high in the former villages, as was the cropping frequency. Higher intensification was also linked to fewer environmental constraints, being closer to markets (Dacca) and having received more government assistance.

Finally, in a study on three districts of Tanzania, Meertens *et al.* (1996) observed changes over a century based on historical records and recent field studies. They concluded that increasing population densities have led to intensification, but that the process has been shaped by agroecological conditions (such as differences in rainfall and type of land). They also noted that the land frontier was exhausted first, followed by out-migration, before intensification occurred.

Population growth and out-migration

Among the several recent studies on Africa, that of Netting *et al.* (1989) stands out. They looked at how the Koyfar of the Jos Plateau in Nigeria adapted to population growth between the 1950s and the 1980s. For decades before, much of the population migrated seasonally to the nearby Benue valley (lowlands) to grow yams to supplement their subsistence production in the plateau homeland. The rapidly growing internal market for yams led to this migration eventually becoming permanent for over half the families. Another demographic response then occurred, with the need for labour leading to increases in both polygamy and multiple household units (fertility was already high). The serendipitous linkages are directly related to the atypical advantages of having a large area of unused, unowned, fertile land nearby, as well as a growing market for yams. Finally, it was noted that certain cultural factors also played important roles in the positive adaptation – household autonomy, work ethic and lack of local autocratic rulers and institutions.

Population growth and multiphasic responses

Research which has attracted considerable attention recently is that of Tiffen and colleagues on the Machakos district, near Nairobi, Kenya (Tiffen and

Mortimore, 1994; Tiffen, 1995). Even as population grew fourfold over the period 1930–1990, the value of agricultural output per capita tripled, mainly from huge increases in the production of cash crops in the 1960s–1970s and in horticulture in the 1980s, while livestock production declined. This reflects a more intensive use of land, as expected by Boserup (1965). Meanwhile, environmental conditions actually improved, with reforestation and the creation of broad labour-intensive bench terraces, both of which saved water and reduced erosion. Still, other changes also occurred, together comprising a multiphasic response, with the first in the 1930s being extensification of agriculture to inferior land and substantial out-migration of males for wage labour, both prior to intensification. Indeed, remittances and savings of the out-migrants later financed an increase in education (human capital) and the creation of non-farm businesses, which probably contributed to the later intensification. More recently, increases in local employment opportunities (including non-farm) have led to a decline in male out-migration. Finally, a demographic transition now appears to have begun since the 1980s, which may reflect increasing difficulties in achieving further extensification (there being no unclaimed land by 1980 (Tiffen, 1995)) or intensification (the value of output per hectare failed to rise as much as population during the 1980s (Tiffen, 1995)). Nevertheless, it should be noted that in Kenya, generally, fertility is now declining, as policies initiated in the 1970s begin to find 'fertile ground'.

Summary, Gaps in Knowledge and Research Needs

The data presented here on linkages between population dynamics and changes in land use in Latin America and other developing regions of the world suggest that neither neo-Malthusian nor Boserupian models offer satisfactory explanations of population–environment linkages. Complex processes are difficult to disentangle on a country scale. Subnational data for provinces or even smaller political units are better suited for testing the interrelationships because they permit taking into account additional factors, including contextual factors that differ from one area to another.

Evidence of linkages between population dynamics and the various economic, economic–demographic and demographic responses appears stronger in other regions than in Latin America for the 1960s to the mid-1990s. This is because, unlike in the other regions, rural population density in Latin America rose only slightly during the first half of that period and actually fell during the latter half. Thus, at a continent-wide level, it is not plausible to expect a relationship between changes in rural density and induced responses, since there was nothing inducing the response. However, this does not mean that such relationships did not obtain for some countries in the region, as discussed above.

Unfortunately, a major shortcoming of cross-continent and cross-country (Latin America) studies, such as those summarized here, is the egregious lack of reliable and comparable data, with the result that many of the data used are based only on 'guestimates' by officials in the countries or at the FAO. For example, few countries in Latin America or Africa carried out national agricultural censuses in the 1980s or 1990s. Although studies of the same relationships (using appropriately different measures) based upon farm-level data could be even more useful than those based on census data for testing population–extensification/intensification linkages, few data sets are available.

With these caveats in mind, let us briefly review some patterns that emerge on population and land-use/land-cover patterns during recent decades (1966–1994) in Latin America, focusing on the differences between Central and South America. Rural population densities in South America were already starting to decline by the early 1980s or before. This was due both to the opening up of new lands for agriculture in frontier areas (mainly forests) and also to declines in fertility and out-migration from rural areas by people searching for better opportunities in urban areas and frontier regions. In Central America, in contrast, the rural population continued to grow throughout the 1966–1994 period in virtually all countries. Rural Central America was seven times more densely populated than rural South America in 1966, but 11 times more in 1994. As a result, the connection between population dynamics and land-use change should be greater there.

However, even in countries such as Brazil, where rural population density is declining, demographic factors are important underlying factors in the expansion of the agricultural frontier and deforestation. One key factor is rural–rural migration, rather than high or rising rural population density due to high population growth and high fertility. Thus, it is crucial to ascertain the origins of rural–rural migrants – and conditions in the places of origin – to determine the ultimate causes of that migration and hence of the forest degradation that occurs in other parts of the country.

The fact that extensification has continued unabated where land is available is sobering, given diminishing forest reserves worldwide and the ecological imperative of tropical forest conservation. Economic and demographic changes have contributed to the widening gap between those regions moving towards rapid rural and overall development (China, South-east Asia and some countries in Latin America) and others which are stagnant (Africa and much of the rest of Asia and Latin America). While demographic responses, especially further fertility decline towards replacement fertility, may help in confronting these problems, and while some further extensification will prove necessary, mechanisms need to be found to achieve much more successful intensification of agriculture, especially in the poorer regions of the Third World.

Unfortunately, there are growing signs of limitations to intensification. The FAO estimates that two-thirds of the global increase in agricultural production from 1990 to 2010 will need to come from increasing yields, with only 21% from expanding the land area and 13% from increasing crop intensity in the sense of shorter fallow periods (WRI, 1996). But the cost of expanding production is rising. Of the remaining potentially usable land, half is already in protected forests, while three-quarters has serious soil or terrain (topography, drainage, etc.) constraints; meanwhile 10% of current agricultural land is already moderately or severely degraded (WRI, 1996). By the early 1990s, three-quarters of the world's wheat and rice acreage and half the land in maize were already in high-yielding varieties (HYV). And irrigated land, accounting for one-third of the world's agricultural production (with 60% of the irrigated land in Asia, and only 5% in Africa and 4% in Latin America), is increasing at a slower rate, growing 2% per annum in the 1970s and 1980s but at less than 1% in the 1990s (WRI, 1996). The main inputs expected to contribute to further intensification are chemical fertilizers, the usage of which is expected to double during the period 1990–2010. Chemical fertilizer usage levels in Africa are currently only one-seventh of those in Asia. Still, increases in fertilizer use have led to smaller increments in production in recent decades (Brown, 1997).

With continued population growth, diminishing available land and future intensification constraints, policy-makers at all levels will be challenged to improve the agriculture–population nexus in developing countries. Successful policies may include: (i) helping rural farmers intensify production through more technical assistance and credit targeted to raising crops rather than cattle; (ii) improving access to reproductive health and family-planning services in rural areas; and (iii) better conserving what is left of the precious vestiges of tropical forest through higher royalties for logging concessions, more rational road-building policies and involving local populations in protecting conservation areas.

Notes

1 Among the nations of the temperate regions, only Finland and Japan (at approximately 0.2% annually) reported positive deforestation rates in the 1990s (World Bank, 1998b). In contrast, all tropical nations are experiencing deforestation at a rate of 0.5% or greater annually.

2 A more extreme version of Boserup's hypothesis, which is discredited by empirical evidence, is that population growth is automatically a boon to the developing world by inducing technological innovation, enabling output to rise faster than population (Simon, 1977, 1981).

3 The use of the same five regions continues to present difficulties for the Near East, which we intended to comprise all countries from Morocco across northern Africa and western Asia to Afghanistan. However, because of the almost total dependence on oil of many countries and/or the lack of adequate data (e.g. Syria, Algeria, Morocco, Tunisia), the final sample comprised only four countries – Egypt, Iran, Iraq and Jordan. In these countries, a couple of values seemed implausible and were omitted, namely, the increase in agricultural productivity in Iraq of 158% in 1980–1995 implied by FAO data.

4 Chen *et al.* (1998) find that rural–urban migration accounts for about two-fifths of urban population growth in the region, with the rest due to natural population growth (the difference between the fertility and mortality of the urban population).

5 The standard definition was used. Land in use one year followed by 4 years in fallow would yield an intensification measure of 20 (20% of the time), while land used for double-cropping in a year leads to a measure of 200.

4 The Economics of Biodiversity Loss and Agricultural Development in Low-income Countries

CHARLES PERRINGS

Environment Department, University of York, York, UK

Introduction

Biodiversity conservation has traditionally been seen as a problem of protecting genetic diversity. It has had two dimensions: *ex situ* germ plasm preservation in zoos, aquaria and arboreta (and, by extension, seed banks, tissue cultures and genomic libraries), and *in situ* species preservation in refugia, especially in megadiversity areas involving high levels of endemism. Increasingly, however, biodiversity conservation is being taken out of zoos and protected areas. It is recognized that biodiversity is important for the functioning of all ecosystems, and that excessive loss of biodiversity imposes real costs on resource users (Heywood, 1995). It is therefore interesting to consider the problem of biodiversity loss not just in refugia, but also in managed ecosystems. These are ecosystems from which some species have been deleted in order to enhance the productivity of others. The problem of biodiversity conservation in such cases does not therefore involve preservation of all existing species. It involves the maintenance of sufficient interspecific and intraspecific diversity to protect the productivity of the system. Put another way, the problem of biodiversity conservation in managed systems requires us to think about the optimal or efficient level of species deletion. The main question posed in this chapter is whether current rates of biodiversity loss are efficient.

This is not an uncontroversial way of looking at the problem. It implies that it is reasonable to apply conventional economic tests to biodiversity loss, and many regard such an approach with repugnance. Wilson (1984, 1993), for example, argues that humans have an inherent inclination to affiliate with life and life-like processes and these innate tendencies form a basis for an ethic of care and conservation of the diversity of life. But to say that there may be an efficient level of biodiversity implies that it may be optimal to drive some species to extinction (if only locally). While most people have little difficulty with this suggestion when the species at issue are, say, the acquired immune deficiency syndrome (AIDS) or smallpox viruses, there is less consensus about endemic agricultural pests or competitors. Nevertheless, this is the approach taken here: to consider whether current rates of biodiversity loss in agroecosystems are efficient.

This chapter does not report original research results, but uses existing literatures in ecology and economics to consider three aspects of the problem. The first, addressed in the following section, is to identify the external costs of biodiversity loss in agroecosystems in developing countries. The biodiversity implications of agricultural growth are also considered. Agricultural growth takes two forms: extensive growth and intensive growth. Extensive growth leads to land conversion, and this is associated with both habitat destruction and habitat fragmentation. It is generally seen as the major proximate cause of biodiversity loss. Intensive growth leads to an alteration in the mix of species due to changes in cropping or livestock regimes or to pest management practices. Intensification affects the combination of crops, livestock, symbiotics, competitors and predators. I argue that a reduction in the diversity of species in the system due to intensification may, in some cases, make agroecosystems more susceptible to exogenous shocks or changes in environmental conditions, and that this effect is not captured in market prices (Conway, 1993; Perrings *et al.*, 1995).

The second issue, addressed in the subsequent section, is the relationship between market failure and income. The costs of biodiversity loss turn out to be sensitive to the distribution of income and assets (Perrings *et al.*, 1994). The Brundtland Report (WCED, 1987b) is best known for the proposition that poverty is causally linked to environmental degradation. Dasgupta, in a number of contributions (1993, 1995a, 1996), has since provided a convincing explanation for why poverty, working through population growth, should cause increasing pressure on the environment. At the same time, there is a growing empirical literature on the so-called environmental Kuznets curve (EKC) (reviewed in Barbier, 1997). This literature suggests a more complex relation between per capita income and environmental change. In this chapter, I consider both the empirical evidence for a relation between indices of poverty and proxies for biodiversity loss, and the behavioural link between rural poverty and the underlying causes of biodiversity loss. More particularly, the discussion addresses how rural incomes may be related to the market failures that drive biodiversity loss in low-income countries.

Finally, Article 11 of the Convention on Biological Diversity (UNEP, 1993) calls on the contracting parties to 'adopt economically and socially sound measures that act as incentives for the conservation and sustainable use of components of biological diversity'. If market failures are driving biodiversity loss beyond the level that is justified by the gains from extensive and intensive growth of agriculture, what is the scope for developing instruments that will work in a developing-country context? I want to consider what can be done through market-based instruments and through institutional and property-rights reforms to address the problem.

The final section offers my conclusions. In summary, these may be stated as follows. Biodiversity loss matters in agroecosystems for a number of reasons, the most important of which is that it reduces the capacity of farmers to cope with external shocks (whether market or environmental). These costs of biodiversity loss are external to the market; they involve market failure. The problem is most severe in low-income countries, where mechanisms for private and social insurance against the risks to agricultural incomes are limited. Governments frequently act as insurers of last resort, distributing famine relief when farm incomes fail. Given the limited resources of governments in low-income countries, however, and given the fact that the risks to farm incomes are often highly correlated within such countries, this is seldom an effective solution. In the absence of effective private or social insurance mechanisms, the best way to deal with biodiversity externalities may be through the private costs of different farming systems. Where biodiversity-poor farming systems involve greater social costs, they should also involve greater private costs. Some concrete proposals are presented which address this matter.

The External Costs of Biodiversity Loss in Agroecosystems

The focus of this chapter is the local costs of the deletion of species. In the case of endemics, the local deletion of some species may well imply global extinction, but even in such cases, the costs to local resource users may be significant. This is rather different from much recent work on biodiversity loss, which tends to focus on the global value of local conservation and the scope for local capture of global values (see, for example, Pearce and Moran, 1994; Pearce, 1999b). My concern, however, is with the local efficiency of biodiversity loss and the scope for developing local incentives for biodiversity conservation.

A local focus draws attention to the role of biodiversity in the provision of locally valuable ecological services. Biodiversity supports a range of ecological services, including, at the global level, the maintenance of the gaseous quality of the atmosphere and amelioration of climate. But it also supports a number of much more localized services: the operation of the hydrological cycle, including flood control and water supply, waste assimilation, recycling of nutrients, conservation and regeneration of soils, pollination of crops and so on (Daily, 1997). The local value of biodiversity derives from the value of these services. In agroecosystems, for example, the most important ecological services are those influencing the productivity of the system and its capacity to maintain productivity over a range of environmental conditions. These comprise both on- and off-farm services. Watershed protection, for example, offers a range of off-farm services to agriculture, including regulation of surface runoff, ground-water recharge, erosion control and localized climatic effects. The conversion of watersheds as a by-product of extensive agricultural growth often means the loss of these services. In what follows, I consider the available evidence on the value of changes in the mix of species as a result both of habitat conversion and of alteration in the intensity of farming activity.

Habitat conversion and degradation: the extent of the problem

The main proximate cause of biodiversity loss is the habitat loss associated with the processes of deforestation and desertification. Both processes are associated with areas where a high proportion of output and/or employment derives from agriculture. Specifically, biodiversity loss due to agricultural growth at the extensive margin is associated with regions of low population density but high population growth (sub-Saharan Africa and Latin America). Biodiversity loss due to agricultural growth at the intensive margin is associated with regions with high rural population density and growth (South Asia and South-east Asia). This said, as the productivity gains of agricultural intensification have faltered in Asia, pressure in that region has been turned to remaining forested areas, and recent rates of deforestation are higher in South Asia and East Asia than elsewhere.

Consider first the process of deforestation. Table 4.1 reports deforestation in selected subregions for the period 1980–1990. Not only did deforestation accelerate in regions where the process was already under way at the beginning of the decade, but also, afforestation turned to deforestation in other regions. In sub-Saharan Africa, the highest rates of forest loss occurred in West Africa – Ghana and Togo, in particular. But the annual rate of loss in these countries, 1.3–1.4%, was still low compared with regions where forest stocks are more depleted. Four countries in Latin America – Costa Rica, El Salvador, Honduras and Paraguay – were converting remaining forests at more than 2% year^{-1} during the 1980s, while Bangladesh, Pakistan, Thailand and the Philippines were all converting remaining forest resources at 2.9–3.0% year^{-1}.

Compare this rate of habitat loss and fragmentation in forested areas with rates of change in arid and semi-arid areas. Desertification is the term most frequently used to describe environmental damage in arid, semi-arid and dry

Table 4.1. Forest resources and deforestation, 1980–1990 (from WRI, 1994).

	Extent of natural forest (1000 ha)		Annual deforestation (1981–1990)	
Region	1980	1990	(1000 ha)	(%)
West Sahelian Africa	43,720	40,768	295	0.7
East Sahelian Africa	71,395	65,450	595	0.8
West Africa	61,520	55,607	591	1.0
Central Africa	215,503	204,112	1,140	0.5
Tropical southern Africa	159,322	145,868	1,345	0.8
Insular Africa	17,128	15,782	135	0.8
South Asia	69,442	63,931	551	0.8
Continental South Asia	88,377	75,240	1,314	1.5
Insular South-east Asia	154,687	135,426	1,926	1.2
Central America and Mexico	79,216	68,096	1,112	1.4
Caribbean subregion	48,333	47,115	122	0.3
Tropical South America	864,639	802,904	6,174	0.7

subhumid areas. Like deforestation, desertification implies a reduction in the vegetative cover of land and tends to be associated with the expansion of agricultural output. Desertification currently affects some 3.6 billion ha and some 900 million people (Tolba *et al.*, 1992). This makes it a more extensive problem than that of deforestation (Table 4.2).

Desertification includes a range of different types of damage. In irrigated lands, for example, desertification is a consequence of the salinization and alkalinization of soils and aquifers. Annual losses due to these causes in the early years of this decade were running at about 1.5 Mha. In rain-fed croplands, the dominant manifestation of land degradation is soil erosion and the loss of soil organisms, which account for at least 3.5 Mha annually. But, in rangelands, degradation takes the form of loss or alteration of vegetation, loss of soil moisture and soil organisms and soil erosion. Annual losses at the beginning of the decade were estimated to be between 4.5 and 5.8 Mha (Tolba *et al.*, 1992).

The value of biodiversity

The costs of the biodiversity loss associated with deforestation and desertification include the loss of globally important services, such as carbon fixation and sequestration, together with the loss of genetic information. There are few estimates of the value of these costs but all indicate that the sums involved are not trivial (Pearce and Moran, 1994; Heywood, 1995). The point has already been made, however, that these two processes also lead to changes in ecological services that are of primarily local importance. These include watershed protection and the derivative services of flood control, water supply and storage already mentioned. But they also include the amelioration of microclimate and effects on soil conservation, nutrient cycling, and timber and non-timber forest production. These services have both current use value and option value – the potential value of such services in the future.

Table 4.2. Extent of desertification of drylands (from Tolba *et al.*, 1992, pp. 137–139).

	Moderately, severely and very severely desertified land[a]					
	Irrigated areas		Rain-fed croplands		Rangelands	
Region	(1000 ha)	(% affected)	(1000 ha)	(% affected)	(1000 ha)	(% affected)
Africa	1,902	18	48,863	61	995,080	74
Asia	31,813	35	122,284	56	1,187,610	75
Australia	250	13	14,320	34	361,350	55
Europe	1,905	16	11,854	54	80,517	72
N. America	5,860	28	11,611	16	411,154	75
S. America	1,517	17	6,635	31	297,754	76

[a]Includes all lands at least moderately damaged in one of the ways encompassed by the term 'desertification' (see text).

It has been recognized for some time that the local value of such services is relevant to the analysis of investments in natural resource-based sectors. Anderson (1987), for example, argued that inclusion of indirect environmental benefits substantially improved the economic rate of return estimated for forestry investments. Illustratively, an estimate of the indirect benefits of forest conservation in Korup National Park, Cameroon (Ruitenbeek, 1989), found the net benefits of watershed protection, flood control and soil fertility maintenance to be roughly comparable to the foregone benefits from timber production. Estimates of the net benefits of habitat conservation in Costa Rica suggest a range between US\$102 and US\$214 ha^{-1} $year^{-1}$ (in 1995 dollars), indicating a net present value for investment in conservation of between US\$1278 and US\$2871 ha^{-1} at an 8% discount rate (Heywood, 1995).

In some cases, the indirect value of conserved tropical forests has been shown to exceed the private value of the converted land (see, for example, the valuation of the conservation of Khao Yai Park, Thailand, by Kaosard *et al.* (1994)). This is true of marginal, steeply sloping lands in watersheds, where soil erosion is a major cost of land clearance. In many cases, however, the indirect value of ecological services from tropical forests may not dominate the private net benefits of conversion to commercial arable or livestock production, but may be sufficient to favour investments that conserve key ecological services over investments which do not.

As a hypothetical illustration, Peters *et al.* (1989) used a productivity/earnings method to estimate the value of Peruvian forest at Mishane, Rio Nanay, under alternative uses. They obtained a net present value (in 1989 dollars) of US\$6300 ha^{-1} for non-timber forest products (fruit and latex) harvesting, US\$490 for sustainable timber production, US\$1000 for timber clear-cutting, US\$3184 for plantation harvesting and US\$2960 for cattle ranching. In this case, non-timber forest products clearly dominate other activities. 'Sustainable forestry' is the least desirable option. But, if the ecological services of forests were valued at levels estimated for Costa Rica, it would be sufficient to reverse the rankings between sustainable forestry, clear-cutting, plantation harvesting and cattle ranching. The point here is that the indirect value of forest conservation is generally such that ignoring it leads to inefficient outcomes.

Another example is to be found in the use made of the gum arabic tree (*Acacia senegal*) in the Sudano-Sahel region. This tree has a number of direct uses: the gum is widely used as an emulsifier in confectionery, beverages, photography and pharmaceuticals, and the tree provides fodder for livestock, fuelwood and shade. But it also offers a number of indirect benefits. The most important of these are that its extensive lateral root system reduces soil erosion and runoff and, as a leguminous tree, it fixes nitrogen, which encourages grassy growth for livestock grazing.

A study of the economics of gum arabic in Sudan (Barbier, 1992) found that, because of a decline in the real producer price relative to other crops (sesame, groundnuts, sorghum and millet), its private profitability was lower than that of other crops, except in the Tendelti system of the White Nile, where field-crop damage occurs frequently. But it also concluded that inclusion of the indirect benefits of gum arabic (control of erosion/runoff, wind-breaks, dune fixation, nitrogen fixation) and the role of the gum belt in controlling desertification meant that its social profitability was much higher than its private profitability. In other words, the estimated private rate of return understated the value of *A. senegal* precisely because it does not take into account the indirect value of the resource in the provision of a range of ecological functions.

More recently, the link between the diversity of species and the disruption of local ecological services has been clarified. Many ecological functions are supported at any one moment in time by a relatively small number of species (Holling, 1992). The removal of these species can cause a fundamental transformation of the ecosystem – whether from forest to grassland or from grassland to a shrubby semi-desert. Where a forest or savannah is transformed in this way, most of the main ecosystem functions are disturbed. Examples of such relations between species and ecological disruptions include the insectivorous birds that mediate budworm outbreaks in the eastern boreal forest (Holling, 1973, 1986) and the grass species that maintain productivity in savannahs (Walker and Noy-Meir, 1982).

Consider the latter case. Increasing livestock diversity has led to two major changes in the biodiversity of savannah rangelands. The first involves the loss of perennial grasses and their replacement by animals. These vary far more in response to fluctuations in rainfall. The second involves the reduction in phenological diversity in the grass sward. This moderates interannual variation in production. Rangelands subject to light grazing pressure tend to have roughly even amounts of early-, mid- and late-season growing grasses, implying that there is roughly the same amount of grass to respond to rainfall whenever it occurs in the season (Walker, 1988). But in rangelands subject to heavy grazing pressure, the loss of early-season palatable species implies an increase in the relative amount of later-growing species (Silva, 1987). There is a reduction in the functional diversity of the range, and this leads to an increase in interannual variation in fodder production. McNaughton (1985) showed that Serengeti grasslands in which there was greatest variation in biomass also varied least in annual production.

In both forests and savannahs, there is an element of redundancy in the system under given environmental conditions. Species redundancy, in this context, means that their removal has relatively few implications for the functioning and productivity of the system – at least under those environmental conditions (Walker, 1993). There is always a threshold of diversity below which their various functions cannot be maintained. Any self-organizing living system requires a minimum diversity of species to capture solar energy and to develop the cyclic relation of fundamental compounds among producers, consumers and decomposers on which biological productivity depends. If the level of biodiversity drops below such a threshold, the productivity of the system will fall. However, since not all species play a vital part at all times, it is often possible to delete some species without immediately affecting the functioning of the system.

The important point, though, is that, even if the biodiversity needed to maintain ecological services under one set of environmental conditions is low relative to the actual level of biodiversity, a change in environmental conditions can alter the mix of organisms, populations and communities needed to maintain those services. That is to say, biodiversity has value both in supporting the flow of ecological services under given environmental conditions and in assuring those services over a range of environmental conditions. Thus, it has an option or insurance value (Perrings, 1995).

The level of biodiversity in an agroecosystem determines its capacity to respond to external shocks, whether market or environmental. This is measured by its resilience. From an ecological perspective, biodiversity protects ecosystem resilience by underwriting the provision of ecosystem services over a range of environmental conditions (Holling *et al*., 1995). Certain species have greater ecological value under one state of nature than others, but species that are 'passengers' under one state of nature may have a key structuring role to play under other states of nature. The ecological impact of biodiversity loss depends on the link between the species and the functions of the system. Whether the deletion of some species affects a given function depends on the number of alternative species that can support the function if the ecosystem is perturbed (Schindler, 1990).

The insurance value of biodiversity has accordingly been argued to lie in its role in protecting ecosystem resilience (Perrings *et al*., 1995; Heywood, 1995). In the rangelands example, the market value of the live-weight gains secured by having a combination of grass species is a measure of the insurance value of that combination. From an economic perspective, biodiversity has insurance value because of its role in protecting the productivity of agroecosystems over a range of both environmental and market conditions.

The genetic diversity of cultivated species and environmental risks

Genetic resources are used in agriculture in two ways: selection of species for domestication and cultivation and genetic 'improvement' of domesticated species. The vulnerability of agriculture due to the narrow genetic base of most major crop plants is a well-recognized source of risk. More than 90% of the food supply derives from a handful of grasses (wheat, rice, maize and oats), *Solanaceae* (tomato and potato), mammals (cattle, sheep and pigs) and birds (chickens and ducks). The narrow genetic base of these species is a cause of disease and pest epidemics, leading to sharp variation in agricultural yields. Wild or traditional genetic stocks have value because they offer scope for breeding or engineering resistance to crop pests and pathogens. Most cultivated crop varieties and many livestock strains already contain genetic material incorporated from wild relatives or from crops and livestock still used by traditional farmers. Indeed, it has been estimated that at least half of the increase in agricultural productivity this century is directly attributable to artificial selection, recombination and intraspecific gene-transfer procedures (Heywood, 1995). There are many examples. The use of Turkish wheat to develop genetic resistance to disease in western wheat crops was valued in 1995 at US$50 million year^{-1}. Ethiopian barley has been used to protect Californian barley from dwarf yellow virus, saving damages estimated at US$160 million year^{-1}. Mexican beans have been used to improve resistance to the Mexican bean weevil, which destroys as much as 25% of stored beans in Africa and 15% in South America. Wild rice strains have improved protection of cultivated varieties against grassy stunt virus. The susceptibility of sugar cane to mosaic virus was solved by introducing resistance to it from wild relatives (*Saccharum spontaneum*) (Heywood, 1995).

The value of biodiversity in cases such as these lies in the fact that it provides the raw material for desirable genetic traits in crops. Genetic resources have been used to improve productivity, induce better resistance to pests and improve adaptation to harsh environments. This value depends on the technology of genetic transfer. Changes in biotechnology have altered both the use value of genetic resources and the potential indirect consequences or external costs of their use. The most important of these changes is that biotechnology allows gene transfer both within species and between unrelated species, increasing the flexibility of genetic resources. At the same time, however, the introduction of biological control agents or genetically modified organisms may have indirect effects that give rise to external costs and benefits. The best-understood indirect ecological effects are those relating to species introductions, and include impacts on the phenotypic characteristics of individuals; the genetics, abundance and dynamics of populations; and the structure and functioning of communities and ecosystem processes, such as nutrient cycling and disturbance regimes (Parker *et al*., 1999).

These ecological effects include the deletion of indigenous species through predation, browsing or competition; genetic alteration of indigenous species through hybridization; and the alteration of ecosystem structure and function, including biogeochemical, hydrological and nutrient cycles, soil erosion and other geomorphological processes. The associated external costs include the health control costs associated with the introduction of pathogens and their vectors; productivity costs of changes in population abundance and species compositions; pesticide costs in agroecosystems; and defensive expenditures and control costs of invasive species, such as the water hyacinth (*Eichhornia crassipes*) (Williamson, 1996).

Potentially more important still are the effects of biotechnology on the incentives to conserve wild relatives and traditional crops. Productivity increases attributed to the Green Revolution have been associated with the abandonment of traditional varieties that have been bred over thousands of years. These landraces have been a major source of genetic diversity in agriculture, but many have disappeared with the Green Revolution. Not only has the total number of rice varieties planted been substantially reduced, but also the proportion of the total crops accounted for by ten or less varieties has substantially increased. Current trends in biotechnology threaten to displace even more traditional varieties and to increase the vulnerability of farmers who choose to adopt genetically uniform crops. There are no realistic estimates of the insurance value of the displaced landraces. This would, in any event, be location- and risk-specific. But, if assessed by the variance in farm incomes attributable to crop failures, it is reasonable to suppose that it is highest for farmers in low-income countries without access to alternative methods of insuring against crop failure. I turn now to the relation between biodiversity and income.

Biodiversity, Agriculture and Poverty

The Brundtland Report (WCED, 1987b) argued that one of the main driving forces behind the degradation of agroecosystems was poverty-induced pressure on forests and rangelands in order to meet basic needs. In recent years, however, an empirical relation has been observed between per capita income and certain indicators of environmental quality that appears to tell a different story. First applied to sulphur dioxide by Grossman and Krueger (1993), it has since been observed that a number of other indicators of environmental quality first deteriorate and later improve as per capita incomes rise. Since this trend mimics the relation between income distribution and per capita income identified by Kuznets in the 1950s, it has been labelled the 'environmental Kuznets curve'. A Kuznets relation has been found between per capita income and, *inter alia*, emissions of sulphur dioxide, particulates and dark matter, nitrogen oxides and carbon monoxide, carbon dioxide, chlorofluorocarbons (CFCs), various indicators of water quality, including faecal coliforms, biological and chemical oxygen demand and arsenic (for a review, see Barbier, 1997).

In addition, a Kuznets relation has been found for at least one of the proxies used for biodiversity loss induced by agricultural growth and deforestation. Panayotou (1995), Antle and Heidebrink (1995) and Cropper and Griffiths (1994) all identify an inverted U-shaped relation between deforestation rates and per capita income. This suggests that, in low-income countries, falling average incomes are associated with increasing habitat conservation. Interestingly, the same relation has not been found between habitat loss and more direct measures of poverty. The International Fund for Agricultural Development's (IFAD) integrated poverty index (IPI),[1] for example, has little power to explain variation in habitat loss (Perrings and Ansuategi, 2000). It is worth emphasizing, however, that the EKC results on habitat loss are generated by single-equation models based on cross-sectional data that ignore feedbacks between the economy and its environment. Furthermore, the chosen measures of environmental quality – rates of deforestation – are not only highly suspect, but are unrelated to the size of the forest stock remaining (see Stern *et al.* (1996) and Stern (1998) for an assessment of the data and estimation problems in this literature).

There are, in fact, two main patterns of deforestation, each involving a rather different relation between habitat conversion and rural poverty. Each pattern has been encouraged by infrastructural development (usually roads) and by government policies on migration and settlement. In one pattern, conversion of forestland for agriculture is due to the actions of large numbers of usually landless individuals. This is the case in Indonesia, for example. In the outer islands of Sumatra, Kalimantan, Maluku and Irian Jaya, the resettlement programme known as 'transmigration' brought in some 3.7 million people from the most populated islands of Java and Bali in the 1960s, 1970s and 1980s. In most cases, migrants were settled in what were termed 'conversion forests'. In countries such as Brazil, on the other hand, deforestation is mostly due to the actions of larger-scale ranchers. While forest conversion by small pioneer farmers using slash-and-burn techniques has been part of the problem, most habitat loss has been due to the development of ranches in response to tax incentives and subsidies (Seroa da Motta, 1997). It is not therefore surprising that broad-brush empirical studies of the EKC type should be inconclusive.

To identify the relation between poverty, population growth and environmental change implicit in the Brundtland hypothesis requires a different approach. Dasgupta (1993) argues that poverty drives fertility through the microeconomic decisions of rural households. Where conditions are such that people are locked into particular technologies, population growth then leads directly to pressure on the resource base. This may be exacerbated by the failure of both markets and policies. In sub-Saharan Africa, for example, a combination of cultural and institutional rigidities and policy-induced declining real producer prices is argued to have led to a positive-feedback loop between rural poverty, population growth, landlessness and pressure on soils, vegetation and water (Pearce and Warford, 1993).

The connections between income, asset holdings, population growth and resource use are clearly complex. However, certain linkages stand out. The first is that poverty constrains the information available to decision-makers. This obviously applies to education and training, but it also applies to the evaluation of new options. The poor command less information than do the rich. In the case of new crops or crop varieties in agriculture, information about environmental risks depends on local trials. But local trials are not costless – whether to private farmers or to agricultural departments. The quality of information available to farmers is sensitive to the resources available for trials. Similarly, estimation of the local impacts of other technological innovations, including pesticides or biocontrol agents, requires research into the effects of such agents elsewhere. Once again, this is not without cost.

A second important linkage lies in the fact that people's preferences are sensitive to both their income and their assets. More particularly, the rate at which people discount the future costs and benefits of present actions turns out to be sensitive to their income. People in extreme poverty effectively discount the future costs of resource use at high rates. Because what matters is consumption today, they frequently neglect the future effects of their actions. Perrings (1989) showed that rates of time preference vary with the level of real income and that people in subjective poverty who choose not to save in order to maintain real consumption levels implicitly discount at very high levels.

Empirical studies of the links between rates of time preference, income and wealth and investment in conservation technology confirm this. In a study of rural India, Pender (1996) found that rates of time preference were inversely related to wealth. Holden *et al.* (1998) investigated the relationship between rates of time preference, poverty and conservation in Indonesia, Zambia and Ethiopia. They found that rates of time preference amongst rural households are generally high and increase with poverty in both assets and income. They concluded that poverty is a disincentive to invest in environmental protection. Taken together with the problems of landlessness and lack of access to rural credit, these factors explain the propensity of the poor in some markets to exploit insecure holdings of marginal land and to respond in a frequently perverse way to economic incentives.

More recently, Perrings and Stern (2000) found a similar relation between discount rates and income in a study of the long-run productive potential of a semi-arid rangeland in Botswana. They used a bioeconomic rangeland model of the livestock sector and assumed a Box–Cox transformation utility function. This allowed estimation of the degree of risk aversion and the farmers' implicit rate of discount. Farmers were shown to be both risk-averse (the risk-aversion coefficient was estimated to be –0.48) and to discount the future heavily (the discount rate was estimated to be 16.5%).

Incentives for Biodiversity Conservation

An understanding of the way that rural poverty conditions farmers' responses is crucial to developing incentives to encourage the optimal level of conservation in low-income countries. Current rates of biodiversity loss in agroecosystems in developing countries are inefficient. Many of the social costs of biodiversity loss or the benefits of biodiversity conservation are not taken into account by resource users. The evidence suggests that developing countries should be investing more in conservation. More particularly, they should be developing incentives under Article 11 of the Convention on Biological Diversity (UNEP, 1993) to encourage resource users to pay due attention to the social costs of their activities.

The inefficiency of current levels of biodiversity loss implies market failure. In many cases, this is due to agricultural policies. Examples include administered prices in agricultural markets that distort the private cost of biological resources, destumping subsidies that encourage clearance of ever more marginal land for agricultural purposes, and stumpage fees or royalties in forestry that encourage excessive harvesting rates in timber concessions (Panayotou and Ashton, 1992). These effects are exacerbated by the fact that small farmers in most developing countries face credit constraints. Indeed, in the early part of this decade, only 5% of farms in Africa and 15% in Asia and Latin America had access to formal credit (Hoff *et al.*, 1993). Pearce (1999a) estimates that worldwide subsidies to biodiversity-degrading activities are currently US$684–808 billion and that, outside the Organization for Economic Cooperation and Development (OECD) countries, subsidies to biodiversity-degrading activities are US$151 billion.

As a reaction to this, a number of environmental economists have argued for the environmental benefits of liberalization. Munasinghe and Cruz (1995), for example, have claimed that: (i) removal of market price distortions, such as agricultural or energy subsidies, will both improve the efficiency of economic activity and reduce the impact of that activity on the environment; (ii) enhancing macroeconomic stability will encourage investment and persuade resource users to take a longer-term view of their decisions; and (iii) economic liberalization will reduce poverty and hence pressure on open-access environmental resources. In addition, they argue that improving the security of land tenure by assigning private property or use rights promotes investment in land conservation and environmental stewardship.

However, the environmental effects of liberalization are typically more ambiguous than this suggests. On the positive side, liberalization may lower the costs of environmental protection, but if liberalization stimulates demand for the products of environmentally damaging activities, it will increase environmental damage (Anderson and Blackhurst, 1992). The evidence is that liberalization has occurred alongside a marked increase in price risk. The coefficient of variation of detrended prices for the major food products rose sharply between the mid-1960s and the mid-1980s. In addition, these prices became positively correlated, thus reducing the value both of diversification within agriculture and of export-earning stabilization schemes (Hazell, 1987). The 'Brundtland hypothesis' holds that countries locked into products for which the terms of trade decline will tend to increase exports of those products just to maintain foreign exchange earnings (Pearce and Warford, 1993). Many primary producing countries have faced a secular fall in world commodity prices. The barter terms of trade of these countries have declined sharply over the last decade. However, the response has not been a reduction in primary commodity production, but an increase in the volume of exports.

The environmental implications of this are at present unclear. Most of the increase in agricultural output in sub-Saharan Africa, to take the worst-affected region, is argued to have derived from intensive rather than extensive growth. The increase in land allocation to crop production in arid and semi-arid areas is smaller than the increase in output. Similarly, the increase in land allocated to pasture is trivial compared with the increase in herd sizes. But data on agricultural inputs do not suggest that increased yields reflect a significant change in agricultural technology in the arid and semi-arid areas, with the possible exception of draught power. The impression is that increased agricultural output and exports have been achieved by putting existing agroecosystems under greater pressure with no change in agricultural technology.

Indeed, there seem to be five recurring effects of liberalization and other elements of a structural-adjustment package. First, the stimulation of export-oriented primary commodity production increases pressure on the resource base. Secondly, the real income of consumers tends to fall and there is disemployment in both the public and private sectors. This leads to a reduction in demand for domestically produced goods and services, which worsens the condition of the poor. Thirdly, the reduction in public expenditure programmes reduces the budgets of agencies protecting the environment. Fourthly, the reduction of credit to small rural investors leads to lower on-farm investments and declining agricultural yields, countering efforts to stabilize the agricultural frontier (particularly in the absence of effective land-tenure systems). Finally, deregulation makes it harder to correct price distortions in the forestry, agriculture and energy sectors.

Because of these effects, there is a growing argument that any adjustment package or liberalization policy should be accompanied by a set of environmental reforms designed to minimize adverse environmental impacts. However, this is not just a matter of setting environmental user fees to match the social opportunity cost of resource use or of creating private property rights. If the aim of a strategy for the conservation of biodiversity is to preserve enough species to guarantee the resilience of local ecosystems, then the system of incentives should encourage appropriate behaviour. Holden *et al.* (1998) conclude that poverty and liquidity constraints in low income rural economies mean that assignment of property rights in land, water or other environmental resources is not sufficient to reduce pressure on environmental resources to sustainable levels. They argue for additional interventions both to alleviate poverty and to regulate resource use.

This brings us back to the conditioning effect of income. The common intuition behind the Brundtland hypothesis is straightforward. If farmers are impoverished by the removal of subsidies and/or the imposition of charges for environmental resources, they will focus their attention on 'free' or open-access resources, and their decisions will become increasingly myopic. The economic intuition is equally plain. A change in relative prices induces both substitution and income effects. For the poor, the income effects of price changes may be very strong. This may not only weaken the effectiveness of economic incentives; if income effects are strong enough, it can lead to perverse results.

The study by Perrings and Stern (2000) identifies the impact of price changes on the resilience of agroecosystems working through their effects on farmers' resource-allocation decisions. Changes in cattle prices affect optimal offtake. This, in turn, affects the growth of the herd, grazing pressure and growth of the range and therefore the resilience function. In the case of Botswana, Perrings and Stern find that the elasticity of supply has the opposite sign to that which would be expected in commercial ranching. Raising prices results in less offtake. This is mainly driven by risk aversion. Under risk neutrality, the sign of the elasticity depends on whether there are increasing or decreasing returns to the size of the herd. Under risk aversion, however, price increases have a strictly limited effect on offtake.

Conclusions

Biodiversity matters in agroecosystems because it determines both the actual and the potential productivity of those systems. Determining the optimal level of biodiversity is, however, a far from trivial exercise. The elimination of pests, competitors and pathogens may increase the productivity of cultivated or husbanded species, but it may also adversely affect the capacity of agroecosystems to respond to environmental stresses and shocks. Experimental studies of the functionality of plant species diversity in pastures have, for example, shown that more diverse grass communities are better able to exploit variation in plant-available moisture than less diverse communities (Tilman and Downing, 1994; Naeem *et al.*, 1995). To the extent that these experimental studies can be generalized, they imply that management options that reduce diversity will also decrease long-run productivity and resilience.

A second point is that the long-run loss of potential productivity involves market failure. They are external to the market. In many cases, biodiversity externalities have been exacerbated by government macroeconomic and microeconomic policies and by the tendency to use the agricultural pricing, tax and subsidy regime to meet governments' distributional objectives. The net effect of market and policy failure is that farmers may be encouraged to adopt management strategies that yield relatively high returns under average market or environmental conditions, but which are highly susceptible to a change in those conditions. Monocultures supported by the heavy use of pesticides and fertilizers provide the best examples of these strategies.

The external costs of such strategies tend to be most severe in low-income countries, where mechanisms for private and social insurance against the risks to agricultural incomes are limited. Governments typically act as insurers of last resort, distributing famine relief when farm incomes fail. However, the resources available for famine relief in low-income countries are often limited. Moreover, the risks to farm incomes from market and environmental shocks alike tend to be highly correlated within individual countries, so this is seldom a completely effective solution. Insurance against highly correlated local income risks typically takes the form of grain reserves. In such cases, there is an argument for the development of 'regional' social insurance structures large enough for the risks to producer incomes to be uncorrelated or only weakly correlated. There already exist some regional groupings with a limited commitment to disaster relief, but there is scope for the development of more systematic contributory schemes.

More importantly, there is scope at both national and regional levels for reducing the correlation between (and hence improving the insurability of) agricultural risks. The correlation between agricultural risks typically varies inversely with crop genetic diversity. Since a reduction in genetic diversity in agroecosystems can decrease both resilience and productivity, conservation of biodiversity helps to ensure a flow of ecological services required for agricultural production over a range of environmental conditions. While there are as yet no estimates of the insurance value of conservation in particular agroecosystems, it is clear that this function is potentially highly useful.

What are the implications of all this? The first thing to note is that it is not an argument against the intensification of agriculture. The alternative to intensification involves encroachment on ever more marginal land and the destruction and fragmentation of ever more scarce habitat. It is, however, an argument against intensification strategies that ignore the costs of a change in the mix of species in the system. It is also an argument for the development of private or social insurance mechanisms that address the change in risks to farm incomes associated with intensification. Finally, it is an argument for the appropriate pricing of resources. In some cases, farmers are offered inducements to maintain production of drought- or disease-resistant crops or livestock. If one crop mix is less susceptible to drought or disease than another, then its social value (in terms of averting expenditures on famine relief) should be reflected in its relative private cost. Whether this implies taxation of the high-risk components or subsidy of the low-risk components depends on local circumstances and the international trading regime.

Note

1 The IPI (for 114 developing countries) is based on Sen's composite poverty index. It is increasing in poverty. It is calculated by combining a head-count index of poverty, the income-gap ratio, life expectancy at birth and the annual rate of growth of per capita gross national product (GNP). The head-count index is simply the percentage of the population below the poverty line. The income-gap ratio is the difference between the highest per capita GNP in the sample and the per capita GNP of the country concerned, expressed as a percentage of the former. Life expectancy at birth is included as a proxy for income distribution below the poverty line. Using this measure, countries are classified in three broad groups. An IPI of 40 or less indicates severe poverty; an IPI between 40 and 20 indicates moderate poverty; and an IPI of less than 20 indicates little poverty. The IPI was developed on the basis of data for a number of different years, but notionally describes the situation in 1988.

5 Farm Household Intensification Decisions and the Environment

STEFANO PAGIOLA[1] AND STEIN HOLDEN[2]

[1]*Environment Department, The World Bank, Washington, DC, USA;*
[2]*Department of Economics and Social Sciences, Agricultural University of Norway, Ås, Norway*

Introduction

With the earth's population projected to grow by two-thirds over the next 30 years and with rising incomes leading to a shift in consumption towards foods higher on the food-chain, forecasts indicate that world food production must at least double by 2025 to meet rising demand. The fundamental arithmetic of agricultural production is that total production equals mean yield times area cultivated. The increasing demand for agricultural products, therefore, can only be met by increasing yields on current agricultural land (intensification) or by expanding the area under cultivation (extensification).

Extensive agricultural growth poses serious environmental threats (Pagiola *et al.*, 1997). Already, agriculture is the human activity that affects the greatest proportion of the earth's surface and is the single biggest user of fresh water. Extensive agricultural growth is considered to be a major contributor to loss of habitat and reduction of biodiversity. Extensive agricultural growth can lead to substantial release of greenhouse gases if forests are cut down – agricultural sources are estimated to account for about 30% of total carbon dioxide emissions and deforestation is the single most important source of these emissions (Duxbury, 1995). Loss of forest cover caused by extensive growth can result in substantial downstream damage from sedimentation and changes in the timing and volume of stream flows (Cassells *et al.*, 1987; Chomitz and Kumari, 1996). Since extensification is taking place in increasingly fragile areas, it is also leading to increased land degradation and there are serious doubts as to its long-term sustainability. The factors that lead farmers to grow extensively rather than intensively are still not fully understood, however. A greater understanding of these factors will provide a useful input to efforts to encourage the intensification of agriculture, as well as to work on deforestation problems, on interactions between agricultural development and biodiversity, and on land degradation.

This chapter develops a stylized model of farm household decision-making with respect to intensification or extensification and uses it to examine the factors that are likely to result in intensification decisions. The focus is on frontier areas in which extensification remains an option; in many areas, of course, extensive growth is no longer possible. The analysis presented here focuses on instances in which intensification and extensification cause environmental problems that are external to the farm household and, as such, are not incorporated into farmer decision-making. Instances in which environmental problems are internal to the farm household (such as land degradation causing loss of productivity) are considered in further extensions (Holden and Pagiola, forthcoming). Although the results given below are intuitive in economic terms, the need for a rigorous analytical framework forces the use of a fair amount of mathematical notation. Interested readers may find a more complete formulation of the model and derivations of the main results in Appendix 5.1.

Farm Household Decision-making

To examine farm household decision-making with respect to intensification and extensification, we use a two-period model in which decisions about intensification and extensification are made in the first period and their results are experienced in the second period (which is modelled as representing the present value of all subsequent periods). The farm household is assumed to maximize utility $U[c,L_L]$, which is a function of consumption, c, and leisure, L_L. Consumption in any period is limited by a budget constraint dependent on income from crop production and from uncleared land.[1] The household's rate of time preference is embedded in the discount factor, ρ – the smaller ρ is, the more the household prefers current consumption relative to future consumption.[2] The household is assumed to cultivate an initial area A_0 (normalized to $A_0 = 1$ for convenience) and to have access to an uncleared area, A, which generates an income for the household of κ per unit of land (so that household income from uncleared land is $A\kappa$) where $\kappa \geq 0$. The household also has a labour endowment of L. A key feature of the model is that we assume that the household faces an imperfect labour market and so can neither hire in nor hire out labour. Kaimowitz and Angelsen (1998) show that assumptions about labour markets have a very important effect on a model's predictions (see also Angelsen and Kaimowitz, Chapter 6 of this volume). We adopt this Chayanovian approach with imperfect labour markets (Angelsen, 1999) because we believe it better reflects conditions in frontier areas than the ideal of perfect markets.

In the first period, the household can allocate its labour endowment, L, to one of four uses:

1. Production on current land. The production potential on current land is summarized in the production function $f_0[L_{00}]$, which is a standard well-behaved production function with positive but diminishing returns to labour. Labour allocated to production on current land thus generates an immediate income of $p_0 f_0[L_{00}]$.

2. Improvements to current land (intensification). By allocating labour to improving current land, the household can increase the productivity of this land in future periods to $\theta[L_I]f_0[L_{0t}]$. For example, farmers might install irrigation systems or improve the drainage of their fields. It is assumed that there are diminishing returns to labour in intensification. If no labour is allocated to intensification, productivity on current land remains unchanged, $\theta[0] = 1$.
3. Clearing of additional land (extensification). The household can allocate labour to clearing additional land for cultivation. The area cleared is given by $A_1 = \alpha L_E$, where α is the clearing efficiency.[3] This land will become available for production in the second period and will have the production function $f_1[L_{1t}]$. The same crop may or may not be produced on cleared land as on current land; in general, however, we assume that the value of production for given inputs will be lower on cleared than on current land, since cleared land tends to be less suited to agriculture.[4] Clearing land also means forgoing any income that that land might have generated had it remained uncleared.
4. Leisure.

Because of the labour market imperfections, these four allocations must equal the household's labour endowment: $L = L_{00} + L_I + L_E + L_{L0}$.[5]

In the second and all subsequent periods, the household can allocate its labour endowment, L, to one of three uses:

1. Production on current land. The production potential on current land will depend on the extent to which improvements to current land were made in the previous period, as summarized in the production function $\theta[L_I]f_0[L_{0t}]$. Labour allocated to production on current land thus generates an annual income of $p_1\theta[L_I]f_0[L_{0t}]$.
2. Production on cleared land. This land has the production function $f_1[L_{1t}]$, generating an annual income of $p_1A_1f_1[L_{1t}]$.
3. Leisure.

Once again, these allocations must equal the household's labour endowment: $L = L_{0t} + A_1L_{1t} + L_{Lt}$.

To examine this problem, we develop a multiperiod version of the approach used in de Janvry *et al.* (1991). In this model, presented fully in Appendix 5.1, households must decide how to allocate their labour in the first and subsequent periods so as to maximize the present value of utility they derive from the resulting consumption and from leisure, subject to constraints set by the production technologies they use, the total amount of labour available and the prices they face for inputs, outputs and consumption goods. Since choices in the future periods depend on first-period choices, we focus on the latter.

There are four possible sets of solutions to this problem: (i) farm households do not allocate any labour to either intensification or extensification and continue producing only on current land; (ii) farm households extensify but do not intensify – that is, in the first period they allocate labour to clearing new land but not to improving their current land; (iii) farm households intensify production and do not extensify; and (iv) farm households both intensify and extensify. We focus here on factors that tend to discourage extensification and encourage intensification.

Extensification Decisions

As shown in the appendix to this chapter, equation (9), farm households will allocate no labour to extensification if

$$\sum_{t=1}^{\infty} \rho^{t} \frac{\partial U}{\partial c_{t}} \alpha \frac{(p_{1} f_{1}[L_{1t}] - w_{t} L_{1t} - \kappa)}{p_{c}} < \frac{\partial U}{\partial c_{0}} \frac{(\kappa\alpha + w_{0})}{p_{c}}$$

The left-hand-side term is the present value of the increased utility resulting from increased future consumption that the expansion of cultivated area will make possible. This depends on the extent of the increase in cultivated area (which in turn depends on clearing efficiency, α) and on the returns achievable on the newly cleared area, net of production costs – including the opportunity cost of labour, w – and of any income that this area might have generated in its uncleared state, κ.[6] The right-hand-side term is the short-term cost of allocating labour to clearing, which includes both the opportunity cost of labour, w_0, and any forgone income from the cleared area. This expression thus states the common-sense result that farm households will not allocate labour to extensification if the short-term costs of extensification exceed the long-term gains. More importantly, however, it allows us to examine how these relative benefits are affected by the characteristics of the technologies available to farm households, by their socio-economic environment and by the households' preferences. Thus, we see that farm households are less likely to allocate labour to extensification:

1. The lower the efficiency of land clearing, α;
2. The lower the productivity of the newly cleared land, f_1;
3. The lower the price of products produced on cleared land, p_1;
4. The greater the income from land in its uncleared state, κ;
5. The greater the opportunity cost of labour in other uses in the first period, w_0;
6. The greater the rate of time preference (that is, the lower the discount factor, ρ); and
7. The greater the marginal utility of current consumption, $\partial U/\partial c_0$, and the lower the marginal utility of future consumption, $\partial U/\partial c_t$, $t \geq 1$.

Intensification opportunities affect the likelihood that extensification will be undertaken indirectly, through their effect on the opportunity cost of labour in each period, w_t. As shown in Appendix 5.1, equation (8a), the opportunity cost of labour in the first period, w_0, depends on the marginal benefits that households can derive by allocating an additional unit of labour to intensification. And, as shown in equation (8b), the opportunity cost of labour in subsequent periods, w_t, depends on the marginal benefits of allocating an additional unit of labour to production on improved land. The greater the benefits of intensification, the higher the opportunity cost of allocating labour to extensification and, hence, the less likely it is that households will choose to do so. Returns to production on current land influence extensification decisions in the same way. We see here the importance of the assumption about imperfect labour markets. If labour markets were perfect, decisions to allocate labour to extensification would depend solely on the wage and the marginal return to labour in extensification; returns to intensification would not play any role.

The other interesting factor in extensification decisions is the income that households derive from uncleared land, denoted by κ. The higher this value, the less likely that households will find it optimal to devote labour to extensification, since it would forfeit the corresponding income, in both the short and the long term. Households can derive income from uncleared land, for example, by collecting a variety of non-timber products, such as fuelwood, fruit or fodder. A considerable literature has developed in recent years on the value of non-timber forest products, spurred by a study estimating that products harvested from a tropical forest in Peru could be worth as much as US\$422 ha^{-1} year^{-1} (Peters *et al.*, 1989). Although this particular estimate is almost certainly an overestimate, other studies indicate that the collection of non-timber products can yield non-negligible returns. Lampietti and Dixon (1995) review a number of studies, many of which show values in the US\$50–100 ha^{-1} year^{-1} range. A study of households living near Mantadia National Park in Madagascar found that they obtained goods worth about US\$100 year^{-1} from forest areas (Kramer *et al.*, 1995). Uncleared areas are also likely to generate many environmental services. The benefits of these services, however, are likely to accrue mostly to people downstream of the uncleared areas, and as such would not be included in an individual household's estimate of κ. For this reason, it is likely that extensification will be undertaken at a rate greater than would be socially optimal.

Intensification Decisions

From equation (10) in Appendix 5.1, farm households will not allocate any labour to intensification if:

$$\sum_{t=1}^{\infty} \rho^t \frac{\partial U}{\partial c_t} \frac{p_0}{p_c} \frac{\partial \theta}{\partial L_I} f_0[L_{0t}] < \frac{\partial U}{\partial c_0} \frac{w_0}{p_c}$$

Here, the left-hand-side term is the present value of the future increased utility resulting from the increased consumption that increased production will make possible; the right-hand-side term is the reduction in first-period utility that will result from lower current consumption due to diverting labour away from current production. Again, the result is eminently plausible: if the long-term gains from intensification are lower than the short-term costs, farm households will not allocate labour to intensification. Once again, the model allows us to examine how changes in the technologies or socio-economic conditions facing farm households affect this result. Farm households will be less likely to intensify:

1. The lower the efficiency of the intensification technology, $d\theta[L_I]/dL_I$;
2. The lower the price of crops produced on current land, p_0;
3. The greater the opportunity cost of labour in other uses in the first period, w_0;
4. The greater their rate of time preference (δ);
5. The greater their marginal utility of current versus future consumption.

Here too, extensification opportunities enter indirectly, through the opportunity cost of labour, w_0. It is interesting to note that increasing the price of consumption goods, p_c, does not affect intensification decisions, since this price affects both the long-term benefits and the short-term costs of intensification.

Intensification will occur if the present value of long-term benefits from intensification is greater than that of the short-term costs. Assuming this is the case, farmers must still decide how to allocate available labour in the first period among current production, intensification and leisure. In this case, the following condition must hold (from Appendix 5.1, equation 12):

$$\sum_{t=1}^{\infty} \rho^t \frac{\partial U}{\partial c_t} \frac{p_0}{p_c} \frac{\partial \theta}{\partial L_I} f_0[L_{0t}] = \frac{\partial U}{\partial c_0} \frac{p_0}{p_c} \frac{\partial f_0}{\partial L_{00}} = \frac{\partial U}{\partial L_{L0}}$$

This expression says that households should allocate available labour so as to equate the present value of increased future marginal utility resulting from higher future consumption made possible by intensification (first term) to both the marginal utility of current consumption forgone because less labour is devoted to current production (second term) and to the marginal utility of current leisure forgone (third term).

Figure 5.1 illustrates how this condition leads to labour allocation among the three competing uses in the first period. The total amount of labour available to the household is given by the width of the x axis. Curve A shows how the present value of the household's future marginal utility from intensification is affected by increasing labour allocation. The marginal utility of intensification falls as more labour is allocated to it for two reasons: because of diminishing returns to labour in intensification, and because of the diminishing utility of

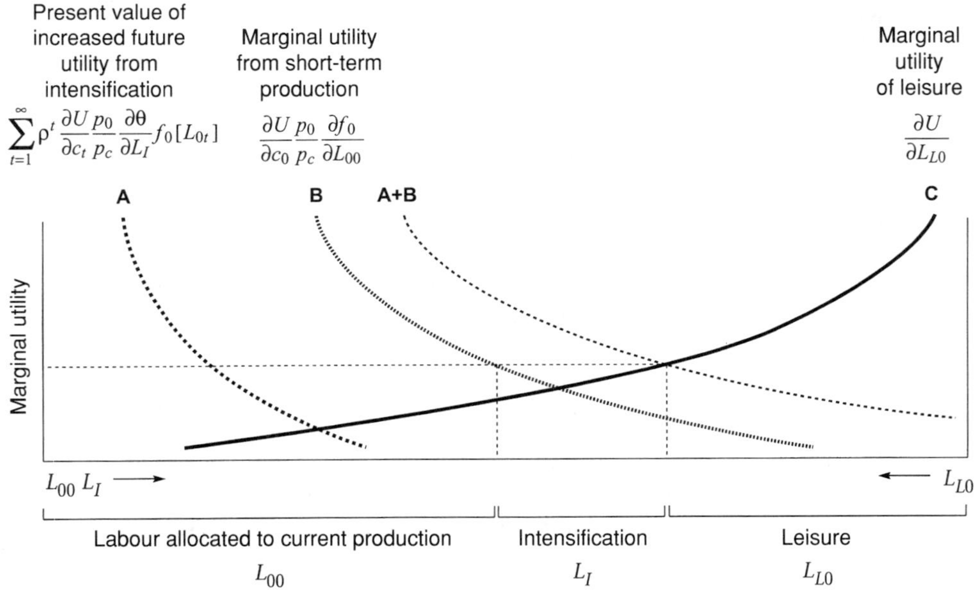

Fig. 5.1. First-period labour allocations when intensification is undertaken.

additional consumption. Likewise, curve B shows that the marginal utility from short-term production also falls as additional labour is devoted to this purpose. Curve C illustrates the marginal utility of leisure; in this case, the amount of labour allocated to leisure is measured from right to left. As with the other curves, the marginal utility falls with increasing labour allocation. Curve A + B combines the two curves showing the marginal benefits of labour allocation to productive activities and is used to find the equilibrium between these uses of labour and leisure. The shape of each marginal utility curve as shown is quite general, but their location is arbitrary. In the analysis that follows, we examine how the relative share of labour devoted to current production, leisure and intensification changes as various factors are varied.

The factor that most directly influences the effort devoted to intensification is the efficiency of intensification, $\partial\theta/\partial L_I$. This may take the form of improved irrigation technologies, for example.[7] Increasing the efficiency of intensification results in the first curve being shifted to the right, from A to A′, as shown in Fig. 5.2. For any given amount of labour allocated to intensification, the household will experience a greater marginal utility, even though the effect is dampened by the diminishing marginal utility of consumption. Under these conditions, it is probable that more labour will be devoted to intensification at the expense of both current production and leisure.[8] Murray (1994) describes a case in the Dominican Republic in which the introduction of a gravity-driven sprinkler irrigation system not only encouraged farmers to intensify production on already cleared land, but actually induced them to allow the more marginal lands which they had cleared to revert to forest cover.

In many developing countries, government policies have tended directly or indirectly to discriminate heavily against agriculture. Resources have been extracted from agriculture in a variety of ways: overvalued exchange rates, protection of competing sectors, price controls and high direct taxation. Analysis of policies in a sample of 18 developing countries found that transfers out of

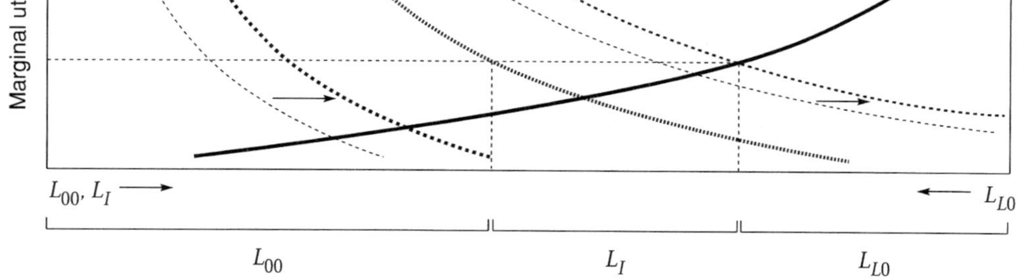

Fig. 5.2. First-period labour allocations: effect of increasing the efficiency of intensification.

agriculture averaged 46% of agricultural gross domestic product (GDP) during 1960–1984 (Schiff and Valdés, 1992). It has often been argued that distortions to economic incentives created by government policies have played an important role in encouraging activities that damage natural resources (World Bank, 1992; Panayotou, 1993). Southgate *et al.* (1999), for example, blame government policies that discriminate against agriculture for expansion of the area cultivated in Ecuador.

Would farm households devote more effort to intensification if policies did not discriminate against agriculture? Figure 5.3 examines this question. Improving the terms of trade for farmers (that is, increasing p_0 relative to p_c) increases both the short-term returns to current production (shifting curve B to B′) and the long-term returns to intensification (shifting curve A to A′). The drop in leisure is unambiguous. The extent to which labour devoted to intensification increases is less clear, since it would mean forgoing some current production, which is now more profitable. Moreover, if the policy change also improves the terms of trade for crops that would be produced on cleared land (that is, increases p_1 relative to p_c), farmers would also face increased incentives to extensify, as noted earlier. It cannot be predicted *a priori*, therefore, whether policy changes that encourage agriculture in general will tend to encourage intensification or not, or, if so, by how much. Much will depend on the specific interaction of conditions in a given case.[9]

Poverty is another factor that has often been thought likely to affect investments (Reardon and Vosti, 1997b). In comparison with better-off households, poorer households are likely to have both a higher marginal utility of consumption and a higher rate of time preference (Holden *et al.*, 1998). As shown in Fig. 5.4, the higher marginal utility of consumption will shift curve B to B′, compared with better-off households. Higher rates of time preference suggest that the poorer households will place relatively less weight on the future benefits of intensification. As a result, it might be expected that poor

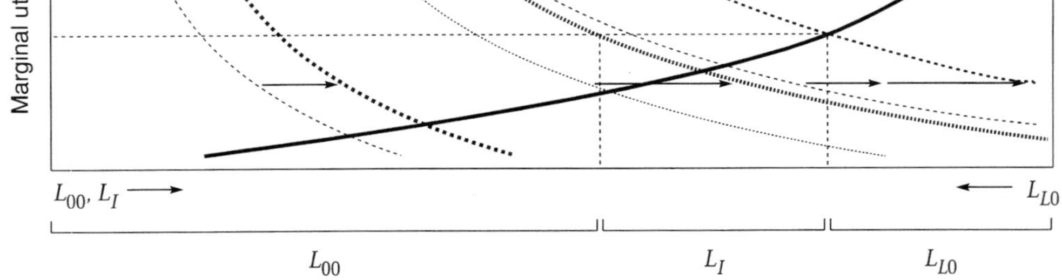

Fig. 5.3. First-period labour allocations: effect of improving the terms of trade for farmers.

households will: (i) be less likely to undertake intensification at all; and (ii) devote proportionally less labour to intensification if they do undertake it. The effect of high rates of time preference on the marginal utility of intensification is offset, however, by the higher marginal utility of consumption. Poorer households may discount the benefits of increased consumption from intensification more than better-off households, but they will also value increments in consumption more. These effects tend to offset each other, so that the net effect on increased future utility from intensification is ambiguous.

Figure 5.4 illustrates the interaction of these factors, based on the assumption that the net effect of higher rates of time preferences and higher marginal utility of consumption offset each other exactly, so that curve A remains unchanged.[10] The increase in labour allocation to short-term production is clear, as is the reduction in labour allocation to leisure. The impact on labour allocation to intensification is unclear, however. It is possible that labour allocation to intensification might actually increase, relative to better-off households, if the marginal utility of increased consumption from increased future production is sufficiently high.[11] Conversely, if the impact of the higher rate of time preferences dominates, it is likely that less labour will be devoted to intensification. Even in this second case, however, the reduction in intensification efforts will be less than might have been expected if higher rates of time preferences alone were at work.[12] It is interesting to note that very similar arguments apply to poor households' extensification decisions. As noted previously, a higher rate of time preference tends to reduce the likelihood that households will find it optimal to devote labour to clearing an additional area, although in the case of poor households this effect is tempered by the higher value such households place on the additional consumption.

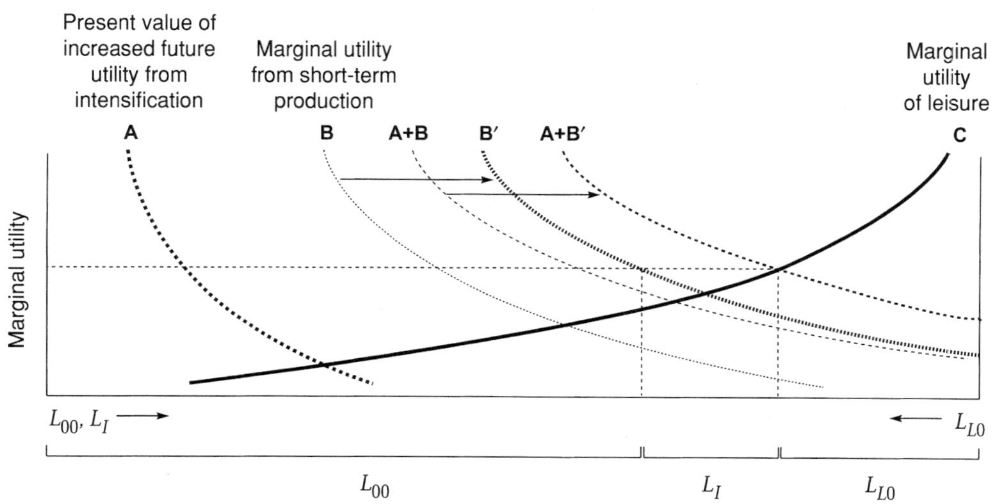

Fig. 5.4. First-period labour allocations: effect of poverty.

Conclusions

The factors that determine farmer intensification and extensification decisions are complex, making it difficult to develop policies that encourage the former and discourage the latter. Moreover, the magnitude, and sometimes even the direction, of the effect of certain factors is likely to vary with site-specific conditions, making it unlikely that general policies will be successful. In particular, it cannot be assumed that removing policy distortions that discriminate against agriculture alone will always tend to reduce pressure to extensify production and encourage intensification.[13] Nor should it be assumed that poorer farmers will always have lower incentives to intensify. Considerable knowledge about the specific conditions of a given case will be required, therefore, in order to be able to reliably design policies that reduce extensification and increase intensification.

Only changes that specifically target intensification or extensification decisions are likely to reliably have the expected effect. Thus, taxing chainsaws and subsidizing improvements such as irrigation systems are likely to result in reduced extensification and increased intensification. Efforts to increase the benefits farm households obtain from uncleared areas are also likely to discourage extensification, whether these take the form of efforts to increase the value of products obtained from uncleared areas or institutional changes designed to ensure that farm households can capture these benefits.

The analysis presented here is clearly dependent on the assumptions made herein, chief among which is that of imperfect labour markets. As Kaimowitz and Angelsen (1998) have shown, changes in this assumption can lead to quite different results. That labour markets do not work perfectly is a reasonable assumption for many frontier areas. Moreover, other constraints could have very similar effects. For example, if a household has a limited ability to supervise hired labour, the allocation of labour to extensification would again be limited by the opportunity cost of allocating the household's available supervision capacity elsewhere. It should also be borne in mind that, while extensification probably poses the graver environmental threat, intensification can also pose threats to the environment (for example, see Crissman *et al.*, Chapter 8 of this volume, on Andean highland agriculture, and Pingali and Rosegrant, Chapter 20 of this volume, on intensive South Asian systems). Intensification often requires greater input use, increasing the risk of contamination of waterways by agrochemicals. Management of water for agricultural purposes can substantially affect the timing, volume and velocity of water flow and groundwater recharge, thereby altering natural lake, riverine, estuarine and marine habitats. Policies should not, therefore, promote intensification indiscriminately.

Acknowledgements

This chapter presents the initial results of ongoing research on the factors affecting farm household choices between intensification and extensification. We are grateful for comments from David Lee, Chris Barrett and Arild Angelsen, and from participants at the American Agricultural Economics Association (AAEA) Conference on Agricultural Intensification, Economic Development and the Environment. The views presented in this chapter are the authors' own and do not necessarily represent those of the World Bank group or of the Agricultural University of Norway.

Notes

1 This formulation allows for the crops produced to be either food crops or cash crops (or some combination), and allows us to examine both households that produce crops either wholly or partly for sale and those that produce solely for their own consumption (in which case, their 'budget constraint' becomes the same as their production).

2 If the farm household's rate of time preference is δ, the discount factor, ρ, is given by $\rho = 1/(1 + \delta)$. Thus, the discount factor will be greater (smaller) when the rate of time preference is smaller (greater). The discount factor is used here because it is easier to manipulate.

3 Assuming linear clearing efficiency is plausible when there is abundant land to be cleared. This assumption could easily be changed to one of diminishing returns to labour in clearing (which would be plausible, for example, if new land to be cleared were progressively further away) without changing the qualitative results in this chapter.

4 In many shifting-cultivation environments, productivity on newly cleared land may initially be higher than on previously cultivated land. Our model focuses on the expansion of the permanently cultivated area, however, and in this context it is plausible to assume that land most suited to permanent cultivation would be cleared first and that expansion would take place into increasingly less favourable land.

5 This formulation ignores the possibility that labour demands for each activity vary during the course of the year. To the extent that they do, the labour constraint will be less binding, since labour use for current production would not necessarily conflict with labour use for intensification or extensification. Introducing this consideration would make the model much less general, however, since the specific details are likely to vary considerably from case to case.

6 Consumption is expressed in physical units, since the value of consumption is divided by its cost; the 'cost' of this additional consumption in terms of lower leisure due to labour demands from the cleared area is subsumed in the shadow wage rate.

7 In a more general model, in which land degradation processes on current land such as erosion or nutrient depletion are taken into account, intensification efficiency could also take the form of improved conservation techniques and/or improved methods to repair the damage of past degradation.

8 The assumption of imperfect labour markets plays an important role in this result. If households can hire additional labour, the higher returns made possible by improved intensification technology may induce them to clear additional areas so as to enjoy these higher returns on a larger scale. Vosti *et al.* (2000a), for example, find that predicted rates of deforestation in the western Brazilian Amazon are very sensitive to assumptions about labour availability.

9 Similarly, Pagiola (1996) finds that, although maize price policy liberalization tends to encourage farm households in semi-arid Kenya to adopt soil conservation measures, it does not always do so, even within the same area.

10 Since the two effects act in opposite directions, any net shift in curve A is likely to be relatively small.

11 This result is similar to that of Pagiola's (1995) analysis of poor farmers' incentives to adopt conservation. He found that poorer households generally have higher incentives than better-off households to adopt conservation, because the penalty for failing to do so (in terms of reduced utility) is so much harsher for the poorer household.

12 It should be stressed that this argument is to be understood relative to a better-off household on the same quality land, using the same technology and facing the same prices. If poorer households differ from better-off households not only in their rates of time preference but also because they have access to poorer land or less productive technology, they may well devote less labour to intensification. But the cause of this shortfall should then be sought in the lower returns of the intensification options open to them, rather than in their rate of time preference alone, as is commonly done.
13 Given the efficiency advantages that are often derived from removal of policy distortions, this should not be interpreted as an argument against policy reform.

Appendix 5.1

Notation

$U[c,L_L]$	Household utility function
c_t	Consumption in period t
L	Household's labour endowment
L_{0t}	Labour allocated to production on current land in period t
L_{1t}	Labour allocated to production on newly cleared land in period t
L_E	Labour allocated to extensification
L_I	Labour allocated to intensification
L_{Lt}	Leisure in period t
δ	Household's rate of time preference
A	Uncleared area
A_0	Current land cultivated (normalized to $A_0 = 1$ for convenience)
A_1	Cleared land
α	Clearing efficiency
κ	Income generated per unit of uncleared land, $\kappa \geq 0$
$f_0[L]$	Production technology on current land, $\mathrm{d}f_0[L]/\mathrm{d}L > 0$, $\mathrm{d}^2 f_0[L]/\mathrm{d}L^2 < 0$
$f_1[L]$	Production technology on cleared land, $\mathrm{d}f_1[L]/\mathrm{d}L > 0$, $\mathrm{d}^2 f_1[L]/\mathrm{d}L^2 < 0$
$\theta[L]$	Intensification technology on current land, $\theta[0] = 1$, $\mathrm{d}\theta[L]/\mathrm{d}L > 0$, $\mathrm{d}^2\theta[L]/\mathrm{d}L^2 < 0$
p_c	Price of consumption goods
p_0	Price of crops produced on current land
p_1	Price of crops produced on cleared land
w_t	Opportunity cost of labour in period t
w_0	Opportunity cost of labour in current period

Model

The household's problem can be summarized as:

$$\max_{\left\{\begin{matrix} c_0, L_{L0}, c_1, L_{Lt} \\ L_{00}, L_E, L_I, L_{0t}, L_{1t} \end{matrix}\right\}} \quad U_0[c_0, L_{L0}] + \sum_{t=1}^{\infty} \rho^t U_t[c_t, L_{Lt}]$$

$$st \begin{cases} p_c c_0 \leq p_0 f_0[L_{00}] + \kappa(A - A_1) & \text{Period 0} \\ p_c c_t \leq p_0 \theta[L_I] f_0[L_{0t}] + A_1 p_1 f_1[L_{1t}] + \kappa(A - A_1) & \text{Period } t,\ t = 1, 2, \ldots \\ L_{L0} + L_{00} + L_E + L_I \leq L & \text{Period 0} \\ L_{Lt} + L_{0t} + A_1 L_{1t} \leq L & \text{Period } t,\ t = 1, 2, \ldots \\ A_1 = \alpha L_E & \text{Clearing efficiency} \end{cases}$$

The Lagrangian for this problem is:

$$\mathbf{L} = U_0[c_0, L_{L0}] + \lambda_0\left(p_0 f_0[L_{00}] + \kappa(A - \alpha L_E) - p_c c_0\right) +$$

$$\mu_0(L - L_{L0} - L_{00} - L_E - L_I) + \sum_{t=1}^{\infty} \rho_t \begin{pmatrix} U_t[c_1, L_{Lt}] + \lambda_t\left(p_0\, \theta[L_I] f_0[L_{0t}] + \right. \\ \left. \kappa(A - \alpha L_E) + \alpha L_E p_1 f_1[L_{1t}] - p_c c_t\right) \\ + \mu_t(L - L_{Lt} - L_{0t} - \alpha L_E L_{1t}) \end{pmatrix}$$

And the first-order conditions are:

$$\frac{\partial \mathbf{L}}{\partial c_0} = \frac{\partial U}{\partial c_0} - \lambda_0 p_c = 0 \tag{1a}$$

$$\frac{\partial \mathbf{L}}{\partial c_t} = \sum_{t=1}^{\infty} \rho^t \left(\frac{\partial U}{\partial c_t} - \lambda_t p_c \right) = 0 \tag{1b}$$

$$\frac{\partial \mathbf{L}}{\partial L_0} = \frac{\partial U}{\partial L_{L0}} - \mu_0 = 0 \tag{2a}$$

$$\frac{\partial \mathbf{L}}{\partial L_{Lt}} = \sum_{t=1}^{\infty} \rho^t \left(\frac{\partial U}{\partial L_{Lt}} - \mu_t \right) = 0 \tag{2b}$$

$$\frac{\partial \mathbf{L}}{\partial L_{00}} = \lambda_0 p_0 \frac{\partial f_0}{\partial L_{00}} - \mu_0 = 0 \tag{3}$$

$$\frac{\partial \mathbf{L}}{\partial L_{\mathrm{I}}} = -\mu_0 + \sum_{t=1}^{\infty} \rho^t \lambda_t p_0 \frac{\partial \theta}{\partial L_I} f_0[L_{0t}] \le 0;\, L_I \ge 0;\, \frac{\partial \mathbf{L}}{\partial L_I} L_I = 0 \tag{4}$$

$$\begin{aligned} \frac{\partial \mathbf{L}}{\partial L_E} &= -\lambda_0 \kappa\alpha - \mu_0 + \sum_{t=1}^{\infty} \rho^t \left(-\lambda_t \kappa\alpha + \lambda_t \alpha p_1 f_1[L_{1t}] - \mu_1 \alpha L_{1t}\right) \\ &\le 0;\, L_E \ge 0;\, \frac{\partial \mathbf{L}}{\partial L_E} L_E = 0 \end{aligned} \tag{5}$$

$$\frac{\partial \mathbf{L}}{\partial L_{0t}} = \sum_{t=1}^{\infty} \rho^t \left(\lambda_t p_0 \theta[L_I] \frac{\partial f_0}{\partial L_{0t}} - \mu_t \right) = 0 \tag{6}$$

$$\frac{\partial \mathbf{L}}{\partial L_{1t}} = \sum_{t=1}^{\infty} \rho^t \left(\lambda_t \alpha L_E p_1 \frac{\partial f_1}{\partial L_{1t}} - \mu_1 \alpha L_E \right) \le 0;\, L_{1t} \ge 0;\, \frac{\partial \mathbf{L}}{\partial L_{1t}} L_{1t} = 0 \tag{7}$$

To simplify the exposition, it is useful to define the shadow wage rates. From equations (3), (1a), (2a) and (4), we have:

$$w_0 = \frac{\mu_0}{\lambda_0} = p_0 \frac{\partial f_0}{\partial L_{00}} = p_c \frac{\partial U / \partial L_{L0}}{\partial U / \partial c_0} = \frac{1}{\lambda_0} \sum_{t=1}^{\infty} \rho^t \lambda_t p_0 \frac{\partial \theta}{\partial L_I} f_0[L_{0t}] \tag{8a}$$

and from (6), (7), (1b) and (2b):

$$w_t = \frac{\mu_t}{\lambda_t} = p_0 \theta[L_I] \frac{\partial f_0}{\partial L_{0t}} = p_1 \frac{\partial f_1}{\partial L_{1t}} = p_c \frac{\partial U / \partial L_{Lt}}{\partial U / \partial c_t} \tag{8b}$$

These terms can be interpreted as the opportunity cost of labour to the household (de Janvry *et al.*, 1991).

From equations (5), (1) and (8), farm households will allocate no labour to extensification if:

$$\begin{aligned}
&-\lambda_0\kappa\alpha-\mu_0+\sum_{t=1}^{\infty}\rho^t\left(-\lambda_t\kappa\alpha+\lambda_t\alpha p_1 f_1[L_{1t}]-\mu_t\alpha L_{1t}\right)<0\\
&\sum_{t=1}^{\infty}\rho^t\alpha\left(\lambda_t p_1 f_1[L_{1t}]-\lambda_t\kappa-\mu_t L_{1t}\right)<\lambda_0\kappa\alpha+\mu_0\\
&\sum_{t=1}^{\infty}\rho^t\lambda_t\left(\alpha\left(p_1 f_1[L_{1t}]-w_t L_{1t}\right)-\kappa\alpha\right)<\lambda_0\left(\kappa\alpha+w_0\right)\\
&\sum_{t=1}^{\infty}\rho^t\frac{\partial U}{\partial c_t}\alpha\frac{\left(p_1 f_1[L_{1t}]-w_t L_{1t}-\kappa\right)}{p_c}<\frac{\partial U}{\partial c_0}\frac{\left(\kappa\alpha+w_0\right)}{p_c}
\end{aligned}\qquad(9)$$

From equations (4), (1) and (8), farm households will allocate no labour to intensification if:

$$\sum_{t=1}^{\infty}\rho^t\frac{\partial U}{\partial c_t}\frac{p_0}{p_c}\frac{\partial\theta}{\partial L_I}f_0[L_{0t}]<\frac{\partial U}{\partial c_0}\frac{w_0}{p_c}\qquad(10)$$

Conversely, if the present value of long-term benefits from intensification are greater than the short-term costs, intensification will occur, and we have:

$$\sum_{t=1}^{\infty}\rho^t\frac{\partial U}{\partial c_t}\frac{p_0}{p_c}\frac{\partial\theta}{\partial L_I}f_0[L_{0t}]=\frac{\partial U}{\partial c_0}\frac{w_0}{p_c}\qquad(11)$$

If intensification occurs, farmers must decide how to allocate the available labour among current production, leisure and intensification. From equations (11) and (8a), we have:

$$\sum_{t=1}^{\infty}\rho^t\frac{\partial U}{\partial c_t}\frac{p_0}{p_c}\frac{\partial\theta}{\partial L_I}f_0[L_{0t}]=\frac{\partial U}{\partial c_0}\frac{p_0}{p_c}\frac{\partial f_0}{\partial L_{00}}=\frac{\partial U}{\partial L_{L0}}\qquad(12)$$

6 When Does Technological Change in Agriculture Promote Deforestation?

ARILD ANGELSEN[1,2] AND DAVID KAIMOWITZ[2]

[1]*Department of Economics and Social Sciences, Agricultural University of Norway, Ås, Norway;* [2]*Center for International Forestry Research (CIFOR), Bogor, Indonesia*

Introduction

The belief that technological progress in agriculture reduces pressure on forests by allowing farmers to produce the same amount of food in a smaller area has become almost an article of faith in development and environment circles (Serrao and Toledo, 1990; Bandy *et al*., 1993; Southgate, 1998). Indeed, many experts believe that yield-improving technological change offers the only hope for meeting the agricultural needs of an expanding population and growing economies without encroaching on primary forests. These people frequently point to the Green Revolution in Asia and the experience of agricultural development in the USA as examples of where new agricultural technologies increased food supplies and pushed down producer prices, which, in turn, discouraged farming in marginal areas. At the more micro level, proponents of this view often advocate projects to improve agricultural productivity in the buffer zones of protected areas to discourage farmers from expanding into the parks themselves.

At the same time, basic economic theory suggests that technological progress makes agriculture more profitable and gives farmers an incentive to expand production on to additional land. This implies that technological change in areas close to forests should increase, rather than reduce, forest clearing (Southgate, 1990; DeShazo and DeShazo, 1995). Moreover, technological change may provide capital-constrained farmers the additional income they need to finance activities involving deforestation (Richards, 1997).

Previous attempts to assess this issue yielded mixed results (Kaimowitz and Angelsen, 1998). Jones *et al.* (1995) used household regression models to analyse data on small farmers in the Brazilian Amazon and found that farmers with higher per hectare values of production cleared less forest. Godoy *et al.* (1997) reached the same conclusion when assessing the relation between physical yields and forest clearing among indigenous rice growers in Honduras. Other studies also associate higher agricultural productivity and investment in research and extension with slower expansion of land in agriculture. These include a paper on Mexico by Deininger and Minten (1996), an analysis of Thailand by Panayotou and Sungsuwan (1994), both based on regional regression models, and multi-country modelling work by Southgate (1994) in Latin America and by Binswanger *et al.* (1987) across different regions. On the other hand, Foster *et al.* (1997) estimated a household regression model using Indian data and found that farmers with higher productivity deforested more. Katila (1995) obtained a positive correlation between higher yields and deforestation in Thailand. Meanwhile, studies by Barbier and Burgess (1996) and Chakraborty (1994), focusing on Mexico and India, respectively, found that productivity did not significantly affect forest clearing.

These apparently contradictory results are not surprising, because the way technological change in agriculture affects deforestation depends on the type of technologies and farmers involved, as well as the conditions under which the change takes place. Important questions regarding the type of technologies are: what agricultural subsectors they apply to; how technology affects factor intensities (land and capital used per hectare); and whether technological change makes agriculture more or less risky. Key farmer characteristics include their objectives, income levels and degree of risk aversion. Important contextual considerations include household labour market conditions, the importance of migration, whether credit rationing limits farmers' ability to expand their area and the elasticity of demand for the products farmers sell.

This chapter uses simple theoretical models and selected empirical evidence to analyse under what conditions technological progress in agriculture can be expected to increase or reduce deforestation. Growth models in the Solow tradition inspire the models presented. These models distinguish between physical and efficiency units of factors of production and describe technological progress as an increase in the efficiency of input use. To capture the spatial dimension of agricultural production and deforestation, we incorporate 'von Thünen-type' features into the models. This means the models take into account transportation costs, and cultivation stops at the point where the costs are so high that further expansion is unprofitable.

Most of the discussion is limited to exogenous technological change. Thus, the analysis starts after farmers have adopted a certain technology and does not address *why* the change takes place in the first place. This issue, however, is addressed briefly towards the end of the chapter in a discussion of the theory of induced innovation.

The chapter deals primarily with labour and output markets and not capital or land markets. This simplifies the analysis and permits some unambiguous results with only a minor loss of practical relevance. In principle, many of the lessons drawn from the analysis of labour markets apply equally to capital markets. A final simplification is to focus on single-product farming systems. Most real-life farmers produce several products simultaneously; however, extending the models to take that into account would unnecessarily complicate them. Fortunately, most of the results from the two sector models discussed below hold when one interprets them as representing a single farmer who produces two different crops.

Technological Change at the Farm Level

Farm-level decisions about how to allocate land – in particular, whether or not to convert forests to cropland or pasture – largely determine how much tropical deforestation occurs. Often, however, it may be more useful to think of deforestation as resulting from a *labour* allocation problem, particularly if labour is the scarcest factor of production and the main input besides land. How do farmers allocate their time between work and leisure, between on- and off-farm work, between different crops and production systems, and between improving existing land and clearing new land?

Technological change clearly affects these decisions, but not always in an obvious manner. It turns out that the answer depends largely on what one assumes with regard to farmers' preferences and the market conditions they face. Farmers may seek to simply attain a minimum predetermined (subsistence) level of consumption or they may attempt to maximize profits. An intermediate situation occurs when farmers want to balance consumption and leisure, but face market imperfections that prevent them from freely substituting family and hired labour and freely choosing between working on the farm or off-farm themselves. This situation is referred to as the 'Chayanovian' model, in reference to the Russian economist who studied this type of behaviour among Russian peasants at the beginning of the century.

Since the models in this chapter focus on the labour and output markets, they are classified based on what each one assumes about how these markets function. We first consider a situation with an imperfect labour market, in which case both the 'subsistence' and Chayanovian approaches are relevant. 'Subsistence' behaviour is more likely to occur in situations where one also has imperfect output markets. In the terminology of the literature on the economics of rural organization, this corresponds to a 'high-transaction-costs economy' (see Hoff *et al.* (1993) for an overview of this literature). After looking at models that assume subsistence-type behaviour, we then examine a situation with perfect output and labour markets. We refer to these latter models as 'open economy' or 'profit-maximizing' models. These models apply more to economies with low transaction costs.

Modelling technological change

Economic theory normally defines technological progress as an increase in total factor productivity: one can produce more with the same amount of inputs. To facilitate the theoretical analysis of technological progress, we distinguish between physical and efficiency units of land and labour. These are the only two factors of production in our model, even though capital enters indirectly, as shown below. An 'efficiency unit' can be thought of as the effective contribution to production made by a physical unit of some factor of production. Technological change can bring about changes in the efficiency of land, labour or the overall production process.

We define[1] $N = \varepsilon L$, $A = \beta H$, where L is the labour input in physical units, which when multiplied by an efficiency factor, ε, gives the efficiency units of labour, N. Similarly for land, H is the physical measurement in hectares, β the efficiency factor and A the efficiency units of land. We assume the production function $Y = \alpha F(N, A)$ has constant returns to scale. This implies that doubling the inputs of both land and labour will double production. This is a fair approximation to reality and simplifies the analysis. We can then define the yield function, where y is output per efficiency (rather than physical) unit of land:

$$y = \alpha f(n),\ y \equiv \frac{Y}{A} = \frac{Y}{\beta H},\ n \equiv \frac{N}{A} = \frac{\varepsilon L}{\beta H} = \frac{\varepsilon}{\beta} l,\ l \equiv \frac{L}{H},\ f_n > 0,\ f_{nn} < 0$$

and where $l = (\beta/\varepsilon)n$ is the labour intensity, defined as physical labour input per hectare, while n is the equivalent measure based on efficiency units, that is, efficiency units of labour per efficiency units of land. Based on this terminology, we specify three types of technological change, depending on which of the three efficiency parameters (α, β, ε) changes.

Pure yield-increasing technological change (α)

An increase in α represents pure yield-augmenting technological progress that does not affect the marginal rate of technical substitution between land and labour. This is referred to in the economics literature as *Hicks-neutral* technological progress. An example of this type of technological progress could be higher-yielding crops or more pest- and weather-resistant plant varieties. It might be tempting to label this type of technological change as labour-neutral, but that would be a mistake. To the farmer, this type of technological progress is exactly the same as an output price increase and will give an incentive to employ more labour per hectare.

Labour-intensive technological change (β)

An increase in the efficiency factor for land, β, will increase labour intensity, and we therefore refer to this as *labour-intensive* technological progress.[2] It can also be labelled *land-saving* technological progress. Factor β will typically be a function of certain kinds of capital inputs, such as fertilizers, which simultaneously increase physical yields and the amount of labour used per hectare.

Labour-saving technological change (ε)

Applying capital inputs, such as chain-saws and tractors, increases the productivity of labour (ε) and allows the same job to be done in less time. We shall therefore refer to this as *labour-saving* technological progress. Note that, more than the two other types of technological change, this change frequently requires large capital investments and/or purchased inputs, and it could thus also be labelled *capital-intensive* technological change. As discussed below, an increase in ε may, in theory, lead to higher labour input per hectare (because labour measured in efficiency units has become cheaper), but that possibility is excluded here. Thus, this can also be thought of as *labour-displacing* technological progress.

A subsistence model

The first model, the household subsistence model, assumes that farmers only wish to produce enough agricultural products to meet their basic needs, at some predetermined subsistence level of production or income, $\bar{I}$. The subsistence model may apply in at least three types of situations: first, when farmers have no desire to consume more than a fixed level of goods and services; secondly, when village norms require farmers to share any production above the subsistence level, thus eliminating their incentive to increase production; or, thirdly, when thin or non-existent output markets (perhaps limited by high transport costs) effectively impede farmers from converting additional agricultural output into other types of consumption goods.

The area dedicated to agriculture forms a circle around the village, with the edge of that circle being the outer limits of cultivation (b^e). The farm-gate price is $(p - tb)$, where b is the distance (km) from the farm to the circle's centre where farmers sell their products and t is a measure of transport costs per kilometre. We assume that labour input per hectare in efficiency units is fixed, $n = \bar{n}$. It follows that agricultural output per efficiency unit of land is also fixed. This simplifies the model without limiting its empirical relevance and still enables us to illustrate the main conclusions.

We further assume that farmers are constrained in the labour market and cannot sell more than a fixed amount of labour $\bar{L}^0$ at a fixed wage rate w. The farmer's problem, then, is to work as little as possible while still producing enough to meet his or her family's subsistence needs:

$$\text{Min} \int_0^{b^e} \bar{n}\beta\varepsilon^{-1}hb\,\mathrm{d}b + \bar{L}^0 \text{ given} \int_0^{b^e} (p - tb)\alpha f(\bar{n})\beta\, hb\,\mathrm{d}b + w\bar{L}^0 = \bar{I}; h \equiv 2\pi / K$$

Here, K is the number of households in the village, which is exogenous, and determines the share of the circle around the village available for each individual household. With n being fixed, the solution to this problem is straightforward. Farmers will work as much off-farm as possible, and then meet the remainder of their subsistence requirements by cultivating the smallest possible agricultural area that will allow them to make up the difference.[3]

How will a new technology influence farmers' land-use decisions? The key is how it affects the output per hectare (physical yield): $\beta\, y = \beta af(n)$. Both pure yield-increasing and labour-intensive technological progress will reduce deforestation, because the same income can be obtained from a smaller agricultural area. Following the assumption of fixed n, labour-saving technological change (ε) will not change the rate of deforestation in this model. Output per hectare will remain the same and so the farmer must cultivate the same amount of land to reach his or her subsistence needs. The only difference is that the farmer does not have to work as much to reach subsistence. Thus, this type of technological change improves people's welfare, but does not help conserve forests.[4]

Risk is not formally included in the model, but we still might consider the effects of risk aversion within the subsistence approach. When farmers decide how much land to farm based on a desire to guarantee a minimum consumption level, even in the worst-case situation, changes that make production less risky reduce deforestation even if they do not change average yields. Since subsistence behaviour seems more likely in contexts where farmers are weakly integrated into output markets, the most important risks are probably production risks related to rainfall and pests. Reducing pest problems should also increase average yield, thus giving a dual benefit for forest conservation.

The pure subsistence model probably does not apply to most tropical rural areas, although it may be relevant for some remote areas with poorly developed markets. Its greatest utility lies in the fact that it makes explicit what assumptions must hold at the household level for technological progress in agriculture to automatically reduce deforestation. This conceptual model also appears to implicitly underlie many of the integrated conservation and development projects (ICDPs) that assume households will deforest less if their agricultural yield rises.

A perfect-market (open-economy) model

At the opposite end of the spectrum in regard to subsistence models are models that assume that farmers: (i) can sell as much as they want of their products at fixed (exogenously determined) prices; (ii) can obtain or sell all the labour and capital they desire at predetermined wage and interest rates; (iii) use family and non-family labour interchangeably (the two are perfect substitutes); and (iv) are indifferent about whether they work on or off their farms.

An important result from standard agricultural household models is that when these conditions apply the models become recursive. We can first analyse household decisions about production (including how much family or outside labour to use on the farm) from a profit-maximization perspective. Then we can look at consumption decisions (including how much the family should work) as utility-maximization problems. This implies that, when utility-maximizing households operate in perfect-market contexts, they base their farming decisions solely on the desire to maximize profits.

The model is also recursive in another sense. The farmer first determines how much labour to apply at a given distance following the standard profit-maximizing maxim that the marginal return to labour should equal the wage. Next, he or she decides how far the outer edge of cultivation (the agricultural frontier) should be from the village centre by continuing out until the profit or land rent is zero. Beyond that point, transport costs are too high to make further expansion profitable. We can define per hectare land rents in terms of either physical or efficiency units. Analytically, it makes no difference. Two equations characterize farmers' behaviour at the point of maximum profit:

$$(p - tb)\alpha\varepsilon f_n = w$$

$$(p - tb^e)\alpha f - wl\beta^{-1} = 0$$

The full derivation of how technological change will affect deforestation under these circumstances is given in Appendix 6.2. The main results, however, follow. Pure yield-increasing technological progress has the same effect as a price increase: it makes agriculture at the frontier (the outer edge of cultivation) more profitable, and farmers clear more forest. Labour-saving technological change will do the same. Since the cost to farmers of doing their tasks has been reduced, it has become cheaper for them to expand their operations.

One surprising result is that labour-intensive technologies have no effect on deforestation. Assume that initially $\beta = 1$ and that, after adopting the new technology, $\beta = 2$. This implies that the production and labour input per hectare would double, and the rent earned before from 1 ha could be earned from just 0.5 ha. But, at the frontier, the rent is zero, and whether it is zero from 1 or 0.5 ha does not matter. The key is the labour costs per efficiency unit of land, and these are only affected by α and ε (and p, w, t and b, of course), not β. Thus, a land-saving (labour-intensive) new technology will not help conserve the forest within this model (nor will it stimulate deforestation).

Two caveats to this result are in order. First, assuming non-constant returns to scale may change the result, although – perhaps surprisingly – it is not obvious in which direction.[5] Secondly, in real life there may be few pure labour-intensive technological changes as defined in this model. Most labour-intensive technologies probably also bring about an increase in α. Nevertheless, the model demonstrates that, in open-economy situations, labour-intensive technological progress is less likely to increase deforestation than other types of technological change, even before labour market effects are taken into account.

How will technological change affect labour intensity (l)? Higher α increases the marginal productivity of labour at the outer edge of cultivation, and farmers therefore want to increase their labour input. Higher β has the same effect. Higher ε generates two contradictory effects: an increase in the efficiency of labour will lower the costs per efficiency unit of labour, but the same amount of work can be done with less physical labour. Although in theory it is indeterminate which effect will dominate, as mentioned previously, we assume the second effect dominates – when mechanization makes labour more efficient, labour intensity declines.[6]

Technological changes that reduce production risks, such as the adoption of chemical pesticides, should encourage deforestation, following the logic of the perfect-market model (although risk is not formally included). As long as farmers are risk-averse, such changes make agricultural production more attractive to farmers. This contrasts with the result of the subsistence model.

The perfect-market (open-economy) model is most likely to approximate situations involving medium- and large-sized farmers, who hire most or all of their labour and have good access to credit. This may apply, for example, to ranchers with large landholdings and mechanized agricultural producers in Latin America and to the large tree-crop, sugar and banana plantations present in all tropical regions. The model also better describes the long-term adaptations where, for example, migration is an option. Under those circumstances, it is more realistic to assume wage rates to be exogenously determined outside the agricultural frontier.

The importance of the market assumptions

The subsistence and open-economy models predict quite different results with regard to how technological change will affect deforestation rates. Whereas, in the former model, technological progress tends to decrease deforestation, in the latter, it generally increases it. A survival logic drives the subsistence model, whereas the open-economy model is all about profitability.

That being said, it should be pointed out that even a subsistence model can lead to rather similar results to the open-economy model with different assumptions about the labour market. The subsistence model described above assumed that farmers could not sell more than a predetermined amount of labour, but that they would have liked to do so. What if they were not constrained and could sell as much labour as they wanted? Then – as in the open-economy model – the farmer would expand his or her land until the returns to labour at the frontier equalled the wage rate:

$$(p - tb^e)\alpha\varepsilon f(\bar{n})\bar{n}^{-1} = w \Leftrightarrow b^e = \frac{p}{t} - \frac{w}{p}\frac{\bar{n}}{f(\bar{n})}\frac{1}{\alpha\varepsilon}$$

Under these circumstances, different types of technological change have exactly the same qualitative effects as in the open-economy model. Thus, what matters is not as much the behavioural assumption ('subsistence' or 'income maximization') but rather the market assumption. To be more precise: where perfect markets exist, assumptions about farmers' objectives will not affect the qualitative results, whereas, in imperfect market situations, they will.

Most smallholders in tropical countries behave neither as pure subsistence-type producers nor as profit-maximizers, nor do they operate in perfect markets. Hired labour requires supervision and is not a perfect substitute for family labour. Farmers may also face difficulties obtaining hired labour or capital at predetermined fixed wages and interest rates, or in selling as much as they want of their own labour.

In each of these cases, farmers perceive a subjective ('shadow' or 'virtual') cost for their labour and/or capital that differs from the market rate (if that even exists). Household consumption and production decisions are no longer separable and farmers' preferences, household size and composition, farm productivity and market prices all influence their farming decisions. This is the Chayanovian case referred to previously. Whereas the open-economy model only considers the *substitution* effect of a technological change (i.e. farming becoming more profitable), the Chayanovian assumptions introduce an *income* effect. Technological progress will improve farmers' income and change the labour requirements of different production options. This will affect farmers' shadow wage rates and therefore the decisions they make.

Consider what happens under pure yield-increasing technological change. The substitution effect leads farmers to clear more land because the additional income they receive from each day they work in the field increases. The income effect leads them to clear less land because the productivity rise has made them wealthier and they now want to dedicate more of their time to leisure. *A priori*, it is impossible to know whether the substitution or the income effects will dominate and thus whether higher yields will raise or lower deforestation. In a previous paper (Angelsen, 1999), it has been shown that income effects are likely to be stronger when farmers find it difficult to obtain or supervise hired labour and have limited opportunities for working off-farm, and weaker when agricultural production constitutes a small portion of household income. It is also argued that the income effect is likely to be stronger among poor households. When household income is close to subsistence requirements, it will respond as in the subsistence model, which is to say that the income effect dominates.

Technological Change at the Macro Level

Macro-level models of deforestation are particularly useful for studying the interaction between different sectors. This interaction can take place through labour migration between the sectors, which is mediated by wage rates, or through output price effects, as production in one sector affects the prices received by the others. The open-economy model just presented can be viewed as a special case, where the changes in labour demand or output supply induced by technological change are too small to affect wages or output prices. In many real-life situations, these conditions do not apply, making it more realistic to use models with endogenous wages and/or prices.

Keeping the focus on labour and output markets, four separate cases can be identified, based on whether output price and/or wages are endogenous or exogenous: (i) when both prices and wages are exogenous, the open-economy model already analysed results; (ii) when output prices remain exogenous but wage rates become endogenous, the first model presented below obtains; (iii) the second model presented below reverses these assumptions and focuses on output market effects; for each of these situations, we analyse how the three types of technological change in either the intensive or extensive sectors influence land clearing in the extensive sector; and (iv) the fourth possibility is that both price and wages are endogenous. This simply combines the two models presented in a straightforward fashion, and the results are not presented here.

A model with endogenous wages

The model developed below belongs to a class of two-sector models with the following basic assumptions: (i) output prices are fixed; (ii) each sector uses two inputs – land and labour; (iii) the total amount of labour is fixed and mobile between the sectors, depending on the endogenous wage rates; and (iv) the other input, land, is fixed in one of the sectors, but not in the other. In the second sector, additional land can be brought into production, although at an increasing cost. Examples of models like this include various contributions by Jones and O'Neill (1992, 1993, 1994) and Findlay and Lundahl (1994). The present model has a structure similar to the Jones and O'Neill basic model, but has a more detailed specification of technology, which allows the analysis of the effects of different types of technological change.

The two sectors of the model are intensive and extensive agriculture (for our purposes, we ignore all other economic sectors). To facilitate analysis, assume that the sectors produce the same product and differ only in that the former uses a higher input (labour)–higher output technology. The difference in the technologies farmers use in each sector results from the fact that land is scarcer in the intensive sector. For simplicity, we assume there are no transport costs in the intensive sector. Thus, the price farmers in that sector receive equals the market price ($p = p^i$).

The model

The first-order conditions for profit maximization in each sector in this model are similar to those for the open-economy farm model. Farmers add labour in farm production as long as the increased output is higher than the costs of labour:

$$p^j \varepsilon^j \alpha^j f_n^j - w = 0; j = i, e$$

where j refers to the sector involved (i = intensive, e = extensive). Since land in the intensive sector is exogenously given, $H^i = \overline{H}^i$, deforestation is related only to the expansion of cultivated land in the extensive sector. In the extensive sector, land is expanded up to the point where the land rent is zero:

$$(p - tb^e)\alpha^e f^e - wl^e \beta^{e^{-1}} = 0$$

As in the open-economy model, this equation allows us to derive the forest frontier (b^e). In this case, however, the equilibrium wage rate is endogenous, thus w, n^e, n^i and b^e are determined simultaneously.

Finally, the total labour supply is fixed and allocated between the two sectors (demand = supply) to ensure full employment and the same wage in the two sectors:

$$\overline{L} = L^e + L^i; L^e = \int_0^{b^e} l^e hb\,db, h \equiv 2\pi; L^i = l^i \overline{H}^i$$

The four key equations are the two conditions for optimal labour use, the definition of the outer edge of cultivation in extensive agriculture and the labour market equilibrium condition. These four determine the labour input units per hectare (l^i, l^e), the wage rate (w) and the outer edge of cultivation (b^e). Using the other equations, one can recursively find the output and labour input in the two sectors and the total land area in the extensive sector. The exogenous variables are the three technology parameters in the two sectors, the output price, the transport costs (t), the land area in the intensive sector and total population.

The effects of different types of technological progress

The mathematical derivation of the effects of exogenous changes is in Appendix 6.3 and the results of three different types of technological change are summarized in Table 6.1. Pure yield-increasing technological change in intensive agriculture makes that sector more profitable than the extensive sector. This bids up wages in the intensive sector and attracts labour from extensive agriculture. The agricultural frontier contracts as a result. The larger the land area of the intensive sector, the bigger the effect.

Table 6.1. Effects of technological change on deforestation in a model with endogenous wages.

Type of technological change	Intensive sector	Extensive sector
Pure yield-increasing technological progress (α)	Decrease	Increase
Labour-intensive technological progress (β)	Decrease	Decrease
Labour-saving technological progress (ε)	Increase	Increase

Pure yield-increasing technological progress in the extensive sector has the opposite effect. The marginal productivity of labour will increase, attracting migrants to the frontier, which, in turn, leads to greater deforestation. Compared with the open-economy case with fixed wages, however, the impact on deforestation will be lower when wages are endogenous. As the marginal productivity of labour rises, wages will be bid up and that will dampen the expansion of the agricultural land by discouraging employment of additional labour in the extensive sector.

Labour-intensive technological change in the intensive sector has a similar effect. The marginal productivity of labour increases, and labour will move from the extensive to the intensive sector. Besides the size of the intensive sector, the magnitude of the effect depends on the intensive sector's capacity to absorb additional labour, or, in technical terms, how rapidly the marginal productivity of labour declines (f^{j}_{nn}).

Perhaps surprisingly, labour-intensive technological changes in the extensive sector lead to less deforestation. One may recall that in the open-economy model this kind of technological change had no effect. Now, however, with endogenous wages, the demand for labour in the extensive sector will increase, wages will be bid up and that will discourage forest conversion. Even if labour is pulled from the intensive to the extensive sector, the increase in labour used per hectare will be large enough to offset this effect and reduce agricultural land area.

In theory, the effect of labour-saving technological change in the intensive sector is ambiguous, as shown earlier in the context of household models. Once again, however, we shall assume that this type of technological change reduces the demand for labour in the sector where the change takes place. Thus, labour-saving technological change in the intensive sector frees labour for migration to the forest frontier, resulting in more deforestation.

Labour-saving technological progress in the extensive sector will certainly boost deforestation, as the unit costs of labour have declined. It is not known, however, whether the labour-market effects will diminish this result. Even though labour intensity has decreased, total area expands, making the net change in extensive-sector labour demand ambiguous. Assume first that demand is reduced. Then wages will decline and further strengthen the effect on deforestation. What if total labour demand rises and pushes up wages? By looking at the effect within a standard supply-and-demand framework, the wage increase will only have a dampening effect, and the final outcome will still be more deforestation (total labour use is up in the extensive sector, and less labour is used per hectare). Further, we know that the forest impact will be larger than in the case of labour-intensive technological change because any dampening effect that might exist in the labour market will be smaller.

The following points summarize the lessons learned from analysis of this model:

1. The effects of technological change on deforestation depend both on the type of technological change and to which sector it applies. Technological progress in the intensive sector will reduce deforestation, except in the important case of labour-saving technologies that expel labour from the intensive sector to the agricultural frontier (extensive sector).
2. Pure yield-increasing and labour-saving technological changes in the extensive sector will both increase deforestation. The effect will be larger in the latter case because not only does labour migrate to the extensive sector as agriculture there becomes more profitable, but also each person working in the extensive sector will clear a larger area of land. On the other hand, labour-intensive technological progress in the extensive sector will reduce deforestation.
3. In reality, it may be difficult to target new technologies to a particular sector. If the same technology applies to both sectors, pure yield-increasing and labour-saving technological changes should stimulate forest conversion, whereas labour-intensive technological changes should promote forest conservation.

A model with endogenous output price

In our final model, we consider a situation where employment in agriculture is small compared with the rest of the economy, and thus the wage rate can be considered exogenous. On the other hand, it is assumed that changes in agricultural output induced by technological changes are large enough to affect prices. Since the two sectors produce for the same market, changes in supply in either sector will affect the price farmers in both sectors receive, and hence the decisions they make.

The model

The first three equations in this model are identical to those of the previous model, with the exception that the wage is exogenous and the price endogenous. The equation for labour-market equilibrium is replaced by the condition for output-market equilibrium, simply stating that supply must equal demand:

$$\alpha^i f^i \beta^i \overline{H}^i + \int_0^{b^e} \alpha^e f^e \beta^e \, hb \, \mathrm{d}b - \gamma E(p) = 0$$

Demand is a function of price, $E(p)$, and γ is a shift parameter, which may be used to study changes in demand.

The effect of different types of technological progress

The effects of technological change on deforestation with endogenous output prices are formally shown in Appendix 6.4. The linkage between the two sectors is through the output price and the model results are more straightforward than in the model with endogenous wages. Any type of technological progress in the intensive sector will increase production in this sector. This creates a downward pressure on the output price, and therefore reduces land expansion in the extensive sector. The effect will be larger the more inelastic the demand for the output (thus small shifts in supply lead to large reductions in price), and the larger the size of the intensive sector relative to the extensive one.

Labour-intensive technological changes in the extensive sector will reduce deforestation. Again, recall that in the model with fixed prices this type of technological change had no effect on deforestation. Now, however, the higher extensive-sector output resulting from the technological change will lower the price, and that will provoke a contraction of the extensive frontier.

Pure yield-increasing and labour-saving technologies in the extensive sector have contradictory effects, and we cannot predict based on theory alone what the net result will be. As in the previous case, both types of technological change raise output and that will push the price down and reduce deforestation. At the same time, the marginal productivity of labour increases and that tends to increase deforestation. If product demand is sufficiently inelastic and the extensive sector has a high share of total output, deforestation will be reduced. Otherwise, it will be increased.

Comparison of Different Models and Working Hypotheses

To summarize, how technological progress affects deforestation depends on the type of technology involved, which agricultural sectors adopt the new technology, and the existing labour and output market conditions (if farmers are constrained, and the price effects). When farmers are not fully integrated into perfect labour and product markets, the results will also depend on their preferences and factors such as their income level. The interaction between these factors is not always obvious, and the four models presented here permit the systematic analysis of the resulting effects.

It is often useful to think of the effects of new agricultural technologies in two stages. First, one might consider how farmers respond, taking all prices as given. This is the partial-equilibrium approach, applied in the first two models. Secondly, one can analyse how the aggregate response from the farmers affects wages and output market prices through changes in supply and demand. These general equilibrium effects, discussed in the third and fourth models, may enlarge, modify, or even reverse the effects of the partial-equilibrium model. Hence the four models yield quite different conclusions. The results are summarized in Table 6.2.

In the subsistence model (model 1), new technologies that increase yield – the typical case – will reduce deforestation. The critical assumptions in this model are that farmers have no desire to increase income and consumption beyond a predetermined level of subsistence and that they are constrained in the labour market. These results are challenged by the perfect-market (open-economy) model, where farmers behave as if they maximize their income. In

Table 6.2. Comparison of the results of the four models.

Type of technological progress (t.p.)	1. Subsistence model with imperfect markets	2. Farm model with perfect markets	3. Macro model with endogenous wages	4. Macro model with endogenous prices
Pure yield-increasing t.p. in intensive sector (α^i)	n.a.	n.a.	–	–
Labour-intensive t.p. in intensive sector (β^i)	n.a.	n.a.	–	–
Labour-saving t.p. in intensive sector (ε^i)	n.a.	n.a.	+	–
Pure yield-increasing t.p. in extensive sector (α^e)	–	+	+	?
Labour-intensive t.p. in extensive sector (β^e)	–	0	–	–
Labour-saving t.p. in extensive sector (ε^e)	0	+	+	?

Sign (+, –, 0) indicates effect on deforestation; n.a., not applicable.

this model, technological progress will generally make forest conversion more profitable. Including a second sector (intensive agriculture) and labour market effects (changes in wages) tends to dampen the effects. Output-market effects (more production depresses output prices) may, however, reverse the conclusions of model 2, which assumed that all prices were given.

These model results lead to the following working hypotheses, which are presented for empirical testing:

1. Technological progress in the intensive sector tends to conserve forests regardless of the characteristics of the labour and product markets involved, except for labour-saving technological change, which expels labour to the extensive sector. The magnitude of the effects depends on market conditions. In the special case where wages and prices are fixed (open economy), technological change in the intensive sector has no effect on deforestation.
2. The effects of technological change in the extensive sector are more complex. In a subsistence economy where farmers are constrained in the labour market, technological progress tends to decrease deforestation, since farmers can generate sufficient incomes from a smaller area. However, when farmers are unconstrained in the labour market and maximize profits, the result will be the opposite; agriculture at the frontier has become more profitable and technological changes will tend to increase deforestation.
3. Including general equilibrium effects in either the output or the labour market will dampen or even change the direction of the results. In particular, making the output price endogenous when demand is inelastic may reverse the effect on deforestation.
4. A robust and significant result is that labour-intensive technological progress in any sector and under any market conditions reduces deforestation (except in the open-economy model, where it has no effect).

5. Pure yield-increasing and labour-saving technological progress in the extensive sector increases deforestation, unless there are large price effects.

Illustrative Empirical Examples

This section seeks to give the reader a sense of the practical implications of the models just presented by providing examples of how technological change has affected deforestation in different situations. The discussion first offers some cases that illustrate issues related to the characteristics of the output and labour markets involved. Then examples are presented that highlight the importance of whether the technological changes involved are labour-intensive or labour-saving.

Most of the cases encountered involve rather major changes, such as the introduction of new crops or animals, large yield increases or the substitution of thousands of workers by labour-saving machinery. That is probably no coincidence; for technological changes in agriculture to have any noticeable impact on deforestation, they probably have to be rather substantial.

Technological change with elastic and inelastic product demand and labour supply

The history of tropical agriculture includes a long litany of situations where the introduction of new cash crops, such as bananas, cocoa, coffee, oil-palm, rubber and sugar cane, led to widespread deforestation (Barraclough and Ghimire, 1995). Since production of these crops was for export and in most cases international demand was growing rapidly, producers could sell as much of these crops as they could produce without greatly depressing the price (as in models 2 or 3). Thus, the areas involved expanded over long periods of time.

The major constraint faced by most colonial powers was how to obtain sufficient labour to keep expanding production. Otherwise the expansion would inevitably bid up the price of labour and that would choke off further growth. To solve this problem they variously imported slaves and wage workers from other locations, imposed poll taxes and vagrancy laws, created debt-peonage systems, monopolized the prime agricultural areas and prohibited unions. When elastic demand for output coincided with an elastic supply of labour, the perfect conditions existed for the introduction of new crops, leading to large-scale deforestation, as predicted in our model with no dampening wage or price effects (model 2).

The spread of cocoa into new areas of Côte d'Ivoire and Indonesia provide some more recent examples of this type of process. In both cases, low-cost production allowed farmers to ignore any international price reductions their activities induced, a constant inflow of migrants from other regions dampened pressure on wages and massive deforestation resulted (Ruf, 1995).

The introduction of rubber in Indonesia started early this century, but planting accelerated in recent decades in response to government programmes and the crop's relatively high profitability. Studies from Sumatra suggest that the replacement of a traditional shifting-cultivation system with the more intensive smallholder rubber has not been sufficient to stop deforestation, but has instead stimulated forest clearing (Angelsen, 1995; Tomich and van Noordwijk, 1995). Since Indonesia sells most of its rubber on the world market, the expansion has not created a large downward pressure on prices (although the country controls about a quarter of the market). A steady flow of migrants, through both state-sponsored transmigration projects and spontaneous migration, has made labour readily available for expansion.

These experiences contrast sharply with those of many developed countries. In the USA, for example, productivity improvements in agriculture, combined with growing employment opportunities in other sectors, which made the supply of agricultural labour less elastic, generated a decline in the area used to raise crops and livestock after around 1920 (Southgate, 1998). Since the elasticity of demand for agricultural products was rather low, prices fell as production grew. Eventually, many farmers in regions with poorer soils or less access to markets decided they could earn more money elsewhere and sold their land. Landowners took large areas that could not compete with prime farming areas out of production and allowed them to return to forest. Mather (1992) describes how a similar process occurred in France. These cases provide an illustration of model 4, where technological change reduces output prices sufficiently to decrease the pressure on forests, in addition to better off-farm opportunities and higher wages pulling labour out of agriculture.

The introduction of high-yielding rice varieties and chemical fertilizers in Asia is another comparable situation. Rice has traditionally been the main staple food product for most of Asia, and its price elasticity of demand in many Asian countries in the 1960s and 1970s was rather low. Thus, *ceteris paribus*, a large increase in rice production, such as occurred during the Green Revolution, could be expected to greatly depress rice prices. If the Green Revolution had not taken place, a persuasive argument can be made that growing populations would have driven rice prices up very sharply. That would have created a major incentive for families to expand their food-crop production into forested areas, particularly in heavily forested countries, such as Indonesia, Myanmar, the Philippines and Thailand. Coxhead and Shively (1995) have used a computable general equilibrium model to simulate this type of process, using data from the Philippines. Their model shows that yield improvements in maize can be expected to depress food prices, which, in turn, should reduce forest clearing for maize and rain-fed rice production.

Labour-intensive technologies

Generally speaking, it is difficult to promote labour-intensive technologies in agricultural frontier areas. Labour is typically the scarcest resource there and farmers therefore prefer technologies that save labour and use large amounts of land, as suggested by induced innovation theory (Holden, 1993b; Richards, 1997). Nevertheless, situations exist where farmers on the frontier choose to produce high-value products, such as perennial crops that can only be grown in labour-intensive systems. There are also cases where the productivity differences between labour-intensive and extensive systems are so great that farmers adopt more labour-intensive systems even on the agriculture frontier. This may occur, for example, when farmers abandon shifting-cultivation systems and adopt annual crop systems that involve shorter fallow periods or eliminate the use of fallow completely.

Perhaps the paradigmatic example of the first situation is the adoption of coca cultivation in the piedmont areas of Bolivia, Colombia and Peru. Coca production provides almost three times more employment per cultivated hectare than shifting-cultivation rice production and ten times more employment per hectare if one includes fallow lands (Sanabria, 1993). On average, each coca producer in the Andes clears much less forest than producers who only grow food crops. Moreover, at least in Bolivia, the evidence suggests that many farmers who might otherwise have migrated to colonization areas where shifting-cultivation systems dominate moved to coca-producing areas instead (Kaimowitz *et al.*, 1999). Thus, the growth of coca production has probably reduced total forest clearing, in line with what one might expect from model 3.

The introduction of more labour-intensive perennial crops in contexts without significant in-migration, where labour supply can be expected to be relatively inelastic, presents a similar situation. This may apply, for example, to the introduction of black pepper and fruit trees in the older agricultural frontier areas of the eastern Brazilian Amazon (de Almeida and Uhl, 1995).

Sub-Saharan Africa provides several examples of shifts from extensive shifting cultivation systems to more intensive crop systems. The introduction of chemical fertilizers and new crops, such as maize, has encouraged the transition from shifting cultivation to more permanent annual cropping systems. Based on analysis of data from northern Zambia using multiple-goal programming models, Holden (1993a) argues that farmers who were given the option of growing maize and using chemical fertilizers could be expected to dedicate less land to crops and fallow. This decline would be even larger if additional technologies were available that reduced the risks associated with producing maize.

Adoption of cover crops, such as velvet bean (*Mucuna* spp.) in Central America, has apparently produced similar results. This technology has been widely adopted by farmers in Nicaragua and along the Atlantic coast of Guatemala and Honduras because it improves soil fertility and yields and reduces weeding requirements. Humphries (1996) found that, on average, farmers in Honduras who adopted this technology planted twice as much maize as those who did not. But, even so, the total amount of land occupied by their cropping system fell (and per hectare labour intensity rose) because they no longer needed large areas of fallow.

However, the Honduran case also illustrates the importance of the more macro-level effects analysed in model 3. Even though technological change led to a decline in the average area each farmer cleared for crop production, aggregate deforestation rates there continued to climb, due to a rapid influx of new migrants from other areas (Humphries, 1996). No evidence suggests that the higher profits generated by adopting cover crops stimulated this migration, but, had the improvement been more marked, that might have occurred.

Our models with two agricultural sectors imply that improvements in production, postharvest handling and processing of labour-intensive horticultural crops, as well as public investment in small-farmer irrigation systems in traditional agricultural regions, should give poor rural families less incentive to move to agricultural frontier areas and clear forest. Shively and Martinez (1999), in a study from southern Palawan in the Philippines, found that lowland irrigation development increased labour demand by some 27%. About half of this increase in labour use was accounted for by employment of upland households living along the forest margin, resulting in reduced forest clearing, charcoal production and collection of fuelwood. It is also suggestive, in this regard, that in many cases limited forest clearing occurs near labour-intensive agroindustrial plantations, presumably because households can earn more money from wage labour on these plantations than by clearing forested areas for small-farm agriculture.

Labour-saving technologies

Labour-saving (and normally capital-intensive) technological changes are the type most likely to promote deforestation. They give farmers an incentive to expand their production since they make agriculture more profitable, and they either expel labour, which might deforest elsewhere, or employ little of it. The introduction of agricultural machinery into traditional agricultural areas often displaces labour. Technological changes that allow the introduction of cattle or mechanized agriculture into new areas tend to increase net employment, but the per hectare demand for labour tends to be low.

More than any other factor, the widespread usage of agricultural machinery and the growth of cattle-based systems probably account for the fact that, on average, farmers in Latin America clear much larger areas than farmers in Africa or Asia. Labour-saving technologies have been less popular in Asia, because wage rates have generally been lower and government exchange-rate, credit and tariff policies have not favoured the importation of agricultural machinery as much. Capital scarcity and weak infrastructure development and consumer demand have limited the widespread adoption of agricultural machinery and large-scale cattle ranching in much of tropical Africa.

The introduction of labour-saving technologies in traditional agricultural areas often displaces farmers, who subsequently migrate to agricultural frontier areas, as seen in model 3. The introduction of new pesticides, crop varieties and agricultural machinery on the Pacific coast of Nicaragua in the 1960s and 1970s, for example, facilitated the expansion of large-scale cotton production. As a result, large numbers of small farmers, who lacked sufficient land, capital and management skills to produce cotton, migrated to Nicaragua's forested areas, where there was less competition for land (Williams, 1986).

The dissemination of new soybean technologies helped encourage farmers in Parana, Brazil, to convert from labour-intensive coffee cultivation to mechanized soybean production. Land tenure became more concentrated, employment fell and, as a consequence, 2.5 million rural people left the state in the

1970s, many of them to the forested areas of Rondônia in the Amazon (Diegues, 1992).

The introduction of improved varieties and other agricultural practices for use in mechanized soybean production greatly facilitated the spread of this production system and provoked major increases in forest clearing in the Brazilian Cerrados, Santa Cruz, Bolivia and parts of Paraguay (Bojanic and Echeverria, 1990; Wallis, 1997). In the Brazilian Cerrados alone, the soybean area rose from nearly zero in 1970 to 4.4 Mha in 1990, largely at the expense of natural vegetation (Wallis, 1997). Since production was capital-intensive and export-oriented and the farmers involved had ample access to capital, neither rising wages, capital constraints nor technology-induced price decreases significantly dampened this expansion.

Before the introduction of African pasture varieties and barbed wire to the Latin American tropics in the early 20th century, the region's low population density limited widespread deforestation, since each household could only manage a relatively small area (Parsons, 1972). These new production systems demanded little labour per hectare, so farmers were able to convert large areas of forest, despite their limited labour supply.

Similarly, the substitution of manual implements for forest clearing by chainsaws dramatically reduces the amount of time required for clearing a hectare of forest and makes it possible for each individual to deforest a much larger area. Thus, not surprisingly, Pichón (1997a) found that small-farm colonists in the Ecuadorian Amazon who owned chainsaws cleared much larger areas.

Conclusions

This chapter attempts to systematically examine some of the main analytical arguments and empirical evidence regarding the impact of technological progress in agriculture on deforestation. Three factors have been highlighted that are important in determining deforestation outcomes: the type of technological change, in which subsector of agriculture it occurs and under what labour-market and product-market conditions. The answer to the question, 'Does technological progress in agriculture lead to more deforestation?' depends critically on these factors.

Three broad conclusions emerge from the theoretical analysis and empirical review. First, technological progress in the intensive sector is generally good for forest conservation, unless it substitutes labour with capital and expels the displaced labour to the agricultural frontier. Technological change in the intensive sector shifts resources away from the frontier by bidding up wages and/or lowering agricultural prices. Secondly, labour-intensive technological progress will tend to reduce forest clearing because it 'mops up' labour, while labour-saving technologies have the opposite effect. Thirdly, the effect of pure yield-increasing and labour-saving technological progress in the extensive sector is sensitive to market assumptions. If we assume either subsistence-type behaviour and imperfect labour markets or endogenous prices and inelastic demand, these types of technological progress may reduce deforestation.

Otherwise, they will tend to encourage it. From a forest-conservation viewpoint, the 'worst' type of technological progress at the frontier is labour-saving (capital-intensive), with an elastic demand for the output (typically export crops with a fixed world market price). If large-scale in-migration is possible, it makes things even worse. The 'best' type of technological change is labour-intensive, especially where it is utilized in the context of limited opportunities for in-migration and inelastic demand for agricultural products. In the latter case, where technological change depresses output prices, the gain to farmers will also be smaller. Labour-intensive technologies, however, represent 'win–win' technologies, as their use will both bid up rural wages and increase farm income, as well as limit the opportunities for expansion of agriculture into forests.

Acknowledgements

We would like to thank Stein Holden, William Sunderlin, Joyotee Smith and the editors for discussions and useful comments on draft versions of this chapter.

Notes

The topic of this chapter is dealt with in much greater detail in our forthcoming book *Agricultural Technologies and Tropical Deforestation* (2001, CAB International).

1 The notation used in the theoretical exposition below is fully defined in Appendix 6.1.

2 This is true as long as n is kept constant, and n will not change if the real wage is kept constant. See below for the case when the real wage is endogenous.

3 We have assumed that the labour-market constraint is binding, which is to say that the wage rate is higher than the return on labour at the frontier.

4 Modifying the assumption of fixed n could change this result; thus we shall not stress it in the overall discussion.

5 Consider the following profit maximization problem with a general production function: Max $p\alpha F(\varepsilon L, \beta H) - wL - cH$. The effect of a change in β on H turns out to be indeterminate. Using a Cobb–Douglas function, the result will be an increase in H, i.e. more deforestation.

6 The net effect depends on the elasticity of the marginal productivity of labour; it is negative if $f_{nn}\ (n/f_n) > -1$.

Appendix 6.1: List of Variables Used

A	$A = \beta H$	Efficiency units of land
H		Land area in hectares (physical units)
I		Income
L		Labour input (physical units)
N	$N = \varepsilon L$	Efficiency units of labour
Y	$Y = \alpha F(N,A)$	Total production
b		Index for distance (e.g. km)
e		Superscript for extensive sector
f		Superscript for farm sector
i		Superscript for intensive sector
l	$l = L/H$	Labour intensity, i.e. labour per hectare, both in physical units
n	$n = N/A= (\varepsilon/\beta)l$	Efficiency units of labour per efficiency unit of land
o		Superscript for off-farm sector
p		Output price
t		Transport costs (per unit of output and kilometre)
w		Wage rate
y	$y = Y/N = \alpha f(n)$	Output per efficiency unit of land
α		Efficiency parameter for overall production (increase in α: pure yield increasing technological progress)
β		Efficiency parameter for land (increase in β: labour-intensive technological progress)
ε		Efficiency parameter for labour (increase in ε: labour-saving (land-intensive) technological progress)
γ		Shift parameter for demand of agricultural product

Appendix 6.2: Comparative Statics of Open-economy Farm Model

The model equations are:

$$(p-tb)\alpha\varepsilon f_n\left(\frac{\varepsilon}{\beta}l\right)-w=0$$

$$(p-tb^e)\alpha f\left(\frac{\varepsilon}{\beta}l\right)-w\frac{l}{\beta}=0$$

Total differentiation yields:

$$\begin{bmatrix}(p-tb)\alpha\varepsilon^2\beta^{-1}f_{nn} & 0\\ 0 & -t\alpha f\end{bmatrix}\begin{bmatrix}dl\\ db^e\end{bmatrix}$$

$$=\begin{bmatrix}-(p-tb)\varepsilon f_n & (p-tb)\alpha\varepsilon\beta^{-1}nf_{nn} & -(p-tb)\alpha(f_n+nf_{nn})-\alpha\varepsilon f_n & 1 & b\alpha\varepsilon f_n\\ -(p-tb^e)f & 0 & -(p-tb)\alpha\beta^{-1}lf_n & \alpha f & l\beta^{-1} & b^e\alpha f\end{bmatrix}\begin{bmatrix}d\alpha\\ d\beta\\ d\varepsilon\\ dp\\ dw\\ dt\end{bmatrix}$$

The coefficient matrix is positive. Note that the elements a_{21} and b_{22} both become zero, using the model's first equation. The sign of the other elements (except b_{13}) are obvious. The effects of technological and price changes on the agricultural frontier and labour intensity are summarized as follows:

$$\frac{db^e}{d\alpha},\frac{db^e}{d\varepsilon},\frac{db^e}{dp}>0;\frac{db^e}{dw},\frac{db^e}{dt}<0;\frac{db^e}{d\beta}=0$$

$$\frac{dl}{d\alpha},\frac{dl}{d\beta};\frac{dl}{dp}>0;\frac{dl}{dw},\frac{dl}{dt}<0;\frac{dl}{d\varepsilon}=?$$

Appendix 6.3: Comparative Statics of a Model with Endogenous Wages

The model consists of four equations:

$$p\varepsilon^i \alpha^i f_n^i\left(\frac{\varepsilon^i}{\beta^i} l^i\right) - w = 0$$

$$(p - tb)\varepsilon^e \alpha^e f_n^e\left(\frac{\varepsilon^e}{\beta^e} l^e\right) - w = 0$$

$$(p - tb^e)\alpha^e f^e\left(\frac{\varepsilon^e}{\beta^e} l^e\right) - w\frac{l^e}{\beta^e} = 0$$

$$\int_0^{b^e} l^e\, hb\, \mathrm{d}b + l^i \overline{H}^i - \overline{L} = 0$$

Total differentiation yields:

$$\begin{bmatrix} p\varepsilon^{i^2}\alpha^i\beta^{i^{-1}} f_{nn}^i & 0 & 0 & -1 \\ 0 & (p-tb)\varepsilon^{e^2}\alpha^e\beta^{e^{-1}} f_{nn}^e & 0 & -1 \\ 0 & 0 & -t\alpha^e f^e & -l^e\beta^{e^{-1}} \\ \overline{H}^i & \int_0^{b^e} hb\,\mathrm{d}b & l^e hb^e & 0 \end{bmatrix} \begin{bmatrix} \mathrm{d}l^i \\ \mathrm{d}l^e \\ \mathrm{d}b^e \\ \mathrm{d}w \end{bmatrix}$$

$$= \begin{bmatrix} -p\varepsilon^i f_n^i & p\varepsilon^i\alpha^i f_{nn}^i n^i & -p\alpha^i(f_n^i + f_{nn}^i n^i) \\ 0 & 0 & 0 \\ 0 & 0 & 0 \\ 0 & 0 & 0 \end{bmatrix} \begin{bmatrix} \mathrm{d}\alpha^i \\ \mathrm{d}\beta^i \\ \mathrm{d}\varepsilon^i \end{bmatrix}$$

$$+ \begin{bmatrix} 0 & 0 & 0 \\ -(p-tb)\varepsilon^e f_n^e & (p-tb)\varepsilon^e\alpha^e f_{nn}^e n^e & -(p-tb)\alpha^e(f_{nn}^e + f_{nn}^e n^e) \\ -(p-tb^e)f^e & 0 & -(p-tb^e)\alpha^e\beta^{e^{-1}} l^e f_n^e \\ 0 & 0 & 0 \end{bmatrix} \begin{bmatrix} \mathrm{d}\alpha^e \\ \mathrm{d}\beta^e \\ \mathrm{d}\varepsilon^e \end{bmatrix}$$

$$+ \begin{bmatrix} -\varepsilon^i\alpha^i f_n^i & 0 & 0 & 0 \\ -\varepsilon^e\alpha^e f_n^e & b\varepsilon^e\alpha^e f_n^e & 0 & 0 \\ -\alpha^e f^e & b^e\alpha^e f^e & 0 & 0 \\ 0 & 0 & l^i & 1 \end{bmatrix} \begin{bmatrix} \mathrm{d}p \\ \mathrm{d}t \\ \mathrm{d}\overline{H} \\ \mathrm{d}\overline{L} \end{bmatrix}$$

The determinant of the coefficient matrix is:*

$$D = -a_{11}a_{22}a_{34}a_{43} + a_{11}a_{33}a_{42} + a_{22}a_{33}a_{41} > 0$$

The effect of parameter changes on b^e is:

$$\frac{\mathrm{d}b^e}{\mathrm{d}\alpha^i} = D^{-1} b_{11} a_{22} a_{34} a_{41} < 0$$

$$\frac{\mathrm{d}b^e}{\mathrm{d}\beta^i} = D^{-1} b_{12} a_{22} a_{34} a_{41} < 0$$

$$\frac{\mathrm{d}b^e}{\mathrm{d}\varepsilon^i} = D^{-1} b_{13} a_{22} a_{34} a_{41} \lesseqgtr 0 \quad \Leftrightarrow \quad f_{nn}^i \frac{n^i}{f_n^i} \lesseqgtr -1$$

$$\frac{\mathrm{d}b^e}{\mathrm{d}\alpha^e} = D^{-1} \left[a_{11} a_{42} \left(b_{34} + b_{24} a_{34} \right) + a_{22} b_{34} a_{41} \right] = D^{-1} a_{22} b_{34} a_{41} > 0^{\dagger}$$

$$\frac{\mathrm{d}b^e}{\mathrm{d}\beta^i} = D^{-1} a_{11} b_{25} a_{34} a_{42} < 0$$

By manipulation and using the model equations, we get:

$$\left(b_{36} + b_{26} a_{34} \right) = \left(p - tb \right) \alpha^e \beta^{e^{-1}} l^e n^e f_{nn}^e < 0$$

which gives the above sign; and

$$\frac{\mathrm{d}b^e}{\mathrm{d}p} = D^{-1} \left[a_{41} a_{22} \left(b_{17} a_{34} + b_{37} \right) + a_{42} a_{11} \left(b_{27} a_{34} + b_{37} \right) \right] > 0$$

To demonstrate the last result, we can show that $(b_{17} a_{34} + b_{37}) \leq 0$ and $(b_{27} a_{34} + b_{37}) < 0$ by using the model equations and the properties of the production function.

Notes to Appendix 6.3

* Following conventional notation, a_{km} refers to the element in row k and column m in the coefficient matrix, whereas b_{km} used below refers to the element in that row k and column m in the matrix for change in the exogenous parameters. Note that, due to space limitations, this matrix is divided into several smaller matrixes, but the reference should be obvious.

† The expression $(b_{34} + b_{24} a_{34})$ becomes zero, using the second and third equation of the model.

Appendix 6.4: Comparative Statics of a Model with Endogenous Output Price

The model consists of the following four equations:

$$p\varepsilon^i\alpha^i f_n^i\left(\frac{\varepsilon^i}{\beta^i}l^i\right)-w=0$$

$$(p-tb)\varepsilon^e\alpha^e f_n^e\left(\frac{\varepsilon^e}{\beta^e}l^e\right)-w=0$$

$$(p-tb^e)\alpha^e f^e\left(\frac{\varepsilon^e}{\beta^e}l^e\right)-w\frac{l^e}{\beta^e}=0$$

$$\alpha^i f^i\left(\frac{\varepsilon^i}{\beta^i}l^i\right)\beta^i\overline{H}^i+\int_0^{b^e}\alpha^e f^e\left(\frac{\varepsilon^e}{\beta^e}l^e\right)\beta^e\, hb\,\mathrm{d}b-\gamma E(p)=0$$

Total differentiation yields:

$$\begin{bmatrix} p\varepsilon^{i^2}\alpha^i\beta^{i^{-1}}f_{nn}^i & 0 & 0 & \varepsilon^i\alpha^i f_n^i \\ 0 & (p-tb)\varepsilon^{e^2}\alpha^e\beta^{e^{-1}}f_{nn}^e & 0 & \varepsilon^e\alpha^e f_n^e \\ 0 & 0 & -t\alpha^e f^e & \alpha^e f^e \\ \alpha^i f_n^i\varepsilon^i\overline{H}^i & \int_0^{b^e}\alpha^e f_n^e\varepsilon^e\, hb\, db & \alpha^e f^e\beta^e & -\gamma E_p \end{bmatrix}\begin{bmatrix} \mathrm{d}l^i \\ \mathrm{d}l^e \\ \mathrm{d}b^e \\ \mathrm{d}p \end{bmatrix}$$

$$=\begin{bmatrix} -p\varepsilon^i f_n^i & p\varepsilon^i\alpha^i f_{nn}^i n^i & -p\alpha^i\left(f_n^i+f_{nn}^i n^i\right) \\ 0 & 0 & 0 \\ 0 & 0 & 0 \\ -f^i\beta^i\overline{H}^i & -\alpha^i\overline{H}^i\left(f^i-f_n^i n^i\right) & -\alpha^i f_n^i l^i\overline{H}^i \end{bmatrix}\begin{bmatrix} \mathrm{d}\alpha^i \\ \mathrm{d}\beta^i \\ \mathrm{d}\varepsilon^i \end{bmatrix}$$

$$+\begin{bmatrix} 0 & 0 & 0 \\ -(p-tb)\varepsilon^e f_n^e & (p-tb)\varepsilon^e\alpha^e f_{nn}^e n^e & -(p-tb)\alpha^e(f_n^e+f_{nn}^e n^e) \\ -(p-tb^e)f^e & 0 & -(p-tb^e)\alpha^e\beta^{e^{-1}}l^e f_n^e \\ -\int_0^{b^e}f^e\beta^e\, hb\,\mathrm{d}b & -\int_0^{b^e}a^e\left(f^e-f_n^e n^e\right)hb\,\mathrm{d}b & -\int_0^{b^e}\alpha^e f_n^e l^e\, hb\,\mathrm{d}b \end{bmatrix}\begin{bmatrix} \mathrm{d}\alpha^e \\ \mathrm{d}\beta^e \\ \mathrm{d}\varepsilon^e \end{bmatrix}$$

$$+\begin{bmatrix} 1 & 0 & 0 & 0 \\ 1 & b\varepsilon^e\alpha^e f_n^e & 0 & 0 \\ l^e\beta^{e^{-1}} & b^e\alpha^e f^e & 0 & 0 \\ 0 & 0 & -\alpha^i f^i\beta^i & E(p) \end{bmatrix}\begin{bmatrix} \mathrm{d}w \\ \mathrm{d}t \\ \mathrm{d}\overline{H}^i \\ \mathrm{d}\gamma \end{bmatrix}$$

The signs of the new elements in the matrix are readily seen. Note in particular for b_{42} and b_{52} that $(f^j-f_n^j n^j)\geq 0$, i.e. the average productivity is higher than the marginal productivity, except at the frontier of the extensive sector (b^e), where it is zero. The determinant of the coefficient matrix is:

$$D = a_{11}a_{22}(a_{33}a_{44} - a_{34}a_{43}) - a_{11}a_{24}a_{33}a_{42} - a_{14}a_{22}a_{33}a_{41} < 0$$

The effect of parameter changes on b^e is:

$$\frac{db^e}{d\alpha^i} = -D^{-1}\left[a_{22}a_{34}\left(a_{11}b_{41} - a_{41}b_{11}\right)\right] < 0$$

$$\frac{db^e}{d\beta^i} = -D^{-1}\left[a_{22}a_{34}\left(a_{11}b_{42} - a_{41}b_{12}\right)\right] < 0$$

$$\frac{db^e}{d\varepsilon^i} = -D^{-1}\left[a_{22}a_{34}\left(a_{11}b_{43} - a_{41}b_{32}\right)\right] = -D^{-1}a_{22}a_{34}\left(p\alpha^{i^2}\varepsilon^i f_n^{i^2}\overline{H}^i\right) < 0$$

$$\frac{db^e}{d\alpha^e} = D^{-1}\left[a_{11}a_{22}\left(b_{34}a_{44} - b_{44}a_{34}\right) + a_{11}a_{42}\left(b_{24}a_{34} - b_{34}a_{24}\right) - a_{14}a_{22}b_{34}a_{41}\right]?$$

$$\frac{db^e}{d\beta^e} = D^{-1}\left[a_{11}a_{34}\left(a_{42}b_{25} - a_{22}b_{45}\right)\right] < 0$$

$$\frac{db^e}{d\varepsilon^e} = D^{-1}\left[a_{11}a_{22}\left(b_{36}a_{44} - b_{46}a_{34}\right) + a_{11}a_{42}\left(b_{26}a_{34} - b_{36}a_{24}\right) - a_{14}a_{22}b_{36}a_{41}\right]?$$

7 Bioeconomic Models and Ecoregional Development: Policy Instruments for Sustainable Intensification

RUERD RUBEN, ARIE KUYVENHOVEN AND GIDEON KRUSEMAN

Development Economics Group, Department of Social Sciences, Wageningen University, Wageningen, The Netherlands

Introduction

Policy analysis to identify the right policies to address the right questions has become a major imperative for the agricultural research community. A growing mismatch has been perceived between agroecological research focusing on technological options to enhance food security and sustainable land use, and socioeconomic research engaged in the analysis of production efficiency and the (non-)adoption of technologies (Anderson, 1994; Ruben *et al.*, 1998b). Where biophysical research indicates that there is still ample room for yield increases through technological change (Rabbinge, 1990), most socioeconomic analyses point towards the structural and behavioural constraints for their adoption (Feder *et al.*, 1985). A new conceptual framework is therefore required that gives priority to: (i) the objectives of the different stakeholders involved in the rural development process; and (ii) the requirements of policy-makers to identify instruments to support sustainable development. It is increasingly recognized that separate research on agroecological and socioeconomic issues can no longer be fruitful. It is for this reason that the concept of ecoregional development has been launched to promote interdisciplinary research efforts that address locally specific rural development issues and offer policy support to the stakeholders involved (Bouma *et al.*, 1995; Reyniers and Benoît-Cattin, 1995). Bioeconomic models have become one of the major vehicles for the operationalization of ecoregional analysis.

In past decades, many developing countries witnessed rapid population growth and deterioration of their natural resource base, together with substantial reforms in domestic markets and institutions. Food security and rural development have been threatened by a policy bias against agriculture, insecure tenure arrangements, limited access to appropriate technologies and a general lack of adequate incentives. Stagnation in yields and per capita agricultural production, combined with further deterioration of natural resources, has occurred in several regions (Pinstrup-Anderson, 1994; Sanders *et al.*, 1996). These problems are especially pressing in fragile areas (semi-arid regions, hillsides, forest margins, tropical lowlands, etc.). This chapter focuses mainly on the sub-Saharan semi-arid regions, where problems of soil degradation and nutrient depletion prevail.

Ever since sustainable development was placed high on the political agenda (WCED, 1987b), various efforts have been made to combine biophysical and socioeconomic components into a coherent analytical and methodological framework. Agrotechnical solutions to problems of unsustainable and stagnating agricultural development have been widely provided (Rabbinge, 1990; Wolf *et al.*, 1991), and policy experiments for economic and institutional reforms have become equally popular (World Bank, 1996b). However, partial solutions are not capable of fully resolving these issues; a combination of appropriate technologies, policy incentives and institutional reforms is required to increase agricultural productivity while maintaining or enhancing the reproductive capacity of the resource base. This is often referred to as sustainable intensification, in which farmers' short-term welfare objectives are brought in line with long-term regional sustainability criteria (Reardon, 1995).

Bioeconomic modelling approaches, the main focus of this chapter, aim to provide a better assessment of the impact of economic incentives on technology choice and land-use practices. These approaches rely on quantitative procedures that specify the interactions between biophysical and socio-economic processes, including feedback mechanisms between farm and regional levels. Bioeconomic models are based on an interdisciplinary approach that specifies both agroecological and behavioural mechanisms and their interactions. Socioeconomic aspects are mainly comprised of farm household reactions to different policy instruments and the market and institutional arrangements that influence supply response. Agroecological components consist of biophysical processes, such as water and nutrient regimes, plant and animal growth, accumulation and leaching of pollutants, and soil erosion or depletion, all of which influence the sustainability of the resource base.

The purpose of this chapter is to provide a comprehensive review of currently available bioeconomic approaches and to discuss their usefulness for policy support of sustainable agricultural intensification. The main questions addressed concern the appraisal of available technological options to foster sustainable land use, the appropriate policy incentives to enhance farm household choices and welfare, and options to reconcile tradeoffs between different policy aims. Procedures to assess the effectiveness of different policy instruments in supporting regional development strategies are urgently required, since the time horizon and budgetary impacts of policy instruments differ widely and the adjustment of land-use practices is costly. Moreover, agricultural research and extension directed toward more sustainable technologies can be enhanced when the effects of policy and technology changes can be evaluated simultaneously in terms of economic efficiency, equity and sustainability.

Role of Bioeconomic Modelling in Ecoregional Analysis

Although agricultural research has offered important contributions to rural development, different types of constraints continue to prevent sustained growth (Anderson, 1994). The adoption of new technologies is limited at the farm household level by purchasing-power constraints and factor-substitution problems (Feder *et al.*, 1985), while market and institutional failures strongly reduce supply response (Askari and Cummings, 1976; Bond, 1983). It is now increasingly understood that the exclusive focus on biophysical crop and animal research needs to be broadened so as to include additional behavioural and structural factors. Thus, the emphasis of agricultural research priorities has shifted towards: (i) a more detailed analysis of the variables that influence the demand for agricultural technologies; and (ii) a thorough appraisal of agricultural policy conditions that enable the adoption of more sustainable production systems.

Ecoregional research programmes are characterized by their focus on interactions between biophysical resource constraints and the organization of farming systems, recognizing that farmers look for local- and household-specific solutions. Rural differentiation is taken into account by identifying different types of farm households that perceive a distinct structure of objectives (e.g. profit, risk, food security). Due attention is paid to external production conditions, including access to markets and property rights, and to tradeoffs between agrotechnical possibilities and socioeconomic incentives. Structural adjustment programmes emphasize the role of market and institutional reforms for enhancing sustained growth. Therefore, policy-makers and analysts ask for decision-support tools that can identify incentives to induce farm households towards higher-productivity, yet more sustainable, practices. The shifts within the agricultural research system towards ecoregional development also require new research methods that recognize the links between farm household preferences, options for technological change, and market and institutional factors that influence changes in farming practices.

Ecoregional research can benefit from incorporating bioeconomic modelling approaches that identify potential farm household and market responses to different types of policy incentives. Major contributions to bioeconomic modelling have been developed at the level of the household (Dalton, 1996; Kruseman and Bade, 1998), village (Barbier, 1994), watershed (Barbier and Bergeron, 1999) and agricultural sector (Deybe, 1994; Schipper, 1996). These models are generally built from separate modules, which specify farm household preferences, technological choice and sustainability parameters. Common characteristics of these bioeconomic models usually include: (i) econometric specification of consumption choice and labour allocation; (ii) discrete definition of technical production coefficients; (iii) use of a linear programming framework for the optimization of activity choice; (iv) specific indicators for sustainability (e.g. nutrient and carbon balances, soil erosion, pesticide emissions); and (v) aggregation through market clearing procedures. Linkages with spatial, intertemporal and more aggregative analyses are sometimes incorporated, but still require further elaboration. Intrahousehold allocation processes are mostly neglected.

Bioeconomic modelling approaches have been shown to be especially relevant to addressing policy issues related to the environmental implications of structural adjustment programmes in sub-Saharan Africa (SSA). The rather fragile agroecological conditions of agricultural production in semi-arid regions require a detailed *ex ante* appraisal of suitable policy incentives that can generate the desired supply response at farm household and regional levels (Kuyvenhoven *et al.*, 1995). The consequences of market reforms (e.g. exchange-rate adjustment, price liberalization) and institutional reforms (e.g. public budget restrictions, land-rights regulation) for technology choice, land-use adjustment, labour allocation and agroecological sustainability can be better understood with the help of these bioeconomic models. Generic policy instruments tend to yield different reactions from different types of farmers, making selective and targeted interventions desirable. Moreover, tradeoffs between the objectives of productivity growth and sustainable land use are likely to appear, and a comprehensive approach combining different types of instruments is therefore required. A careful balance between market incentives and structural policies is generally considered to be important for successfully addressing joint efficiency and sustainability objectives (Ruben *et al.*, 1996; Kuyvenhoven *et al.*, 1998).

Ecoregional development in the semi-arid regions of SSA implies that solutions need to be found to a number of interrelated problems. Soil erosion and organic matter deficits have led to declining yields for arable cropping activities (cereals, cotton) and the subsequent shortening of fallow periods (van der Pol, 1992). Increasing pressure from livestock activities is manifest, since farmers tend to maintain substantial herds for insurance and consumption-smoothing purposes (Udry, 1990). Given the high-risk environment and the presence of substantial market and institutional failures, farmers' willingness to invest in new technology is limited and supply response to market signals is generally low (Bond, 1983). Common technological strategies for addressing these problems focus on such means as improved integration of cropping and livestock activities (Breman, 1990) and higher nutrient applications to improve yield levels (van Keulen and Breman, 1990). This process of moving towards sustainable intensification can only be made economically attractive for farmers if market prices and infrastructure reforms offer sufficient incentives for the adjustment of their land-use practices and farming systems. This implies that farm household short-term welfare objectives need to be balanced with regional sustainability criteria (Reardon, 1995; Sanders *et al.*, 1996).

Different modelling approaches have been used to address agricultural intensification, and policy prescriptions are found to depend critically on the analytical framework and assumptions of underlying models. Exploratory biophysical optimization models tend to be highly optimistic about potential yield improvements and the increased revenues that result from higher input applications (Veeneklaas, 1990; Breman and Sissoko, 1998). However, integrated bioeconomic models based on farm household welfare and risk objectives under fixed prices demonstrate that a fair number of the actual production technologies characterized by negative nutrient balances are none the less maintained.

This can be explained by their positive net contributions to farm household income, while high cash and labour requirements tend to reduce the attractiveness of more sustainable technologies (Ruben *et al.*, 1996, 1998b). Results from aggregate bioeconomic models with endogenous prices for non-tradable goods confirm that substantial structural reforms are required to stimulate the adoption of more sustainable technological options (Kruseman *et al.*, 1997).

Bioeconomic Models: a Review of Approaches

The relationship between the biophysical and social processes can be analysed in different ways. First, agroecological optimization models can be extended with the incorporation of some economic components (de Wit *et al.*, 1988). Secondly, biophysical features can be incorporated into economic models at aggregate (Linnemann, 1979) or farm household levels (McConnell, 1983). Thirdly, modelling procedures can link separate biophysical and socioeconomic models without explicitly defining feedback mechanisms. Fourthly, integrated bioeconomic models can include procedures to link biophysical and socioeconomic information, using feedback mechanisms. We focus our analysis on the latter types of approaches.

A wide range of methods is available to deal with the integration of and interaction between agroecological and socioeconomic processes in the analysis of agricultural development. These approaches rely on different assumptions, analytical frameworks and methodological foundations. Therefore, the application domain of each of these modelling approaches has to be clearly specified. We shall assess different types of bioeconomic models through a classification based on: (i) the purpose of the model, and (ii) the aggregation level (Table 7.1).

A commonly used distinction of land-use studies[1] according to the aim or purpose of the model defines three types of approaches (Rabbinge and van Ittersum, 1994): (i) explanatory studies, which describe current land use;

Table 7.1. Matrix of bioeconomic modelling approaches.*

Application level	Explanatory models	Exploratory models	Predictive models
Plot/field	Production function analysis; activity budgets	Technical coefficient generator (TCG); production ecology	Precision farming
Farm household	Farming systems analysis and research (FSAR)	Farm management analysis; cost–benefit and multicriteria analysis	Farm household modelling
Village/watershed	Village-level SAM models (and CGEs)	Land-use analysis; land evaluation and zoning	Bioeconomic village CGEs
Region/sector	Optimal control models	Multiple-goal linear programming (MGLP)	Bioeconomic multimarket models, CGEs

*Classified according to model purpose and application level.
SAM, social accounting matrix; CGE, computable general equilibrium.

(ii) exploratory studies, which focus on technical options to improve land use; and (iii) planning or predictive studies, which demonstrate the impact of policy interventions. These approaches rely on essentially different assumptions about the role of biophysical or behavioural variables, the dynamics of agrarian change and the time perspective. Explanatory studies are based on actual behaviour and currently available technologies within a given market and institutional environment. Exploratory studies largely disregard farmers' behaviour and offer a long-term perspective on potential technological options for improving land use, as well as the tradeoffs between income and sustainability implications. Predictive or forecasting studies focus on the impact of economic incentives and medium-term market adjustments when new technologies are made available. These modelling approaches can be classified on a continuum that ranges from entirely static to more dynamic agrarian change.

A second classification criterion refers to the level of aggregation of the analysis. A distinction can be made between: (i) field- or plot-level studies; (ii) farm household studies; (iii) watershed- or village-level studies; and (iv) regional studies (Stomph *et al.*, 1995). The aggregation level strongly influences the nature of the focus on biophysical or socioeconomic processes and the importance attached to market-induced adjustments in land use. At the field level, farmers' behaviour is often considered exogenous and attention is focused on technical efficiency and sustainability criteria (plant–soil interactions, macronutrient and carbon balances). At the farm household level, assumptions regarding consumption, investment behaviour and labour allocation are introduced and allocative efficiency becomes important, but prices are still considered exogenous. Village or watershed models include interactions and exchanges between farm households, and consider off-site effects (e.g. runoff, erosion) and market or institutional constraints. Finally, regional models rely increasingly on endogenous market clearance mechanisms for reaching partial equilibrium conditions.

Table 7.1 presents a classification of the major bioeconomic modelling approaches currently used. Strict delineations cannot be maintained, however, since most approaches make use of information derived from analyses at other system levels. For example, regional and village-level bioeconomic models rely on assumptions regarding farm household behaviour, while the latter requires inputs from field-level analyses. Moreover, accurate predictions of future land use requires a good understanding of the forces currently operating that drive land-use dynamics.

Recently, several of these approaches have shown a certain degree of convergence, characterized by three strands of development. In the biophysical realm, the work done by production ecologists and the heirs of early land evaluation studies shows considerable overlap, and both are increasingly vying for attention by economists. Secondly, economists are increasingly making attempts to incorporate biophysical information and processes into their models. Thirdly, at a more aggregate level, there is a tendency to analyse the effects of land degradation on economic development, and vice versa.

We discuss below the similarities and differences between various types of bioeconomic models, paying attention to their historical origins, analytical foundations and integrating methods. Reference will be made to contributions from different traditions, such as production ecology, farming systems research, land evaluation, farm management and planning approaches, natural-resource economics and micro- and macroeconomics. Interestingly, a number of common methods have emerged that enable the linking of information from the biophysical and social sciences.

Explanatory bioeconomic models

Explanatory models generally have the objective of understanding current land use, taking into account biophysical resource constraints (e.g. land quality, climate) and socioeconomic production conditions (e.g. prices, market development). Model results are usually validated against empirical data. Analytical procedures are often largely static and deterministic, although modelling results are sometimes used for extrapolation (assuming unchanged circumstances).

Explanatory models often start from detailed descriptions of current land use and farming practices that originate from farming systems analysis and research (FSAR). In the early 1980s, farming systems research and analysis contributed substantially to a better understanding of the conditions under which small farmers' production took place in the post-Green Revolution period (Tripp *et al.*, 1990). FSAR proved to be highly relevant in specifying diversity in farming practices (Ruthenberg, 1980; Steenhuijsen Piters, 1995) and in explaining yield gaps between experimental and field research. The practical use of FSAR remained limited, however, mainly because results proved to be highly location-specific and/or difficult to quantify. Moreover, FSAR typically lacks a methodology to effectively address policy issues that constrain farming systems (Hebinck and Ruben, 1997; Jones *et al.*, 1997). Therefore, it has been suggested that FSAR models incorporate linkages with socioeconomic models (Anderson, 1985; Dillon and Virmani, 1985) and biophysical models (Dent and Thornton, 1988). Recently, operations research methods have been used for quantifying farming systems analysis, incorporating a stronger economic orientation and a more systematic treatment of biophysical components (McCown *et al.*, 1994; Van Rheenen, 1995).

For a systematic comparison of differences in cropping and production systems between farmers, production function analysis – based on input–output schedules for different cropping and livestock activities – has been widely used, relying on econometric regression techniques and different types of functional forms to represent various possibilities for factor substitution. Agroecological data are usually not directly used in production functions, but can be included in a damage function. Environmental effects are also taken into account through post-modelling analysis (Freeman *et al.*, 1997). Incorporating environmental information directly into production function analysis has also become increasingly popular (Mausolff and Farber, 1995; Ruben *et al.*, 1998a).

At higher levels of aggregation, models based on social accounting matrices (SAMs) have been used to describe, among other things, exchange relations between farm households. Taylor and Adelman (1996) suggest new directions for village-level modelling based on SAMs. This approach allows, in principle, for an appraisal of agroecological sustainability effects when nutrient flows and off-site environmental effects (erosion) are incorporated. Detailed field research is required, however, for a correct specification of the interactions between agents with respect to open access and common property resources (Runge, 1985). Agency theory can provide a suitable analytical framework (Hayami and Otsuka, 1993). SAMs can be combined with computable general equilibrium (CGE) models for the simulation of effects of price policies on resource allocation.

Finally, at the sectoral or regional level, optimal control (OC) models can be used to quantify optimal harvesting rates of renewable resources and identify the investments necessary for restoring resource stocks. Resource economists have used this approach widely for the appraisal of resource management processes and optimal harvesting rates. Applications are also available in the analysis of soil degradation processes (Barrett, 1991; Goetz, 1997; Hu *et al*., 1997) and for aspects of tropical forest management, fisheries and nature reserves (Bulte, 1997; Bulte and van Soest, 1999). The results of these analyses show that resources will be exploited efficiently as long as recurrent income is used for replacement investments to recover resource stocks. Pricing policies can be used to modify the tradeoffs between current and future consumption. Model outcomes of this type prove to be quite sensitive to the assumptions used (especially discount rates). Moreover, OC models fall short in taking into account adequate feedback mechanisms between economic variables and the adjustments in resource management regimes.

We conclude that explanatory research has been dominated by the social sciences (e.g. sociology, economics, anthropology), while contributions from agroecology are mainly recognized within the FSAR tradition. For analyses at higher systems levels such as the village and region, environmental effects are generally considered as an important by-product, but feedbacks between socio-economic and biophysical processes are generally not treated in a systematic manner.

Exploratory bioeconomic models

Exploratory models have the explicit aim of assessing future options for improved resource use under different agroecological and socioeconomic conditions. These models are mainly used for simulation purposes and outcomes cannot be directly compared with the current situation. The results of exploratory models are especially relevant in identifying tradeoffs between the interests of different stakeholders (farmers, state, environmental pressure groups, etc.).

For the exploration of production possibilities at the plot or field level, production ecology approaches offer a wide range of models for developing technical coefficient generators (TCGs) for different land use activities. In the production ecology tradition, these TCGs are mainly oriented towards increasing land productivity through the use of the most technically efficient (and sustainable) options. This approach was stimulated by the first successes of the Green Revolution and often aims at finding solutions to questions related to world food shortages. It has a strong quantitative and exclusively biophysical orientation, and is biased towards technical solutions of societal problems. Attempts to link production ecology models within a macroeconomic general-equilibrium approach (Linnemann, 1979) have proved to be difficult because of conflicting paradigms. The production ecology school has therefore relied more on systems analysis. The systems approach requires well-defined system boundaries and system components, permitting the development of models composed of different elements and their subsequent integration to study the performance of the system as a whole (Rabbinge *et al.*, 1994). From these biophysically based systems approaches has emerged a marked attention to ecoregional development analysis (Bouma *et al.*, 1995).

At the farm level, exploratory approaches are used for farm management analysis. Specific tools such as cost–benefit and multicriteria analysis are used for the evaluation of investments (Van Pelt, 1993; De Graaff, 1996) or for the selection of agricultural research priorities (Alston *et al.*, 1995). These approaches are partial, by nature, and their results are strongly dependent on assumptions regarding prices and discount rates. Choices between different technological options often cannot be satisfactorily explained, since farm household objectives are not fully specified (Heerink and Ruben, 1996). Traditional cost–benefit analysis has been extended to account for environmental effects (Arrow *et al.*, 1996). Multicriteria analysis offers a way of simultaneously considering economic efficiency, social equity and agroecological sustainability, albeit in a rather subjective manner (Mendez, 1995). Both methods offer results regarding economically feasible technologies and thus provide the building-blocks for bioeconomic models.

At the village or watershed level, procedures for land evaluation and zoning approaches have been developed as a framework for linking information from soil sciences (e.g. geographical information systems (GIS), satellite images) with other biophysical and, sometimes, socioeconomic models to assess soil degradation under different regimes of land use (van Staveren, 1980). Land evaluation gained popularity a few decades ago, when a general consensus emerged that land uses could ultimately be planned. In the late 1980s, land evaluation became more oriented towards agroecological sustainability issues and linked with environmental indicators. Simultaneously, successful attempts were made to integrate land evaluation with farming systems analysis into a single framework (Fresco *et al.*, 1992; Alfaro *et al.*, 1994; Schipper, 1996). Most studies still maintain a strong biophysical starting-point, but economic criteria are increasingly incorporated.

Finally, regional-level studies generally use multiple-goal linear programming (MGLP) techniques to explore the long-term prospects of agricultural development in terms of technology choice (de Wit *et al.*, 1988; WRR, 1992; Veeneklaas *et al.*, 1994; Breman and Sissoko, 1998). Economic parameters are used, but often in a rather *ad hoc* fashion. The use of these programming methods – originally developed for farm planning – at the level of regional development can be explained by the general interest in establishing the 'outer boundaries' of a system. Moreover, growing concern over sustainability issues can easily be incorporated by considering environmental amenities as part of the vector of outputs (Knickel, 1994). Economic parameters (prices) and farm household objectives are usually exogenous in these optimization models. However, recent successful attempts to incorporate behavioural constraints and endogenous prices have made some of these models comparable to predictive bioeconomic models (Bouman *et al.*, 1999a).

We can conclude that biophysical approaches have clearly taken the lead in most exploratory research where sustainable land use was of interest. Interestingly, the major analytical procedures used for aggregation towards higher system levels (farm, village, region) are derived from farm management analysis. In principle, this should enable a fruitful cooperation with social science analysis, once the interactions with behavioural aspects can be adequately addressed.

Predictive bioeconomic models

Predictive models are developed for the purpose of decision support at different system levels. These approaches explicitly take into account the behaviour of actors, such as farmers, and their interactions and exchange relations that give rise to changing production conditions (e.g. through induced price adjustments). Moreover, simulations can be made to assess system performance under different types of policy interventions.

At the field level, predictive models provide useful information for the design and operation of precision farming systems. Detailed knowledge on soil conditions and input requirements for spatially defined units and temporally defined operations permit substantial efficiency gains (Roberts *et al.*, 1995). Whereas increasing cost efficiency in input use is attractive from the farmer's viewpoint, reduced emissions are important from a societal perspective. Information from agroecological growth simulation models and georeferenced soil data are used in this type of assessment, in combination with empirical farming systems research.

Bioeconomic models at the farm level make use of farm household modelling procedures, such as those developed by Singh *et al.* (1986). These models explicitly account for natural resource endowments, input and labour allocation decisions, output choice and consumption preferences under different conditions of market development. Biophysical information can be linked to the production side of the farm household model (Altieri *et al.*, 1993; Dalton, 1996) by making use of mathematical programming procedures. Technical production options in the optimization model are derived from production ecology and land-evaluation analysis (Kruseman *et al.*, 1995, Kruseman and Bade, 1998; Hengsdijk *et al.*, 1996). Econometric techniques are applied to specify farm household behaviour concerning consumption and risk. For the purpose of model calibration, information from farming systems research is used. These models enable an appraisal of supply response of farm households to different policy incentives, taking into account varying criteria, such as income and sustainability.

Bioeconomic village models (*modèles villageoises*) have also been developed, notably in the French tradition of analysing watersheds or comparable microregions (Barbier, 1994; Deybe, 1994; Reyniers and Benoît-Cattin, 1995; Deybe and Robilliard, 1996; Barbier and Bergeron, 1999). These models are based on the aggregation of separate models for individual households or typical farm types, and make use of market clearing procedures to adjust the prices for non-tradable commodities. Another approach relies on village-level SAMs and introduces explicit market mechanisms (Taylor and Adelman, 1996). Exchange relations between farmers can also be analysed through a game-theoretical approach to determine pay-offs in both economic and biophysical terms.

Finally, at the regional level, a general-equilibrium framework is required to address interactions between the agricultural sector and the rest of the economy. These interactions encompass issues such as climate change, pollution, off-site erosion damage and salinity problems in large river basins with large irrigation schemes. For this type of analysis, linkages between sector models and biophysical models are clearly required. Most economic studies of this type rely on CGE models, with some econometrically specified linkages to agrotechnical modules (Alfsen *et al.*, 1997; CAMASE, 1998). Another option is to add a formal aggregation module to the above-mentioned bioeconomic farm household or village models (Kruseman *et al.*, 1997; Bouman *et al.*, 1999a,b).

Evolutionary approaches have become available that specify the relationships between population growth, resource scarcity and technological change while paying explicit attention to the endogenous character of these processes. Problems of soil resource degradation are analysed both as a cause and as a consequence of ongoing trends in population growth and changes in the market and institutional environments that farm households are facing (Reardon and Vosti, 1992).

Decision-support systems for policy appraisal call for an explicit treatment of biophysical and socioeconomic processes. Analysis at lower system levels (field, farm) usually has a biophysical focus, while at higher system levels (village, region) socioeconomic relations become more important. Modelling procedures for linking agroecological and socioeconomic information still require separate attention. Major questions arising in the modelling process pertain to the following:

- *Technology choice*. Linking stochastic and deterministic analytical processes implies that farm household models which rely on econometric specifications need to be related to agroecosystem information of a more deterministic nature. Technical input–output coefficients derived from crop growth models and related sustainability indicators cannot be directly fitted into an econometric production function because of synergy effects and limited factor substitution options (de Wit, 1992). Therefore, recursive modelling procedures are typically applied to identify relevant break points.
- *Farm household behaviour*. Farm household models require a detailed specification of different objectives (e.g. profit, risk, leisure) that are perceived simultaneously, while land-use analysis usually relies on a single objective

optimized under certain constraints. A feasible integration can be found when specific weights for various farm household objectives are specified, using dynamic programming methods (van Kooten *et al.*, 1990; Romero, 1993).

- *Micro–macro aggregation.* Farm household technology choice and related resource allocation decisions under given socioeconomic conditions cannot be directly scaled up to the regional level, since price changes (for outputs and inputs) and spillover effects (for labour demand) are likely to occur. Therefore, a new equilibrium has to be established through iterative procedures to adjust response reactions at a higher aggregation level (Bade *et al.*, 1997; Bouman *et al.*, 1999a).
- *Market and institutional constraints.* Farm household models account for market failures by taking into account specific transaction costs and/or conditions of restricted market access (Kruseman and Ruben, 1998). Relations with the (non-farm) labour market and population dynamics (migration) are still kept exogenous, and are better addressed within a multimarket framework. Institutional constraints, especially those related to common property resources, have to be addressed within a contingent choice framework that accounts for expectations of the behaviour of other agents. Implications for the sustainability of resource use can be identified through backward induction procedures that account for discounted welfare effects.
- *Model validation and calibration.* Predictive models mostly rely on fixed coefficients for relevant model parameters (e.g. risk aversion, savings rate, etc.). Explanatory models can be used for empirical specification of these coefficients, thus improving the validity of model outcomes.

Bioeconomic Models: an Empirical Assessment

The relevance and importance of integrated bioeconomic models can be demonstrated by presenting results from different model specifications applied to a similar agroecological environment. For this purpose, we used the case of the semi-arid region of southern Mali, a region characterized by rapid agricultural development based on food crops (millet, maize, sorghum), cash crops (cotton, groundnuts) and livestock production. Population growth, stagnant agricultural yields and increasing stocking rates make further intensification of farming systems inevitable. For this region, we compare three modelling approaches with respect to their derived policy implications for sustainable intensification:

1. A biophysical optimization model, which gives high priority to sustainability objectives.
2. A recursive bioeconomic model, based on farm household welfare objectives with fixed prices.
3. An aggregate bioeconomic model, with endogenous prices for non-tradable commodities (cereals and livestock).

The exploratory biophysical model takes the optimization of net revenues at lowest environmental costs as a single objective, has no behavourial relations and only addresses production issues, such as adjustment in cropping systems and technology choice. The explanatory farm household model includes consumption behaviour and considers utility maximization (of consumption) as the primary objective. In both models, prices are kept exogenous and farm households are price takers. The latter assumption is relaxed in the third model, which maintains the utility maximization objective, but where market prices for cereals and livestock are determined as the outcome of changes in aggregate supply and demand. The model specifications thus start from a technical optimum and gradually include behavioural aspects and market-clearing procedures.

All models use the same database. Agroecological input–output coefficients for all major production activities (e.g. cereals, cotton, pastures and livestock) are specified through crop growth simulation procedures, including nutrient and organic matter balances as joint products (Hengsdijk *et al.*, 1996). Activities include both actual and potential (more sustainable) production technologies and account for crop–livestock interactions (e.g. manure, traction), different livestock feed regimes (e.g. grazing, harvesting of crop residues, fodder crops) and alternative soil-management practices (fallowing, ploughing of crop residues, manuring). Farm household resource endowments for land, labour, animal traction and implements are derived from Brons *et al.* (1994).

The model framework is based on the integration of discretely defined technical production coefficients with continuous functions for consumption behaviour and labour allocation (Kruseman and Bade, 1998).[2] The biophysical optimization procedure only considers productive choice that leads to optimal revenues at lowest environmental costs. The recursive bioeconomic analysis optimizes consumption utility for an individual farm household under given prices, taking into account both productive and consumptive behaviour. The regional analysis also optimizes consumption utility but takes the effect of changes in production on local market prices into account.

Table 7.2 presents the impact of a fertilizer subsidy for each model specification. Results are shown for selected welfare criteria, a number of sustainability indicators (e.g. organic-matter and nutrient balances) and adjustments in land use, labour allocation and prices. The focus is on the price-subsidy instrument, since the high costs of imported fertilizers following structural adjustment are often mentioned as a major limitation for sustainable land use (Kuyvenhoven *et al.*, 1995; Reardon, 1995). Lower fertilizer costs could thus generate higher nutrient applications and a better integration of cropping and livestock activities, enabling a simultaneous increase in yields and soil-nutrient balances.

Model results are expressed in terms of response multipliers, which indicate the percentage change of a variable caused by a 1% subsidy in the fertilizer price, compared with the base run for each of the different model specifications under unchanged conditions. We focus attention on the responsiveness to the fertilizer subsidy and neglect the substantial differences in absolute levels of the outcomes. Model outcomes turn out to be highly sensitive to the assumptions about farmers' behaviour and the level of market and institutional development. Favourable results for income and sustainability derived from the biophysical model cannot be fully maintained when more realistic assumptions on farm household objectives and market clearing are introduced.

Table 7.2. Response elasticities of introducing a 1% fertilizer subsidy.

Indicators	Biophysical optimization (sustainability)	Farm household welfare (consumption utility)	Regional development (consumption utility and market clearance)
	Percentage change compared with base run for different model specifications		
Welfare indicators			
Per capita income	0.15	0.28	0.08
Food production	0.10	−0.85	−0.49
Food consumption	–	0.01	−0.04
Marketed output ratio	0.10	0.47	0.19
Off-farm labour	0.00	−9.62	0.00
Sustainability			
Organic matter	0.68	−0.40	0.63
Nitrogen balance	0.51	−0.94	0.98
Phosphate balance	0.07	−0.99	10.79
Input intensity			
Labour per hectare	0.09	0.07	0.07
Production			
Livestock (no. TLU)	0.00	−0.10	0.03
Cotton	0.98	5.98	0.32
Fallow	0.09	−78.75	0.00
Cereals	−0.35	−0.02	−0.01
Leguminous grains	−0.56	0.03	0.06
Prices			
Cereals	–	–	0.01
Livestock	–	–	2.62

TLU, tropical livestock unit.

Results from the biophysical optimization model indicate that fertilizer subsidies allow for a substantial improvement of sustainability indicators, while introducing major shifts in the land-use pattern. Farmers' income rises as a result of the reliance on more efficient technologies and the expansion of commercial cropping activities. The area devoted to cotton production expands strongly at the cost of cereals (maize, sorghum, millet and cowpea) and leguminous grains, because cash-crop production permits higher profits with better environmental amenities. Food production still increases, since fertilizer subsidies make the selection of higher-yielding cereal technologies feasible. Food consumption remains unchanged because the model only takes production aspects into account. Surplus production is sold at fixed prices. The livestock stocking rate is maintained, but feed rations are increasingly derived from agroindustrial products such as cotton-cake. On-farm labour demand also increases, while engagement in off-farm labour is hardly modified. Consequently, the time available for leisure decreases. The results from this simulation indicate that fertilizer subsidies enable a simultaneous improvement of farmers' income and the sustainability of the resource base.

The farm household model also includes consumption decisions and the labour–leisure choice. In this simulation, fertilizer subsidies have an even stronger effect on the expansion of the cotton area, and growing labour requirements for cotton production can only be satisfied through a reduction in off-farm employment. This can be explained by the fact that household consumption requirements are now satisfied through market exchange, while additional labour demands for cash-crop production lead to reduced leisure. The area devoted to cereals production is only marginally reduced, but far less efficient production technologies are selected, so that total food production decreases. Household food consumption is maintained, however, through reliance on market-purchased cereals. However, the amount of livestock has to be reduced, since fewer crop residues are available for feed rations. Leguminous grains are only partly able to compensate for this loss. Consequently, organic-matter and nutrient balances show a general deterioration, since less manure is available for nutrient recycling and fallow practices are substantially reduced. The rise in farm household income mainly comes from increased specialization in cotton. Given the lower food production and the decrease in off-farm income, these revenues are largely used to maintain food consumption through the purchase of cereals. Consequently, fewer resources are available for investments to improve sustainability. Given the labour, feed and manure constraints, the fertilizer subsidy induces the selection of less efficient and extractive production technologies and soil-mining practices to attain high, short-term returns.

The regional development model relaxes the assumption of fixed prices and introduces interactive procedures for the clearance of cereal and livestock markets. Consequently, changes in production, consumption and market supply lead to price adjustments. Model outcomes differ greatly from those derived from the farm household model. Response elasticities are generally dampened, and the directions of change are sometimes reversed. Positive production effects resulting from subsidies on inputs for cash-crop production are partly outweighed by secondary price and income effects. Rising cereal prices result from reduced cereal production, rising market demand and the substitution of arable land towards cash-crop (cotton) and fodder (leguminous grains) production; reduced cereal production allows for the maintenance of some fallow areas. Livestock prices rise strongly, as farmers want to retain cattle (financed out of cotton revenues) as a means of managing soil organic matter; rising prices also reflect an increased demand for cattle, due to higher household income. Engagement in off-farm employment remains unchanged; this income source is required to finance inputs and food purchases. The most important differences can be observed for the sustainability indicators, especially for phosphate, which show positive changes. This is due to the rising commodity prices, which offer incentives for better input allocation (mainly in cotton), improved animal-feeding practices (e.g. less reliance on crop residues and more use of fodder crops) and the maintenance of fallow areas.

Comparison of the three model specifications shows that fertilizer subsidies enhance technological change and offer large opportunities for land-use adjustment, contributing to both higher incomes and more sustainable resource management. However, individual farm household responses to the subsidy show rather divergent results. Farm household income increases are higher compared with the results of biophysical optimization, but this is achieved through the selection of less efficient food-production technologies (since food requirements can be satisfied through market purchases), and is accompanied by a reduction in livestock (with less manure availability) and fallow. At the regional level, induced price adjustments again modify farm household modelling results. Not only are aggregate responses substantially lower, but fertilizer subsidies (combined with higher commodity prices) are positive for both income and sustainability.

The differences in direction and intensity of the response to fertilizer subsidies depend critically on the assumptions regarding farmers' production and consumption behaviour and on procedures used for the analysis of market performance. Biophysical optimization models tend to be too optimistic about the possibilities for (costless) substitution in crop and technology choice to enhance sustainable land use. Introducing household consumption into biophysically oriented models leads to high income effects via increased specialization, but at the expense of environmental quality. At the regional level, endogenous prices for non-tradables induce modifications in crop and technology choice, which, in this case, result in an improvement of sustainability by using the market mechanism. Regional models with induced price effects thus offer an intuitive balance between reasonable income effects and moderate improvements in the sustainability of resource management.

Although at first sight the responses in the biophysical and regional optimization models point in the same direction, a comparison of the model scenarios shows that these results are derived under totally different mechanisms. In the regional model, improved sustainability is mainly caused by induced market effects and far less by the input subsidy as such. Moreover, the strong changes in crop and technology choice that result from the biophysical optimization model are substantially dampened when price effects are taken into account. Each model specification also reveals interesting biases. In the biophysical model, technical possibilities for improving sustainable land use loom large. The farm household model illustrates how specialization and (over)exploitation of the resource base can go together. At the regional level, incentives for the improvement of land-use practices are mainly channelled through the market. Hence, fertilizer subsidies can enhance sustainable intensification only if the market is functioning well.

Policy Applications of Bioeconomic Modelling

Agrarian policies for ecoregional development need to be based on a thorough understanding of the likely effects of different types of incentives on adjustments in land use, farm household welfare and sustainability. Integrated bioeconomic models provide an important decision-support tool for policy-makers, enabling a simultaneous appraisal of production and consumption decisions, related changes in land, labour and input use and their implications for farm household income and natural-resource balances. Bioeconomic modelling procedures provide for a comprehensive impact analysis at different levels of scale, based on integrated information on soil-nutrient dynamics, farmers' behaviour and market performance.

Bioeconomic models can be particularly useful in the appraisal of policy options for ecoregional development in semi-arid regions, where soil-nutrient balances are fragile and risk-averse farm household behaviour prevails. The interactions between a large number of simultaneously occurring processes (e.g. land-use change, technology choice and input application, marketing and consumption decisions, labour allocation) can be assessed within this integrated modelling framework.

Agricultural intensification is usually considered to be an important strategy for improving both farmers' incomes and soil-nutrient balances. Different technical options are, in principle, available for enhancing sustainable land use (e.g. better integration between cropping and livestock activities, control of soil erosion, crop residues and manure recycling, etc.), but their rate of adoption is co-determined by economic incentives (Kuyvenhoven *et al.*, 1998). In the West African context, fertilizer subsidies are often mentioned as an important instrument for stimulating these changes. The modelling results presented here indicate that fertilizer subsidies are able to promote the process of sustainable intensification only when biophysical criteria are considered as the primary objectives. Subsidies prove to be a disincentive when farm household welfare objectives are included, since higher income and consumption levels are reached with less sustainable resource management. Once market-clearing procedures are introduced at the regional level, the perceived positive effects on income and sustainability are simultaneously realized, but at more modest levels. Besides the availability of technological options, the development of well-functioning local markets and institutions is thus a precondition for enhancing sustainable intensification. Technical options may be made attractive from the perspective of the farm households' short-term welfare objectives by using price and infrastructure reforms as incentives to adjust land-use practices and farming systems.

Besides fertilizer subsidies and market liberalization, other policy instruments are usually considered in discussions of sustainable intensification. Exchange-rate devaluation and deregulation of cotton exports will tend to put upward pressure on real farm-gate prices. Similar income and substitution effects can be expected to those in the regional development scenario, e.g. both farm household welfare and sustainable land use are likely to be improved. Otherwise, different types of land policies (e.g. land tax or head tax for the use of common land) are sometimes advocated to control the degradation of pastures. Policy simulations indicate, however, that these instruments tend to reduce the size of the herd and stimulate the reliance on crop residues as fodder (instead of ploughing under), thus affecting the sustainability of land use and curtailing the opportunities for farm household consumption smoothing (Ruben *et al.*, 1996).

Finally, some suggestions can be offered for extending the bioeconomic framework. Most of the modelling exercises focus on separate policy instruments, while structural adjustment programmes modify a whole range of production conditions, with simultaneous impacts on factor and commodity markets. To capture these effects, multimarket analysis and CGE features can be integrated into the bioeconomic framework. Moreover, relations with the non-agricultural sectors are only taken into account as far as the opportunity costs of labour are concerned. Forward and backward linkages among the agricultural, industrial and services sectors can be made more explicit. Where relevant, spatial specification of bioeconomic models, providing linkages with GIS, are technically feasible and will enable a better appraisal of the desired location of infrastructure investments to improve market performance.

Summary and Conclusions

Policy analysis for the appraisal of ecoregional development alternatives needs to be based on a thorough understanding of farm household and aggregate regional response to different types of incentives. Bioeconomic models have been developed to identify feasible instruments for handling tradeoffs between welfare and sustainability objectives. This analysis is applied to the debate on which policies best enhance sustainable intensification in West Africa. Some general conclusions are drawn with respect to the possibilities for overcoming economy–sustainability tradeoffs under different conditions of market development. While market failures are a recurrent feature in most sub-Saharan countries, suitable incentives for improving current unsustainable land-use systems can only be expected to generate the desired income and substitution effects when market mechanisms are in force.

Policy-makers often ask for advice from the agricultural research community in identifying the right policies for addressing the right questions. The answers they receive are mostly of a partial nature and, therefore, often cannot be directly used for policy analysis. Simultaneously addressing questions of sustainable land use and farm household welfare requires an integrated modelling effort that takes into account the most important dimensions of regional development. We have demonstrated that the precise specifications of such models are of fundamental importance. The procedures used for the functional integration of technical, behavioural and market aspects have a distinct effect on the model outcomes. Changes in input prices (e.g. fertilizer subsidies) and related adjustments in crop and technology choice are far less important than feedback mechanisms (e.g. output price adjustments) caused by market-clearing mechanisms.

The modelling approaches reviewed in this chapter have become major instruments for understanding and operationalizing ecoregional development. The functional integration of technological options, farm household behaviour and market mechanisms into a single analytical framework that recognizes interactions and feedback procedures offers promise in bridging the gaps between different academic disciplines and between scientists and policy-makers.

Comparing different modelling specifications with respect to their policy impacts on sustainable intensification, we observe that the optimistic results that follow from biophysical optimization models with prioritized sustainability objectives are dampened when farm-household welfare objectives and endogenous prices for non-tradables are introduced. Model outcomes prove to be highly sensitive to different assumptions about farmers' behaviour and the level of market and institutional development. In situations where improving sustainable land use in West African dryland areas might be technically feasible, the available technologies are not likely to be adopted as long as local markets remain constrained. Morever, since sustainability is hardly ever a major objective from the viewpoint of the farm household, market price adjustments are required to provide effective incentives for the desired land usage.

Ecoregional research can clearly benefit from understanding bioeconomic modelling, which identifies potential farm household and market responses to policy incentives. Fundamental research on the specification of modelling components and procedures needs to continue, since some major interactions are not yet fully understood. We have shown that results (and thus policy recommendations) are sensitive to model specifications and assumptions regarding technical and economic performance. At the biophysical level, relationships between soil erosion and yield potential are still subject to much debate. In addition, the dynamics of multiple cropping systems cannot be captured adequately in available crop-growth models. Farm household behaviour can be better specified when procedures for modelling risk and adoption behaviour are explicitly incorporated. Integration of bioeconomic models with spatial, multimarket and general-equilibrium analyses can continue to be improved, while linkages with non-agricultural activities should be enhanced to enable a better appraisal of resource allocation procedures.

Notes

1 Land-use studies offer an appraisal of change, in agricultural systems in terms of time and space (Schipper, 1996).

2 A full specification of the linear programming model is available in Kruseman *et al.* (1997).

8 Tradeoffs in Agriculture, the Environment and Human Health: Decision Support for Policy and Technology Managers

CHARLES C. CRISSMAN,[1] JOHN M. ANTLE[2] AND JETSE J. STOORVOGEL[3]

[1]International Potato Center (CIP), Quito, Ecuador; [2]Trade Research Center at Montana State University, Bozeman, Montana, USA; [3]Laboratory of Soil Science and Geology, Wageningen University, Wageningen, The Netherlands

Introduction

The key to designing sustainable agricultural production technologies and identifying policies that promote sustainable systems is in understanding their economic, environmental and human-health impacts. The research reported here supports the development of a methodology designed to quantify such impacts and to represent them as tradeoffs. This tradeoffs methodology addresses two key elements: first, it provides an organizational structure around which to design successful interdisciplinary research that assesses the sustainability of production systems; and, secondly, it provides a successful means of communicating research findings to policy-makers and the public.

The research reported here is a second application of the tradeoffs methodology. The initial application was in a study of the environmental and human-health aspects of pesticide use in potato production in Ecuador, documented in Crissman *et al.* (1998). The current application of the tradeoffs methodology is through case studies of Andean potato and milk production systems in the Andean highlands of northern Ecuador and northern Peru, in which we are again examining pesticide use but also erosion and soil management practices.

Agricultural intensification and land degradation

Farmers are the most numerous and most important soil-resource managers in the Andes. Agricultural technology ranges from traditional, extensive, low-input, low-output systems to modern, intensive, high-input, high-output systems. The traditional systems have to be maintained within their ecological constraints and, as a result, are generally perceived as environmentally friendly and sustainable.

However, due in part to shrinking farm size, traditional systems have proved to be economically and socially non-sustainable. With a closed agricultural frontier in most parts of the Andes, the fundamental option for Andean farmers is to increase the physical and financial output from the existing farm. This inexorable pressure provides a strong incentive to shift to higher-output modern systems. The basic quest of agricultural and environmental research for sustainable farming systems is to match the environmental friendliness of traditional farming systems while reaching the higher outputs and, thus, the economic and social sustainability found in modern farming systems.

Linking national policy with local impact: the role of aggregation and scale of analysis

There is growing recognition of the critical role that (dis)aggregation and scale of analysis play in assessing the interactions between human activity and the environment. Scientific understanding of the impacts of human activity on biological and physical systems typically occurs at the level of some relatively small unit of analysis – an individual organism, or a square metre of land. At this level, the interactions between individual biological and physical units and human activity depend on characteristics of the physical environment that are temporally and spatially heterogeneous. In the inter-Andean valleys of Ecuador and Peru, this heterogeneity is extreme. Consequently, the interactions between human activity and the environment are spatially and temporally variable, and aggregation of these interactions results in a loss of information about them. Yet, for purposes of understanding the economic and social importance of these changes and for analysing policy options to manage these changes, we must work with much larger aggregate units of analysis composed of many individuals.

Making sustainability operational

Because sustainability is a relatively new objective for research, there is still no consensus on methods to quantify the concept and incorporate it into public policy analysis. Within the context of the international agricultural research centres, making sustainability operational calls for new approaches to research priority setting, problem identification and organization. The International Potato Center (CIP) and its fellow institutes in the Consultative Group for International Agricultural Research (CGIAR) have adopted an ecoregional approach as a means of operationalizing sustainability. The CGIAR identifies ecoregions as 'agroecological zones' and defines the role of the ecoregional approach as follows:

> The main role of the ecoregional approach is to contribute to the goal of increasing sustainability of agricultural production by providing: First, a process that identifies the right research content due to its holistic and forward looking perspective which contrasts with traditional disciplinary and commodity approaches to research. Second, a mechanism for partnership, among relevant actors with complementary functions, that contributes to achieving their common and individual institutional goals through applied and strategic research on the foundations of sustainable production systems. Third, a mechanism that develops, tests, and supports effective research paradigms for the sustainable improvement of productivity.
>
> (CGIAR, 1993, p. 4)

The ecoregional approach places emphasis on modelling production systems and their environmental impacts at a small scale, such as the field or watershed scale, and on how those small-scale impacts affect systems at larger scales or higher levels of aggregation. The approach is primarily a systems modelling approach, which emphasizes the importance of economic decision-making models to capture changing priorities in farm households and communities. Other tools important to the ecoregional approach include geographical information systems (GIS) and crop, livestock and soils models (Bouma *et al*., 1995). We should emphasize that these tools build upon the methods and data provided by the traditional experimental approach of agricultural research that is the hallmark of the CGIAR research system (CGIAR, 1995).

The research reported here is compatible with the ecoregional approach and provides a methodology for the implementation of research within this paradigm. In particular, the tradeoffs methodology operationalizes the ecoregional approach by linking disciplinary models of agricultural production, environmental impact and human health to a regional level for technology assessment and policy analysis.

The research sites

The methodological developments reported here are tested in a comparative case study of potato- and milk-based farming systems in Carchi Province in northern Ecuador and Cajamarca Department in northern Peru. These two cool, highland, tropical sites are distinct in the quality of soil and water resources available to farmers. The Carchi site enjoys deep volcanic soils and abundant rainfall. The Cajamarca site, in contrast, has thinner, eroded soil and scarce rainfall. Low soil organic matter and water constraints limit productivity. Both sites are characterized by smallholder hillside agriculture, in which ample use of externally supplied inputs is typical in potato production. Chemical fertilizers and a wide range of insecticides and fungicides are utilized to improve potato yields. Potato yields between the sites and among the survey farmers range from 7 to 25 t ha^{-1} (fresh weight), which, given the climatic conditions, is far below the potential production. Milk production is also low (8 l animal $unit^{-1}$ day^{-1}); mixed-breed cows are milked once a day and have access to only natural pastures. Changing technologies and policies related to potato and milk production are the focus of the policy analysis applications presented here.

The Tradeoffs Model

The ecoregional research approach taken here is integrated systems modelling at the field level, with aggregation to the watershed and region. Utilizing integrated analysis techniques implies moving through different scales and applying varying analytical techniques, utilizing different knowledge and expertise levels. The tradeoffs model developed and presented here starts with a combination of macro- and micro-level factors, moves to an analysis at the micro level and aggregates the results back up to the macro level. The tradeoffs model in its initial application consisted of a set of linked econometric and simulation models to evaluate pesticide impacts. The present model structure begins to generalize the conceptual framework to address distinct agriculture, environment and health issues.[1] The pesticide impacts study was an initial application of the tradeoffs model.

Model purpose

The tradeoffs model aims to provide the basis for a decision-support system for assessing tradeoffs between agricultural production, the environment and human health for different economic, agricultural and environmental policies and for agricultural research. A decision-support system is any arrangement by which data and information are structured so as to be referred to for help in decision-making. These systems can vary from simple empirical models to sophisticated mechanical or statistical models. The model assesses linkages between farmers' cropping decisions, economic variables and natural resources and should be able to:

- quantify the impact of existing and proposed agricultural and environmental policies on the sustainability of selected agroecosystems;
- screen proposed agricultural technologies, such as integrated pest management (IPM) and various types of soil husbandry, for their potential impact on the sustainability of selected agroecosystems;
- generate results than can be utilized to develop recommendations for research priorities for national and international research systems;
- recognize the spatial variation of natural resources and communicate with a GIS in order to be able to deal with this variation.

The decision-support system contains the following key features:

- It links disciplinary data and models in a GIS framework.
- It utilizes the minimum data necessary for decision support and policy analysis.

- It is generalizable – results can be extrapolated to larger geographical regions, using a GIS framework.
- It is transportable – the generic structure of the system can be adapted to other geographical settings and applications.

The research chain for tradeoffs assessment

A central theme of this approach is that quantifying tradeoffs is an essential ingredient in setting research priorities and in designing and implementing sustainable agriculture criteria in agricultural research programmes. Tradeoffs assessment provides an organizing principle and conceptual model for the design and organization of multidisciplinary research projects to quantify and assess the sustainability of agricultural production systems. Input from the general public ('stakeholders'), policy-makers and scientists is used to identify the critical dimensions of social concern, i.e. the sustainability criteria. Based on these criteria, hypotheses are formulated as tradeoffs between possibly competing objectives, such as higher agricultural production and improved environmental quality.

Once the key tradeoffs are identified, research team leaders can proceed with project design and implementation, including identifying the appropriate scientific disciplines to plan and implement the research needed to quantify these tradeoffs. The next step, critical to quantifying tradeoffs, is the identification of disciplinary models and data needed to quantify each sustainability indicator. Attention must also be devoted to assessing the data needs for each of the disciplinary components of the analysis, and how the model outputs can be effectively linked for the analysis of tradeoffs. As we discuss further below, a critical element at this stage is for all of the represented disciplines to agree upon basic spatial and temporal units of analysis. Will analysis be conducted at the field or watershed scale? Will time units be daily, weekly, monthly or yearly? Once these fundamental issues in research design have been resolved, data collection and disciplinary research can proceed. Upon completion of the disciplinary components of research, the respective data sets and models can be linked to test hypotheses about tradeoffs, and the findings can be presented to policy-makers and the general public.

A number of challenges face researchers in implementing this type of research. Despite widespread acceptance of the goal of sustainable agricultural systems and recognition of significant tradeoffs associated with the regulation of technologies such as pesticides, a scientific consensus is lacking on how the economic, environmental and public-health impacts of agricultural technologies can be quantified and assessed (D'Souza and Gebremedhin, 1998). Analysis of these complex, interrelated issues raises difficult theoretical and methodological problems for researchers. Environmental, agricultural and health characteristics of farmers, farmland and farming technologies vary over space and time. The problems that concern the public are multidisciplinary and thus require a multidisciplinary approach. Overcoming disciplinary biases and establishing effective interdisciplinary communication is a continuing challenge for a research team.

Units of analysis and scale issues

As noted above, one of the practical methodological challenges is the choice of the unit of analysis. Research and modelling in the biological and physical sciences typically deal with a unit of analysis – whether it is at the cellular, soil unit, plant, animal or field scale – that is different from the farm or sectoral scales relevant to policy analysis. Policy analysis is typically concerned with a larger unit of analysis, usually defined in relation to a geographical or political region that contains a population of the units addressed by biological and physical sciences. The aggregation problem, i.e. the problem of combining heterogeneous small units into a larger unit for policy analysis, must be addressed by all researchers if their data and results are to be useful for policy analysis.

The fact that the various scientific disciplines use different units of analysis frequently means that the data and methods developed for disciplinary research are of limited value for policy research. Disciplinary research typically operates in a format dictated by disciplinary orientation and generates data intended to satisfy disciplinary objectives. This disciplinary orientation of research leads to a situation in which various pieces of the scientific puzzle are investigated with techniques or models designed for single-aspect analysis – without regard to the conjoining of those pieces into the larger picture that is required for policy analysis. Thus, the disciplinary component of research intended to support the assessment of tradeoffs must be planned at the beginning of the research effort to produce methods and data that are required for disciplinary analysis, but which can also be utilized across disciplines to assess tradeoffs. The planning, in advance, of coordinated disciplinary research is one of the key benefits of the tradeoffs assessment methodology.

Tradeoffs associated with agricultural production systems can be defined across several dimensions at a point in time, and can also be defined in one or more dimensions over time. Likewise, in evaluating the long-term sustainability of a production system, economic and environmental indicators can be used to quantify the productivity and other attributes of a system over time. These indicators include measures of economic returns, soil erosion, chemical leaching, nitrate movement through soil profiles and the organic content of the soil. Measuring tradeoffs in multiple dimensions requires site-specific data and models, since the impacts of different production systems may vary across different environmental dimensions at different sites. Thus, any attempt to rank production systems according to sustainability criteria needs to account for spatial variability in economic, environmental and health outcomes.

The larger the spatial or temporal scale, the more complex becomes the process of quantifying tradeoffs for the analysis of agricultural sustainability. Analysis at the regional or national scale is even more difficult than analysis at smaller scales, such as a watershed. Attempts to develop quantitative indicators of the sustainability of the US farming sector (USDA, 1994), or the farming sectors of member countries of the Organization for Economic Cooperation and Development (OECD, 1994), have relied on aggregate data on production, input use and resource degradation. These data do not provide a scientifically defensible foundation for policy formation because production cannot be linked to environmental and health impacts on a site-specific basis.

The research methods applied in the tradeoffs model make use of an alternative approach to addressing regional policy concerns in the area of sustainable agriculture and technology evaluation that is based on solid scientific foundations. The approach followed here is to develop data and related disciplinary models that link the site-specific management decisions of producers with environmental and health impacts, and then to utilize a statistical representation of the relevant human and physical populations to statistically aggregate those impacts to a regional level for policy analysis.

Political pressure to identify a set of sustainable production technologies implies that there must be some means of ranking the importance of the various impacts. Ranking technologies according to multiple criteria requires a method of converting these criteria to a common unit of analysis. The economic approach to this problem is to convert all impacts to monetary terms and to use this information to conduct a benefit–cost analysis. However, despite decades of research on the valuation of environmental and health outcomes by environmental and health economists, there is no scientific or public consensus on valuation methods or their public acceptability, and data for the valuation of most environmental and health impacts are lacking. For this reason, we advocate the view that agricultural sustainability research should focus on establishing a sound scientific basis for quantifying tradeoffs between ecological and economic objectives that exist with alternative production systems, without attempting to value impacts for benefit–cost analysis. The valuations of the impacts will be determined anyway during the political negotiation process.

Assessment of tradeoffs

The approach described here is compatible with the ecoregional approach and provides a methodology for the implementation of research within this paradigm. In particular, the methodology developed here can be viewed as a way to operationalize the ecoregional approach by linking disciplinary models of agricultural production, environmental impacts and human health at a regional level for technology assessment and policy analysis. The methods for quantifying and assessing tradeoffs presented here provide an explicit framework for setting research priorities and organizing research within the framework of the ecoregional approach.

The conceptual framework for disciplinary integration and policy analysis developed for the application of the tradeoffs model is illustrated in Fig. 8.1. This framework is designed to address the methodological issues raised by disciplinary integration and aggregation from the field scale where modelling is valid to the level appropriate for policy analysis (e.g. the watershed or larger scale). Moving from top to bottom, the framework captures the logical sequence of how macro-level policy affects farming decisions that result in micro-level impacts, and how those impacts should be aggregated back up to units useful for macro-level policy analysis. This sequence crosses several levels of analysis and,

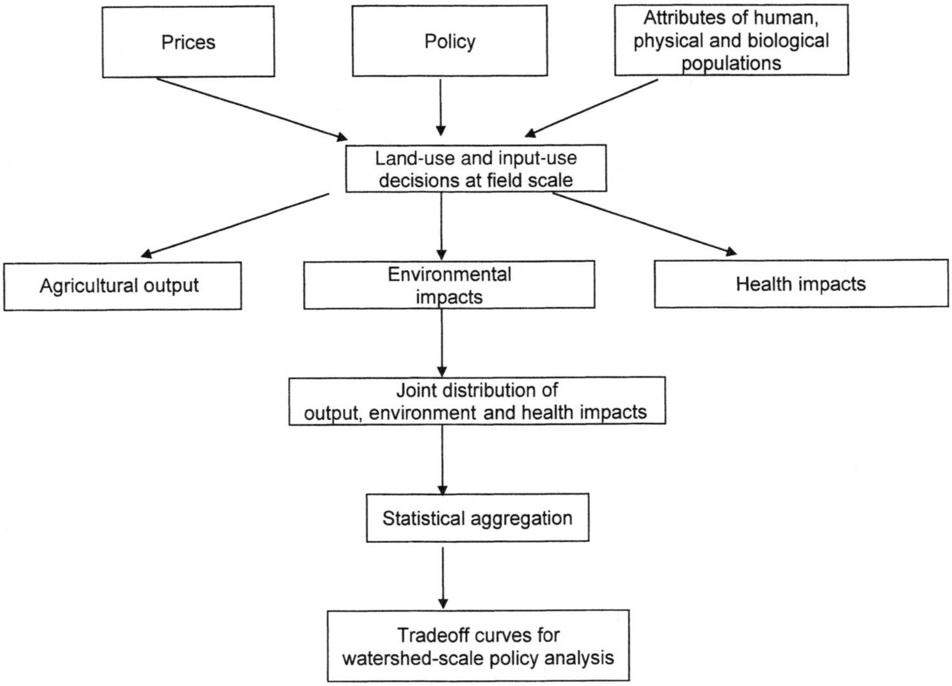

Fig. 8.1. Conceptual framework for disciplinary integration and policy analysis.

because it is statistically based, provides a basis for aggregation. The farm-level component of the model represents farmers' decision-making. By incorporating the decision-making process of the land manager, the model provides a link to the available set of policy instruments and regulations. The 'what if' questions needed for policy analysis can be explicitly incorporated into the model.

Starting at the top of Fig. 8.1, using a parcel of land as the unit of analysis, the model shows that prevailing policies and market prices, technologies, farmer characteristics, and the physical attributes of land affect farmers' management decisions in two ways. The farmer makes land-use decisions on how to use the parcel, plant a crop, leave the land in pasture or plant something else, such as trees. With the decision to plant an annual crop, for example, the farmer then makes management decisions on how to plough and what inputs to use when and in what quantities. Physical relationships between the environmental attributes of the land in production and management practices then jointly determine the agricultural output, environmental impacts and health impacts associated with a particular unit of land in production.

The farm-level decision models show that each unit of land that is in production has management and environmental characteristics, which, in turn, are functions of prices, policies, technology and other farm-specific variables. As shown in the lower part of Fig. 8.1, the individual probability distributions of technology, farmer and environmental characteristics in the region induce a joint distribution of management practices, environmental characteristics and health outcomes for each land unit in production as a function of prices and policy parameters. This joint probability distribution provides a statistically valid representation of the outputs, inputs, environmental impacts and health impacts for the population. Recall that, in this case, the population is the farm field and the farmer of that field. Therefore, these individual outcomes can be 'added up' to produce an aggregate distribution of impacts. In a statistical framework, the aggregation problem is defined as the adding up of individual characteristics to obtain summary statistics for the group.

In production analysis, Antle and McGuckin (1993) show that aggregate production functions and aggregate input demand functions can be interpreted as population means. They also show that the aggregate functions for output, input and contamination depend on the underlying distributions of characteristics of the disaggregate populations. Thus, the distribution of individual farmer's usage of individual fields produces agricultural output and environmental contamination at the same time – a joint distribution. The aggregation results show that these joint results per field can be added up to yield summary statistics for the group. The aggregate outcomes – measured in terms of agricultural output, environmental quality indicators and health indicators – are used to construct tradeoffs for policy analysis. This information can be utilized in several ways. If monetary values can be assigned to all impacts, a benefit–cost analysis of policy alternatives can be conducted. However, since monetary values are usually not available, the more useful approach is often to present information about tradeoffs directly to decision-makers and policy-makers.

The structure of the tradeoffs model as used to examine pesticide leaching and health effects is presented in Fig. 8.2. Given the modular nature of the model construction, this structure changes with consideration of the different scenarios. Thus, for example, if the tradeoffs model is run to evaluate erosion impacts, the outputs of the economic model include land-preparation practices and times of soil exposure. Instead of a leaching model, an erosion model would be utilized.

A Pesticide Use Scenario

In the case study areas, pesticide contamination of groundwater and occupational exposure to pesticides are priority problems. Farmers use a wide array of highly toxic organophosphates and carbamates as an effective first option for pest management in potato production. Pesticides are not used in dairy production, the principal alternative to potatoes (although per hectare income from potatoes greatly exceeds that of dairy). The agrochemicals have potential for accumulation in the environment and subsequent contamination of groundwater. The products are also neurotoxins, and acute or chronic exposure (principally dermal) can cause peripheral and central nervous system damage. Pesticide leaching was simulated with a mechanistic simulation model based on the LEACHP model (Wagenet and Hutson, 1989). Health effects of human exposure were measured with a World Health Organization (WHO) battery of neuropsychological tests (Crissman *et al.*, 1994). The tradeoffs examined compared agricultural production and pesticide loadings or neurological damage. In

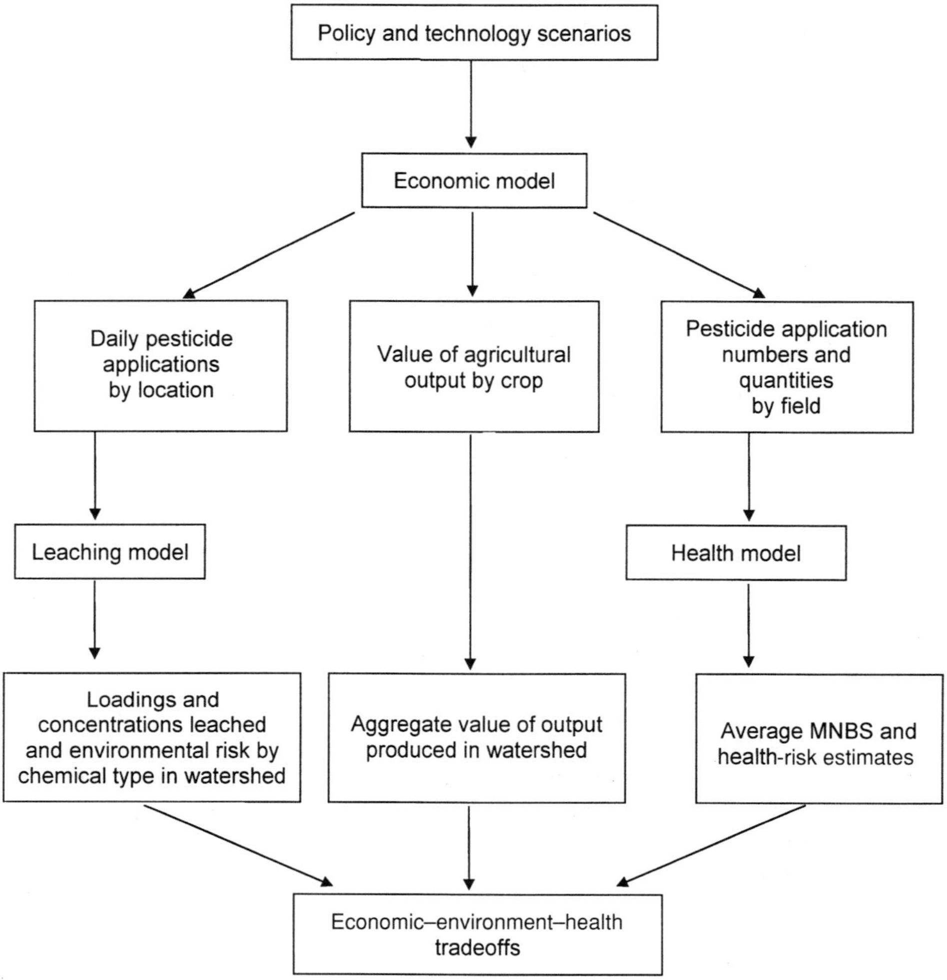

Fig. 8.2. Integrated model for tradeoff analysis. MNBS, mean neurobehavioural score.

the context of sustainable agricultural systems, these three factors can be defined as sustainability indicators for agriculture in the case-study areas.

A policy or technology scenario – e.g. increased pesticide prices or IPM technologies – is input into the economic model. The economic model randomly samples from distributions of physical and biological populations to select a field. This field carries with it soil, slope, climate and planting-date characteristics. By comparing the results of a net revenue function, a land-use choice is made. In this case, the choice is pasture for milk production or potatoes. The typical rotation is two 6-month potato crops followed by 2 years of pasture. If the choice is potatoes, a dynamic production function analysis then generates three types of output, which are used subsequently. First, pesticide applications are generated at a daily time step by field to be input into the pesticide leaching model. The pesticide leaching model generates outputs in the form of loadings into the environment (the total mass leached into groundwater) and water concentrations leached below the root zone. Secondly, the value of agricultural output by field is generated and saved so that it can be aggregated and used to construct aggregate tradeoffs for policy analysis. Thirdly, numbers and quantities of pesticide applications by field are generated to be input into the health component of the simulation model. The health simulations generate estimates of the effect of pesticide exposure on the farm population in terms of mean neurobehavioural score (MNBS) averaged over the population, and in terms of the risk that the MNBS exceeds a critical value. The model can follow the same field through numerous cropping cycles and select different fields in numbers determined by the analyst. The last step in the simulation model aggregates the economic, environmental and health outcomes to the watershed level so that aggregate tradeoffs across those outcomes can be analysed.

The statistical stochastic structure of the model is key to capturing the spatial variability in both biophysical processes and human behaviour. Neither representative farm optimization models nor aggregate models can capture the linkage between these spatial phenomena.

Tradeoff curves

To facilitate interpretation of the tradeoffs model results, the tradeoffs between agricultural, health and environmental criteria are presented in a series of pair-wise comparisons illustrated with tradeoff curves, a concept similar to the production possibilities frontier familiar to students in introductory economics courses. The hypothetical tradeoff curves in Fig. 8.3 illustrate the tradeoffs between agricultural output and environmental impact. A tradeoff curve represents all the possible pairs of outcomes for a given technology; thus different curves are available for different technologies. The movement from T1 to T2 shows a change in technology that, for each point on the *x* axis, maintains

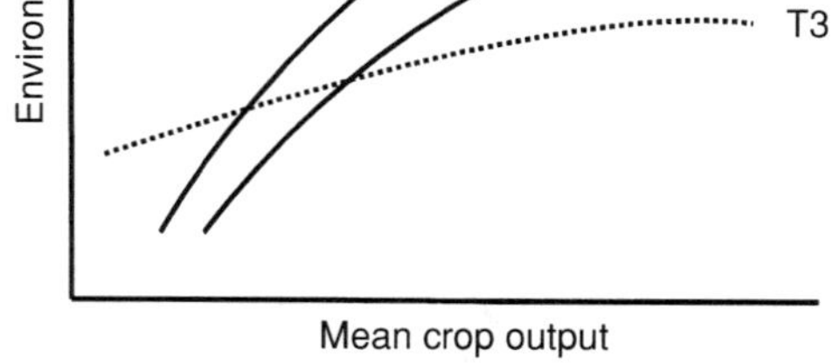

Fig. 8.3. Output–environment tradeoffs associated with alternative technologies.

output while reducing environmental impact. The slope of the tradeoff curve provides information about the opportunity cost of environmental quality in terms of forgone output. As curve T3 shows, different technologies may be more or less damaging at different levels of output. The factors on the axes are identified based on their relevance and importance to analysts and stakeholders and thus can be interpreted as indicators of sustainability.

Politicians implicitly use tradeoff curves every day. By the very nature of their job, they are concerned with winners and losers resulting from policy decisions. The tradeoff curve is simply a concrete expression of what is usually a mental calculation. In the example, the politician or analyst can readily see if the sacrifice of a single unit of environmental quality will result in a gain of a single unit or five units of agricultural production. Politically determined weights would then guide the decision as to whether the size of the sacrifice is acceptable.

An empirical curve is constructed by imposing different policy scenarios on the model. Recall that the model is stochastic; thus the outcomes of the model can be plotted as a scatter diagram and a curve fitted through the points. Figure 8.4 shows two tradeoff curves generated by the tradeoffs model. The curves show the relationship between the value of production and carbofuran leaching beyond the root zone in potato fields for the existing technology and for a projected IPM technology that reduces carbofuran use. Each separate point represents the average outcome of 30 different fields, each with five production cycles measured in kilograms of active ingredient leached and value expressed in Ecuadorian sucres, measuring total production value from the field. These values are presented in log form and the fitted curve shows a curvilinear response. The characteristics of each field are drawn from a joint distribution of climate and biophysical factors. This is repeated 30 times, giving the observed variability in response. A particular field may carry high or low agricultural

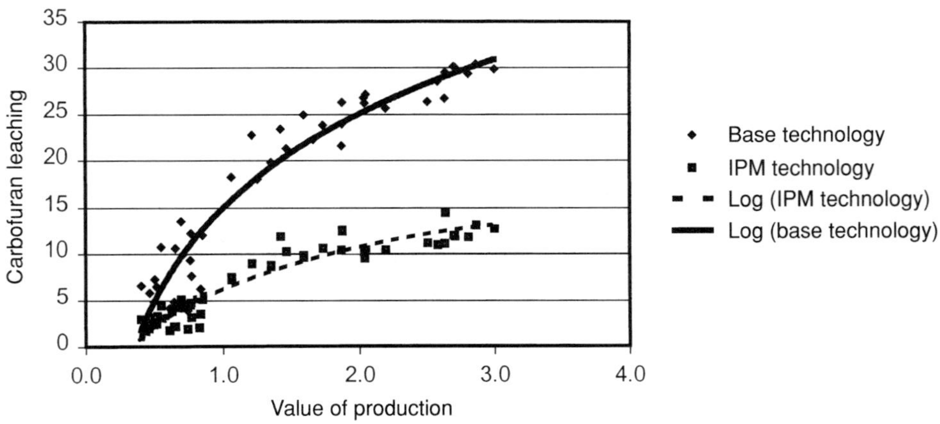

Fig. 8.4. Carbofuran leaching – value of production tradeoffs under present technology and projected IPM technology. Value of production is measured by log (1000 sucres) of total production value over five crop cycles and replications; carbofuran leaching is measured by log (kg of active ingredient) over five crop cycles and replications.

potential combined with high or low contamination potential. Note the spread of leaching at given levels of production. The randomly selected fields, combined with variability of climate information, produce results that capture the variability in real life. Recall that changing relative prices causes movements along a curve. Different levels of insecticide tax are imposed on the model. This tax causes a reduction in carbofuran use, with a consequent reduction in leaching and the value of agricultural production. The reduced value of production comes from two sources. Incomes are reduced because costs increase and because farmers dedicate a smaller area to potatoes. Increases in potato prices were also imposed on the model. The price rises increase the profitability of production and cause farmers to dedicate a greater area to potatoes and to make more profitable extra applications of carbofuran, so both leaching and the value of production increase. The regression fits a line through this scatter of points to provide a summary indication of the trend.

Figure 8.5 illustrates an *ex ante* analysis produced by the tradeoffs model of alternative policies and technological change, comparing human health risk and the value of production. The same set of scenarios of a changing tax on pesticides and prices on potatoes generates a scatter of points, with the base technology summarized in the uppermost line. A set of technology scenarios is then used in the tradeoffs model to create the set of lines depicted in the table in which the observed technology can be compared *ex ante* with possible technology alternatives. Available IPM technologies can significantly reduce pesticide use while maintaining low levels of crop losses. Safe handling practices, such as wearing impermeable capes and gloves, can significantly reduce dermal exposure and reduce adverse health effects. The impacts of different IPM technologies and safe handling practices shift the tradeoffs curve downward, reducing health risk while maintaining value of production. At high levels of agricultural production, the combination of IPM and improved safety can

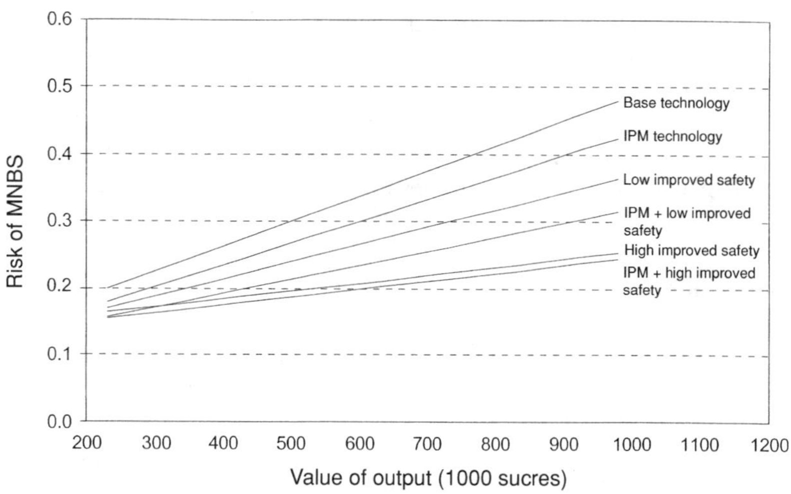

Fig. 8.5. Health–output tradeoffs for Andean weevil IPM and improved safety practices. MNBS, composite score of neurobehavioural performance. Risk is the chance of an individual scoring > 1 SD below the control population.

reduce the health risk by more than half. The analysis provides support for the role of research and education in comparison with simply taxing pesticides. The technology changes and safe handling practices are superior to taxes in reducing health risk while preserving the value of production.

The current version of the tradeoffs model consists of economic decision models linked to leaching models, as well as to crop growth simulation models in a Windows user interface (Antle *et al.*, 1999). The structure of the current version is described in Antle *et al.* (2000). The interface allows the user to set scenario levels. This software integrates field-scale GIS-based soils and climate data with the Decision Support System for Agrotechnology Transfer (DSSAT) suite of crop growth simulation models (IBSNAT, 1990), econometric-based economic simulation models of land use and management decisions, and environmental process models (leaching, runoff and erosion). The software provides the basis for drawing a statistically representative sample of fields in a region such as a watershed, conducting integrated analysis and statistically aggregating the results to a scale relevant to policy decision-making. Tradeoffs can then be displayed between competing or complementary policy objectives in simple two-dimensional graphs, and changing tradeoffs can be examined under alternative policy and technology scenarios. A long-term objective of the project is to make this model general enough for it to be applied to the analysis of tradeoffs in any agroecosystem. A related long-term objective is to develop the software and documentation for distribution to and use by analysts worldwide.

Methodological Issues

Linkages to geographical information systems (GIS)

The tradeoff model makes use of spatial data in the construction of the distributions of field characteristics; these data are then arranged using GIS techniques. The use of GIS aids in the systematization of relevant spatial data and permits statistically valid stratification of these data. This, in turn, creates the possibility of examining further 'what if'-type questions for land-use planning and for refining and targeting policy or technology. The model results can also be displayed with GIS techniques for additional illustrative impact.

Alternative approaches to model integration

Figures 8.4 and 8.5 show that economic, environmental and health processes need to be linked to quantify the relevant tradeoffs and to measure how these tradeoffs change in response to changes in technology or policy. In principle, this means that two or more sets of complex relationships – economic, behavioural, biophysical, geophysical, toxicological, etc. – must be integrated for the analysis. No matter how simplified and parsimonious these disciplinary models are, truly integrating them into one model that is larger and more complex poses serious design and implementation problems. The result can be difficult for disciplinary specialists to understand and apply, and can be time-consuming and unwieldy to use. Clearly, the principle of parsimony indicates that the simplest models adequate for the problem at hand should be used.

Our experiences with this type of analysis suggest that several different levels of model integration are possible, depending on the features of the economic and physical processes involved and on the types of technology or policy scenarios to be analysed. Here we provide examples of three levels of model integration that are possible in the context of agricultural–environmental interaction and discuss hypothetical economic and physical models.

Level I: economic and physical models are simulated independently; outputs are combined to infer environmental impact

Under certain conditions, environmental or health impacts can be represented as a function of management decisions, but the properties of this function are independent of any specific management decisions. This type of model integration is compatible with certain types of physical conditions, such as chemical leaching, where the physical processes do not feed back to affect production in future periods. It is also compatible with policy scenarios where changes in environmental conditions do not feed back to affect future production conditions.

Level II: an economic model is simulated for each policy scenario, and output of the economic model is used as input into the physical simulation model, but there is no feedback from physical processes in one production period to crop production or management decisions in future periods

In some cases, it is not reasonable to assume that environmental impacts can be accurately modelled independently from the management decisions. For example, the form of the relationship describing pesticide leaching beyond the root zone is generally dependent on the amount of pesticide applied and the soil surface moisture conditions. As in level I, this type of simulation modelling is compatible with physical and policy conditions where there is no dynamic feedback from environmental outcomes to future production decision-making.

Level III: an economic model is formally linked to the physical model, and the two are simulated jointly to represented dynamic feedback from environmental conditions to production

When either physical conditions or policy do dictate a feedback from environmental conditions to production, it is not possible to conduct economic simulations independently of physical simulations. For example, crop choice, fertilizer use, tillage or other management practices affect soil organic matter and soil productivity in future production periods. Under these conditions, full model integration is required to account for dynamic interactions.

Antle *et al.* (1998) illustrate how feedback occurs between pesticide application decisions and farmer health. In that analysis, pesticide use affects productivity and also affects farmer health, and farmer health affects productivity. The analysis considered the short-term interrelationship between certain acute and chronic health effects and productivity. In the longer term, there would be dynamic interactions between health and productivity. For example, some health effects of pesticides (such as neurological or dermatological effects) may be reversible, so a reduction in pesticide use may lead to an improvement in health over time, which would not be captured in a static analysis. Nor did the analysis consider longer-term effects of pesticides on reproduction or immunity, with consequences such as malformations or cancer.

Summary and Conclusions

The central focus of the research reported here is the development of a decision-support system that can provide information for policy-makers and others that improves their ability to assess tradeoffs among competing goals and how those tradeoffs might be changed through policy and technology interventions. The tradeoffs model provides a framework in which to structure input from stakeholders, such as farmers, research administrators, and local and national policy-makers. This information is used to define the indicators that are quantified as tradeoffs, and to define policy and technology scenarios to be evaluated.

Initial applications show that tradeoffs in issues important to sustainable agriculture systems can be empirically estimated. We show that tax or price policy more typically causes movement along a given curve and thus retains the 'winner–loser' setting. We show that technological solutions frequently provide at least an improvement in one factor, while the other factor is held constant. It is theoretically possible to show 'win–win' solutions with the tradeoffs analysis. One such example would be to compare the improvement in groundwater quality with the improved neurobehavioural function – the two factors measured on the vertical axes in Figs 8.4 and 8.5. By demonstrating the effects of the adoption of IPM practices, improvement in these two sustainability indicators was shown as possible. If demonstrated to a Minister of the Environment or Minister of Health, this 'win–win' solution could easily convince these political leaders to lobby the Minister of Agriculture to emphasize this technological alternative. Through its utilization of GIS-based information organization and the extrapolations that are available, the model can be used at various scales of policy analysis from community, to watershed, regional, national and transnational scales. We expect the principal clients of the model to be technology managers in agricultural and environmental research institutes and policy analysts in governmental and non-governmental organizations.

Note

1 Accessibility in this application is enhanced by the development of a Windows-format user shell and modular structure, where different modules can be inserted or removed depending on the issues of interest.

9 Balancing Regional Development Priorities to Achieve Sustainable and Equitable Agricultural Growth

PETER HAZELL AND SHENGGEN FAN

Environment and Production Technology Division, International Food Policy Research Institute (IFPRI), Washington, DC, USA

Introduction

Past agricultural development strategies have emphasized irrigated agriculture and 'high-potential' rain-fed lands in an attempt to increase food production and stimulate economic growth. This strategy has been spectacularly successful in many countries and was responsible for the Green Revolution. At the same time, however, large areas of less-favoured lands have been neglected and lag behind in their economic development. These lands are characterized by lower agricultural potential, often because of poorer soils, shorter growing seasons and lower and uncertain rainfall, but also because past neglect has left them with limited infrastructure and poor access to markets. Despite some out-migration to more rapidly growing areas, population size continues to grow in many less-favoured areas, and this growth has not been matched by increases in yields. The result is worsening poverty and food insecurity problems, and widespread degradation of natural resources (e.g. mining of soil fertility, soil erosion, deforestation and loss of biodiversity) as people seek to expand cropped areas.

Less-favoured lands are extensive in the developing world. According to a report prepared by the Consultative Group of International Agricultural Research/Technical Advisory Committee (CGIAR/TAC, 1998), 'marginal' and sparsely populated arid lands account for 75% and 85%, respectively, of the total agricultural area in Asia and sub-Saharan Africa. Their shares in total agricultural production are lower but still large. In China and India, for example, we estimate that less-favoured lands account for about one-third and 40% of total agricultural output, respectively. Globally, some 500 million poor people live in less-favoured lands (Hazell and Garrett, 1996).

It is becoming increasingly clear that, on poverty and environmental grounds alone, more attention will have to be given to less-favoured lands in setting priorities for policy and public investments. Social and environmental crises are already common in many less-favoured areas, and some governments

(eds D.R. Lee and C.B. Barrett)

and donors are spending more resources on crisis relief than on development in these areas (Owens and Hoddinott, 1998). This leads to two key policy questions. First, what level of investment can be justified in less-favoured lands? Secondly, how should the resources allocated to less-favoured lands be used to promote development that is beneficial to the poor and environmentally sustainable? This chapter addresses these two issues.

Returns to Public Investments in Less-favoured Areas

The amount of public investment allocated to any region should be guided to the greatest extent possible by the net social returns that are realized through productivity growth, poverty reduction and the containment of environmental degradation. While 'win–win–win' investments are usually more desirable, an investment that involves some tradeoff between these social goals may also be attractive, providing any sacrifice of one goal is adequately compensated by gains in achieving other goals.

Conventional wisdom suggests that the productivity returns to investment are highest in irrigated and high-potential rain-fed lands and that growth in these areas also has substantial trickle-down benefits for the poor, including those residing in less-favoured areas. Even though investing in less-favoured lands might have a greater direct impact on the poor living in those areas, it is argued that investments in high-potential areas give higher social returns for a nation than investments in low-potential areas. The logic behind this position is as follows. Investment in high-potential areas generates more agricultural output and higher economic growth at lower cost than in less-favoured areas. Faster economic growth leads to more employment and higher wages nationally, and greater agricultural output leads to lower food prices, both of which are beneficial to the poor. Less-favoured areas will benefit from cheaper food, from increased market opportunities for growth and from new opportunities for workers to migrate to more productive jobs in the high-potential areas and in towns. Fewer people will try to live in less-favoured lands, and this will help reduce environmental degradation and increase per capita earnings. Migrants may also send remittances back to relatives in less-favoured areas, further increasing per capita incomes there, especially for the poor.

Many of the expected benefits arising from rapid agricultural growth in high-potential areas have been confirmed (Pinstrup-Andersen and Hazell, 1985; Hazell and Ramasamy, 1991; David and Otsuka, 1994). Nevertheless, the rationale for neglecting less-favoured areas is being increasingly challenged by: (i) the failure of past patterns of agricultural growth to resolve growing poverty, food insecurity and environmental problems in many less-favoured areas; (ii) increasing evidence of stagnating levels of productivity growth and worsening environmental problems in many high-potential areas (Pingali and Rosegrant, 1994; Pingali *et al.*, 1997); and (iii) emerging evidence that the right kinds of investments can increase agricultural productivity to much higher levels than previously thought in many less-favoured lands (Scherr and Hazell, 1994; Reil *et al.*, 1996; Conway, 1997; Pender *et al.*, 1999a). It now seems plausible that increased public investment in many less-favoured areas may have the potential to generate competitive, if not greater, agricultural growth on the margin than comparable investments in many high-potential areas, and that these investments could have a greater impact on the poverty and environmental problems of the less-favoured areas in which they are targeted. If so, additional investments in less-favoured areas may actually give higher aggregate social returns to a nation than additional investments in high-potential areas. In fact, they might even offer 'win–win–win' possibilities.

To test this hypothesis, we analysed the growth and poverty alleviation impacts of alternative types of investments in high- and low-potential areas in India over recent decades.[1] Unfortunately, the available data only permit a more speculative assessment of the environmental impacts of public investments in rural India. India is a good example because, like many other Asian countries, past public investments have been biased towards high-potential areas, and the remarkable productivity gains achieved in those areas (which led to national food surpluses) can now be juxtaposed against the lagging productivity and widespread poverty, food insecurity and environmental degradation that exists in many less-favoured rain-fed areas. The results provide strong support for our hypothesis that greater levels of investment in less-favoured lands are warranted, at least based on growth and poverty-alleviation grounds.

Contrasting Returns to Public Investments in India

Land classification

Our analysis is based on three categories of land in India: irrigated and high- and low-potential rain-fed areas. Following Kerr (1996), we classified districts as irrigated if more than 25% of the cropped area (averaged from 1970 to 1995) is irrigated and as rain-fed if the irrigated share is less than 25%. We further subdivided rain-fed areas into high- and low-potential areas according to their agroecological characteristics. For this purpose, we used the classification scheme of the Indian Council of Agricultural Research (NBSS and LUP, 1992), which divides India into 20 agroecological zones, based on soils and climate. The district data available to us cover 18 of these zones; the remaining zones are mostly mountain, island and desert areas, which account for only a small share of national agricultural output. We considered all rain-fed districts falling in zones with poor soils, short growing periods and low rainfall as low-potential in this study and all other rain-fed districts as high-potential areas. Defined in this way, irrigated and high- and low-potential rain-fed areas account for 46, 17 and 35% of the total cropped area and 56, 17 and 27% of total agricultural output, respectively.

Table 9.1. Production and productivity growth.

Year	Production growth index			Land productivity (Rs ha^{-1}, 1990 prices)			Labour productivity (Rs $worker^{-1}$, 1990 prices)			Total factor productivity		
		Rain-fed			Rain-fed			Rain-fed			Rain-fed	
	Irrigated	High potential	Low potential	Irrigated	High potential	Low potential	Irrigated	High potential	Low potential	Irrigated	High potential	Low potential
1970	100	100	100	5046	3478	2287	3645	3337	2881	100	100	100
1971	100	102	96	5037	3513	2202	3613	3367	2746	100	102	96
1972	93	101	78	4735	3475	1819	3366	3306	2224	93	101	78
1973	96	96	104	5015	3340	2474	3437	3112	2910	95	95	102
1974	96	99	99	4994	3539	2233	3431	3168	2733	95	98	97
1975	107	112	117	5620	3986	2657	3802	3528	3186	106	110	114
1976	105	100	111	5545	3552	2519	3700	3105	2977	104	98	107
1977	116	115	119	6140	4088	2713	4037	3526	3146	114	112	115
1978	122	119	121	6474	4174	2748	4198	3590	3149	119	116	116
1979	103	98	114	5447	3433	2592	3517	2917	2917	101	95	109
1980	120	124	118	6462	4471	2713	4047	3626	2949	116	119	111
1981	127	128	129	6860	4568	2942	4217	3683	3150	122	123	121
1982	129	130	126	6855	4674	2864	4189	3674	3023	123	124	117
1983	141	154	144	7531	5518	3292	4507	4266	3394	134	146	133
1984	143	152	136	7571	5504	3084	4460	4103	3121	134	143	125

1985	151	179	120	8115	6471	2744	4632	4751	2703	142	167	109
1986	146	162	120	7927	5772	2775	4395	4214	2635	136	150	108
1987	149	171	126	8160	6145	2932	4371	4345	2710	138	157	112
1988	172	183	157	9609	6630	3782	4952	4554	3305	159	168	139
1989	169	185	156	9101	6637	3520	4765	4508	3203	155	169	137
1990	170	195	148	9209	6992	3338	4688	4651	2954	155	178	129
1991	170	194	154	9154	7198	3470	4588	4507	2996	155	175	134
1992	173	207	168	9290	7699	3787	4585	4707	3178	157	186	145
1993	178	222	166	9439	8223	3693	4605	4934	3051	160	199	142
1994	189	233	171	9963	8469	3845	4770	5069	3057	169	208	146
Annual growth rate												
1970–80	1.88	2.15	1.67	2.50	2.54	1.72	1.05	0.83	0.23	1.53	1.79	1.09
1980–90	3.49	4.68	2.26	3.61	4.57	2.10	1.48	2.52	0.02	2.93	4.05	1.48
1990–94	2.67	4.47	3.77	1.99	4.91	3.60	0.44	2.18	0.86	2.13	4.03	3.06
1970–94	2.68	3.58	2.26	2.88	3.78	2.19	1.13	1.76	0.25	2.21	3.10	1.58

Agricultural growth

As a result of the rapid adoption of new technologies (particularly high-yielding varieties, fertilizers and irrigation) and increases in rural infrastructure, agricultural production and factor productivity have grown rapidly in Indian agriculture in recent decades (Table 9.1). Five major crops (rice, wheat, sorghum, pearl millet and maize), 14 minor crops (barley, cotton, groundnut, other grain, other pulses, potato, rapeseed, mustard, sesame, sugar, tobacco, soybeans, jute and sunflower) and three major livestock products (milk, chicken, and sheep and goat meat) are included in our measure of total output. Unlike traditional measures of production growth, which use constant output prices, we used the more appropriate Törnqvist–Theil index (a discrete approximation to the Divisia index) to calculate output.[2]

For the period 1970–1994, agricultural production grew fastest in the high-potential rain-fed areas (3.58% year^{-1}), followed by irrigated areas (2.68% annually) and then low-potential rain-fed areas (2.26% annually). This may reflect a 'catching-up' effect, since irrigated production grew rapidly prior to 1970 as a result of the Green Revolution, and the use of high-yielding varieties and fertilizers spread more slowly to rain-fed areas. Production growth in irrigated and high-potential rain-fed areas slowed in the early 1990s, whereas it increased in the low-potential rain-fed areas to 3.77% per year, more than double the rate of growth achieved in the 1970s.

Land productivity, measured as the gross value of output in rupees (1990 prices) per hectare of net cropped area, also grew fastest on average in the high-potential rain-fed areas, but the most rapid growth in these areas occurred after 1980, when it increased to 4.5–5.0% year^{-1} (Table 9.1). Average growth in land productivity has also been quite high in irrigated areas (2.88% year^{-1} since 1970), though it slowed to less than 2% year^{-1} after 1990. In contrast, low-potential rain-fed areas experienced the slowest average growth in land productivity between 1970 and 1990, but growth has accelerated to 3.6% year^{-1} since then. Measured land productivity remains more than twice as high in irrigated and high-potential rain-fed areas than in low-potential areas, and this gap has widened since 1970.

Growth in labour productivity has been consistently low in all types of Indian agriculture since 1970, averaging only 1.13, 1.76 and 0.25% year^{-1} in irrigated and high- and low-potential rain-fed areas, respectively (Table 9.1). It has accelerated a little in the low-potential rain-fed areas since 1990 (only to 0.86% year^{-1}), but has slowed in irrigated and high-potential rain-fed areas. Given that the welfare of the rural poor can be expected to be closely linked to labour productivity in agriculture, these trends should be a matter of considerable concern to Indian policy-makers.

Total factor productivity (TFP), a measure of the return to all direct and indirect inputs used in agriculture, grew fastest in high-potential rain-fed areas during 1970–1994 (3.1% year^{-1}), followed by irrigated areas (2.21% year^{-1}) and then low-potential rain-fed areas at 1.58 % year^{-1} (Table 9.1).[3] TFP growth has slowed in irrigated areas since 1990, remained unchanged at nearly 4% year^{-1} in high-potential rain-fed areas and accelerated to 3.06% $year^{-1}$ in low-potential rain-fed areas.

Rural poverty

Table 9.2 shows the incidence of rural poverty for the three land types defined above in 1972, 1987 and 1993. Poverty has been measured as the percentage of the rural population falling below the official poverty line (Rs 15 per capita per month, calculated at 1960/61 prices). In 1993, there were 184 million rural poor in the areas covered by our data set, representing about two-thirds of India's rural poor. This total had hardly changed since 1972, when there were 192 million rural poor. Of the 184 million rural poor in 1993, 154 million (84%) lived in rain-fed areas. These were distributed about equally between high- and low-potential rain-fed areas, a feature that has also not changed since 1972. The density of poor people is highest in the high-potential rain-fed areas: 1629 poor people 1000 ha^{-1} of geographical area in 1993, compared with 705 in irrigated areas and 599 in low-potential rain-fed areas. The percentage of the rural population living in poverty is also highest in the high-potential rain-fed areas (44% in 1993), and lowest in the irrigated areas (28%). These patterns reflect interregional migration over time and regional differences in per capita resource availability and labour productivity. The poverty shares declined by 25–30% between 1972 and 1993 in all three types of areas, but more because of population growth than because of any decline in the number of rural poor. The large number of rural poor remaining in rain-fed areas represents a continuing

Table 9.2. Changes in rural poverty by type of region in India (authors' calculations, based on data from Dreze and Srinivasan (1996) and the Government of India).

			Rain-fed areas		
		Irrigated areas	Total	High potential	Low potential
Percentage of poor in total population (%)	1972	39	52	59	47
	1987	32	46	48	44
	1993	28	39	44	36
Number of poor (millions)	1972	37	155	80	75
	1987	35	167	79	88
	1993	30	154	78	76
Number of poor per 1000 ha geographical area	1972	862	880	1680	583
	1987	813	951	1660	688
	1993	705	878	1629	599

Data are only available for 47 agroclimatic zones (of total 65). An agroclimatic zone is defined as rain-fed if less than 40% of the total cropped area is irrigated, and otherwise as irrigated.

challenge for India's policy-makers, and highlights the importance of investing more in these areas.

The model

The model used to estimate the effects of rural infrastructure and technologies on production growth and rural poverty is given in equations (1)–(7).

$$Y_{i,t} = f(LABOUR_{i,t}, LAND_{i,t}, FERT_{i,t}, MACH_{i,t}, ANIMALS_{i,t}, HYV_{i,t}, GIRRI_{i,t}, PIRRI_{i,t}, LITE_{i,t}, ROAD_{i,t}, MKT_{i,t}, ELECT_{i,t}, T), \quad (1)$$

$$HYV_{i,t} = f(ROADS_{i,t}, LITE_{i,t}, GIRRI_{i,t}, PIRRI_{i,t}, ELECT_{i,t}, ATT_{i,t}), \quad (2)$$

$$GIRRI_{i,t} = f(\mathrm{ROADS}_{i,t}, LITE_{i,t}, ELECT_{i,t}, ATT_{i,t}), \quad (3)$$

$$PIRRI_{i,t} = f(ROADS_{i,t}, LITE_{i,t}, ELECT_{i,t}, GIRRI_{i,t}, ATT_{i,t}), \quad (4)$$

$$P_{i,t} = f(Y_{i,t}, WAGE_{i,t}, TT_{i,t}, T), \quad (5)$$

$$WAGE_{i,t} = f(Y_{i,t}, ROADS_{i,t}, LITE_{i,t}, ELECT_{i,t}) \quad (6)$$

$$TT_{i,t} = f(Y_{i,t}, Y_{n,t}) \quad (7)$$

Equation (1) is the production function. The dependent variable is the Törnqvist–Theil index of agricultural output ($Y_{i,t}$), while explanatory variables include traditional farm inputs, such as labour ($LABOUR_{i,t}$), land ($LAND_{i,t}$), fertilizer ($FERT_{i,t}$), machinery ($MACH_{i,t}$) and draught animals ($ANIMAL_{i,t}$); technology inputs like the percentage of high-yield varieties (HYVs) in total cropped area ($HYV_{i,t}$), the percentage of the total cropped area that is under canal irrigation ($GIRRI_{i,t}$) and private irrigation ($PIRRI_{i,t}$); infrastructure variables, such as road density ($ROAD_{i,t}$), development of rural markets ($MKT_{i,t}$) and rural electrification ($ELECT_{i,t}$); literacy rate of the rural population ($LITE_{i,t}$); and a time-trend variable that is intended to capture anytime-related changes not captured elsewhere.

Our specification of technology and infrastructure variables is incomplete, but these are the key ones for which we have district-level data. Due to the endogeneity of HYVs and irrigation variables in the production function, irrigation (public and private) and the adoption of high-yielding varieties are also modelled as endogenous variables in equations (2), (3) and (4). The explanatory variables here are lagged output prices ($ATT_{i,t}$), measured as the previous 5-year moving average of agricultural prices divided by a relevant gross national product (GNP) deflator, and social and infrastructure variables, such as rural education and road density. The estimation of these equations also enables us to measure the indirect impact of rural infrastructure on agricultural production through improved technologies.

Poverty ($P_{i,t}$), measured as the percentage of the rural population falling below the official poverty lines, is modelled in equation (5) as a function of agricultural production, wages and a terms-of-trade variable (TT), measured as agricultural prices divided by a relevant GNP deflator. Increases in agricultural output or wages should help the poor to increase their incomes and hence reduce poverty. Improvements in the terms of trade for agriculture can be expected to benefit many farmers, but by raising the price of food they can also increase poverty in the short run. However, in the long run, improvements in the terms of trade should stimulate greater investment in irrigation and HYVs, which in turn can be beneficial to the poor.

Equation (6) models wages as a function of agricultural output, roads, electrification and rural literacy. All four variables are expected to contribute to higher wages. Roads have an indirect impact on wages through agricultural growth (equation 1), but may also contribute directly by promoting growth of the rural non-farm economy (not modelled explicitly here) and through enhancing commuting opportunities.

The agricultural terms of trade are modelled in equation (7) at the district level. They are expected to decline with increases in agricultural production at the national ($Y_{n,t}$) and district ($Y_{i,t}$) levels. We initially included some demand-side variables (population size and per capita income), but these turned out not to be statistically significant and were dropped from the model.

As specified above, estimation biases could arise if government investments in technology and rural infrastructure are systematically targeted at different kinds of regions (e.g. if the government gives higher priority to high-potential areas). In this case, the investment variables may be correlated with the error terms in the production and poverty equations (Binswanger *et al.*, 1989). We do not have enough exogenous variables to fully resolve this problem, and have attempted to reduce the size of the possible estimation biases by including annual rainfall in many of the equations. We expect rainfall to act as a proxy for agricultural potential, which is presumed to drive any selection bias in government decisions. We also added district dummy variables in all equations to capture any remaining fixed effects of agroecological characteristics.

Estimates of the production and poverty equations

The estimation for all the equations except the poverty equation was based on a pooled time series (25 years: 1970–1994) and cross-sectional (250 districts) data set. Because we are interested in differences in the impacts of infrastructure and technology variables across different types of lands in India, a variable coefficients model was estimated. This is equivalent to adding slope dummies for each variable. Since our poverty data are only available for 47 agroclimatic regions and for 3 years (1972, 1987 and 1993), a two-step procedure was used in estimating the full equation system. The first step involved estimating all the equations without the poverty equation, using the district-level data from 1970 to 1994. Then the values of *Y*, *WAGE* and *TT* at the agroclimatic zone level were predicted using the estimated parameters. The second step involved estimation of the poverty equation, using the predicted values of the independent variables at the agroclimatic zone level and the available poverty data for 1972, 1987 and 1993.

The results reported here were derived from double-log form specifications of all the equations. As such, all coefficients are in elasticity form. For brevity, we only report the results of the production and poverty equations (see Fan and Hazell (1999) for a full report on the analysis); these are reported separately for irrigated and high- and low-potential areas.

The results for the production function estimation in Table 9.3 are satisfactory from a statistical perspective: most of the positive coefficients are significant at the 5% confidence level or better; only two of the negative coefficients obtained are statistically significant, and the R^2 value for the pooled regression is good at 0.972.

The coefficients in the poverty equation in Table 9.4 are all signed according to expectations. Higher production growth and higher wages both contribute to reductions in poverty (hence the negative signs), while increases in the agricultural terms of trade are harmful to the poor in the short run.

Table 9.3. Estimated production functions by type of region, India.

	Irrigated areas		High-potential rain-fed areas		Low-potential rain-fed areas	
	Coefficient	*t* value	Coefficient	*t* value	Coefficient	*t* value
Labour	0.254	7.16*	0.373	8.40*	0.089	3.59*
Land	0.470	9.46*	0.097	4.86*	0.278	6.51*
Fertilizer	0.002	0.20	0.004	0.32	0.039	2.36*
Machinery	0.008	1.48	0.016	2.34*	0.025	4.03*
Animals	0.037	2.81*	0.053	3.28*	0.080	4.81*
HYV	0.005	0.43	0.013	2.33*	0.061	5.54*
Roads	0.091	3.69*	−0.004	−1.02	0.167	5.50*
Market	−0.001	−0.15	−0.016	2.11*	−0.019	−1.46
Canal irrigation	0.028	2.09*	0.071	3.97*	0.002	0.52*
Private irrigation	0.039	2.25*	−0.042	−0.69	0.089	4.47*
Electrification	−0.069	−4.69*	0.007	1.33	0.032	1.86
Education	−0.025	−1.05	−0.029	−1.25	−0.003	−1.05
Time trend	0.025	14.41*	0.019	12.88*	0.042	3.96

R^2 = 0.972 and * indicates statistically significant at the 5% confidence level.

Table 9.4. Estimated poverty functions by type of region, India.

	Irrigated areas		High-potential rain-fed areas		Low-potential rain-fed areas	
	Coefficient	*t* value	Coefficient	*t* value	Coefficient	*t* value
Production growth	−0.160	−1.83	−0.170	−1.75	−0.310	−2.18*
Wage	−0.157	−1.99*	−0.161	−2.01*	−0.022	−1.41
Terms of trade	0.258	1.48	0.268	1.41	0.123	1.97*

R^2 = 0.757 and * indicates statistically significant at the 5% confidence level.

Marginal returns

The estimated elasticity coefficients in Tables 9.3 and 9.4 give the direct impact of each infrastructure and technology variable on agricultural production and poverty. But the full model captures indirect as well as direct impacts. For example, roads not only contribute directly to agricultural production in equation (1), but also affect the adoption of HYVs (equation 2) and the levels of investment in irrigation (equations 3 and 4); these variables in turn also have an impact on agricultural production. To capture the full impact of each infrastructure and technology variable on production and poverty requires totally differentiating the full equations system with respect to each variable of interest. The marginal impacts obtained from these calculations are shown in Table 9.5, where the derivatives were evaluated using the 1994 values for all relevant variables.

For every investment, the highest marginal impact on production and poverty alleviation occurs in one of the two rain-fed lands, while irrigated areas rank second or last. Moreover, many types of investments in low-potential rain-fed lands give some of the highest production returns, and all except markets and education have some of the most favourable impacts on poverty. These results provide strong support for the hypothesis that more investment should now be channelled to less-favoured areas in India.

The marginal impact of HYVs on production is much larger in high- and low-potential rain-fed areas than in irrigated areas, and they contribute much

Table 9.5. Marginal returns to infrastructure and technology inputs by type of region, India, 1994.

		Irrigated areas	High-potential rain-fed areas	Low-potential rain-fed areas
Economic returns to production (1990 prices)				
HYV	Rs ha^{-1}	63	243	688
Roads	Rs km^{-1}	100,598	6,451	136,173
Markets	Rs number^{-1}	(276,745)	7,808,112	(4,794,073)
Canal irrigation	Rs ha^{-1}	938	3,310	1,434
Private irrigation	Rs ha^{-1}	1,000	(2,213)	4,559
Electrification	Rs ha^{-1}	(546)	96	1,274
Education	Rs labour^{-1}	(360)	571	102
Returns to poverty reduction				
HYV	Persons ha^{-1}	0.00	0.02	0.05
Roads	Persons km^{-1}	1.57	3.50	9.51
Markets	Persons number^{-1}	(2.62)	537.79	(313.72)
Canal irrigation	Persons ha^{-1}	0.01	0.23	0.09
Private irrigation	Persons ha^{-1}	0.01	(0.15)	0.30
Electrification	Persons ha^{-1}	0.01	0.07	0.10
Education	Persons labour^{-1}	0.01	0.23	0.01

The numbers in parentheses are negative; in most cases they are not statistically significant.

more to poverty alleviation in the rain-fed areas. Roads have a large productivity effect in all three types of lands, but a much larger impact on poverty alleviation in rain-fed areas, particularly the low-potential rain-fed areas. Rural electrification and education have their largest productivity impacts in rain-fed areas, and they also have a favourable impact on the poor in these areas. Their impacts in irrigated areas are very small. Canal irrigation has its largest productivity and poverty impacts in high-potential rain-fed areas, while private irrigation has its most significant impacts in low-potential rain-fed areas. Market development has a huge marginal impact on production and poverty alleviation in the high-potential rain-fed areas, but not in irrigated and low-potential rain-fed areas.

It should be noted that the marginal impacts of different investments reported above are gross of their costs. It could be argued that some investments are more expensive to undertake in less-favoured areas because of their diverse and generally less favourable agroecological conditions. For example, the development of HYVs may be more difficult for less-favoured areas, and their widespread adoption may be more constrained by the diversity of growing conditions. Investments in roads and other infrastructure may also be more costly in many less-favoured lands because of difficult topographical conditions or remoteness from major population centres or markets. Data that we have obtained at the state level for India (Fan *et al.*, 1998) suggest that the unit costs of key investments are not all that different across states, despite considerable diversity in the proportions of their irrigated and rain-fed areas. But further analysis at more disaggregated levels is required before we can be sure that our marginal results hold when measured net of investment costs.

Another possible limitation of our analysis is that we have not allowed for the potential spillover effects of agricultural growth in one type of land on others. For example, agricultural production growth in any one type of area can contribute to poverty reduction in other areas by helping to lower national food prices. Also, agricultural growth in one area that generates additional employment may attract migrants from other areas, with potential benefits to the poor in those other areas. These spillover effects can be ignored if they are similar for all regions per unit of investment (they simply offset each other). In our results, the rain-fed areas give the greater production returns per unit of investment, and hence are likely to generate larger spillover benefits than irrigated areas.

It should again be noted that our econometric analysis has ignored the environmental impacts of public investments, due to the lack of comparable district-level data. Resource degradation problems are widespread in rain-fed areas in India, and have already reached the point where productivity has fallen in many areas. In a recent review of the evidence, Kerr (1996) identifies soil erosion, falling groundwater levels and the degradation of community forests and grazing pastures as the most widespread problems for rural people. Soil erosion is depressing yields in many areas, or yields can be maintained only by using more fertilizers. Falling groundwater levels are increasing the costs of well irrigation and the likelihood of water shortages. Less than half of the officially forested land still bears trees and pastures have been seriously overgrazed. The resulting shortage of fuel and biomass is contributing to worsening soil fertility problems, as well as to the destruction of remaining forest.

Environmental degradation in less-favoured areas is driven by many factors, but the more important are population growth, poverty, increasing landlessness and near-landlessness, ineffective regulation and management of common properties (including groundwater) and inadequate tenure security over rented lands. Tenancy is even prohibited in some states, so that, while it continues to exist, it remains illegal and provides insufficient incentives for tenants for sustainable land management. Ongoing solutions to these problems include public investments in soil and water conservation, watershed development, social forestry, rehabilitation of wastelands and the creation of more effective institutions for managing groundwater, forests and pastures (Kerr, 1996; Kerr *et al.*, 1998a). Many of these investments increase productivity even as they resolve environmental problems, and they reach many of the poorest communities in India. They are therefore largely complementary to other public investments designed to increase growth and reduce poverty.

Strategies for Developing Less-favoured Lands

Many past attempts to develop less-favoured lands (e.g. integrated rural development projects) were not very successful, including the results of some of the research undertaken by the international agricultural research centres. The reasons for this are complex, but include: inappropriate macroeconomic, trade and sector policies that penalized agriculture; insufficient levels of investment in agricultural research in less-favoured areas and research of the wrong type; insufficient investments in rural infrastructure and human capital targeted on less-favoured areas; inappropriate development strategies that neglected smallholders; and poor performance by and coordination among many public-sector institutions working in less-favoured areas. Structural adjustment programmes are creating a more enabling economic environment for the development of many less-favoured lands, although drastic reductions in government budgets are also constraining the needed expansion of public investment in these lands. But future investments in less-favoured lands need to be based on new or improved paradigms for sustainable development.

Less-favoured areas are very diverse in their agroclimatic conditions and hence in their potential for agricultural growth. In some areas, agricultural development may not be an economically viable alternative, and solutions will have to be sought through development of the rural non-farm economy and through accelerated out-migration. Possibilities for achieving these alternatives are most promising when the national economy is growing rapidly and when agriculture has become a relatively minor share of national income and employment. Prospects are much less promising in stagnant and predominantly agrarian economies, since the rural non-farm economy is then constrained by local demand for its output, which in turn is constrained by the level of per capita incomes. Without agricultural growth, incomes in these areas remain low and the demand for rural non-farm goods and services remains stagnant (Hazell and Reardon, 1998).

Migration and non-farm diversification will also have to play an important role in the long run for many types of less-favoured areas if their per capita incomes are to keep pace with increasing national living standards. This longer-term view needs to be kept in mind when developing strategies for the short to medium term, particularly when those strategies are focused on alleviating poverty and environmental problems. Policy-makers need to avoid inadvertently locking too many people into marginal areas where their long-term prospects are limited.

However, for many less-favoured lands, agricultural intensification must be a key component of their development strategy, particularly over the next few decades while their population densities continue to grow. But, because of poor infrastructure, low to moderate yield potential, fragile soils and high climate risk, the strategy will typically need to be different from the Green Revolution approach adopted in irrigated and high-potential rain-fed areas. Agricultural development strategies need to be tailored to local agroclimatic conditions and to the type of development pathway that local communities are following (see Pender *et al.*, Chapter 10 of this volume). Nevertheless, some key elements of appropriate agricultural intensification strategies for many less-favoured areas can be given.

Promote broad-based agricultural development

Broad-based agricultural development that reaches small and medium-sized family farms, as well as larger commercial farms, should be promoted. There are few economies of scale in agricultural production in developing countries (unlike many agricultural processing and marketing activities); hence targeting family farms is attractive on both equity and efficiency grounds. Broad-based development strategies require that small and medium-sized farms receive priority in publicly funded agricultural research and extension and that they obtain adequate access to markets, credit and input supplies. These requirements demand special attention at a time when markets and agricultural services are being privatized, since the high transport costs and thin markets of many less-favoured areas do not make them attractive to private agents. Special attention must also be given to women farmers, who have traditionally been discriminated against in their access to resources and improved technologies, credit and farm inputs.

Improve technologies and farming systems

Because of their poor infrastructure, low yield response and high climate risk, the intensive use of modern inputs, such as fertilizers, is unlikely to be economic in many less-favoured lands. Moreover, monocrop farming systems can be environmentally destructive, as well as too risky. Agricultural researchers and farmers need to step back from narrow commodity approaches and take a more holistic approach to improving resource management practices at the farm and landscape levels. These practices may need to include: management at the watershed level of water catchment and use and soil erosion control; improved soil moisture and fertility management, including improved crop rotations and intercropping, and better integration of farm trees and livestock into cropping systems to generate and recycle plant nutrients; and more rational exploitation of favourable niches in the landscape for production of high-value crops and trees. This will require research to be more multidisciplinary, more site-specific and more responsive to farmer and community needs.

Ensure secure property rights and effective institutions

Farmers need assured long-term access to land if they are to pursue sustainable farming practices and to make long-term investments in improving and conserving resources (e.g. tree planting, continuous manuring, and terracing and contouring for soil and moisture conservation). Many of the indigenous land-tenure systems to be found in less-favoured lands do provide farmers with adequate tenure security, and they seem to evolve quite efficiently towards greater privatization of rights as population and commercialization pressures increase (Migot-Adholla *et al.*, 1991; Place and Hazell, 1993). In these cases, the appropriate role for governments is not to abruptly replace the indigenous systems, but to seek ways of strengthening them and facilitating their adaptation to changing circumstances. Legal registration of blocks of community or village-held land and simple voluntary systems for recording land transactions and resolving disputes may sometimes help increase security by reducing land disputes between and within communities. In contrast, land registration may only be economically worthwhile in areas of high population density and/or commercialized agriculture, particularly when formal lending institutions are also well developed and land is already effectively privatized. It may also be required in newly settled areas where there are no indigenous land tenure systems and disputes over ownership and boundaries are common.

Many resources are owned and managed as common properties in less-favoured lands (e.g. wood lots, grazing areas and wetlands), because this provides a more effective way to share risks and to ensure equitable access to resources by all members of the community. If these resources are to remain common properties and are not to be privatized or overexploited, effective local organizations are needed to manage them. Often, governments have undermined indigenous institutions by nationalizing important common-property resources, such as forests and rangelands. Public institutions have then failed to manage these resources effectively and they have degenerated into open-access areas. The most successful institutions for managing common properties are local organizations dominated by the resource users themselves.

Conserving or improving natural resources often requires collective action by groups of users, even when the resources are not commonly owned. Examples include organizing adjacent farmers in a landscape to invest labour in land terracing, bunding or water catchment or for biological pest control. Organizing farmers into effective and stable groups for collective action is difficult and success is conditioned by a range of physical, social and institutional factors (Uphoff, 1986; Ostrom, 1994; Rasmussen and Meinzen-Dick, 1995). Collective action is facilitated by a smaller number of users, by homogeneity of members in terms of shared values and economic dependence on the activities of the group and if the net benefits from group membership are substantial and equitably distributed. Institutional design is also important. Ostrom (1994) has identified seven design principles for effective local organizations: (i) there must be a clear definition of the members and the boundaries of any resource to be managed or improved; (ii) there should be a clear set of rules and obligations that are adapted to local conditions; (iii) members should collectively be able to modify those rules to changing circumstances; (iv) there should be an adequate monitoring systems in place, with (v) enforceable sanctions, preferably graduated to match the seriousness and context of the offence; (vi) there should be effective mechanisms for conflict resolution; and (vii) the organization, if not empowered or recognized by government authorities, should at least not be challenged or undermined by those authorities.

Ensure that risks are managed effectively

Risks of crop disaster due to bad weather or pests can discourage investments by farmers in land improvements and their adoption of yield-improving technologies. Agricultural research can help reduce risk, for example, by improving drought resistance in crops or developing better ways to conserve soil moisture. Additionally, governments may need to assist farmers in coping with catastrophes and to provide effective safety-net programmes and credit and insurance markets. However, care should be taken in designing such interventions, for, if subsidized, they can easily lead to changes in farming practices that increase the dependence of the beneficiaries on subsidized assistance in the future. Subsidized drought insurance, for example, increases the profitability of risky farming practices beyond their true economic value, and encourages their adoption even though this may lead to greater financial exposure in future droughts and to resource degradation. Subsidized fodder programmes in drought years for rangeland users can also encourage overstocking, which, over time, degrades the range.

Agricultural insurance has often appealed to policy-makers as an instrument of choice for helping farmers and agricultural banks manage climate risks, such as drought, and many countries spend large sums of public money each year on such insurance. But the experience has generally not been favourable (Hazell *et al.*, 1986; Hazell, 1992). Publicly provided crop insurance has, without exception, depended on massive subsidies from the government and, even then, its performance has been plagued by moral-hazard problems associated with many sources of yield loss, by high administration costs, by political interference and by the difficulties of maintaining the managerial and financial integrity of the insurer when the government underwrites all losses (Hazell, 1992). Nor has crop insurance been able to reach the poorer farmers or to assist non-farm members of rural communities, who also suffer in catastrophic agricultural years (among them landless labourers, agricultural traders and shopkeepers). Area-based yield insurance may offer a better alternative and has recently been tried in India with some success (Mishra, 1996). Unfortunately, it remains very costly to the government because the premium rates are set far too low in relation to costs. It is also unnecessarily restricted to farmers who grow the insured crops. A more promising approach would be area-based insurance, based on rainfall rather than yield. This could be a useful risk-management aid to all kinds of rural households and could be simpler and cheaper to operate than area-yield insurance schemes (Hazell, 1992).

Invest in rural infrastructure and people

Less-favoured areas are often poorly placed to compete in liberalized markets, because of their restricted access to markets and high transport and marketing costs. The public sector has an important role to play in building and maintaining roads in these areas and in promoting expansion of private transport, marketing, input supply and financial services that are competitively priced. Investments in electricity and telecommunications are also needed if the private sector is to grow. Investments in clean water and the education and health of local people not only increase their productivity in agriculture, but enhance their opportunities to diversify into non-farm activities, including out-migration to better-paying jobs. The results from our India study show that these kinds of public investments in less-favoured lands can yield favourable growth, as well as poverty alleviation pay-offs. As such, these investments do not have to be a net drain on the national economy. However, full-cost pricing of publicly provided services in less-favoured areas may not be warranted, since their favourable impacts on growth and on the poor have the potential to make substantial future savings in the costs of government relief and safety-net programmes.

Provide the right policy environment

Market reforms, including price and trade liberalization, are necessary to ensure that prices provide the right production signals for farmers and that production and input markets can be competitive and work well. Available evidence suggests that, prior to recent reforms, less-favoured lands were typically penalized along with the rest of the agricultural sector by distorted macroeconomic, trade and sectoral policies (Awudu and Hazell, 1998). As a result, many of the ongoing policy reforms have improved the terms of trade for less-favoured areas and have increased their market opportunities. However, in order to take advantage of these new opportunities, adequate investments in rural infrastructure are needed to improve market access and to reduce transport and marketing costs. If market reforms are not matched by appropriate levels of investment in local infrastructure, they can be quite destructive for many rain-fed areas. For example, market reforms have reduced the availability of inorganic fertilizers and increased their costs in many areas in Africa, and the resultant reduction in their use (often from modest levels to begin with) is now contributing to worsening soil-fertility problems. The associated reduction in food production also adds to the food-insecurity problems of the poor. Transitional policies, sometimes including targeted subsidies, may be necessary in some less-favoured areas to manage some of the negative impacts of market reforms, at least until such time as the required infrastructure investments have been made.

Strengthen public institutions

Many of the public institutions that service agriculture and rural areas have tended to neglect less-favoured areas, and they are often poorly positioned to address the unique problems of these regions. For example, agricultural research and extension systems have been structured to serve the needs of irrigated and high-potential rain-fed areas and, while reasonably efficient at promoting Green Revolution technologies in these areas, they are much less able to deliver the kinds of multidisciplinary, farmer-oriented, natural resource management approaches needed in most less-favoured areas. Similar biases have existed in rural credit and insurance institutions. On the other hand, public agencies with resource mandates (e.g. forestry and rangeland departments) have often been very active in less-favoured lands, but have taken top-down approaches to the management of these resources. Not surprisingly, by excluding local users from any real stake in the ownership and management of these resources, resources are exploited and degraded, while the relevant public departments are hamstrung by their inability to effectively regulate resource use on the ground. The development of less-favoured lands will require significant changes in the objectives and operational modalities of many public agencies. More participatory approaches that build on the interests and abilities of local people to manage resources are needed, and this will require very different incentive structures within public institutions, with greater accountability to intended beneficiaries.

Another problem that has plagued the effectiveness of public institutions has been their seeming inability to coordinate relevant activities in rural areas. Key functions are compartmentalized within different ministries and at different levels (local, regional and national) of government, and there is rarely an effective institutional mechanism for coordinating their plans and activities. Integrated development projects attempted to overcome this problem, but, with few exceptions, they were never successful in moving the coordination beyond the planning stage. Coordination units at the highest level of government have rarely worked in practice, and more effective solutions probably require greater devolution of authority to local governments.

Conclusions

In order to promote economic growth and to redress poverty and environmental problems, policy-makers in developing countries will need to pursue appropriate and sustainable methods of agricultural intensification for both high- and low-potential regions. This dual strategy will be particularly challenging if government budgets for investment in agriculture and rural areas continue to remain tight, and striking the right investment balance between irrigated and rain-fed regions and between high- and low-potential rain-fed areas will be particularly important. Investments in irrigated and high-potential rain-fed areas cannot be neglected because these areas still provide much of the food needed to keep prices low and to feed growing livestock and urban populations.

On the other hand, the poverty, food security and environmental problems of many low-potential areas are likely to remain serious in the decades ahead, as populations continue to grow. While out-migration and economic diversification should become increasingly important in the development of most low-potential areas, agricultural intensification will often offer the only viable way of raising incomes and creating employment on the scale required in the near future. Even when the investments needed to achieve this growth yield lower economic returns than investments in high-potential areas, they might still be justified on the basis of their significant social benefits in the form of poverty alleviation and improved environmental management. Moreover, with worsening income disparities between many high- and low-potential areas, policy-makers are likely to come under increasing pressure to invest more in low-potential areas.

The potential tradeoffs between investing in high- and low-potential areas have yet to be widely quantified, and it is possible that they may be changing. Productivity levels in many high-potential areas have reached a plateau, while at the same time recent agricultural research in some low-potential rain-fed areas is suggesting new avenues for increasing their productivity (Scherr and Hazell, 1994). Our analysis of investments in irrigated and high- and low-potential rain-fed areas in India suggests that investments in rural infrastructure, agricultural technology and human capital are now at least as productive in many rain-fed areas as in irrigated areas, and they have a much larger impact on poverty. These results raise the tantalizing possibility that greater public investment in some low-potential areas could actually offer a 'win–win' strategy for addressing productivity and poverty problems.

The successful development of less-favoured lands will require new and improved approaches, particularly for agricultural intensification. These will require stronger partnerships than are needed in high-potential areas between agricultural researchers and other agents of change, including local organizations, farmers, community leaders, non-governmental organizations, national policy-makers and donors. It will also require time and innovation; new approaches will need to be developed and tried on a small scale before they are scaled up, and their testing will take time to assess and evaluate, particularly given the noise introduced by climatic variability. All this will require patience and perseverance on the part of policy-makers and donors, perhaps more than the current aid culture allows.

Notes

1 A more detailed report of the study is available in Fan and Hazell (1999).

2 The formula for the Törnqvist–Theil index of aggregate production is:

$$\ln YI_t = \Sigma_i 1/2^*(S_{i,t} + S_{i,t-1})\ ^*\ln(Y_{i,t}/Y_{i,t-1})$$

where $\ln YI_t$ is the log of the production index at time t, $S_{i,\,t}$ and $S_{i,t-1}$ are the share of output I in total production value at time t and t_{-1}, respectively; and $Y_{i,t}$ and $Y_{i,t-1}$ are quantities of output I at time t and t–1, respectively. Farm prices are used to calculate the weights of each crop in the value of total production. The index is desirable because of its invariance property; if nothing real has changed, the index itself remains unchanged.

3 Total factor productivity is defined as aggregate output minus aggregated inputs. Again, a Törnqvist–Theil index is used to aggregate both inputs and outputs:

$$\ln TFP_t = \Sigma_i 1/2^*(S_{i,t} + S_{i,t-1})^*\ln(Y_{i,t}/Y_{i,t-1}) - \Sigma_i 1/2^*(W_{i,t} + W_{i,t-1})^*\ln(X_{i,t}/X_{i,t-1})$$

where $\ln TFP_t$ is the log of the total factor productivity index; $W_{i,t}$ and $W_{i,t-1}$ are cost shares of input i in total cost at time t and t–1, respectively; and $X_{i,t}$ and $X_{i,t-1}$ are quantities of input i at time t and t–1, respectively. Five inputs (labour, land, fertilizer, tractors and buffalo) are included. Labour input is measured as the total number of male and female workers employed in agriculture at the end of each year; land is measured as gross cropped area; fertilizer input is measured as the total amount of nitrogen, phosphate and potassium used; tractor input is measured as the number of four-wheel tractors; and bullock input is measured as the number of adult bullocks. The wage rate for agricultural labour is used as the price of labour; rental rates of tractors and bullocks are used for their respective prices; and the fertilizer price is calculated as a weighted average of the prices of nitrogen, phosphate and potassium. The land price is measured as the residual of total revenue net of measured costs for labour, fertilizer, tractors and bullocks.

10 Pathways of Development in the Hillside Areas of Honduras: Causes and Implications for Agricultural Production, Poverty and Sustainable Resource Use

JOHN PENDER,[1] SARA J. SCHERR[2] AND GUADALUPE DURÓN[3]

[1]*Environment and Production Technology Division, International Food Policy Research Institute (IFPRI), Washington, DC, USA;* [2]*Department of Agricultural and Resource Economics, University of Maryland, College Park, Maryland, USA;* [3]*International Center for Women, Washington, DC, USA*

Introduction

In Central America, rapid population growth and agricultural expansion and intensification in steep hillside areas have led to growing concerns about deforestation, soil erosion, watershed degradation, loss of soil fertility, sedimentation of reservoirs and other resource problems (Leonard, 1987; Neidecker-Gonzales and Scherr, 1997). Despite the common perception that population growth and agricultural intensification are primary causes of land degradation and related problems, evidence from many parts of the world suggests that sustainable use of resources in fragile lands can be achieved at much higher population densities than exist in most of Central America (Turner *et al.*, 1993; Tiffen *et al.*, 1994; Templeton and Scherr, 1999). Whether such pressures lead to resource and human welfare degradation or improvement depends upon a complex set of factors, which can vary substantially from one community to the next; these include population density, access to markets and infrastructure, the development of local markets and institutions, the nature and fragility of local resources, and the local incidence of government policies and programmes (Scherr *et al.*, 1996; Lopez, 1998).

Accounting for such a complex set of factors and their impacts on resource management in the diverse circumstances of hillside areas is a major challenge for policy research. To address this challenge, we employ the concept of 'pathways of development'. A development pathway represents a common pattern of change in resource management, associated with a common set of causal and conditioning factors.[1] For example, two of the pathways in central Honduras identified in this study are expansion of basic grains production in areas far from

(eds D.R. Lee and C.B. Barrett)

roads and the urban market, and intensification of horticultural production in areas close to roads and urban markets. The nature, causes and consequences of such pathways are different, and many of the appropriate technology and policy strategies to achieve productivity, resource and welfare improvement in these different pathways will also differ.

The principal research questions addressed by this study are:

- What are the major pathways of development in the central region of Honduras, their causes and their implications for agricultural productivity, natural resource sustainability and poverty?
- How can policies and technologies facilitate more productive, sustainable and poverty-reducing pathways of development in this region?

Conceptual Framework and Methodology

The conceptual framework for this study is shown in Fig. 10.1. In this framework, pressures from population growth, markets, new technology or other external factors induce changes in local markets, prices and institutions within individual communities. These shifts are conditioned by community characteristics, such as human and natural resource endowments, infrastructure, market linkages and the local knowledge base. Community changes may induce responses in agriculture and natural resource management (NRM) strategies at both household and collective levels, such as changes in land use, land

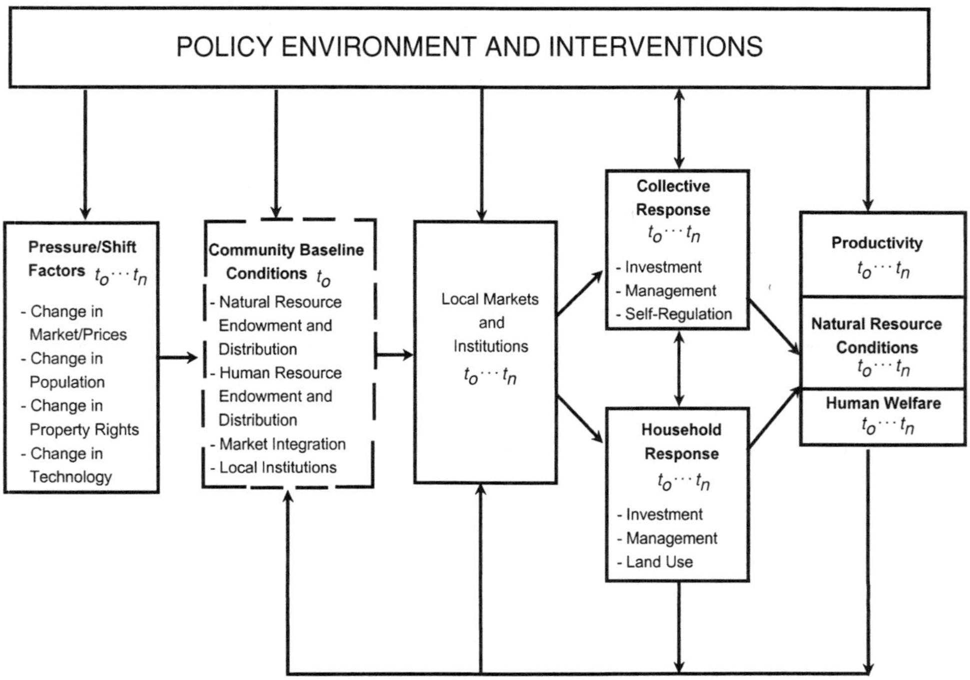

Fig. 10.1. Conceptual framework (from Scherr *et al.*, 1996).

investment, land management intensity, input mix, conservation practices and collective action. Changes in NRM then affect natural resource conditions, economic conditions and human welfare, and these in turn have feedback effects on local conditions, institutions and NRM decisions.

Public policies can influence the process at various levels. For example, agricultural research and price policies affect shift factors. Resource regulations and infrastructure investment affect community conditions. Land titling and credit programmes affect local markets and institutions. Technical assistance influences response patterns. Nutrition programmes and state forest management also directly affect outcomes. Policy research must consider which types and sequences of policy action are likely to be most effective in different circumstances.

Research hypotheses

Several general hypotheses linking the driving forces of change to the nature and outcomes of change can be suggested:

1. *Impacts of population growth.* Population growth induces expansion of production into more marginal and fragile lands, particularly where unsettled land is available and property rights are not well enforced. Where land is limited, intensification of labour per unit of land will occur, including shortening of fallow cycles and adoption of more labour-intensive products and practices (Boserup, 1965). These changes increase land productivity but, holding the state of technology and market development constant, may reduce labour productivity (Salehi-Isfahani, 1988) and per capita income (Pender, 1998).[2] Effects on resource management and conditions may be mixed. Expansion into marginal lands can cause deforestation and land degradation, particularly where property rights are not well defined, while population growth can induce labour-intensive land improvements where land tenure is secure (Scherr and Hazell, 1994; Tiffen *et al.*, 1994). In addition, population growth may promote institutional changes, such as the development of individual property rights, which contribute to improved resource management (Boserup, 1965). Thus, there may be a U-shaped relationship between population density and resource conditions, with resource conditions first worsening and later improving as population rises (Scherr and Hazell, 1994; Pender, 1998; Templeton and Scherr, 1999).
2. *Impacts of improved market or technological opportunities.* Increases in the profitability of agricultural products, whether resulting from infrastructure investment, market development, changes in market prices, technological innovation or government policies affecting these factors, will promote expansion of agriculture into marginal areas if the costs of productive factors are unaffected by the change (Angelsen, 1996). However, if the costs of factors rise, a reduction in agricultural area is possible, as productive factors are concentrated on the most profitable lands (Angelsen, 1996). If the expansion of agricultural land is limited, increased profitability will cause intensification of labour and/or capital per unit of land, though the effects on capital relative to labour depend on the

nature of factor markets and the nature of the change. Market or technology development may promote a shift to cash crops and will tend to increase farm incomes (unless offset by falling farm prices). The implications for resource management and environmental conditions may be mixed. For example, changes in commodity prices have a theoretically ambiguous effect on soil conservation investments (LaFrance, 1992; Pagiola, 1996). Market or technology development may increase the externalities associated with demand for water and agricultural chemicals.

3. *Pathways of development.* Although many pathways of change are theoretically possible, we hypothesize that a relatively small number of pathways represent much of the variation in the processes of change occurring in hillside areas of Central America. These pathways are expected to be determined primarily by differences in comparative advantage, which is largely determined by agricultural potential, market access and population density (Pender *et al.*, 1999a). More commercially oriented pathways (such as intensive production of vegetables) are expected to be found closer to urban markets, or where a comparative advantage in high-value non-perishable commodities exists (such as for coffee and some forest products). In areas of high market access, non-farm development is also likely to be important. In areas more remote from markets and lacking comparative advantage in high-value commodities, subsistence production of cereals and livestock is likely to continue to be important. Where population density is low, there may be greater comparative advantage in more extensive cereal and livestock production and/or in forestry activities. The problems, constraints, opportunities and responses to change may differ substantially across pathways; thus different technology, policy and institutional strategies may be required for more productive, sustainable and welfare-improving development to occur in different pathways.

Methodology

This study is based primarily on a survey of communities in the central region of Honduras. The central region, defined to include 31 *municipios* (counties) in the departments of Francisco Morazan and El Paraíso, was selected as the study region, due to the predominance of hillsides, the presence of several important watersheds, concerns about poverty and resource degradation in the region and the presence of the capital city of Tegucigalpa and major road networks within the region. A stratified random sample of 48 of the 325 *aldeas* (villages) in the region was selected for the survey. The stratification was based upon the 1974 rural population density of the *municipio* in which each *aldea* was located and the distance by road of the *municipio* county seat to Tegucigalpa.[3]

The analytical approach used in this study seeks to address the complexity of factors and diversity of situations influencing agricultural resource management and NRM in central Honduras by identifying the broad development pathways in the region and their determinants and implications. The causes and consequences of different development pathways are different, and the opportunities and constraints affecting NRM decisions may also differ across the pathways. For example, labour-intensive land management technologies are likely to be more suited to high-population-density areas, where subsistence grain production is dominant, compared with low-population-density areas or where off-farm income opportunities are high.

To identify the pathways of development, we adopted a simple classification system based on primary and secondary occupations and land use in the communities studied and changes since the mid-1970s. Six pathways were identified (described below). The analysis was based upon comparisons of descriptive statistics across the pathways, econometric analysis and qualitative information from the survey. The general form of the econometric analyses, based on the conceptual framework in Fig. 10.1, is as follows:

1. Pathway = *f* (driving factors, conditioning factors, policies/programmes).
2. Household-level responses = g (pathway, driving factors, conditioning factors, policies/programmes).
3. Collective responses = *h* (pathway, driving factors, conditioning factors, policies/programmes).
4. Outcomes = *k* (pathway, driving factors, conditioning factors, policies/programmes).

Many of the dependent variables in this system are measured as discrete variables, including pathways (categorical), household responses (ordinal),[4] collective responses (binary) and some of the outcome variables (ordinal). Several different types of regression models are used, depending on the nature of the dependent variable.[5] In some regressions, the dependent variable is measured at a single point in time (e.g. use of conservation practices in 1996), and in some the dependent variable is measured as a change (e.g. perceived change in resource conditions since the mid-1970s).

The explanatory variables used to explain the determinants of pathways include factors affecting agricultural potential (altitude and number of rainfall days), population density, access to markets (distance to the urban market and to the nearest road) and access to technology (the presence of a technical assistance programme). The cross-sectional regressions for household responses and outcomes include as explanatory variables the pathways, population density and density-squared (the squared term included to check for possible U-shaped relationships), the distance to the urban market and to the nearest road, adult literacy and the presence of programmes of technical assistance, credit, agrarian reform and land titling.[6] In the regressions for changes in the dependent variables, we replaced population density and density-squared with the change in these measures between 1974 and 1988 (two census years), and replaced the distance measures of market access with indicators of whether a road had existed in the community since the 1970s or whether one had been constructed since then.

The endogeneity of some explanatory variables – particularly population growth and the pathway variables – may be a problem. In all regressions, including these explanatory variables, we ran the regressions twice, using actual and predicted values of these variables, to test the robustness of the findings. We discuss only significant and robust findings in the results below.

Pathways of Development in Central Honduras

Basic grain production (maize, beans, sorghum) is the most or second most important economic activity (occupation) of farm households in all but one of the sample communities. Other factors are therefore more important in distinguishing the pathways.

Classification of the pathways

The pathways of development are labelled 'basic grains expansion', 'basic grains stagnation', 'horticultural expansion', 'coffee expansion', 'forestry specialization' and 'non-farm employment' pathways. In the basic grains expansion communities, basic grains production is the primary activity, livestock production is the secondary activity (in almost all cases) and basic grains production has been expanding since 1975. Basic grains and livestock production are also the dominant activities in the basic grains stagnation communities, but basic grains production has been stagnating or declining in these. In the horticultural expansion communities, horticultural crops production (especially of vegetables) is the primary or secondary economic activity and horticultural area has been increasing in all cases. In coffee expansion communities, coffee production is the primary or secondary activity and the importance of coffee has been increasing in all cases. In forestry specialization communities, forestry activities were of primary or secondary importance. In non-farm employment communities, off-farm employment was first or second in importance and has been increasing in importance in almost all cases.

Characteristics of the pathways

The six pathways are similar in some respects and significantly different in others (Table 10.1).[7] The basic grains pathways are at lower altitudes and have lower rainfall than the other pathways. Population density was lowest in the forestry, horticultural and basic grains expansion pathways in the mid-1970s, but rapid population growth in the horticultural pathway caused this pathway to be among the most densely populated by the late 1980s. Access to the urban market and roads is lowest in the basic grains pathways and highest in the non-farm employment and horticultural pathways. In the case of the horticultural pathway, this has been a result of road construction since the mid-1970s. Access to technical assistance and other government programmes has been lowest in the basic grains expansion pathway and generally highest in the coffee and forestry pathways. Literacy is lowest in the basic grains and horticultural pathways.

Land-use change has been moderate overall since the 1970s, but substantial changes are apparent in a few pathways. The area in annual crops grew in all pathways, but most rapidly in the horticultural and basic grains expansion pathways.[8] The expansion in annual crops occurred at the expense of tree cover in all pathways, but especially in the basic grains expansion and horticultural

Table 10.1. Development pathways in central Honduras.

Variable	Basic grains expansion	Basic grains stagnation	Horticultural expansion	Coffee expansion	Forestry specialization	Non-farm employment
Number of communities in sample	5	15	5	10	3	10
Sample represents:						
% of population of region	5	20	5	31	11	28
% of area of region	6	15	4	34	22	20
Dominant economic activities						
Basic grains	×	×	×	×	×	×
Livestock	×	×				
Horticultural crops			×			
Coffee				×		
Forestry					×	
Non-farm employment						×
Change in economic activities since 1975						
Basic grains	↑	0↓	0↓	0↓	↓	↑↓
Livestock	0↓	0	0	0↓	0↑	0↓
Horticultural crops	0	0↑	0↑	0↑	0↑	↑↓
Coffee	↑↓	0	0	↑	↑	0↓
Forestry	0↓	0↓	0↓	0↓	0↑	↑↓
Non-farm employment	0	0↑	0↑	0↑	0	0↑
Index of proportion sold outside[a] (mean, robust standard error in parentheses)						
Basic grains	3.4	1.9	0.0	0.6	2.0	1.1
	(0.3)	(0.2)	(0.0)	(0.3)	(0.0)	(0.3)
Cattle/cattle products	2.9	2.6	2.7	1.4	2.0	2.2
	(0.8)	(0.5)	(0.9)	(0.5)	(1.4)	(0.8)
Vegetables	0.0	2.0	4.0	1.9	0.0	1.1
	(0.0)	(0.4)	(0.0)	(0.7)	(0.0)	(1.0)
Coffee	1.1	0.0	0.0	4.0	2.0	1.4
	(0.5)	(0.0)	(0.0)	(0.0)	(1.4)	(1.1)
Pine resin	6.0	0.0	0.0	6.0	6.0	6.0
	(0.0)	(0.0)	(0.0)	(0.0)	(0.0)	(0.0)

[a]Values of index: 0 = none; 1 = less than 10%; 2 = minority; 3 = half; 4 = majority; 5 = more than 90%; 6 = all.
× indicates a dominant economic activity.

pathways. The area of devegetated land increased in the basic grains expansion pathway but declined in the others.[9] Changes in area of pasture and fallow land were generally relatively small, with the exception of fallow in the non-farm employment pathway, which has increased significantly since the 1970s. Survey respondents attributed land-use changes to population growth and burning practices (especially in the basic grains communities), increased knowledge about land-management options and improved market access (especially in the horticultural pathway), increased profit potential of commercial crops (especially in the coffee pathway), overexploitation of forests by loggers and community members (in the forestry pathway), poor soils (forestry pathway), decline in farming (non-farm employment pathway) and other factors.

Crop production practices vary substantially across the pathways. Continuous (no fallow) crop production is most common in the forestry pathway, despite low population density, probably due to the limited availability of arable land. Continuous cultivation has increased in all pathways, but especially in the forestry and coffee pathways. Burning to prepare fields is most common in the basic grains pathways, but has declined significantly in all pathways. Crop production is most technified in the horticultural pathway, in which the use of chemical fertilizer, insecticides, herbicides, improved seeds and irrigation is most common. Use of modern inputs has increased in most communities since the mid-1970s, but especially in the horticultural pathway.

The most common livestock are poultry, pigs and cattle. While chickens are common in all pathways, pigs are most common in basic grains communities and cattle in forestry and basic grains expansion communities. Use of all types of livestock has generally declined, although declines were more common outside the basic grains pathways.

Use of organic inputs and annual conservation practices is generally uncommon in all pathways. The most common annual conservation practice is contour planting, which is practised by significant fractions of farmers (but less than half), mainly outside the basic grains pathways. The most common land-improving investments are live barriers, stone walls, tree planting and terraces; all of these, however, have been adopted by less than half of the households. Adoption varies by type of investment and pathway: live barriers and terraces are most common in the forestry and coffee pathways, stone walls most common in the forestry and basic grains stagnation pathways, and tree planting most common in basic grains expansion and coffee pathways.

Local collective action related to natural resources includes formal or informal regulation of use of common-property resources, such as forests and water, and collective investments, such as planting trees near water sources or building stone walls to control runoff. Local regulation of common resources is strongest in the forestry and coffee communities, where the external institutional presence has been the greatest.[10] Collective investments in tree planting and runoff control have been greatest in the basic grains expansion and non-farm employment pathways, and no such investments were reported in any of the horticultural or forestry communities.

In terms of outcomes, maize yields are lowest in the basic grains pathways, but improvement in maize yields is more common in the basic grains expansion pathway than in other pathways. Agricultural wages and wage improvement have been lowest in the basic grains pathways, and real wages have fallen in the basic grains stagnation pathway since the mid-1970s. Measures of poverty are high but declining in all pathways, with the greatest improvement occurring in the non-farm employment pathway. Deforestation on steep slopes is greatest in the basic grains stagnation pathway, while there is also significant cultivation on steep slopes in the horticultural pathway. Perceptions of changes in resource conditions (cropland quality, forest area, forest quality, water availability and water quality) indicate a decline in most measures in all pathways, but resource degradation is most common in the basic grains and forestry pathways.

These indicators suggest that the different pathways have distinct causes or conditioning factors and different implications for resource management decisions, agricultural productivity, poverty and resource sustainability. Below we investigate these relationships using econometric analysis, controlling for the explanatory factors mentioned previously.

Determinants of the pathways

A multinomial logit model was used to investigate the determinants of the pathways (Table 10.2).[11] The results (together with the descriptive statistics discussed above) imply that different determining and conditioning factors are critical for different pathways. In general, the factors determining comparative advantage are important distinguishing factors. Agroecological characteristics, especially altitude and rainfall, distinguish the basic grains pathways from most other pathways, while horticultural communities are at higher elevations than others. The basic grains expansion communities have lower population densities, are more remote from roads and markets and are less well served by external programmes than all other pathways. Horticultural communities have relatively good access to roads and markets, but this access has come relatively recently, in contrast to most communities outside the basic grains pathways. Technical assistance does not appear to have been a major driving factor behind the basic grains expansion pathway; indeed, there appear to be opportunities for more external presence in this pathway. In contrast, involvement of external programmes appears to have been important in the coffee and forestry communities. Forestry communities are also distinguished from the other pathways by their sparse population density, relatively good access to markets and high presence of forestry department officials. Non-farm employment communities have the best access to the urban market, roads and public services.

Agriculture and Natural Resource Management

In this section, we consider the factors determining the patterns and changes in agriculture and NRM, including household-level crop-production practices, conservation measures and collective action to manage common resources.

Table 10.2. Determinants of the pathways.

	Multinomial logit regressions[a] (standard errors in parentheses)				
Coefficient	Basic grains expansion	Horticultural expansion	Coffee expansion	Forestry specialization	Non-farm employment
Altitude at mid-point (m a.s.l.)	–0.0187	0.0187***	0.00807	0.00309	0.00191
	(0.0127)	(0.0052)	(0.00500)	(0.0105)	(0.00303)
Average number of rainfall days per year	0.1300**	0.0499	0.1556**	0.1360	0.0657***
	(0.0627)	(0.0308)	(0.0673)	(0.0963)	(0.0210)
1974 population density (persons km^{-2})	–0.2263*	–0.1429**	0.0049	–0.1269	–0.0236
	(0.1142)	(0.0626)	(0.0495)	(0.1870)	(0.0190)
Distance to Tegucigalpa (km)	–0.0969*	–0.0605	0.1810	0.1425	–0.0469
	(0.0511)	(0.0557)	(0.1111)	(0.1461)	(0.0327)
Distance to nearest road (km)	1.133***	–69.03***	–0.975**	–0.7625	–79.86***
	(0.347)	(4.76)	(0.450)	(0.4728)	(3.15)
Presence of technical assistance programme	–35.35***	–50.85***	15.89	10.08	–53.22***
	(11.65)	(9.02)	(20.68)	(35.71)	(9.20)
Mean predicted probability of correct pathway[b]	0.83	0.71	0.68	0.14	0.79

[a]Reference category is basic grains stagnation pathway. Only 47 observations were used due to missing data on population density for one community. Coefficients and standard errors are adjusted to account for sampling weights, stratification and finite population size. Intercept not reported.

[b]For example, the predicted probability that a community is in the basic grains expansion pathway is 0.83 for basic grains expansion communities. The mean predicted probability of the correct pathway for basic grains stagnation communities is 0.71.

*, ** and *** indicate statistical significance at the 10%, 5% and 1% levels.

m a.s.l., metres above sea level.

Crop production

Econometric analysis reveals significant variation among the pathways in crop-production practices in 1996, even after controlling for other factors, such as population density, access to markets, technical assistance and other programmes (Table 10.3).[12] Consistent with the descriptive statistics, the use of continuous cropping is most common in the forestry pathway, while burning is most common in the basic grains stagnation pathway. Use of fertilizer, herbicides, improved seeds and irrigation is more common in the horticultural pathway than in most other pathways.

Population density has a significant impact only on pesticide use. The impact is positive, with higher population density associated with increased use, but at a diminishing rate.[13] Somewhat surprisingly, continuous cropping, insecticides and herbicides are more commonly used further from the urban market. Burning to clear fields is more common further from a road (probably because it is illegal), while use of fertilizers, insecticides and improved seeds is more common close to a road, as one would expect. Technical assistance programmes are associated with less burning and greater (i.e. more common) use of insecticides and herbicides. Agrarian reform programmes are associated with greater use of burning (probably because smallholder beneficiaries are oriented towards basic grain production), while land-titling programmes are associated with greater use of fertilizer, herbicides and improved seeds. Literacy is associated with greater use of burning, possibly because the opportunity cost of labour is greater where literacy is higher.

The preceding results are based on cross-sectional regression analysis and may be affected by many possible sources of omitted-variable bias.[14] Fortunately, we are also able to investigate determinants of change in some cropping practices, and these regressions will be less affected by omitted variables (at least those which do not change over time).[15] These results are presented in Table 10.4.

Some of the results explaining changes in practices confirm results in the cross-sectional analysis. We find that burning has declined more commonly in the horticultural and non-farm employment pathways than in the basic grains stagnation pathway. Population growth has a positive but diminishing impact on insecticide use. Road access and land-titling programmes contribute to the increasing use of insecticides. Technical assistance is associated with reduced burning, while agrarian reform programmes are associated with increased burning. However, some of the factors that have a significant impact in the cross-sectional regressions have an insignificant impact in the analysis of changes, reducing our confidence in those cross-sectional results. In some cases, we find significant impacts in the analysis of changes that we did not find in the cross-sectional analysis. For example, we find that the increase in continuous cropping is more common in the basic grains expansion pathway and where technical assistance has operated. Such impacts could easily be missed in cross-sectional regressions.

These results imply that changes in market access, technical assistance and other programmes have been important factors influencing the intensification of

Table 10.3. Determinants of cropping practices.

Variable	Ordered probit regressions – signs of significant coefficients[a]						
	Continuous cropping	Burning	Fertilizer	Insecticide	Herbicide	Improved seeds	Irrigation
Basic grains expansion	0	– – – (R)	+++ (R)	+++ (R)	0	0	0
Horticultural expansion	++	– – – (R)	++ (R)	++	++ (R)	+++ (R)	++ (R)
Coffee expansion	0	– – – (R)	+ (R)	0	– (R)	+	0
Forestry specialization	+++ (R)	– – – (R)	0	– – –	– – –	++ (R)	0
Non-farm employment	++	– – –	0	++ (R)	0	0	0
Population density, 1988 (per km^2)	0	0	0	+++ (R)	– – –	0	0
Population density squared	0	0	0	– – – (R)	+++	0	0
Distance to Tegucigalpa (km)	+++ (R)	0	0	+++ (R)	+++ (R)	0	0
Distance to nearest road (km)	0	+++ (R)	– – – (R)	– – – (R)	0	– – (R)	0
Technical assistance programme	0	– – – (R)	0	+++ (R)	+++ (R)	0	0
Credit programme	0	– – –	0	0	0	0	0
Agrarian reform programme	0	+++ (R)	+ (R)	0	0	0	+ (R)
Land-titling programme	0	0	++ (R)	0	++ (R)	++ (R)	0
Proportion of adults literate, 1988	0	++ (R)	– (R)	0	0	0	0

[a]Based on ordered probit regressions with coefficients and standard errors adjusted for sampling weights, stratification and finite population. Dependent variables take integer values from 0 (no one uses practice) to 6 (everyone uses).

+, ++ and +++ indicate a positive coefficient statistically significant at the 10%, 5% and 1% level, respectively; –, – – and – – – indicate a negative coefficient statistically significant at the 10%, 5% and 1% level, respectively; 0 indicates that the coefficient is not statistically significant at the 10% level; (R) indicates that the coefficient is the same sign and significant at the 10% level when pathways are replaced by predicted pathways.

Table 10.4. Determinants of changes in crop practices.

Variable	Ordered probit regressions – signs of significant coefficients[a]		
	Continuous cropping	Burning	Insecticide
Basic grains expansion	++ (R)	0	++
Horticultural expansion	0	– – – (R)	0
Coffee expansion	0	0	0
Forestry specialization	0	– –	0
Non-farm employment	0	– – – (R)	0
Change in population density, 1974–1988 (per km^2)	0	0	0
Change in (population density squared)	0	0	– – (R)
New road access since 1975	0	0	+++ (R)
Already had road access in 1975	0	0	+++ (R)
Technical assistance programme	+++ (R)	– – (R)	0
Credit programme	0	– – – (R)	0
Agrarian reform programme	0	+++ (R)	+++
Land-titling programme	0	0	+++ (R)
Change in proportion of adults literate, 1974–1988	0	0	– – (R)

[a]Based on ordered probit regressions with coefficients and standard errors adjusted for sampling weights, stratification and finite population. Dependent variables take values of +1 (increase), 0 (no change), and –1 (decrease).
+, ++ and +++ indicate a positive coefficient statistically significant at the 10%, 5% and 1% level, respectively; –, – – and – – – indicate a negative coefficient statistically significant at the 10%, 5% and 1% level, respectively; 0 indicates that the coefficient is not statistically significant at the 10% level; (R) indicates that the coefficient is the same sign and significant at the 10% level when pathways are replaced by predicted pathways and population growth replaced by predicted population growth.

crop production, both directly (i.e. controlling for the pathways) and via their influence on development of the pathways themselves. Market access has had the most influence on the adoption of purchased inputs, while programmes have affected the use of labour-intensive practices as well. Population growth appears to be of less importance in directly causing these changes. However, by influencing the choice of development pathway, differences in population density (and other initial conditions) still have an important influence.

Use of organic inputs and conservation measures

Different pathways are associated with different types of conservation measures. The use of contour planting is more common in the horticultural and forestry pathways than in the other pathways, controlling for other factors (Table 10.5). Minimum tillage is most common in the coffee pathway. Mulching is least common in the horticultural and non-farm employment pathways. Terraces are most common in the coffee and non-farm employment pathways, while live

Table 10.5. Determinants of annual conservation or organic practices.

Variable	Ordered probit regressions – signs of significant coefficients[a]					
	Contour planting	Minimum till	Mulching	Incorporation of crop residues	Cattle manure	Chicken manure
Basic grains expansion	0	0	0	0	0	+++
Horticultural expansion	+++ (R)	–	– – – (R)	0	0	+
Coffee expansion	++	++ (R)	0	0	0	0
Forestry specialization	++ (R)	+++	0	0	0	+++
Non-farm employment	0	0	– – (R)	0	0	0
Population density, 1988 (per km^2)	0	0	0	0	++ (R)	0
Population density squared	0	0	0	0	– – (R)	0
Distance to Tegucigalpa (km)	– – – (R)	– – – (R)	– – (R)	– – – (R)	0	0
Distance to nearest road (km)	–	+++ (R)	0	+	0	– – – (R)
Technical assistance programme	++	++	–	++(R)	0	0
Credit programme	–	0	0	0	0	0
Agrarian reform programme	0	+	++	0	0	0
Land-titling programme	++ (R)	0	0	0	0	0
Proportion of adults literate, 1988	+++ (R)	0	– –	– –	++ (R)	0

[a]Based on ordered probit regressions with coefficients and standard errors adjusted for sampling weights, stratification and finite population. Dependent variables take integer values from 0 (no one uses practice) to 6 (everyone uses).

+, ++ and +++ indicate a positive coefficient statistically significant at the 10%, 5% and 1% level, respectively; –, – – and – – – indicate a negative coefficient statistically significant at the 10%, 5% and 1% level, respectively; 0 indicates that the coefficient is not statistically significant at the 10% level; (R) indicates that the coefficient is the same sign and significant at the 10% level when pathways are replaced by predicted pathways.

barriers are also most common in the coffee pathway (Table 10.6). These differences suggest that the potential for adoption of particular conservation measures varies across the pathways, and suggests the importance of a targeted approach to the promotion of conservation measures in different pathways.

Population density has an insignificant impact on most conservation or organic measures, except the use of cattle manure and tree planting. Both of these are more common at higher population density, but at a diminishing rate. This may be because the relative returns to such investments are higher at higher population densities and/or the labour costs of the measures are lower (particularly for manuring).

Several conservation measures are less common further from the urban market, including contour planting, minimum tillage, mulching and incorporation of crop residues. This could be because the returns to such efforts are lower where returns to agriculture may be lower (due to lower market access), but it may also be due to lower intensity of involvement of programmes promoting conservation in more remote areas. Surprisingly, minimum tillage is more

Table 10.6. Determinants of land-improving investments.

	Ordered probit regressions – signs of significant coefficients[a]			
Variable	Terraces	Live barriers	Stone walls	Tree planting
Basic grains expansion	++	++	0	0
Horticultural expansion	0	0	+	0
Coffee expansion	++ (R)	+++ (R)	0	0
Forestry specialization	+++	+++	0	0
Non-farm employment	+++ (R)	0	0	0
Population density, 1974[b] (per km^2)	0	++	0	+++ (R)
Population density squared	0	0	0	– – (R)
Distance to Tegucigalpa (km)	0	0	0	+
Distance to nearest road (km)	0	– – (R)	0	– –
Technical assistance programme	+++ (R)	+++ (R)	+++ (R)	0
Credit programme	+++ (R)	0	– – (R)	0
Agrarian reform programme	+++ (R)	0	0	++
Land-titling programme	+ (R)	+	– – – (R)	0
Proportion of adults literate, 1974[b]	++ (R)	0	0	0

[a]Based on ordered probit regressions with coefficients and standard errors adjusted for sampling weights, stratification and finite population. Dependent variables take integer values from 0 (no one uses practice) to 6 (everyone uses).

[b]1974 population density and proportion of adults literate used because some investments may have occurred before 1988.

+, ++ and +++ indicate a positive coefficient statistically significant at the 10%, 5% and 1% level, respectively; –, – – and – – – indicate a negative coefficient statistically significant at the 10%, 5% and 1% level, respectively; 0 indicates that the coefficient is not statistically significant at the 10% level; (R) indicates that the coefficient is the same sign and significant at the 10% level when pathways are replaced by predicted pathways.

common further from a road, while use of chicken manure and live barriers is more common close to roads. The effect of road access on the use of chicken manure (which is marketed by commercial poultry operations) is similar to the effect of road access on inorganic fertilizer use.

As one would expect, several conservation measures are positively associated with technical assistance programmes, including incorporation of crop residues, terraces, live barriers and stone walls. Credit, agrarian reform, land-titling programmes and literacy are also positively associated with the adoption of terraces. Credit and titling programmes are negatively associated with the use of stone walls. Perhaps farmers feel less need to build stone walls around their fields for tenure security if they have a title to the land and access to credit based on land ownership.

Collective investment in natural resource management

Collective investment in tree planting and to control runoff is higher at moderate rates of population growth than at low or very high rates.[16] This may be because low and high rates of population growth are associated with high rates of emigration and immigration, respectively, which may reduce the ability to achieve collective action by reducing the stability and homogeneity of the community (Baland and Platteau, 1996). Higher initial population density reduces collective investments, possibly because the number of required participants and, hence, the organizational costs of collective investments are higher where population density is greater. The presence of local organizations involved in NRM appears to stimulate collective investments in NRM, while external organizations are negatively associated with local collective investment. This latter finding suggests that external organizations may undermine local collective efforts, and implies that such organizations should take a cautious approach when intervening in local communities, to avoid such displacement.

Development Outcomes

The above patterns of resource management are the proximate causes of change in the 'critical triangle' of development outcomes: economic productivity or growth, sustainability of the natural resource base and poverty alleviation (Vosti and Reardon, 1997). These outcomes thus also vary by development pathway and other factors.

Productivity

In the econometric analysis, we did not find significant and robust differences in land productivity (as measured by maize yields) among the pathways (Table 10.7). The only factor found to be significantly associated with maize yields was population density, with higher density associated with lower yields, though at

Table 10.7. Determinants of productivity, resource use and poverty outcomes.

	Least-squares regressions – signs of significant coefficients[a]						
	Productivity, 1996		Resource conditions, late 1970s[b]			Poverty, 1988	
Variable	High maize yield	High male wage	Forest on steep land	Devegetated steep land	Cultivation on steep land	Houses with dirt floor	Households whose last child died
Basic grains expansion	0	0	+	– – (R)	0	0	0
Horticultural expansion	0	+++ (R)	++ (R)	– – – (R)	++	0	0
Coffee expansion	0	+++ (R)	+++ (R)	– – – (R)	0	0	0
Forestry specialization	0	+ (R)	+++ (R)	– – (R)	– – –	+ (R)	0
Non-farm employment	0	+++ (R)	0	0	0	0	0
Population density (per km^2)	– – – (R)	0	– –	0	++	0	0
Population density squared	++ (R)	0	++ (R)	– – (R)	–	0	0
Distance to Tegucigalpa (km)	0	++	0	0	0	++	0
Distance to nearest road (km)	–	– – – (R)	0	0	– –	0	0
Technical assistance programme	0	++	NE	NE	NE	+	0
Credit programme	0	0	NE	NE	NE	0	0
Agrarian reform programme	0	0	NE	NE	NE	0	+++ (R)
Land-titling programme	0	+ (R)	NE	NE	NE	0	0
Proportion of adults literate, 1988	0	0	0	0	– – (R)	– – – (R)	0
Proportion of land with > 30% slope	NE	NE	0	0	– (R)	NE	NE

[a]Based on least-squares regressions with coefficients and standard errors adjusted for sampling weights, stratification and finite population.
[b]Population density and proportion of adults literate are for 1974 in these regressions, for 1988 in the other regressions.
+, ++ and +++ indicate a positive coefficient statistically significant at the 10%, 5% and 1% level, respectively; –, – – and – – – indicate a negative coefficient statistically significant at the 10%, 5% and 1% level, respectively; 0 indicates that the coefficient is not statistically significant at the 10% level; (R) indicates that the coefficient is the same sign and significant at the 10% level when pathways are replaced by predicted pathways; NE means not estimated.

a diminishing rate. This suggests that population pressure is causing a decline in soil productivity, perhaps because of shortening fallow periods without sufficient use of organic or inorganic sources of soil fertility. This hypothesis could not be confirmed in the regression explaining changes in maize yields, since the population growth variables were found to have an insignificant effect (Table 10.8). The only significant and robust finding in that analysis was that growth in maize yields was more likely in the horticultural pathway than in any others. This is consistent with the finding above of increasing use of modern inputs in the horticultural pathway. Such inputs are not restricted to horticultural crops in this pathway, but are also used in maize production.

Labour productivity, as measured by wage rates, is highest outside the basic grains pathways (Table 10.7). Real wages have also been growing more rapidly in horticultural and non-farm employment communities (Table 10.8). Wages are lower further from a road, as one would expect. Surprisingly, however, we find that wage growth has been slower in communities that have access to a road than in those that do not.

Natural resource conditions

Natural resource conditions vary substantially across the development pathways. In the late 1970s, forests were most likely to be found on steep lands (with slopes over 30%) in the horticultural, coffee and forestry pathways (Table 10.7).[17] Devegetation of steep lands is more common in the basic grains stagnation pathway than in most other pathways, controlling for other factors, as was found in the descriptive statistics. Higher population density is associated with greater cultivation and less forest on steep lands, but at a diminishing rate. Literacy is associated with less cultivation on steep lands.

Perceived changes in natural resource conditions also varied across the development pathways. Cropland quality was most likely to improve (or least likely to decrease) in the horticultural pathway (Table 10.8). Forest area was most likely to increase in the horticultural and non-farm employment pathways. However, water availability was more likely to decline in both of these pathways, as well as in the basic grains expansion pathway, than in the other pathways. These factors also affect resource degradation. Population growth is associated with a greater likelihood of deforestation, though at a diminishing rate. Road access and land titling are both associated with deforestation, as one might expect. Water quality was more likely to decline where there had been an agrarian reform programme, while the presence of national or municipal efforts to protect forests was associated with improvement (or less reduction) in water availability and quality.

Poverty

By most available measures, socioeconomic conditions improved in all pathways between 1974 and 1988. The measures that we examine through

Table 10.8. Determinants of changes in outcomes.

Variable	Least-squares and ordered probit regressions – signs of significant coefficients[a]								
	Productivity, 1975–1996		Perceived natural resource conditions, 1975–1996					Poverty, 1974–1988	
	Maize yield	High male wage	Cropland quality	Forest area	Forest quality	Water availability	Water quality	Houses with dirt floor	Households where last child died
Basic grains expansion	–	0	0	– – –	– – –	– – – (R)	0	0	0
Horticultural expansion	+ (R)	+++ (R)	++ (R)	++ (R)	0	– – (R)	0	0	0
Coffee expansion	0	0	0	+++	+ (R)	–	0	0	0
Forestry specialization	– – –	0	– – –	0	– – –	– – –	– –	0	0
Non-farm employment	0	+++ (R)	0	++ (R)	0	– – (R)	0	0	–
Change in pop. dens., 1974–1988	0	++	0	– – – (R)	0	++	++	0	0
Change in (pop. dens. squared)	0	– – –	0	+++ (R)	0	0	0	0	0
New road access since 1975	0	– – – (R)	0	0	0	0	0	–	0
Already had road access in 1975	0	– – – (R)	0	– – – (R)	0	0	0	0	0
Technical assistance programme	0	0	0	–	0	0	0	+++ (R)	0
Credit programme	0	0	0	0	0	0	0	0	0
Agrarian reform programme	0	– – –	0	0	0	0	– – – (R)	0	0
Land-titling programme	0	0	0	– – (R)	0	0	++	0	0
Change in proportion of adults literate, 1974–1988	0	0	0	0	0	– (R)	0	0	0
Presence of national or municipal forest regulation	NE	NE	0	0	0	+++ (R)	+++ (R)	NE	NE

[a]Based on ordered probit regressions for change in maize yield and change in resource conditions, and least squares for other regressions. Coefficients and standard errors adjusted for sampling weights, stratification and finite population. Dependent variables for ordered probit regressions take the values +1 (increase), 0 (no change) and –1 (decrease).

+, ++ and +++ indicate a positive coefficient statistically significant at the 10%, 5% and 1% level, respectively; –, – – and – – – indicate a negative coefficient statistically significant at the 10%, 5% and 1% level, respectively; NE means not estimated; 0 indicates that the coefficient is not statistically significant at the 10% level; (R) indicates that the coefficient is the same sign and significant at the 10% level when pathways are replaced by predicted pathways and population growth replaced by predicted population growth.

econometric analysis include the percentage of houses with a dirt floor and the percentage of households whose last child died.[18] We do not find statistically significant and robust differences among the pathways in these measures of poverty or changes in these measures, controlling for other factors (Tables 10.7 and 10.8). The only factor that we find significantly associated with the proportion of houses with a dirt floor is literacy (a negative effect). We find that the presence of an agrarian reform programme is associated with child mortality, while technical assistance programmes are associated with an increase (or less decrease) in the proportion of households having a dirt floor. These last two findings may be a result of such programmes focusing on areas where poverty is severe and perhaps worsening.

Overall, these results do not show strong differences across the pathways in terms of changes in poverty, which has declined fairly broadly in the region. Perhaps public and non-governmental organization (NGO) investments in social services have been more powerful determinants of changes in poverty than differences in comparative advantage and changes in agriculture occurring across the different pathways of development.

Summary and Conclusions

The general pattern of agricultural change in the central region of Honduras is toward increased specialization and commercialization of production based upon comparative advantage. Differences across the pathways in comparative advantage and other factors led to differences in land use, agriculture and resource management decisions. Given such differences, it is not surprising that we also find significant differences across the pathways in their outcomes for agricultural productivity and natural resource conditions. In general, productivity outcomes were more favourable in the horticultural, coffee and non-farm employment pathways, while the implications for resource conditions were more mixed, with some resource conditions improving in these pathways and others becoming worse.

While many natural resource and environmental problems were getting worse in the central region, many aspects of human welfare were improving over the study period. In general, the changes in agricultural productivity and resource conditions were more mixed, and more often negative, than changes in measures of poverty. A major reason for this seems to be that the factors influencing these different outcomes are different. The factors influencing agricultural productivity and natural resource conditions are very location-specific. In contrast, welfare conditions are largely affected by provision of services by public agencies and NGOs, and the impacts of these interventions were fairly broad. In addition, migration has tended to reduce differentials in wages across communities (although large disparities remain), while general growth in economic opportunities resulting from market liberalization and changes in market conditions have led to increasing real wages in almost all pathways.

Although changes in natural resource conditions and human welfare are moving in opposite directions in much of the central region, this does not imply that there must be large tradeoffs between these objectives. As mentioned above, the differences in these outcomes appear to result from different causal factors, suggesting that these causes can be addressed separately.[19] In some cases, direct tradeoffs do exist – for example, the increased use of irrigation and agrochemicals in the horticultural pathway leads to rising productivity and incomes, but also to problems of water scarcity and contamination. In these cases, careful consideration of the impacts on farmers' welfare is needed when considering measures to address the resource and environmental issues, and vice versa.

Findings relative to the research hypotheses

The preceding results adequately demonstrate the validity of hypothesis 3 regarding the determinants and impacts of the pathways. Here we recapitulate our findings with regard to hypotheses 1 and 2 concerning the impacts of population pressure and access to markets and technology.

Population pressure

The results of the regression analysis and the explanations provided by survey respondents support the hypothesis that population growth induces deforestation and expansion of agricultural area at relatively low levels of population density. Survey responses also support the expected impact of population growth on reducing fallow periods, though, surprisingly, this was not supported by the regression analysis. In general, we found few significant and robust impacts of population density or population growth on agricultural and resource management practices. Agricultural change in central Honduras appears to have been more market- and technology-induced than population-induced.

Population pressure was found to be associated with deforestation on steep lands, as expected. This relationship had the hypothesized U shape, although the predicted turning-point is at levels of population density above those found in central Honduras. Population pressure was also associated with lower maize yields, contrary to the prediction that population pressure induces more intensive land use and thus greater land productivity (if not labour productivity). This result may be due to population-induced land degradation.

Population pressure was not significantly associated with wages or measures of poverty, perhaps because populations respond to differences in wages and poverty via migration. Large differences in the rate of population growth across the pathways are indicative of this. Thus, the potentially negative welfare effects of population growth appear to have been mitigated, or at least shared among communities to some extent, as a result of migration.

Overall, these results support the concerns raised by many observers about the negative implications of population growth for agricultural productivity and the environment, although the impacts are not all negative or large. While we do find some evidence that population-induced investments in land improvement are occurring, these responses are not sufficient to compensate for all the negative effects, particularly on soil fertility and forest resources.

Market access

Access to markets, as measured by road access, is associated with deforestation and expansion of agricultural area. As expected, road access is also associated with greater use of purchased inputs. The impacts of market access on changes in resource conditions are mixed, as hypothesized. As mentioned above, deforestation is greater where there is good road access. On the other hand, the use of burning is lower in such areas, which will tend to lead to better resource conditions. Communities closer to the urban market were more prone to adopt several conservation measures; however, this may be a reflection more of access to information and technical assistance than access to markets or economic opportunities. The impacts of road access on measures of welfare are not as favourable as we expected to find. However, wage growth has been higher in the horticultural and non-farm employment pathways, indicating that the indirect effects of market access (as a determinant of these pathways) must also be considered. Road and urban-market access did not have significant direct effects on measures of poverty.

Overall, these results confirm our expectations that market access should lead to intensified use of agricultural inputs, while having mixed impacts on natural resource conditions. The findings do not show the impacts of market access on changes in wages and poverty that we expected, but this may be because these effects are mitigated by migration.

Access to technology

The presence of technical assistance programmes in agriculture has contributed significantly to several aspects of agricultural intensification and natural resource management, including reduced use of fallow and burning and the adoption of many conservation measures. We found no significant impacts of technical assistance on most outcome measures. Overall, the results suggest that technical assistance has been most effective in promoting more labour-intensive practices, but that these changes have not produced large measurable impacts on agricultural productivity, resource conditions or poverty.

Policy implications

These research findings suggest several important implications for policy-makers seeking to increase agricultural productivity, ensure resource sustainability and reduce poverty in hillside areas of Honduras. The findings with regard to poverty suggest that basic infrastructure and public services are critical and badly needed throughout most of the central region and that interventions to promote sustainable agricultural production are likely to be insufficient to address the problems of poverty in the region. Efforts to provide these services should be continued and expanded to poorly served areas, particularly more remote areas, regardless of what is done to address resource management issues.

Our findings with regard to technical assistance programmes indicate the importance of increasing productivity as a primary objective if such programmes are to have a substantial and long-term impact. The low adoption of most conservation measures despite substantial efforts to promote them is often due to their high labour costs and limited short-term economic benefits, according to survey respondents. This has often been the case with similar measures promoted elsewhere (for example, see Lutz *et al.*, 1994).

Part of the problem with technical assistance appears to be a lack of targeting of different types of measures to different situations. A 'one-size-fits-all' approach is unlikely to be successful, given the pathway-specific nature of opportunities and constraints. For example, production and conservation practices that are appropriate for labour-intensive agriculture, such as in forestry communities, may not be appropriate in more extensive production systems, such as in the basic grains expansion pathway or in more external input-intensive production, such as in the horticultural pathway. Less labour-intensive practices with income-earning potential need to be developed and promoted in such pathways. For example, fruit production appears promising as a way to improve income and conserve soil, especially where population pressure is lower, as in the basic grains expansion pathway.

Credit, agrarian reform and land-titling programmes have had a limited impact in central Honduras, probably because of their limited presence. The importance of education as a conditioning factor in the adoption of some soil conservation measures suggests that educational improvement may have important 'spin-off' benefits for resource conservation, in addition to its direct impacts on reducing poverty.

The widespread decline in the use of burning is an example where policies and technical assistance programmes seem to have had a large beneficial impact for environmental objectives. The increase in continuous cropping, access to alternative techniques (particularly use of herbicides) and shifts into cash-crop production were undoubtedly also important contributing factors to this change.

In conclusion, there are many ways to promote more productive, sustainable and poverty-reducing development in the central region of Honduras. While some of these can be applied across the board, most will need to be tailored more specifically to the particular problems and opportunities available in different development pathways.

Acknowledgements

The authors gratefully acknowledge the financial support of the Swiss Development Cooperation Agency and the Inter-American Development Bank for this research, and institutional support from the International Food Policy Research Institute, the Inter-American Institute for Cooperation in Agriculture and the University of Maryland. We are grateful to the International Centre for Tropical Agriculture (CIAT) for providing access to census and other secondary data for Honduras. We are especially grateful to the rest of the study team – Fernando Mendoza, Carlos Duarte, Juan Manuel Medina, Oscar Neidecker-Gonzales and Roduel Rodriguez – and to the many farmers and others in Honduras who generously agreed to respond to our many questions.

Notes

1 This concept was inspired by the work on comparative patterns of economic development by Morris and Adelman (1988).

2 Krautkraemer (1994) offers a contrary view of the implications of population-induced intensification for labour productivity, based on assuming that the production set is non-convex.

3 Details on the study sample and the survey are available in Pender and Scherr (1997).

4 Most of the household-response variables (such as adoption of various conservation practices) were measured by an ordinal index from 0 to 6: 0 = no households used the practice; 1 = fewer than 10%; 2 = minority; 3 = about half; 4 = majority; 5 = more than 90%; 6 = all.

5 For example, we use multinomial logit to estimate determinants of the pathways, binary probit to estimate determinants of collective action, and ordered probit to estimate determinants of household responses and some outcome variables. All of these models are based on maximum likelihood estimation. The multinomial logit model is generally used to predict the probability of selecting a particular option from a set of mutually exclusive alternatives. Binary probit is used to predict the probability of selecting one of two options from a binary choice set. Ordered probit is used to predict the probability of selecting one option from an ordered choice set. Further details are available in Greene (1990).

6 In regressions explaining changes in resource conditions, we also included efforts to regulate forest use by the national or municipal governments.

7 Due to space constraints, further descriptive statistics cannot be included here, but they are available from the authors.

8 This is based on comparisons of aerial photographs in the late 1970s and early 1990s. Aerial photos were available for both dates for only 23 communities, including only one basic grains expansion and one forestry community and no basic grains stagnation community.

9 Devegetated land includes areas that are rocky or overgrazed or which have recently been logged or cleared for cultivation. The latter is probably more common in the basic grains expansion pathway than in the others.

10 In Pender and Scherr (1999), we report a positive association between the presence of external organizations and local organizational development in the study communities.

11 The reference (omitted) category in the regression results is the basic grains stagnation pathway. The coefficients reported in Table 10.2 represent the effect of the indicated variable on the (logarithm of the) ratio of the probability of the community being in the indicated pathway relative to the probability of the community being a basic grains stagnation community.

12 In the remainder of this chapter, we present tables containing only the qualitative results of the econometric analyses. The actual regression results are available in Pender *et al.* (1999b).

13 The relationship reaches a maximum at a population density of 100 persons per km^2, which is greater than the maximum population density of communities in the sample. The same is true in all subsequent analyses in which the coefficients of population density and population density squared are statistically significant.

14 For example, the surprising positive association between distance to the urban market and adoption of some purchased inputs may be due to omitted land quality or climate characteristics that are correlated with distance to the market.

15 Unfortunately, our measures of change are relatively rough, in most cases only indicating whether a practice had increased, remained constant or decreased. For some types of practices, such as use of fertilizers and herbicides, it was so common for use to increase that regression analysis was not possible. For other practices, such as use of improved seeds and irrigation, changes were so uncommon that, again, regression analysis was not possible.

16 These results are reported in Pender and Scherr (1999). The probit model was not estimable with the pathway variables included.

17 Cross-sectional regressions explaining land use on steep slopes were not possible for recent land-use because of the limited number of communities for which aerial photos were recently available (23). The regressions for land use in Table 10.7 are based on land-use data from 37 communities.

18 These measures are plausibly more affected by agricultural and resource management decisions and productivity than other available measures from the census data (such as availability of sanitation, water and electricity), which are more related to public services.

19 Of course, to the extent that public expenditures are required to finance both efforts to improve public services and improvements in NRM, there will be tradeoffs in the use of such expenditures.

11 Implications of Resource-use Intensification for the Environment and Sustainable Technology Systems in the Central African Rainforest

JAMES GOCKOWSKI, G. BLAISE NKAMLEU AND JOHN WENDT

Humid Forest Ecoregional Centre of the International Institute of Tropical Agriculture (IITA), Yaoundé, Cameroon

Introduction

The rainforest of the Congo River basin of Central Africa is being rapidly degraded and deforested, contributing to global climate change and threatening unique plant and animal diversity, including the lowland gorilla, chimpanzee and African elephant.[1] Addressing these great environmental concerns must, however, focus on alleviating the grinding poverty that plagues the region. Since the Central African population is still largely rural-based and uses mostly underdeveloped agricultural technologies, this in turn means increasing the productivity of farming systems in the Congo basin.

Finding development pathways that can lead to sustainable livelihoods while minimizing or arresting declines in environmental services is a daunting task for stakeholders and policy-makers, but is critical to relieving the poverty and environmental stress that characterize much of the region. The principal goal of the Alternatives to Slash-and-Burn (ASB) system-wide programme of the Consultative Group on International Agricultural Research (CGIAR) is to identify and develop agricultural land-use systems and supportive policies and institutions that can best balance growth, poverty alleviation, carbon sequestration and biodiversity objectives (see Lee *et al.*, Chapter 1 of this volume). This chapter is concerned with assessing the extent to which agricultural intensification is compatible with these multiple objectives in the Congo basin.

To develop a policy-driven strategy for increasing productivity, it is important to first understand the process of autonomous agricultural intensification and build on its positive implications for both the environment and livelihood strategies (Lele and Stone, 1992). The Forest Margins Benchmark area provides a unique 'laboratory' for studying these processes over a gradient

(eds D.R. Lee and C.B. Barrett)

of population- and market-driven intensification of agriculture (Fig. 11.1). Ecoregional research at the Humid Forest Ecoregional Centre of the International Institute of Tropical Agriculture (IITA) in Cameroon seeks to understand the evolution of rural development processes (including the intensification of land use) under different levels of resource-use pressure. This understanding is then used to develop crop and resource-management technologies specifically targeted to development domains defined by varying endowments of land, labour, rural institutions and market access.

In this chapter, we examine: (i) the environmental consequences of both population-driven and policy-led intensification; (ii) tradeoffs in economic performance, carbon stocks and biodiversity for the dominant land-use systems; and (iii) the underlying factors leading to intensification at both the household and community levels. These findings have implications for targeting policies and research to address the difficult goals of poverty alleviation and improvement in the environment. The main target of policy-driven intensification in southern Cameroon since the 1930s has been the cocoa agroforest. We examine briefly the mostly unsuccessful attempts to intensify this land-use system and recent policies that led to extensification of the sector, with important negative environmental consequences. For national and international research institutes interested in developing resource-management technologies under conditions faced by farmers, the delineation of development domains becomes very important. Using an area-based sampling frame, household endowments of land, labour and field-management practices, as well as commercialization and diversification strategies, are examined. An analysis of the spatial patterns of intensification and commercialization strategies also contributes to domain definition.

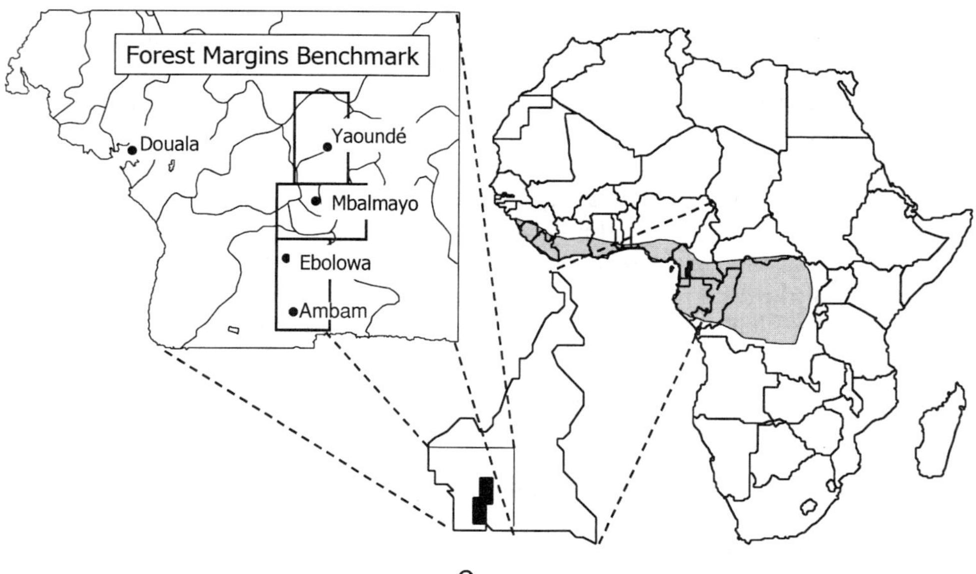

Fig. 11.1. ASB forest margins ecoregional benchmark area and the humid forest of West and Central Africa.

Agricultural Intensification in the Congo Basin

While agricultural intensification has led to increasing per capita food supplies in many areas of the world, Malthusian concerns are still raised for Africa (Lele and Stone, 1992; Pinstrup-Andersen *et al.*, 1997). Low productivity growth has resulted in stagnant rural incomes and incessant pressure on the forest margin. Boserup (1965) countered the arguments of classical economists, such as Malthus, who foresaw population growth leading to declining natural resources and threatening food security, by pointing to the role of growing populations in stimulating technical change and land-use intensification. Empirically, she concluded that labour productivity in shifting cultivation using unimproved techniques depends on the length of the fallow period. As population pressures increase and fallow periods shorten, declining productivity leads to a shift towards more labour-intensive agricultural techniques, such as manuring, improved fallow techniques, composting, ridging and fertilizer use.[2] In instances where households may not have access to these techniques or the knowledge to manage them, yields will typically decline and land resources may be exhausted.[3] This latter scenario is a particular worry for the Congo basin, where the infrastructure and extension services needed to rapidly diffuse technologies are underdeveloped.

A corollary to Boserup's central hypothesis is that farmers will adopt a labour-intensive system once knowledge of that system becomes available and if the returns to labour are higher than for existing technologies. This corollary implies an important but largely unachieved research and extension objective – that is, to develop and diffuse technology systems capable of increasing both labour and land productivity in land-abundant environments.

Schultz (1964) prescribes investments in agricultural research and industrial capacity to produce modern inputs (pesticides, fertilizers, etc.) and, perhaps most importantly, in the capacity of farmers to use these improved inputs effectively. The success of the scientific community in developing the dwarf wheat and rice varieties of the Green Revolution in Asia and Latin America has lent credence to these prescriptions (Hayami and Ruttan, 1985). Application of the Green Revolution paradigm to the land-surplus economies of sub-Saharan Africa has been questioned, as has the appropriateness and profitability of inorganic fertilizer use on the acidic soils of Central Africa (Lele, 1975, 1989; Eicher and Baker, 1982; Spencer, 1985; Delgado and Pinstrup-Andersen, 1993; Spencer and Badiane, 1995). In contrast, Sanchez (1976; see also Sanchez *et al.*, Chapter 17 of this volume) concludes that long-term experiments conducted throughout Africa demonstrate the agronomic sustainability of continuous cultivation with the combined use of fertilizers and manure. The profitability of fertilizer use is location-specific and depends on transportation costs, pricing policies, crop responses and internal fertilizer production capacity.

Conceptually, agricultural intensification consists of increasing all other factors of production relative to the scarcest factor, with scarcity defined in terms of social opportunity cost (Delgado and Pinstrup-Andersen, 1993). Evaluating land scarcity is complicated by difficulties in measuring the potential outputs and services provided by the tropical rainforest. Plant nutrients generated by slashing and burning the forest biomass, sustainable timber harvests, non-timber products, biodiversity values, watershed functions, impacts on local and global climate change – all of these are among the outputs and services of the tropical rainforest. Some of these outputs are valued privately by the farmer and others by society at large. The increasing depletion of these services has led to increases in both the private and social opportunity cost of a hectare of tropical forest.[4]

However, where population densities are still low, as in many parts of the Congo basin, labour scarcity (evaluated at private costs) results in labour-extensive, land-intensive cropping systems that seek to maximize the returns to labour. Labour scarcity is also likely to be subject to seasonal variation according to the agricultural calendar. Under such conditions, technologies and innovations that are land-saving will only be adopted if they are also labour-saving and increase the farmer's returns to labour. When the factor endowments of these agricultural systems are evaluated at their private costs, generally labour appears to be the scarce resource. However, from a policy perspective, we are concerned with the social costs of cropping a hectare of rainforest, and we therefore posit that land is the scarce factor (measured at social cost). We then define resource use intensification as an increase in the quantity and/or quality of inputs used in conjunction with agricultural land (inputs per hectare per unit of time).[5] Inputs are defined broadly to include seeds, scientific knowledge, management practices, labour, irrigation, soil amendments, pesticides and capital inputs (including human investment).

The ASB Forest Margins Benchmark Area and Resource-use Intensification

Population-driven intensification has proceeded farthest in the northern portion of the benchmark region, referred to as the Yaoundé block (see Fig. 11.1). Here, population densities exceed 150 persons km^{-2} in some localities and the fallow period has been reduced to 2 or 3 years on average. In contrast, most of the southern benchmark site (Ebolowa block) is characterized by low population densities (fewer than 10 persons km^{-2}) and extensive crop–fallow rotational systems. While Ebolowa farmers fallow on average for 7–8 years, they have the option of resting the land much longer if they begin to encounter problems due to declining soil fertility, pests or disease.

Farms in the benchmark area are generally small, estimated to be 2.6 ha in the Yaoundé block, 2.4 ha in the Mbalmayo block and 3.6 ha in the Ebolowa block, and use few capital inputs (Gockowski *et al.*, 1997). The most important and frequently encountered food cropping system is the groundnut/cassava-based mixed system that is responsible for household food security (Westphal, 1981; Gockowski *et al.*, 1997). In areas with access to markets, this system also generates surplus revenues for the women farmers who manage it. The next most frequent system, and that with the largest land area, is the cocoa agroforest. Managed by men and, in some instances, widowed female household heads, this system has been the foundation of the rural cash economy in southern Cameroon since the 1940s. The third most frequently encountered field system is the plantain/cocoyam system, which often also includes the cultivation of melon seed (*Cucumeropsis mannii*), where population densities are low.[6] These systems are generally targeted to longer-period fallow fields and secondary forests. High-value monocrop horticultural systems are by far the most intensive annual cropping system, and are found predominantly in the Lékié division of the densely populated Yaoundé block. First introduced in the 1970s, diversification into horticulture has expanded rapidly in response to growing urban demand, with little support from public research or extension.

Policies and institutional development directed at the cocoa sector have played a significant role in its intensification since the early 1920s. When the French were entrusted with western Cameroon following the First World War, they undertook a concerted effort to develop the smallholder cocoa sector, reflecting a shift in French colonial policy favouring the intensification of indigenous export agriculture (Bonneuil and Kleiche, 1993). Factors influencing the spread of cocoa included farmers' cash requirements to pay poll taxes, cocoa's role in establishing land-tenure rights and the development of new demands through merchandise trading and a nascent rural goods market (Myint, 1965; Assoumou, 1977; Weber, 1977; Guyer, 1984).

The cocoa stabilization fund, established in 1956, set producer prices on average at 50% of the Douala free-on-board (f.o.b.) price. The significant surpluses accruing to the fund were to support producer prices in periods of low world prices and for cocoa-sector development. In the late 1980s, the Cameroon economy, in the throes of its most severe recession since independence, was further pummelled when world cocoa prices plunged. By 1990, the government was no longer able to maintain official producer prices in the cocoa and coffee sectors and cut cocoa prices from 420 to 250 CFA francs kg^{-1}. In response, cocoa producers shifted labour and capital into other employment. The impacts on the environment of this policy-led extensification in the cocoa sector are explained below.

The success of efforts to intensify the agricultural sector in southern Cameroon has been mixed. Cocoa yields are low, ranging from 100–200 kg ha^{-1} in the land-abundant southern portion of the benchmark area to 300–500 kg ha^{-1} in the more intensified systems of the Yaoundé block (MINAGRI, 1987; Losch *et al.*, 1990; SODECAO, 1995; Gockowski *et al.*, 1997). With the exception of improved open-pollinated maize varieties developed by the cereal grains project of the US Agency for International Development, IITA and Cameroon's Institut National de Recherche Agricole pour le Développement (IRAD) in the 1980s, almost no policy/research-led intensification of staple-food cropping systems in southern Cameroon has occurred. Efforts to intensify agriculture in the benchmark area have proceeded along two fronts: policy-led intensification of the cocoa sector through an emphasis on research and extension (with mixed results), and the autonomous intensification of both cocoa and annual food-crop systems. The latter are mostly confined to areas of high population pressure and good market access (i.e. the Yaoundé block), with some localized intensification in the urban peripheries of the other two major towns in the benchmark, Ebolowa and Mbalmayo.

The Environmental Effects of Agricultural Intensification

The effects of intensification on carbon stocks

Tropical forests constitute a major sink in the global carbon dioxide (CO_2) balance sheet (Tinker *et al.*, 1994). The more CO_2 that is sequestered in tropical forests and cropping systems, the less is available for global warming. Collecting data on carbon dynamics across tropical agroecosystems and land-use systems was one of the objectives of the ASB programme. The measure of carbon stocks developed in ASB research for a given land-use system is the average annual carbon stock measured in tons per hectare over one cycle of the system. Mathematically, it is the time integral of the equation describing carbon sequestration divided by the number of years in the cycle (see Appendix 11.1). Above- and below-ground carbon was measured in six land-use types: annual crop fields, short fallow fields of 1–4 years dominated by the woody herbaceous species *Chromolaena odorata*, medium fallow fields of 7–9 years, older fallow fields of 13–18 years, forest plots of 50–100 years and mature cocoa perennial crop systems of 25 years. The Ruthenberg ratio, $r = t_{crop}/(T_f + t_{crop})$, where t_{crop} is the period of cropping and T_f is the fallow period, is a useful descriptive index of cropping intensity in fallow rotational systems (Ruthenberg, 1980). When r increases, it indicates an intensification of cropping over time. When $r = 1$, there is no longer a fallow period and the land is in continuous cultivation. An increase in the Ruthenberg index in the Cameroon context, where agricultural mechanization is not a factor, is also associated with increasing average annual labour input per hectare.

Total time-averaged carbon stocks[7] for the four principal administrative divisions of the benchmark area and for the various land-use systems are reported in Table 11.1.[8] In terms of agricultural intensification, these divisions can be ranked as follows: Lékié > Mefou > Nyong et So'o > Ntem. Overall, we estimate that the effects of agriculture have reduced the total carbon pool in the benchmark area to 80% of its primordial level. Carbon stocks decline with agricultural intensification, from 89% of the primordial level in the Ntem division to 36% in the intensely cultivated Lékié division (Table 11.1). The share of total carbon that is in perennial crop systems increases with overall intensification. In the Lékié division, it is estimated that there are 2.5 Mt more carbon in cocoa agroforests than in the remaining forested land.

The carbon stock in crop–fallow rotational systems is an indicator of intensification and system sustainability. In moist tropical forests, the ratio of nutrients in the biomass to those in the soil is far higher than that in temperate forests.[9] Total carbon stock is proportional to total biomass, while total biomass is positively correlated with total nutrient stocks (although this relationship is non-linear). The carbon stock is thus an approximate indicator of the total nutrient stock of a land-use system. If carbon levels are high, the elevated

Table 11.1. Carbon stocks (Mt) across land-use categories and administrative divisions of the Forest Margins Benchmark area of southern Cameroon.

	Lékié		Mefou		Nyong et So'o		Ntem		Subtotal	
Land-use system	Mt	% total	Mt	% total	Mt	% total	Mt	% total	Mt	% total
Annual food	16.5	50	10.0	9	4.3	5	12.9	3	42.8	6
Biennial food	6.2	19	8.3	8	4.4	5	25.4	6	42.3	6
Cocoa	6.8	21	4.1	4	1.9	2	6.0	1	19.0	3
Forested land	3.7	11	85.3	79	83.9	89	390.7	90	563.6	84
Divisional total	33.2		107.7		94.5		435.0		667.7	
Average (t C ha^{-1})	11		22		26		27		24	
Primordial state (Mt)	91.8		144.7		109.9		491.2		837.4	
Percentage current to primordial	36		74		86		89		80	

quantity of nutrient-rich ash following burning will produce sufficient returns to labour to maintain rural livelihoods and enable the system to be reproduced. However, once system carbon declines and fertility is reduced, intensification, in the form of new management techniques (improved planted fallows, new crops, purchased inputs and new cropping rotations), is required in order to sustain rural livelihoods.

The estimated carbon stock in an intensive crop–fallow rotational system with a Ruthenberg ratio (r) of 0.5 at the time of slashing (T) is 90 t ha^{-1}, versus 212 t ha^{-1} for an extensive system with an r value of 0.11 (Table 11.2). Assuming that the composition of plant nutrients is roughly the same for the two different fallow periods, available plant nutrients in the intensive short-fallow system will be considerably lower. Empirically, crop systems of rural households in the Lékié division are most likely to have Ruthenberg ratios approaching or exceeding 0.5. Thus, these households have intensified the field management of their food-crop systems to a much greater degree than other areas of the benchmark area (Gockowski *et al.*, 1997).

The carbon stocks sequestered in cocoa agroforests may help reduce global warming and should be considered in discussions of carbon sinks and emissions trading (Newmark, 1998). The conversion of 1 ha of a short-fallow–crop rotational land use to a cocoa agroforest could sequester up to 72 t of additional carbon, depending on the fallow length and the cocoa production cycle (Table 11.2). Cocoa agroforests can also play an important role in maintaining hydrological functions. Deforestation can disrupt the energy/water balance, local precipitation patterns, drainage, runoff and water yield (Tinker *et al.*, 1994). The scale at which these effects occur is not easily discerned because of the difficulty in obtaining data for large areas. The substantial area still in perennial crop systems in the Lékié division (an estimated 13% of total area) may help maintain the hydrological functioning of the environment and precipitation patterns.

Table 11.2. Dynamics of carbon stocks for various land-use systems in southern Cameroon.

	Crop period/ age of tree stock (years)	Fallow period (years)	Total carbon at slashing (t C ha^{-1})	Time-averaged carbon (t C ha^{-1})	Ruthenberg ratio
Short fallow–intercrop	1.5	1.5	90	82	0.50
Short fallow–intercrop	1.5	3.0	104	87	0.33
Short fallow–intercrop	1.5	4.0	115	91	0.27
Medium fallow–intercrop	1.5	5.0	127	96	0.23
Medium fallow–intercrop	1.5	7.0	153	106	0.18
Medium fallow–intercrop	1.5	9.0	178	117	0.14
Long fallow–intercrop	2.0	12.0	206	132	0.14
Long fallow–intercrop	2.0	14.0	215	142	0.13
Long fallow–intercrop	2.0	16.0	212	150	0.11
Cocoa plantation	15.0	–	143	111	–
Cocoa plantation	25.0	–	190	132	–
Cocoa plantation	40.0	–	190	154	–
Primary forest	–	–	307	307	–

The environmental importance of maintaining cocoa agroforests in the landscape has not been fully recognized by national and international decision-makers. Evidence for this is found in policies undertaken in the name of structural adjustment, which severely undermined cocoa-producer incentives in the late 1980s and early 1990s and permitted Cameroon's smallholder rural economy to decline.

Policy-led extensification and carbon stocks

From the 1940s to the present, cocoa has been the major source of revenue and employment in the rural economy of southern Cameroon. When official prices were abruptly reduced by 40% in 1989/90, sliding to 50% of their former levels by 1993/94, producers responded by shifting labour into extensive slash-and-burn plantain and cocoyam production systems. Assessing the environmental impacts of this policy-led extensification of the cocoa sector depends on resolving empirical questions – in particular, the magnitude of the resulting shift in land use (specifically, the supply elasticity of land for plantain/cocoyam production) and the type of land converted (i.e. short-fallow fields versus secondary forest). A simple partial-equilibrium model of plantain/cocoyam acreage response to changes in cocoa prices is used to analyse the effects of this land-use change on carbon stocks.

In 1994, an ASB characterization survey in the four principal benchmark divisions estimated that 24,630 ha were planted in long-fallow plantain/cocoyam systems and 114,700 ha were planted in shaded cocoa agroforests (Gockowski *et al.*, 1997). In 1993/94, when these fields were planted, the official price of cocoa was 250 CFA francs, significantly below the official price of 420 CFA francs in 1988/89. The quantitative response of producers as measured by the change in plantain/cocoyam acreage depends on the cross-price elasticity of cocoa.[10] For our purposes here, we assume constant-elasticity functional forms and competitive products in production. The acreage response models were parameterized using a range of plausible elasticities and changes were estimated given the price changes from 1988/89 to 1993/94 (Table 11.3). The estimated increases in plantain-based cropping systems range from 1240 ha to 5630 ha.

The net impacts of these cropping-system changes on carbon stocks and deforestation depend on the type of land conversion. The conversion of forested land will have a larger negative impact on carbon stocks than will the conversion of 14-year-old fallow land. Farmers can expand plantain-based systems by bringing currently existing long-fallow fields into production, resulting in an increase in the Ruthenberg cropping intensity ratio (i.e. the fallow period shortens). Alternatively, they can extend the forest margin and convert forestland to plantain/cocoyam production. There is no information available at this level of detail, however. To look at these possibilities, we examine three scenarios. In scenario 1, we assume that all of the expansion occurs in existing long-fallow rotations. In scenario 2, the expansion occurs by converting only forested land. In scenario 3, the expansion is distributed equally across the existing long-fallow fields and forested land.

The results are presented in Table 11.4. The estimated net decline in the total carbon stock of the agroecosystem ranges from 0.21% to 2.14%, depending on the elasticity assumption and land-conversion scenario. For a given elasticity, the lowest loss in carbon stock occurs under scenario 1, where production is expanded by targeting already existing fallow lands. While this type of expansion is most desirable from an environmental perspective, from the farmer's perspective, it may not be optimal. It yields a shorter fallow period and lower accumulated biomass at the end of the fallow; thus, there is likely to be a decline in output if no additional inputs are substituted. The most serious loss in carbon stocks occurs when forested land is converted and the supply elasticity is high. In addition to the impact on carbon stocks, biodiversity would certainly be affected by the substitution of a plantain/cocoyam intercrop for a relatively diverse land use consisting of long fallow and secondary forests.

Table 11.3. Estimated changes in plantain-based cropping systems as a result of 1989/90 cocoa-price decline in southern Cameroon.

Cross-price elasticity	Area change in plantain-based system (ha)	Assumed factor demand
–0.1	+1240	$L_{\text{plant}} = 42{,}782\ P_c^{-0.1}$
–0.2	+2430	$L_{\text{plant}} = 74{,}311\ P_c^{-0.2}$
–0.3	+3550	$L_{\text{plant}} = 129{,}076\ P_c^{-0.3}$
–0.4	+4620	$L_{\text{plant}} = 224{,}202\ P_c^{-0.4}$
–0.5	+5630	$L_{\text{plant}} = 389{,}434\ P_c^{-0.5}$

L_{plant}, area in plantain-based cropping system; P_c, producer price of cocoa.

Table 11.4. Estimated changes in carbon stocks as a result of cocoa price decline and expansion of slash-and-burn plantain systems in southern Cameroon.

	Scenario 1: Fallow conversion			Scenario 2: Forest conversion		Scenario 3: 50% fallow–50% forest conversion		
Increase in plantain-based systems (ha)	Change in fallow length (years)	Change in carbon stocks (Mt)	Change relative to total stock (%)	Change in carbon stocks (Mt)	Change relative to total stock (%)	Change in fallow length (years)	Change in carbon stocks (Mt)	Change relative to total stock (%)
1240	0.8	−3.8	0.21	−8.6	0.47	−0.39	−6.2	0.34
2430	1.6	−7.6	0.42	−16.7	0.92	−0.78	−12.1	0.67
3550	2.5	−11.1	0.61	−24.4	1.35	−1.16	−17.7	0.98
4620	3.5	−14.3	0.79	−31.8	1.75	−1.55	−23.1	1.28
5630	4.4	−17.0	0.94	−38.7	2.14	−1.93	−28.2	1.56

The likelihood of the type of policy shock (e.g. a cocoa price decline) examined here has been mitigated with the liberalization of cocoa export marketing. However, this analysis illustrates how policy distortions can have serious environmental effects and, alternatively, how price policy might be used to correct for environmental externalities. In the context of evolving discussions on carbon trading and greenhouse-gas reductions under the Kyoto Protocol, the seemingly small changes estimated above can, in fact, be quite significant. To illustrate, if a price of US$5 t^{-1} is assumed for carbon released into the atmosphere, the model simulations presented in Table 11.4 represent environmental costs ranging from US$19 million to US$193.5 million.

The effects of intensification on plant biodiversity

This research only measured plant diversity at the plot level. The biodiversity measure used in ASB research incorporates the notion of plant functional attributes developed by A. Gillison of the Centre for International Forestry Research (CIFOR) with standard measures of species richness. Six plant functional-attribute classes are used: leaf size, leaf inclination, chlorotype, leaf type, life form and root type. Each attribute class is composed of a subset of elements. For instance, the three possible elements for the leaf inclination attribute class are vertical, lateral, pendulous and composite. The combination of all classes and their descriptive elements is referred to as the 'plant modus'. Each species is associated with one modus. The ratio of species to plant modi measured in a 40 m × 5 m transect is used as an indicator of plant diversity. This ratio, by combining the more commonly used measure of species richness with plant functions, provides a measure of the robustness of the ecological functions of the system. Land uses with a higher ratio (i.e. more species for a given number of modi) would be expected to have greater ecological resilience in the face of change. Biodiversity in crop–fallow rotational systems was evaluated by comparing points in the time sequence having the highest species/modi ratio. The relationships between these measures of plant biodiversity and the biophysical functioning of agricultural land-use systems were not addressed.

Surprisingly, significant differences in plant biodiversity as the fallow period shortens were not observed (Table 11.5). Substantial differences do seem to exist between the annual crop systems and cocoa agroforests. The species/modi ratio and species richness were highest for the cocoa agroforest and approached 80% of the indicated plant diversity in forestland. The overall importance of the biodiversity preserved in cocoa agroforests increases with population pressure. In the intensely cropped Lékié division, the most important stocks of biodiversity are found in the lands accounted for by cocoa agroforests and secondary forests (13% and 7% of total area, respectively).

The cocoa agroforest is particularly effective in maintaining the biodiversity most valued by rural populations. These complex multistrata systems serve as biological reserves for many forest products used and traded. Over the last 60 years, where land pressures have increased and the forest has largely disappeared, farmers have nurtured and transplanted wild seedlings of indigenous

Table 11.5. Measures of plant biodiversity by land-cover type in southern Cameroon.

Land cover	Modi	Species richness	Species/modi ratio
Raffia forest	29	57	1.97
Secondary forest	39	76	1.95
Cocoa plantation	38	63	1.66
Bush fallow (8–15 years)	35	53	1.51
Chromolaena fallow (2–4 years)	44	64	1.45
Annual food-crop field	32	42	1.31
Forest food-crop field	12	14	1.17

fruit and timber trees into their cocoa agroforests. In the subdivision of Sa'a, one of the most heavily populated areas in the benchmark region, the production and marketing of the widely consumed oilseed *ndjansan*, from the forest tree species *Ricinodendrom heudelotii*, is an important commercial activity. Most of this production originates from cocoa agroforests, where this species also serves as shade for the cocoa understorey.

While the international community is greatly concerned with the maintenance of the environmental services discussed above, poor farmers eking out a marginal existence can only be concerned with the capacity of their farming systems to maintain livelihoods. This naturally leads to concerns about tradeoffs between environmental and economic indicators.

Tradeoffs between economic growth and the environment

The policy analysis matrix (PAM) approach (Monke and Pearson, 1989) was used to evaluate the economic returns and policy distortions affecting the major land-use systems and hybrid oil-palm production system. The net present value (NPV) of costs and returns over a 30-year period was calculated for these systems, using a 10% discount rate and an opportunity cost of family labour of US\$1.21 person^{-1} day^{-1}. Six perennial crop systems were evaluated:

1. Intensive cocoa with mixed fruit-tree shade canopy planted into short fallow.
2. Intensive cocoa planted into short fallow.
3. Extensive cocoa with mixed fruit-tree shade canopy, planted into forestland/long fallow.
4. Extensive cocoa planted into forestland or long fallow.
5. Improved *Tenera* hybrid oil-palm system planted into short fallow.
6. Improved *Tenera* hybrid oil-palm system planted into forested land or long fallow.

These were compared with the two dominant slash-and-burn food-cropping systems in the benchmark area:

7. Intercropped groundnut/cassava-based food field planted into a short fallow.

8. Intercropped melon/plantain/cocoyam food field planted into a long fallow.

Together, these last two cropping systems and the cocoa agroforest account for an estimated 87% of all cropland in the benchmark area (Gockowski *et al.*, 1997). The four variants of cocoa agroforest reflect differences in the intensification and commercialization of these systems across the benchmark region. As mentioned above, the cocoa agroforests in the Yaoundé block are more intensively managed than those found elsewhere. A distinction is also drawn between agroforests with urban market access for secondary fruit-tree production and those without a commercial fruit-tree component.

The domestic resource cost ratio (DRC) was estimated to be less than 1 for all systems (Table 11.6).[11] The intensive cocoa system with fruit and the hybrid oil-palm system in forested land had the lowest DRCs (0.49 and 0.48), while the extensive plantain-based system, the extensive cocoa system without fruit and the short-fallow rotation had the highest estimated DRCs. Effective protection coefficients (measuring the degree of taxation or subsidy) for the perennial crop systems were in the range of 0.89 to 0.93.[12] Taxation on cocoa (mainly consisting of import tariffs on pesticides and a 10% excise tax on production) was much lower in the late 1990s compared with when the national marketing board was operating and setting producer prices by presidential decree. The liberalization of export markets in 1994, particularly with increased world market prices in the late 1990s, has provided greater incentives for cocoa and coffee producers. Internal food crop markets are not distorted, as indicated by the estimated effective protection coefficients (EPCs) of 1.0 for the rotational fallow systems.

The returns to labour are a critical variable for determining the adoption potential of a given technology system, particularly in land-abundant, labour-scarce economies, such as that of the Ebolowa block of the benchmark region. A grid search was conducted to determine the wage rate at which the NPV was equal to zero; this was used as the measure of the returns to labour. The intensive fruit–cocoa system and the oil-palm system planted into forestland had the highest returns (US\$2.36 and US\$2.44 person^{-1} day^{-1}, respectively) and the extensive cocoa and long-fallow plantain systems the lowest (Table 11.6).

The results for biodiversity and social profitability show that the two intensive cocoa systems and the extensive cocoa system with fruit-trees offer relatively high profitability while maintaining a high level of biodiversity. In contrast, the crop–fallow rotational systems and the short-fallow oil-palm systems perform rather inadequately on both counts. The possibility of augmenting the biodiversity in these latter systems is low (unlike the potential for increasing their productivity and profitability).

Both above-ground carbon stocks and social profitability were high for the intensive cocoa and extensive cocoa with fruit systems, and for hybrid oil planted into forestland. Oil-palm systems, while low in biodiversity, are similar to cocoa in terms of carbon sequestration. The long-fallow intercrop rotation is also comparable to the perennial tree-crop systems in terms of time-averaged carbon; however, the social profitability of this system is low. The short-fallow–intercrop rotation performs poorly in both dimensions.

Table 11.6. Selected results for ASB land-use systems.

	Domestic resource cost	Effective protection coefficient	Social profitability (US$ ha^{-1})	Private returns to labour (US$ day^{-1})	Time-averaged carbon stock (t C ha^{-1})	Biodiversity (species/modi ratio)
1. Int. cocoa w/fruit	0.49	0.90	175	2.36	154	1.66
2. Int. cocoa	0.58	0.89	123	1.95	154	1.66
3. Ext. cocoa w/fruit	0.68	0.90	113	2.13	154	1.66
4. Ext. cocoa	0.54	0.93	61	1.63	154	1.66
5. SF–oil-palm	0.62	0.90	98	1.81	153	1.18
6. Forest–oil-palm	0.48	0.92	165	2.44	153	1.18
7. SF–intercrop	0.69	1.00	64	1.79	82	1.45
8. LF–intercrop	0.73	1.00	28	1.70	142	1.51

Int., intensive; Ext., extensive; SF, short fallow; LF, long fallow.

Carbon stocks in short-fallow systems can be increased through agroforestry techniques. Four-year *Chromolaena* fallow–intercrop rotations were estimated to have a time-averaged value of above-ground carbon equal to 4.53 t ha^{-1} and an above-ground accumulation rate of 2.90 t C ha^{-1} $year^{-1}$ during the fallow period. IITA on-station work with improved fallow interventions has shown above-ground accumulation rates of up to 10 t C ha^{-1} $year^{-1}$ using leguminous tree species, such as *Calliandra cathyothryus* (S. Hauser, Yaoundé, Cameroon, 1998, personal communication). A 4-year fallow–intercrop rotation with *Calliandra* accumulating at this rate would increase above-ground time-averaged carbon accumulation to 14 t ha^{-1}, sequestering nearly 10 t more carbon than the natural fallow system.

Resource-use Intensification and Technology Targeting

Thus far, we have focused on the environmental aspects of the intensification process. We now turn our attention to household objectives and the process of intensification across the benchmark. A principal components analysis (PCA) using survey data from 225 households in 15 villages across the benchmark region was used to determine relationships between cropping intensification, various field-management practices, field differentiation and agricultural commercialization strategies (see Appendix 11.2). This analysis seeks to uncover patterns and associations by looking at the loadings on variables across the principal components of the variation. Questions one might address include: Are particular agricultural commercialization strategies associated with field differentiation or increased field management? The analysis can also be utilized spatially to find out where a majority of households are exhibiting similar responses to variables. This information is useful for targeting on-farm interventions that are compatible with household-factor endowments and strategies.

The variables included in the analysis are given in Table 11.7. ALLMGT is the sum of enumerated (or 'count') variables for various field-management

Table 11.7. Variables used in principal components analysis.

	Mean	Standard deviation
ALLMGT = sum of field-management practices	3.086	2.839
GNUT = 1 if groundnuts commercial strategy, 0 otherwise	0.201	0.368
CASSAVA = 1 if cassava commercial strategy, 0 otherwise	0.455	0.457
COCOA = 1 if cocoa commercial strategy, 0 otherwise	0.550	0.457
PLANTAIN = count of plantains/cocoyams among top three earners	0.899	0.624
HORT = count of tomatoes/peppers/green maize among top three earners	0.364	0.579
OTH_LAB = count of non-family labour tasks	4.431	4.382
NOTYPE = count of different field types	4.631	2.075
TLAB = number of adult labour equivalents at the household level	4.297	2.867
PRESSURE = fallow fields/annual-crop fields cultivated	1.688	1.714

practices. Included in this measure are counts of fields tilled before planting, crops for which ridges are used, crops planted in rows, the number of improved varieties cropped, and fertilizer and pesticide use on food crops. CASSAVA, PLANTAIN, COCOA, HORT and GNUT are qualitative indicators of commercialization strategies. NOTYPE is a count of the number of different field types cropped by the household differentiated over time and space. This variable is often associated with the intensification process, as households modify their cropping systems in response to resource-use intensification in order to exploit resource niches and market opportunities more efficiently (Baker and Dvorak, 1993). The variables assumed to be exogenous in the short run are household endowments of labour and land (TLAB and PRESSURE). The variable PRESSURE is a proxy for the inverse of the Ruthenberg ratio at the household level. The rotated loadings for the five principal components with eigenvalues greater than 1.0 are presented in Table 11.8. Together, the five components account for 71.3% of the total variation among the set of ten variables.

The high positive loadings on HORT and ALLMGT and the negative loading on PLANTAIN for the first linear variate, CROPSYS, distinguish households pursuing commercial horticultural production (associated with intensive management practices) from those with commercial plantain and cocoyam production (using extensive farming practices). Spatially, the horticultural strategy is concentrated in the villages of the Yaoundé block, while households targeting extensive plantain systems are found mainly in the Ebolowa and Mbalmayo blocks, where population pressures are lower and woody biomass resources are more abundant. This pattern is, in part, related to the better-developed input markets in the Yaoundé block. In Yaoundé alone, there are approximately 30 retail shops selling the agrochemicals essential for horticulture production. There is also a significant market for chicken manure in the

Table 11.8. Principal components analysis of resource-use intensification and commercialization processes in southern Cameroon.

	Five principal (rotated) components				
	CROPSYS	INTENSI	COCOASYS	AFUBO	HHLABOR
ALLMGT	0.510	0.429	0.133	0.327	−0.058
GNUT	0.034	−0.224	0.245	−0.729	0.020
CASSAVA	0.107	−0.098	0.192	0.807	0.016
COCOA	−0.016	−0.091	−0.901	0.027	0.037
TLAB	0.202	0.176	−0.335	−0.130	−0.691
OTH_LAB	0.090	0.133	−0.344	−0.124	0.783
HORT	0.858	0.122	0.126	−0.046	−0.110
PLANTAIN	−0.864	0.098	0.194	−0.087	−0.010
PRESSURE	0.091	−0.782	−0.075	−0.005	−0.117
NOTYPE	−0.024	0.821	0.139	0.100	0.093
Percentage of total variance explained	18.120	16.110	12.355	13.426	11.307

peri-urban agricultural sector. In contrast, input access in other parts of the benchmark area is much more difficult.

The high positive loadings on ALLMGT and NOTYPE and the negative loading on PRESSURE for the second linear variate, INTENSI, suggest that, as land pressures increase, farmers respond both by augmenting the number of field-management interventions and by differentiating their field systems to exploit new commercial and ecological niches. Spatially, households in the upper tail of the distribution for this variate were concentrated in the Yaoundé block, where population pressures are highest. Low scoring villages were mainly found in the Mbalmayo and Ebolowa blocks.

The third linear variate (COCOASYS) is somewhat unidimensional, with a high positive loading only on COCOA and marginal loadings on the labour variables, TLAB and OTH_LAB. Spatially, cocoa production decreases in commercial importance for households residing in villages around the Yaoundé urban periphery. High transport costs in the Forest Margins Benchmark region confer a von Thünen-type pattern of comparative advantage for food commodities, in which the importance of cocoa diminishes in the urban periphery.[13]

The high positive loadings on CASSAVA and ALLMGT and the negative loading on the fourth linear variate, AFUBO, distinguish women's commercialization and management strategies emanating from the mixed cassava–groundnut field. Women who produce marketable surpluses of cassava using intensive management practices are at one end of the distribution of this linear variate; while those marketing commercial quantities of groundnuts with relatively extensive management practices are at the other. Spatially, evidence of a von Thünen-type pattern of intensification is again exhibited, with cassava production most important in villages within the Yaoundé block, where transport costs are relatively low. Consumption preferences for fresh boiled tubers (rather than processed forms of cassava) are in part responsible for high transport-cost margins. In contrast, groundnuts, which have a much higher value-to-weight ratio, constitute a commercial strategy in villages more removed from major urban markets.

The high positive loading on OTH_LAB and the negative loading on TLAB for the fifth linear variate (LABOR) indicate that households which have higher levels of family labour availability tend to use other resources less frequently than those households which are less abundantly endowed with labour.

Summary and Conclusions

Agricultural intensification is inevitable for Central Africa. Given the rapid population growth of the region, the question is not whether intensification will proceed, but rather where, when, how, and with what impact on the environment and rural development. This analysis has shown that, in the Forest Margins Benchmark region of southern Cameroon, substantial environmental costs are associated with endogenous intensification driven by declining fallow periods and nutrient depletion of soils, as described by Boserup (1965). Carbon stocks have declined significantly and biodiversity has been lost. Across land-use systems, the importance from both environmental and rural development perspectives of maintaining perennial tree-crop systems in the landscape is evident. Mature cocoa agroforests (> 25 years) maintain approximately 60% of the carbon stock of primary forest, are important genetic reserves for indigenous fruit-trees and preserve important habitats for avian populations of threatened species, such as African grey parrots and hornbills. The cocoa and oil-palm perennial crop systems, however, were the most profitable production systems examined, now that major policy distortions, such as overvalued exchange rates and fixed-price marketing regimes, are no longer factors.

Policy change has been described as a pillar of the strategy for agricultural development in sub-Saharan Africa (Cleaver, 1993). The example of related markets for cocoa and plantain/cocoyams shows the importance of paying attention to policy linkages not only between sectors, but, perhaps even more importantly, between commodities within a sector. In southern Cameroon, most of the price distortions have been eliminated, as evidenced by EPCs close to 1, although perennial tree-crop producers are taxed more heavily than annual food-crop producers. Historically, this disparity has been much wider. An argument could be made that, if the full range of outputs from the cocoa agroforests (biodiversity, carbon sequestration and watershed functions) could be valued, a policy subsidizing the creation of new cocoa agroforests would be socially desirable. There is, of course, a major caveat: perennial tree crop systems generate net environmental benefits only when replacing degraded short-fallow lands. If these systems were established in primary or secondary forest, there would be a net loss to the environment, although conversion to multistrata agroforestry systems still maintains more environmental services than any other type of land conversion.

From both an agronomic and an economic perspective, perennial crops are well adapted to the conditions of the Congo basin. However, with the exception of cocoa and robusta coffee in southern Cameroon and robusta coffee in the Democratic Republic of the Congo, smallholder perennial tree-crop systems are not widespread (Manyong *et al.*, 1996). Encouraging the establishment of perennial tree-crop systems in the Congo basin should be a development priority. A strategy to encourage their establishment in already deforested fallow land offers a potential 'win–win' situation for both development and the environment. However, farmers will normally choose to establish their plantations in long bush and forest fallows when this type of land is at their disposal, in order to capture the fertility rent (Ruf, 1998). To counter this rational economic behaviour, new planting subsidies could be targeted for planting in degraded short-fallow situations, perhaps as part of a still-to-be-developed carbon emissions/sinks trading scheme. Such a strategy presumes either increased productivity of the remaining short-fallow lands or the substitution of market food purchases for on-farm production.[14]

In much of Central Africa, rural markets are not well developed, compared with West Africa. Without their development, households have no choice but to rely on own production for the bulk of their food consumption. In the short and medium term, while better infrastructure and market institutions are being developed, heavy emphasis must be placed on food-crop intensification. Thus a two-pronged strategy at the household level is envisioned: first, the substitution of perennial tree-crop systems for short-fallow annual cropping systems; and, secondly, the increased intensification of annual food-crop production systems.

A strategy to intensify annual crop systems requires paying attention to institutions, infrastructure and factor endowments. In areas such as the Yaoundé block of the benchmark site, where fallows are short but market access for outputs and inputs is high, 'Green Revolution'-type seed–fertilizer technology systems have a role to play in the intensification process. In areas where medium- and long-fallow annual food-crop systems are still in place and infrastructure and institutions are less developed, new varieties and cropping techniques are called for. In particular, varieties tolerant of the many mineral deficiencies that characterize the soils of the basin need to be developed. At the same time, cropping systems and land-husbandry techniques that are more efficient at capturing the high flush of nutrients following burning are needed.

Most of the staple crops of the region – cassava, plantain, yams, taro and cocoyams – are vegetatively propagated (VP). This is both a blessing and a curse in terms of achieving impacts from plant-breeding efforts. It is a blessing in that, once farmers have new VP varieties, they do not need to repeatedly purchase the same variety, as is the case with hybrid maize or horticultural crops. On the other hand, this means that the private sector is unlikely to play a role in research and multiplication. This places the onus on national researchers, extensionists, non-governmental organizations (NGOs) and farmer organizations to find, develop, adapt and distribute these varieties. Multiplication of VP material is generally more time-consuming than seed multiplication and often results in the transmission of pests and disease. Tissue culture speeds up the process and can control pests and disease, but this capacity needs to be built up in the region. Perhaps the most promising vehicle for diffusing improved VP materials is farmer groups and farmer federations, which have both flourished in southern Cameroon, following legislation in 1992 that allowed farmers to freely associate and form legally recognized village-based organizations. These organizations are proving to be an important means for the diffusion of varieties using rapid 'on-the-shelf' multiplication techniques for plantain, cassava, yams and cocoyams.

The allocation of political resources is relatively inefficient in Central Africa. The restrictions placed on political organization have limited political development in rural areas and biased the allocation of state resources in favour of the urban-based political élite (Bates, 1981). The institutionalization of public-sector capacity to provide a continuous stream of technology consistent with resource endowments has generally been most effective when the political environment has encouraged the development of farmer organizations (Binswanger and Ruttan, 1978). The development of these organizations in southern Cameroon is a step in the right direction, but much more needs to be accomplished. These grassroots initiatives are potentially important vehicles for accomplishing the bottom-up institutional changes so desperately needed to effect agricultural intensification in the Congo basin.

At the national level, policy-makers are concerned about food-security issues and maintaining adequate food supplies in urban areas. Interregional trade and trade liberalization should be encouraged, particularly across agro-ecological zones, to address these concerns. All countries in the Congo basin have significant land areas of moist savannahs. The productive potential of this zone for annual food-crop production can be even higher than the humid forest zone due to increased solar radiation. Furthermore, the environmental costs of bringing this land into production are likely to be considerably lower than in the humid forest. However, a large portion of the population of the Congo basin lives in urban centres with extremely poor linkages to these potentially productive savannahs. Developing transport corridors would significantly reduce the von Thünen-type intensification pressures around urban centres and enhance urban food supplies.

It is also important to consider the impacts of rapid urbanization in West and Central Africa on consumption patterns. Urban household consumption of imported rice and wheat flour is many times higher than that in rural areas. According to the last census, 52% of the population residing in the humid forest zone of Cameroon was urban-based in 1987. Countries across West and Central Africa are experiencing similar demographic trends, with many predicted to be predominantly urbanized within a short time. These factors lead to questions about the strategic comparative advantage of the humid forest zones of West and Central African countries. Would these countries be better off concentrating food production in their savannah areas, while promoting diversified perennial tree-crop systems in the humid forest zone to generate foreign exchange? The areas in the world capable of successfully producing cocoa, coffee, rubber, oil-palm and other tree crops are limited compared with those areas that can grow maize, wheat, rice and other staple grains. Further analysis of this issue is warranted.

Acknowledgements

The authors would like to thank Lindsey Norgrove, Stefan Hauser and the editors of this volume for their comments and suggestions on earlier drafts.

Notes

1 Annual estimated forest loss in the Congo basin was 11,420 km^2 from 1990 to 1995, equivalent to 0.6% $year^{-1}$. In comparison, estimated forest losses for Indonesia and Brazil (locations of two other ASB sites) were 10,840 km^2 (or 1.0%) $year^{-1}$ and 25,540 km^2 (or 0.5%) $year^{-1}$, respectively (FAO, 1997b).

2 Unfortunately, most of Central Africa has yet to achieve this endogenous intensification. There is no fertilizer production capacity in the region and in the largest country, the Democratic Republic of the Congo, average fertilizer application is less than 0.5 kg ha^{-1}. This low usage reflects high costs, a lack of availability and a lack of responsive crop varieties at farmers' disposal (World Bank, 1996a).

3 Improving farmers' access to technology and knowledge is a major institutional challenge for Central Africa. Development investment is low, with health services, agricultural research, extension, rural roads and communication infrastructure all woefully underfunded (Bosc and Hanak-Freud, 1993; Cleaver, 1993; Kesseba, 1995).

4 Globally, the Food and Agriculture Organization (FAO, 1997b) estimates that the total tropical forest area declined by 9.7% from 1980 to 1995, representing a loss of 163 Mha.

5 Use of this unit assumes that intensification is time-dependent. Intensification usually first means a shortening of the fallow period, implying that the amount of labour per ha per unit of time will increase.

6 This is a nutrient-demanding system, requiring large amounts of woody biomass and organic matter.

7 Per hectare carbon stocks were estimated using reported mean fallow periods and cropping periods for the different land uses, based on unpublished ASB survey results.

8 The four divisions account for 85% of the benchmark area. The Lékié and a small portion of the Mefou divisions correspond to the Yaoundé block; the Nyong et So'o and the major portion of part of the Mefou division correspond to the Mbalmayo block; and the Ntem division corresponds to the Ebolowa block of the benchmark area.

9 Nutrient measures from a mature forest on an alfisol in Ghana found 27% of total nitrogen, 94% of phosphorus, 57% of potassium and 50% of calcium and magnesium in the biomass (Greenland and Kowal, 1960).

10 This is defined as the percentage change in area planted to plantains divided by the percentage change in relative cocoa prices.

11 The DRC is used for comparing efficiency or comparative advantage across different production enterprises. Minimizing the DRC is equivalent to maximizing economic efficiency (Monke and Pearson, 1989, p. 27). It is defined as the ratio of domestic factor prices to value-added measured in economic (i.e. social) prices. Value-added is the difference between the value of output and the cost of tradables.

12 The EPC is the ratio of private returns to the domestic factors of production to the social returns to domestic factors of production. If the EPC is greater than 1, the domestic factors are being subsidized. If less than 1, they are being taxed.

13 At a distance of 100 km, the cost of transporting bulky fresh products to Yaoundé is estimated to be the equivalent of a 41% production tax. The stronger price incentives for bulky commodities, such as cassava or fresh fruits, as one draws nearer Yaoundé lead to intensification and specialization in the production and commercialization of these products in the urban periphery. This type of production–market linkage was first described by von Thünen.

14 It would be self-defeating if, after establishing cocoa agroforests on degraded lands, farmers cleared more forest for annual food-crop production.

Appendix 11.1: Time-averaged Carbon Stocks

The data for carbon pools in various natural fallow cover types as reported in Kotto-Same *et al.* (1997) were used to estimate the empirical relationship between time and total system carbon (equations 1 and 2). The functional relationship between fallow length and carbon was combined with that for the cropping phase to determine the overall carbon–time relation $f(x_{cf})$, during the crop–fallow rotation, where:

$$f(x_{cf}) = 79.8 \quad \text{for } -t_{\text{crop}} \leq x_{cf} < 0 \tag{1}$$
$$= 79.8 + 4.90\, x_{cf} + 1.25\, x_{cf}^2 - 0.065\, x_{cf}^3 \quad \text{for } 0 \leq x_{cf} \leq T$$

and where x_{cf} are years in the crop–fallow cycle, t_{crop} are years cropped, year 0 is the start of the fallow period and T are the years of fallow.

For perennial crop systems common to southern Cameroon (shaded cocoa, shaded coffee and fruit tree-based home gardens), the estimated functional form $f(x_p)$ was linear, where:

$$f(x_p) = 79.8 + 4.216\, x_p \quad \text{for } 0 \leq x_p \leq 25 \tag{2}$$
$$= 185 \quad \text{for } 25 < x_p < T$$

x_p are years in the perennial crop system, year 0 is the start of the establishment period and T is the age of the plantation. As all perennial crop measurements were taken in 25-year-old plantations, there are no regression properties associated with this function, which is simply the line fit between mean carbon estimates for $T = 0$ and $T = 25$. Plantations older than 25 years were assumed to be in a steady state at 185 t ha^{-1}. Forested land is also assumed to be in a carbon steady state at 307 t ha^{-1}.

The time-averaged carbon stock (L_C) for annual and biennal crop–fallow rotational systems is the mean value of the integral of $f(x_{cf})$ over the interval $[-t_{\text{crop}}, T]$:

$$L_C = \left(\int f(x)\,d_x\right) / \left(T + t_{\text{crop}}\right) \tag{3}$$

Appendix 11.2: Principal Components Analysis of Commercialization/ Intensification

Consider the random matrix of household variables $\mathbf{X} = (X_1, X_2, \ldots, X_p)$ with mean vector $E(\mathbf{X}) = \mu$ and positive definite covariance matrix $\text{var}(\mathbf{X}) = \mathbf{S}$. We wish to find arbitrary linear combinations, $Y = a_1X_1 + a_2X_2 + \ldots a_pX_p$ of the original variables that have the largest variation, where $\text{var}(Y) = \text{var}(\mathbf{a}'\mathbf{X}) = \mathbf{a}'\mathbf{S}\mathbf{a}$. The orthogonal transformation from $X_1, X_2, \ldots, X_p$ to principal components $Y_1, Y_2, \ldots, Y_p$ is a transformation to p random variables with zero covariances, where each successive principal component has maximum variance (Karson, 1982). Thus the proportion of the total variation attributed to each principal component is largest for the first and successively smaller for the second, third, etc. principal components. As principal components analysis (PCA) of a random matrix $\mathbf{X}$ is sensitive to scale, analysis is typically conducted on standardized variables, with the result that the covariance matrix is also the matrix of simple correlations. Under these circumstances, the relative importance of the random variables of $\mathbf{X}$ in relation to the principal component, Y_j, is determined by a comparison of the relative sizes of the coefficients a_{ij}.

In our application, the linear variates are interpretable based on the relative sizes and signs of the coefficients (loadings) of the original random variables. For instance, the first linear variate, labelled CROPSYS, distinguishes households with a horticultural commercialization strategy (HORT = 0.858), using an increased number of field-management practices (ALLMGT = 0.510) from households pursuing a plantain/cocoyam commercialization strategy (PLANMAC = –0.864) using extensive field-management practices.

12 Agricultural Intensification, Deforestation and the Environment: Assessing Tradeoffs in Sumatra, Indonesia

THOMAS P. TOMICH,[1] MEINE VAN NOORDWIJK,[1] SUSENO BUDIDARSONO,[1] ANDY GILLISON,[2] TRIKURNIATI KUSUMANTO,[1] DANIEL MURDIYARSO,[3] FRED STOLLE[1] AND ACHMAD M. FAGI[4]

[1]International Centre for Research in Agroforestry (ICRAF – SE Asia), Bogor, Indonesia; [2]Centre for International Forestry Research (CIFOR), Bogor, Indonesia; [3]BIOTROP–GCTE Impacts Centre for South-east Asia (IC-SEA), Bogor, Indonesia; [4]Alternatives to Slash-and-Burn Programme, Agency for Agricultural Research and Development, Indonesia

Introduction

The conditions necessary for increased productivity of agroforestry and other land uses to jointly reduce poverty and deforestation are not well understood. The global Alternatives to Slash-and-Burn (ASB) Programme was founded to provide scientific insights and workable innovations for the simultaneous pursuit of precisely these two goals: poverty reduction and rainforest conservation in the tropics. This chapter summarizes results from study sites on the island of Sumatra, Indonesia, which were chosen to represent the lowland humid tropical forest zone in Asia for the global ASB project.[1]

A clear gradient in population density in Sumatra occurs from Lampung Province at the southern tip (174 people km^{-2} in 1993) to Jambi Province in the middle of the island (39 people km^{-2} in 1993). Most of the ASB work in Sumatra has concentrated on study sites in Jambi and Lampung Provinces, both of which are located in Sumatra's broad peneplain agroecological zone. This region is almost flat land, less than 100 m above sea level, consisting of about 10% river levees and flood-plains with fertile alluvial soils and 90% uplands with a gently undulating landscape and mostly red-yellow podzolic soils.

The peneplains have been the focus of government-sponsored settlement schemes (called transmigration), large-scale logging and various large-scale public and private land development projects since the 1970s. Because of these activities, little natural forest remains. The process of deforestation, which is almost complete in lowland Sumatra, seems likely to be repeated elsewhere in Indonesia. By understanding the factors contributing to this process and its consequences, ASB researchers hope to identify policies and technologies that can ameliorate the effects of deforestation and contribute to conservation of the remaining rainforests in Asia.

Conflicting interest groups

All ASB research seeks to integrate social and economic analyses with measurements of biophysical outcomes of land-use change, in order to understand conflicts and complementarities between development and resource-conservation objectives. The work in Indonesia is distinguished by including analysis of large estates as well as smallholder activities. This was necessary because Sumatra's dualistic development pattern – a continuing legacy of the colonial period – means that various actors' interests in land conversion, as well as the social, economic and environmental consequences of deforestation, differ substantially by scale of operating unit.

At least six distinct interest groups have a stake in the trajectory of land-use change in Sumatra, but there are crucial differences among them in the weights they place on the various economic and environmental outcomes of forest conversion:

- The international community, concerned with global climate change, extinction of species and loss of distinctive ecosystems.
- Several thousand hunter-gatherers,[2] who continue their traditional migratory lifestyle within remaining forest fragments and national parks in Jambi Province and elsewhere in central Sumatra.
- Millions of small-scale farmers – including local people, spontaneous migrants and government-sponsored settlers (transmigrants) – who depend primarily on land converted from forest in order to make a living.
- Large-scale public and private estates, operating forest concessions and plantations of 100,000–300,000 ha or more. Like smallholders, these large operators currently receive few, if any, incentives or sanctions regarding the environmental impacts of their activities. But large estates and smallholders compete for a limited area of land, which causes pressure for forest conversion. Moreover, the land uses and management strategies of large-scale estates differ significantly from those of smallholders.
- Absentee farmers with medium-sized holdings of 10–25 ha or more. These operators use similar technologies to smallholders, but are able to exert substantial influence, especially on local officials.
- Public policy-makers, who have a range of public policy concerns (discussed below) and whose official salaries are grossly inadequate, making them susceptible to influence by private interests, such as large-scale operators.

ASB 'meta' land uses and major land uses in Sumatra

'Meta' land uses were identified to facilitate cross-site comparisons among the national ASB research agendas. Because deforestation is among the primary concerns of this research, natural forests are the reference point here for global environmental concerns. Grasslands and pastures are included as reference points at the opposite ecological extreme. In between, a representative range of five generic upland, rain-fed land-use systems was selected for cross-continent comparisons of alternatives. These include: extraction of forest products; complex multistrata agroforestry systems, also known as 'agroforests' (Michon and de Foresta, 1995); simple tree-crop systems, including, but not limited to, monoculture; crop–fallow systems, which include the textbook version of 'shifting cultivation' or slash-and-burn agriculture; and continuous annual cropping systems, which may be monocultures or mixed cropping. This analytical scheme was chosen to cover the spectrum of land-use intensification alternatives and to provide counterpart land-use types that can be found in the other ASB sites.

Table 12.1 gives the general specifications for the six Sumatran land uses. Pastures are almost non-existent and, apart from wet rice – which is limited by soil, water supply and topography – continuous annual cropping is rare except in transmigration settlement sites. At the transmigration study site in Lampung, continuous monoculture of cassava and maize and rotations of cassava and maize are common. These fields are often plagued by *Imperata cylindrica*, which covers about 8.6 Mha in Indonesia (Garrity *et al.*, 1997). Estimates for continuous cassava monoculture degrading to *Imperata* are reported here for comparison with other ASB sites, where continuous annual cropping and grasslands are more prevalent than in Sumatra.

Criteria and Indicators for Assessment of Land-use Alternatives

Measurements of field-level differences in the economic, agronomic and global environmental consequences of the various land uses provide a starting-point for quantifying the tradeoffs involved in land-use change and for identifying alternatives that provide an attractive balance among competing objectives. The 'ASB matrix' approach was developed to link global benefits with sustainable alternatives that are adoptable by farmers (Tomich *et al.*, 1998b; Vosti *et al.*, 2000b). The matrix provides a framework for organizing data for the assessment of possible tradeoffs and complementarities across specific indicators representing broad criteria, which are discussed in the next section. The rows of the matrix correspond to specific land uses found in Sumatra, as described above. The columns correspond to various environmental, agronomic, economic and social criteria. Further details on ASB methodologies and data are presented elsewhere (Gillison, 1999; Palm *et al.*, 1999; Swift, 1998; Weise, 1998; Vosti *et al.*, 2000b).

Table 12.1. Specifications for major land uses at the forest margins of the peneplains of Sumatra, Indonesia.

'Meta' land use	Corresponding land use in lowland Sumatra	Type/scale of operation	Landscape mosaic context	Description
Natural forest	Natural forest	25 ha fragment within a logging concession	Forest	Reference point: primary baseline for assessment of land-use alternatives. Undisturbed for at least 100 years
Forest extraction	Community-based forest management	Common forest-land of 10,000 ha to 35,000 ha	Indigenous smallholder landscape	Reference point/possible ASB 'best bet': products are honey (every 3 years), fish, petai
				Reference point/possible ASB 'best bet': products are honey (every 2 years), fish, petai, rattan and songbirds
	Commercial logging	Logging concession of 35,000 ha or more	Forest	Reference point/'best bet' from official perspective: simulation of Indonesian 'sustainable logging system'; 40-year cycle
				Reference point: based on estimates of actual harvesting behaviour for a recently renewed concession; 20–25-year cycle
Complex, multistrata agroforestry systems	Rubber agroforests	Smallholders' plots of 1–5 ha	Indigenous smallholder landscape	Indigenous system: forest clearing followed by upland rice and planting of 'unselected' rubber seedlings, with natural regeneration of forest species. This is the dominant smallholder land use
	Rubber agroforests with improved planting material	Smallholders' plots of 1–5 ha	Indigenous smallholder landscape	Possible ASB 'best bet': forest clearing followed by upland rice and planting of rubber clones, with natural regeneration of natural forest species

Simple tree-crop systems	Rubber monoculture	Smallholders' plots of 1–5 ha	Indigenous smallholder landscape	(Formerly) 'best bet' from official perspective: upland rice and planting of rubber clones, with intensive use of inputs and labour to prevent regeneration of natural forest species
	Oil-palm monoculture	Large-scale private estate of 35,000 ha or more	Monoculture plantation	'Best bet' from official perspective: plantation oil-palm grown in close association with processing mill (processing not included in the economic analysis)
Crop/fallow systems	Upland rice/bush fallow rotation (shifting cultivation)	Smallholders' plots of 1–2 ha $year^{-1}$, often located on community land	Indigenous smallholder landscape	Reference point: 1 year of upland rice followed by bush fallow of 10 years or more. The dominant smallholder land use of 100 years ago, now rare
				Reference point: 1 year of upland rice followed by a short bush fallow of 5 years or less. Now found only in isolated areas
Continuous annual crops/ grasslands	Continuous cassava degrading to *Imperata cylindrica* grassland	Smallholders' plots of 1–2 ha within large-scale settlement project	Large trans-migration project divided into small plots	Reference point: monocrop cassava with little use of purchased inputs
				Reference point: monocrop cassava with intensive use of purchased inputs

Global environmental criteria

Quantification of at least three indicators of the global environmental consequences of deforestation and other land-use changes is essential for formulating sound policy responses – or even for knowing whether intervention is needed. Two of these indicators are linked to global climate change: carbon (C) stocks and net absorption of greenhouse gases, including carbon dioxide, methane and nitrous oxide. Carbon stocks were measured for sample plots in natural forests, shifting-cultivation sites and five other major land-use alternatives in Sumatra's peneplains. The point data from the samples were used to estimate the 'time-averaged C stock' for major land-use systems (see Appendix 11.1 of Chapter 11). Land-use change can thus be translated into a net release or net sequestration of C.[3] The third indicator is biodiversity. Above-ground measurements were made for plant functional groups, as well as the more conventional taxonomic approach. The number of plant species is used as an overall indicator of biodiversity richness that is suitable for cross-continent comparisons. Indicators of above- and below-ground biodiversity were studied for the same land uses where the C stocks and greenhouse-gas emissions were measured.[4]

Agronomic sustainability criteria

Agronomic sustainability refers to the long-term production capacity at the plot level, but researchers and farmers may differ in their assessment of what 'sustainable' means. Soil scientists and agronomists collaborating in ASB research identified a minimum set of seven components of agronomic sustainability; these include soil bulk density (indicating compaction), soil C saturation, active soil C, soil exposure to erosion, nutrient balance (nitrogen, phosphorus, potassium (N, P, K)), potential for weed problems and potential for pest and disease problems (Weise, 1998). Although it has not been possible to arrive at a single summary indicator for agronomic sustainability, it has been possible to use this mix of indicators to assess Sumatra's major land-use systems.[5]

Policy-makers' criteria

Before the severe economic setbacks of recent years, Indonesia's development strategy had simultaneously pursued growth, equity and stability – called the 'development trilogy' – with considerable success for over 30 years. Each of these broad goals yields criteria for the assessment of land-use alternatives, emphasizing the policy objectives most affected by land-use change.

Growth

What is the potential profitability of the activity, and does the country have a comparative advantage in it? If so, expansion of this activity can contribute to economic growth. Since many of the land-use alternatives in Sumatra involve perennials, the appropriate measure of profitability is the estimated net present value (NPV) – the present discounted value of revenues less costs of tradable inputs (fertilizer, fuel, etc.) and of domestic factors of production (land, labour, management) over the full 25-year period considered in our analysis. The policy analysis matrix (PAM) technique provided the framework for estimating profitability indicators, as well as the indicators of labour requirements and cash-flow constraints discussed below. The PAM is a matrix of information about agricultural and natural resource policies and factor market imperfections, which is created by comparing multiyear land-use system budgets calculated at private and social prices (Monke and Pearson, 1989).

Social profitability, calculated at economic (shadow) prices – e.g. the NPV at social prices – is an indicator of potential profitability (or comparative advantage). If land is scarce, the NPV estimates returns to land. (To the extent that management is a scarce factor, it would also be included in the residual.) Land is certainly a constraint that should be considered by policy-makers in choices regarding development of large-scale estates versus smallholders (and there are other reasons for believing that these development strategies are mutually exclusive (Tomich *et al.*, 1995)). Returns to land valued at social prices will be used as the indicator for potential profitability evaluated from the policy-makers' perspective.

There is a long list of potential corrections necessary to arrive at social prices. The adjustments in these analyses focus mainly on policy distortions arising from trade restrictions. We also used a lower real discount rate (15% instead of 20%) to capture a rough approximation of the impact of capital market imperfections on the private cost of capital. We have used the same wage rate in both sets of calculations, implicitly assuming that there are no imperfections in the market for unskilled labour. While this is not completely true, it also seems that these imperfections do not have a significant effect in the unskilled labour market (see discussion of labour markets below). The main omission here is that, due to lack of data, prices are not adjusted to reflect costs and benefits of environmental externalities – such as smoke, ecological changes and loss of watershed functions – arising from agricultural and forestry production activities. In addition, a complete economic picture should also include assessing the private and social profitability of 'downstream' processing activities, especially for timber, rubber, cassava and palm oil.

Indonesian teams undertook studies of the six Sumatran land-use systems selected for study (Aliadi and Djatmiko, 1998; Arifin and Hudoyo, 1998; Budidarsono, 1998; Hadi and Budhi, 1998; Machfudh and Endom, 1998; Maryani and Irawanti, 1998). All of these studies use the macroeconomic parameters prevailing when the data were collected in July 1997, including a wage rate for unskilled labour of Rp 4000 day^{-1}. At that time, the exchange rate was about Rp 2400 per US dollar.[6] To assess land-use alternatives over the longer term, the macroeconomic parameters existing in July 1997 are probably a better guide than those prevailing during the subsequent economic crisis of 1997/98.

Real interest rates – that is, interest rates net of inflation – are the discount factors used to value future cash flows in current terms. As in most developing countries, capital markets in Indonesia are fraught with imperfections, some of which have been manifested in the financial crisis. Private interest rates (at least for smallholders, if not for large corporations, which could secure subsidized credit) have been very high in real terms. In July 1997, formal-sector lending rates were almost 30% year^{-1} and annual inflation was under 10%. Thus, the private interest rate of 20% used in these analyses is a lower bound for the actual cost of capital for smallholders. The real social interest rate is less than the private rate and 10% is probably too low. So, somewhat arbitrarily, a rate of 15% has been used for the real social cost of capital, both as the interest rate and as the discount rate for calculating NPV at social prices.

This difference between private and social interest rates is the main cause of divergences between calculations at private and social prices for many of the land-use alternatives. The analyses are quite sensitive to the choice of discount rates, which unfortunately involves considerable uncertainty. Particularly for the private cost of capital, the subjective discount rate may be much higher (or lower) than the 20% real rate used here. Interest rates in the informal sector often exceed 100% year^{-1}. Holden estimated that the average subjective discount rate (rate of time preference) among transmigrants in Riau exceeded 90% (A. Angelsen, Bogor, Indonesia, 1998, personal communication). On the other hand, as Angelsen has pointed out, 'the desire to claim or secure land rights may modify the effect of high discount rates'.

Equity and stability

Would expansion of a given economic activity create employment opportunities, especially for unskilled rural workers? Or would it displace these workers, forcing more to migrate to Indonesia's cities? If an economic activity is profitable, is it adoptable by smallholders? If so, it may have the potential to contribute to poverty alleviation. Equity and stability are both affected, at least in part, by employment opportunities. The indicators of adoptability presented below are also relevant to poverty alleviation. From the perspective of policy-makers concerned with employment generation, total time-averaged labour requirements are good indicators and are also related to equity and stability criteria. Note, however, that, while labour-intensive alternatives should be attractive for policy-makers who are concerned with job creation, these alternatives will only be attractive to households if they provide attractive returns to labour. Neither this indicator nor other available data can shed much light on potential seasonal bottlenecks in labour demand.

Smallholders' socioeconomic criteria

Alternative systems and technologies must be profitable and socially acceptable for smallholders; if not, they have little prospect for adoption and, hence, impact. A minimum set of three quantifiable socioeconomic objectives were judged necessary for the assessment of land-use alternatives from smallholders' perspectives (Tomich *et al.*, 1998b; Vosti *et al.*, 2000b).

Smallholders' production incentives

Does it pay smallholders to invest in a particular production alternative compared with other options? Production incentives (financial profitability) for smallholders are measured as returns to labour valued at private prices. Private prices are the prices that households and firms actually face, so private profitability – e.g. the NPV at private prices – is a measure of production incentives. We also introduce a measure of returns to labour: the wage rate that sets the NPV equal to zero. This calculation converts the 'surplus' to a wage, after accounting for purchased inputs and discounting for the cost of capital; no surplus is attributed to land. This measure of returns to labour is valid when land is abundant and labour is scarce. Returns that exceed the wage (of Rp 4000 day^{-1}) mean the activity will be attractive to family members or would justify hiring labour.[7] Returns to labour valued at private prices were selected as the indicator of profitability for smallholders' production incentives.

Household labour constraints

Is it feasible for these households to supply the necessary labour themselves or to hire workers? As shown in Table 12.2, two of the systems with the highest potential profitability – smallholder rubber agroforests planted with clones, and large-scale oil-palm – both have high labour requirements. However, each system also has high returns to labour. Thus, although problems in the labour market or credit market could impose a serious barrier to adoption, returns to labour itself are not a problem as long as they exceed the opportunity cost of family labour, taken as the market wage. More generally, returns to labour valued at private prices, which were selected above as an indicator of smallholders' production incentives, are also a good indicator for smallholders' concerns with labour constraints, if combined with assessments of institutional barriers in markets for labour and capital.

Household food security

Even if the alternative is profitable and feasible given household labour constraints and labour-market conditions, is it so risky (either in terms of variance in food yields or as a source of income) that adoption would jeopardize household food security? To accommodate land-use alternatives that do not involve food crops, our food security indicator is based on Sen's (1982) concept of risk of food-entitlement failure, which encompasses trade- and production-based entitlements to food, as well as security of property rights over productive assets (inheritance and transfer entitlements). One of the key dimensions of this analysis is the 'path' of food entitlement – is it derived from consumption of one's own food production, exchange of one's own production for food or working for wages to buy food?

Institutional and policy barriers to adoption

Quantitative measures of the concerns of smallholders and policy-makers need to be supplemented by assessment of institutional endowments as they affect land, labour, capital and commodity markets, as well as the availability of information on production technologies. In turn, markets and other institutions affect the feasibility of adoption of technological innovations by smallholders. Problems in input supply, output, labour and capital markets are indicated respectively by I, O, L and K, respectively, in Table 12.2. The specific non-market problems and issues considered here are access to non-market information (N), regulatory issues (R), local environmental issues (E), insecure property rights (P), equity biases (B) and need for social cooperation (C). The use of upper-case letters in Table 12.2 denotes a serious constraint and lower-case letters denote a constraint that may be overcome by individuals and communities on their own.

Table 12.2. ASB matrix for the forest margins of Sumatra.

Land use		Global environment		Agronomic sustainability: plot-level production sustainability	
Description	Scale of operation/ evaluation	Carbon sequestration: time-averaged (Mg ha^{-1})	Biodiversity: plant species per standard plot	Overall rating	Main sustainability issues[a]
Natural forest	25 ha fragment/ 1 ha	254	120	1	
Community-based forest management	35,000 ha common forest/1 ha	176	100	1	
Commercial logging	35,000 ha concession/ 1 ha	150	90	0.5	C
Rubber agroforest	1–5 ha plots/ 1 ha	116	90	0.5	C
Rubber agroforest with clonal planting material	1–5 ha plots/ 1 ha	103	60	0.5	C, K, W, P
Rubber monoculture	1–5 ha plots/ 1 ha	97	25	0.5	C, W, P
Oil palm monoculture	35,000 ha estate/1 ha	91	25	0.5	C, Fert
Upland rice/bush fallow rotation	1–2 ha plots/ 1 ha	74	45	0.5	Fert, P
Continuous cassava degrading to *Imperata*	1–2 ha plots within settlement project/1 ha	39	15	0	C, Fert, W

[a]Plot-level production sustainability: C = soil compaction; K = potassium balance; Fert = fertilizer cost; P = pest or disease problem.
[b]Market imperfections: I = input market problem; O = output market problem; K = capital market problem.

The ASB Matrix for the Forest Margins of Sumatra

The complete ASB matrix for Indonesia, as presented in Table 12.2, is the basic tool for an integrated assessment of options to balance environmental benefits with sustainable agricultural development. An alternative approach to analysing multiobjective criteria is to derive a set of weights for various indicators, often valued in terms of currency, as a basis for developing an index to rank alternatives. While it is possible to take this approach to estimating returns to land and imputed values for carbon sequestration services (for relevant examples, see Tomich *et al.*, 1997), it is problematic for other objectives (especially measuring biodiversity). The matrix of land-use alternatives (Table 12.2) is, then, a

Table 12.2. *Continued.*

National policymakers' concerns		Adoptability by smallholders			
				Institutional and policy issues	
Potential profitability: returns to land (Rp 1000 ha^{-1}) at social prices	Employment: time-averaged labour input (days ha^{-1} $year^{-1}$)	Production incentives: returns to labour (Rp day^{-1}) at private prices	Household food security: food entitlement via:	Market imperfections[b]	Other institutional problems[c]
0	0	0	n/a		
9.4 to 18	0.2 to 0.4	11,000 to 12,000	Own production and exchange	o	N, R, P, C
(32) to 2102	31	(17,349) to 2008	Wages	O, K	N, R, E, P, B, C
73	111	4000	Exchange		P, b, c
234 to 3622	150	3900 to 6900	Exchange	I, k	N, P, b, c
(993)	133	3683	Exchange	I, k	N, P, b, c
1480	108	5797	Wages	I, o, K	N, R, e, P, B, c
(180) to 53	15 to 25	2700 to 3300	Own production		n, P, c
(315) to 603	98 to 104	3895 to 4515	Own production and exchange	o, K	n, E, p, c

[c]Other institutional problems: N = non-market information problem; R = regulatory problem; E = local environmental problem; B = equity biases (gender or distributional); C = social cooperation required.

For market imperfections and other institutional problems: upper-case letters indicate more serious problems.

pragmatic attempt to circumvent this valuation problem and to organize information that incorporates competing objectives, wherein various groups can assign their own assessments of relative weights. With this array of indicators, it is possible to examine tradeoffs and complementarities across the various criteria.

Global environmental concerns and agronomic sustainability

Carbon sequestration

Carbon stocks of tree-based land-use systems depend largely on the typical cycle length of these systems, as annual C increments are similar. Thus, time-averaged C stocks are roughly similar for long-rotation tree-based systems, which are superior to all other land uses in this regard, except for natural forests themselves.

Biodiversity

Alternative land uses at the forest margins differ significantly in their potential for conservation of above-ground biodiversity. A range of alternatives are shown to fall between the extremes of smallholders' complex, multistrata agroforestry systems (agroforests) and continuous food-crop monoculture, as measured by the number of plant species per standardized plot.

Plot-level agronomic sustainability

All the tree-based systems included in Table 12.2 (smallholder agroforests and rubber monoculture, as well as large-scale plantation monoculture) are agronomically sustainable. On the other hand, the shortening of fallow rotations from 10 years or more to less than 5 years, along with rising land scarcity, is undermining the sustainability of shifting cultivation, which has been disappearing anyway as population pressure increases in Sumatra (van Noordwijk *et al.*, 1995). Prior to the monetary crisis that began in Indonesia in 1997, unsustainable shifting cultivation was not financially profitable in much of Sumatra. This appears to have changed during the collapse of the Indonesian currency in 1997/98, which (temporarily) may have reversed the long-term decline in shifting cultivation (Tomich *et al.*, 1998a, Part VI). Indeed, because of the collapse of the *rupiah*, the potential profitability of many tree-based systems has increased substantially, which may have boosted incentives for forest conversion by smallholders and large-scale operators alike.

Continuous cultivation of cassava does not appear sustainable on this land because of the depletion of nutrients and of soil organic matter. On these soils, marginal revenues from fertilizer applications to cassava do not cover fertilizer costs at current prices, which were near the world market price for most nutrients at the time of these studies, except nitrogen, which is subsidized in Indonesia. Moreover, extensive cassava systems mine nutrients, exhausting the soil and reducing the number of options for future land use. Local, regional and national environmental problems linked to land-use change still need to be addressed to fully understand the sustainability of these land-use alternatives beyond the plot level.

Concerns of smallholders and national policy-makers

Potential profitability and smallholders' production incentives

Estimates of returns to land valued at social prices (potential profitability or comparative advantage) and returns to labour valued at private prices (smallholders' production incentives) are presented in Table 12.2. The upland rice/bush fallow rotation stands out as being unprofitable, in terms of either potential profitability or smallholder production incentives. The high end of the range of estimated returns are for the fallow of 10 years or more, which is no longer feasible because of population pressure. The low end of the range corresponds to short-fallow shifting cultivation. These results are consistent with the disappearance of shifting cultivation in most of Sumatra's peneplains.

Sustainable forms of continuous food-crop production may be technically feasible in Sumatra's peneplains, but often are not financially attractive because they require too much labour and too many purchased inputs. Here, we focus on cassava, which may be among the most profitable of the continuous food-crop alternatives for the peneplains. Profitability estimates for two cassava systems are included in Table 12.2, one with fertilizer applications from the first year and one with fertilizer beginning in the seventh year after forest clearing. The latter system is shown to produce relatively attractive returns to land and labour, but its longer-run sustainability requires further study.

Estimated returns to labour are highest for community-based forest management (including extraction of non-timber forest products (NTFPs)), but these high returns are dependent on the existence of some mechanism to exclude outsiders. This alternative can thus play an important role for those communities which can regulate access to forestlands. If, on the other hand, communities cannot regulate access to their forests, one would expect the returns to labour from extraction of forest products to decline toward the wage rate. Even under 'open access', however, one would still expect returns to labour to exceed the wage rate by some margin equal to a risk premium. The risks involved include the possibility of failure to find products to extract and the risk (and associated costs) of detection by officials, since many of these activities are prohibited.

The relatively low returns to land – well below those in rubber agroforest systems – suggest that NTFP extraction is not a feasible alternative for large numbers of people, because there is not enough land for everyone to practise this livelihood strategy. These results must be interpreted with some care, however, for several reasons. First, it was not possible to include all the myriad commodities collected from the forest by local villagers. Researchers focused on the commodities that villagers reported were most important to them. Secondly, because restrictions banning logging by villagers are enforced actively, it was not possible to obtain data about villagers' timber extraction from the forest, and it is likely that this is significant. Thirdly, extractive activities are highly site-specific, and it may be that the study site was not representative. Finally, at least part of the forestland claimed by long-established communities is on designated government land and it is not clear how this problem of tenure insecurity might bias these results. Long-run profitability may be overstated because of unsustainable harvesting. However, if the community or individual members had secure property rights, this might induce investment to increase productivity. Two estimates of profitability – one based on sustainable harvests only and a second higher estimate including products that may be depleted (songbirds and rattan) – are reported in Table 12.2.

A pair of profitability estimates are reported for commercial logging in Table 12.2; the lower estimate represents complete compliance with government regulations, the higher is closer to actual practice. The sustainable-logging regulations, if they are really followed, reduce profitability, mainly by slowing timber extraction and because high export taxes (effectively an export ban) for logs and sawn timber depressed the domestic prices of logs by 50–70%. However, timber companies can get around both of these problems. First, many companies circumvent regulations on timber extraction. Secondly, these are typically vertically integrated firms, producing products like plywood for the export market. Therefore, the best indicator of profitability of these activities is probably the figure of just over Rp 2 million ha^{-1}, valued at social prices reflecting world prices of forestry products.[8]

Oil-palm is widely viewed as the most profitable alternative for Sumatra's peneplains and Indonesia's oil-palm producers have the lowest unit costs in the world. Thus, it is no surprise that large-scale oil-palm monoculture is among the most profitable alternatives, evaluated either in terms of returns to land, valued at social prices, or returns to labour, valued at private prices. The latter measure is of limited relevance, however, because the official wages for plantation workers are well below these estimates of returns to labour. As occurred earlier in Malaysia (Barlow, 1986), plots of 2–5 ha of oil-palm planted by independent smallholders began to appear in Sumatra beginning in the 1980s. These merit study for their possibility to combine high potential profitability with attractive returns to smallholders' labour. For the time being, however, government development strategies discriminate against the emergence of independent smallholder oil-palm producers. For example, some provinces will not license palm oil mills unless the enterprise also has its own oil-palm plantation or smallholders are organized in nucleus estate/smallholder (NES) schemes. This is intended to prevent NES participants from selling their produce outside the project (as happened in the case of rubber) in order to avoid repayment of loans. But this practice also retards the development of the market for independent smallholder oil-palm producers.

The three contrasting rubber systems produce a wide range of results.[9] First, as already noted, it is encouraging that returns to labour at private prices are virtually identical to the market wage for rubber agroforests planted with seedlings. Although these smallholders are the lowest-cost producers of natural rubber in the world (Barlow *et al.*, 1994), returns to land at social prices are not much above upland rice with a long bush fallow rotation and are well below oil-palm monoculture.

Perhaps the most striking result seen in Table 12.2 is the estimated returns to land at social prices for rubber agroforests planted with PB 260 clones, which rival large-scale oil-palm monoculture. This system also produces attractive returns to labour at private prices. These data must be treated with caution, since they are based on projections from farmer-managed trials and have not been verified through broader experience by smallholders. The top of the range of profitability estimates might actually be attained by 10–25% of smallholders. However, the lower figure in the range represents an expert's best guess about a 'worst-case' scenario for yields in this system for the bottom quartile. A central question is where the middle of the profitability distribution would be for this system; this can only be answered through farmers' experience. But these results support the idea that the potential profitability of rubber agroforests planted with clonal material (and other smallholder agroforests planted with appropriate, higher-yielding germ plasm) may be comparable to that of large-scale oil-palm plantation monoculture.

The profitability estimates for smallholder rubber monoculture planted with GT 1 seedlings provide a cautionary tale to balance the encouraging projections for rubber agroforests planted with PB 260 clones. These monoculture plots were part of a government-sponsored rubber replanting project, which was undertaken with high expectations. But disappointing yields were obtained because of institutional shortcomings involving supplies of planting material (participants received seedlings instead of clones), technical information and credit, and these yields could not offset high project costs. Contrasted to this high-cost approach, the strategy to introduce clones into smallholders' agroforests seeks a moderate increase in yields at minimal incremental costs. Yet the costly lessons of earlier failures in smallholder rubber development should be borne in mind (Tomich, 1991).

Employment

For the rubber and oil-palm systems that were evaluated, total time-averaged labour requirements are similar, ranging between 100 and 150 person-days ha^{-1} annually. Labour for harvesting is the largest component in these systems. Because of a lack of pronounced seasonality in much of Sumatra, the harvesting of rubber and oil-palm can proceed roughly 10 months $year^{-1}$. The two extractive activities – community-based forest management and commercial logging – fall at the opposite extreme. Neither of these extractive activities nor the upland rice/bush fallow rotations can provide many employment opportunities.

Household food security

Calculations presented in Tomich *et al.* (1998a) indicate that production risk for rubber agroforests may be less than for the upland rice/bush fallow rotation. The terms-of-trade risk for rubber, however, is about double the production risk, as measured by their respective coefficients of variation. Although these measures suggest that upland rice/bush fallow is, overall, less risky than rubber, the superior production incentives for rubber agroforests are the reason why rubber has displaced upland rice over the past century. Note that the large-scale alternatives – commercial logging and large-scale plantations of oil-palm or industrial timber – imply a shift to wage labour or migration for displaced local households.

Tradeoffs and Complementarities

Because of the multiple criteria regarding the production and environmental services of forests, 'deforestation' must be viewed as a multidimensional phenomenon. Sometimes the policy problems may be simplified to a few key dimensions (tradeoffs), which are highlighted here.

Is there a tradeoff between growth and poverty alleviation?

If they are really more profitable than smallholder tree-based alternatives, all the large-scale systems will involve tradeoffs with household food security, since such projects often displace local smallholders with little or no compensation. In the case of large-scale logging, there is also a tradeoff with employment creation in smallholder tree-based systems. The potential profitability of some tree-based alternatives for smallholders (namely, rubber agroforests planted with clones) appears to be comparable to that of large-scale estates and logging. However, this requires further verification through additional studies of smallholder rubber and other alternatives, such as smallholder timber and smallholder oil-palm. This result holds promise for complementarity between policy-makers' concerns with potential profitability and smallholders' production incentives. It also suggests that policy concerns regarding equity and mounting concerns about social and political instability can be addressed through a smallholder-based development strategy without a significant reduction in economic growth. If they can be adapted for smallholders, the treecrop-based systems offer attractive production incentives. Since labour markets appear to work well, labour should not be a serious constraint to adoption. Thus, smallholder tree-crop systems also offer complementarity with employment creation objectives.

Are there tradeoffs between global environmental benefits and local objectives?

The relationship between potential profitability and C sequestration is dominated by cycle length (frequency of clear-felling for rejuvenation). Where tree-crop systems can be rejuvenated without clear-felling, a substantial increase in C stock may be possible. Moreover, there do not appear to be large differences between forest extraction and the other tree-based systems regarding C stocks and greenhouse gases. Thus, as far as agronomic sustainability and climate-change objectives are concerned, tree-based systems dominate among the alternatives.

Raising the productivity of rubber agroforests, which span millions of hectares, offers a promising pathway in Sumatra. There appears to be great potential for raising the profitability of these systems through adaptation of existing higher-yielding clones within existing smallholder systems, which would also enhance household food security and expand employment opportunities. It may be possible to combine these potential benefits with significant biodiversity conservation, because the mix of planted species is augmented by natural regeneration of forest species (Michon and de Foresta, 1995; van Noordwijk *et al.*, 1997). Indeed, these agroforests may approximate a number of forest functions, thereby providing the technical foundation for sustainable community-based forest and watershed management. But it must be emphasized that agroforests are not perfect substitutes for biodiversity conservation in natural forests. Indeed, conversion of natural forests to agroforests involves a significant reduction in species richness.

A key unresolved question is whether the potential development of smallholder rubber agroforests can compete with the (private and social) profitability of large-scale land uses, including oil-palm plantations, industrial timber estates and logging concessions. These large-scale alternatives are viewed as 'best bets' for economic development by many policy-makers and donors, in large part because of the conventional wisdom regarding economies of scale in plantation development. If it turns out that large-scale development alternatives are more profitable – recall from Table 12.2 that this is not a foregone conclusion – an important tradeoff between global environmental benefits arising from biodiversity conservation and national economic objectives based on large-scale plantation development will have to be faced.

Even if further analysis shows that the large-scale schemes hold no advantages in terms of private and social profitability compared with smallholder schemes, a potential tradeoff between profitability and biodiversity conservation remains to be addressed concerning smallholder systems (van Noordwijk *et al.*, 1997). Farmer management aimed at increasing the productivity of systems often decreases biodiversity. Whether or not this apparent trade off between productivity and biodiversity is inescapable is the subject of debate and further research.

Barriers to Adoption of Land-use Alternatives

Cash-flow constraints

Because perennials are so important among the Sumatran alternatives, our analysis of cash-flow constraints focused on multiyear (rather than seasonal) cash-flow constraints in order to assess whether the investments required by these systems represent barriers to adoption by smallholders. Table 12.3 takes two approaches to assessing multiyear cash-flow constraints: the number of years to positive cash flow and the NPV of establishment costs (defined as costs prior to positive cash flow). The imputed value of family labour is included in establishment costs, because these labour inputs presumably represent forgone earnings in other activities, even if they do not require cash outlays.

By either measure, community-based forest management is the only profitable system without any multiyear cash-flow constraints. For the other systems, years to positive cash flow range from 2 years for logging to 6–10 years for smallholder rubber and 10 years for large-scale oil-palm. Time is not a constraint by itself, as evidenced by almost 3 million ha of rubber agroforests

Table 12.3. Cash-flow constraints.

Land-use system	Years to positive cash flow at private prices (years)	NPV of establishment costs at private prices (*rupiah* ha^{-1})	Years to positive cash flow at social prices (years)	NPV of establishment costs at social prices (*rupiah* ha^{-1})
Community-based forest management	n/a	n/a	n/a	n/a
Commercial logging	2	820,669–869,199	2	716,917–764,238
Rubber agroforest (seedlings)	10	1,305,536	10	1,477,735
Rubber agroforest (clones)	6–7	2,593,458–2,862,422	6–7	2,950,338–3,303,338
Rubber monoculture	10	2,085,257	10	2,192,584
Oil-palm monoculture	10	8,041,847	9	8,182,015
Upland rice/bush fallow rotation	Never	n/a	Never	n/a
Cassava/*Imperata cylindrica*	2	n/a	2	n/a

that have been planted by smallholders without any formal credit. The NPV of establishment costs at private prices is probably the best indicator of cash-flow constraints for smallholders. In interpreting these estimates, bear in mind that the existing rubber agroforests are evidence that the estimated Rp 1.3 million ha^{-1} required for establishment has not been an insurmountable barrier for smallholders.[10] These estimates suggest that replacing seedlings with higher-yielding clones in rubber agroforests more than doubles investment costs to roughly Rp 2.6–2.9 million ha^{-1}. Since there is no long-term institutional credit for smallholders in Sumatra, whether these investment requirements are barriers to adoption depends in large part on the divisibility of the activity (e.g. planting a bit at a time).

Investment costs for large-scale oil-palm plantations are the highest of all, at over Rp 8 million ha^{-1}. Investments of this magnitude would be difficult for many smallholders. But capital costs for large-scale plantations may be inflated for at least two reasons. First, these plantations formerly received heavily subsidized credit from the government, which tended to make them artificially capital-intensive. Secondly, it is our experience that respondents tend to overstate investment costs to mask profitability. None the less, adapting high-yielding oil-palm systems as alternatives for smallholders will require research to develop options that are less capital-intensive.

Market imperfections

Input-supply markets

Markets for supplying planting materials for clonal rubber and oil-palm represent a significant barrier to the adoption of profitable alternatives by smallholders. Farmers have little access to improved rubber planting material. The

Treecrops Advisory Service, virtually the sole provider of rubber budwood, has focused its efforts on supplying planting materials to project participants in the past and has largely ignored the much larger number of non-participants (Tomich, 1991). Except in a few areas of Sumatra, the private nursery industry has only begun to develop. For public and private sources alike, there are serious problems of reliability regarding the quality of planting material, which is difficult to assess until several years after planting. Current delivery pathways for improved planting material (and the information needed to use it) seem inadequate, but direct government intervention to supply germ plasm may be neither feasible nor desirable. For example, subsidizing germ plasm would hamper development of a private nursery industry.

Output markets

Government restrictions on marketing and international trade are perhaps the greatest barrier to the development of smallholder timber-based alternatives and also hinder community-based forest management. Beginning in 1998, the government agreed to deregulate exports of timber, abolish joint-marketing associations (which functioned as cartels) and end export quotas and numerous other restrictive marketing arrangements. Previous restrictive marketing practices damaged the marketing capacity of most timber companies by inhibiting the development of marketing networks that could respond to buyers' needs. The situation is particularly bad for rattan, since the export ban on raw rattan destroyed overseas markets and induced importers to seek alternative supplies. There is also concern that old rent-seeking practices (such as the plywood and clove cartels) will re-emerge under new guises. These risks are increased by a lack of market information on these commodities. The lack of information is probably worst for NTFPs, especially those occupying narrow market niches. Oil-palm has also been subject to export taxes and, at times, export bans (Tomich and Mawardi, 1995), which seriously depress farm-gate prices. For oil-palm and cassava, there are also some concerns about the development of local markets, which can link smallholders with processors. However, these markets seem to be emerging. Local markets for natural rubber have functioned for a century or more. Although they are characterized by distortions reflecting product quality and the effects of national policy, and the international buffer stock has at times depressed prices, local markets do transmit world price changes to the farm gate, with marketing margins reflecting transportation and other costs.

Labour markets

Although the complete analysis also includes skilled labour requirements, the summary analysis presented here focuses on unskilled labour. Instead of hiring permanent skilled workers, smallholders are more likely to develop certain technical skills themselves. So the relevant barrier is the acquisition of technical information, rather than the market for skilled labour. Although labour markets in Sumatra fall short of the theoretical 'ideal' of economics textbooks, recent empirical studies linked to the ASB project (Suyanto *et al.*, 1998a,b) indicate that these labour markets function reasonably well. None of the land-use alternatives face problems regarding markets for unskilled labour. It is worth noting that casual markets for skilled labour (e.g. chainsaw operators) are also emerging.

Capital markets

No long-term institutional credit is available in rural Sumatra. Yet household savings, which have financed investments in existing smallholder agroforestry systems, such as rubber agroforests, are often underestimated. In rural Indonesia, farmers are able to receive considerable credit from informal sources (e.g. relatives, moneylenders). However, current economic hardships – especially rising food prices – may be straining these resources. Capital market imperfections (lack of credit and interest rates well above the social price of capital) may constrain smallholders' nutrient purchases for cassava production and the use of clonal rubber planting material, and are probably a barrier to smallholder oil-palm production. Whether or not smallholder timber extraction is constrained by capital market imperfections depends, in part, on the development of contract markets for chainsaw services and log transport.

Other institutional problems

Non-market information

Information acquired from research (e.g. new technologies) comes primarily from the government, and existing research facilities are inadequate to meet the research needs presented by diverse production conditions. This constraint is particularly severe for alternatives such as NTFPs and smallholder timber, which are not high priorities for the government, especially compared with rice, the main staple food. This bottleneck in technical information is a concern for all systems, except rubber agroforests using seedlings, where indigenous knowledge is well developed.

Regulatory issues

As discussed above, policies that restrict access to markets are a particular concern for timber and non-timber forest products and for oil-palm. This problem is compounded for timber and NTFPs by policies that attempt to restrict access to state forestland, even if it has been used by local people for generations (see below). Thus, especially for timber and NTFPs – but to a lesser extent for oil-palm – success in these land-use alternatives requires considerable investment of time (and often money) to 'work the system' under current policies.

Local environmental issues

Based on available data, the production of most of these systems seems benign. However, there may be water- and air-quality concerns arising from the processing of rubber, oil-palm and cassava. The exceptions are large-scale logging and continuous cassava cultivation, which are susceptible to erosion.

Further work is needed to assess the environmental impacts of expansion of particular alternatives, including air quality, landscape biodiversity and watershed functions.

Property rights

This is a highly charged political issue for all systems except the continuous production of food crops at a transmigration site; even here there can be problems of tenure conflicts with indigenous groups whose presence predates the settlement. In most cases, the tenure status of lands at the forest margins (and the products derived from those lands) needs to be clarified between the government and local communities.

Equity biases

The primary concern is that potential economies of scale will lead to concentration of land under commercial logging, for which scale economies have been documented elsewhere, and for oil-palm, where scale economies are probably not intrinsic but may result from current development policy. Despite the conventional wisdom, the prevailing faith in scale economies in the production of so-called 'plantation' commodities receives little (if any) support from agricultural economics (Hayami, 1994; Tomich *et al.*, 1995). This is, nevertheless, an empirical question that requires further investigation. Unlike production, the marketing and processing of primary products are often characterized by increasing returns to scale. This is the case for three of the most important land-use alternatives in Sumatra – rubber, pulp and oil-palm. The natural rubber industry in South-east Asia provides an excellent example of the efficiency with which markets can integrate low-cost production by smallholders with processing in factories that achieve economies of scale; similar marketing arrangements should work for pulp. Oil-palm has conventionally been viewed as an estate crop in South-east Asia (but not in Africa) because of its perishability. None the less, oil-palm production on independent plots as small as 1 ha began to emerge in Sumatra in the 1980s. Outgrower schemes, contract farming and other institutional arrangements can all help reduce transactions costs in linking efficient smallholder producers with efficient large-scale processors. There is some cause for concern regarding gender bias in tree-crop production, since recent studies have shown that tree planting induces a shift from matrilineal inheritance to patrilineal inheritance for some categories of trees in some areas of Sumatra (Suyanto *et al.*, 1998a).

Social cooperation

The main need for social cooperation concerns the two forest extraction alternatives, community-based extraction of NTFPs and logging. In each case, the sustainability of the land use is in doubt if communities cannot manage a system to restrict access to their common-property resources. Indigenous communities with their customary laws intact appear to have this capacity; communities of recent settlers may not. Collective action is also required for fire and pest control, and may be an emerging constraint in many agricultural systems.

Summary and Conclusions

The key hypothesis underlying ASB research in Indonesia has been that intensifying land use as an alternative to slash-and-burn can simultaneously reduce deforestation and reduce poverty. Poverty reduction in most of the tropics depends on finding ways to raise the productivity of labour and land through intensification of smallholder production systems. Thus, the intensification hypothesis hinges on the existence of opportunities to raise the productivity of smallholder systems at the forest margins without degrading forest functions. Although there may be (a few) 'win–win' opportunities to alleviate poverty while conserving tropical forests, it is naïve to expect that productivity increases necessarily slow forest conversion or improve the environment. Indeed, quite the opposite is possible, since increasing the productivity of forest-derived land uses also increases the opportunity cost of conserving natural forest. These increased returns to investment can spur an inflow of migrants or attract large-scale land developers and thereby accelerate deforestation.

Estimates of returns to land and labour presented in this chapter indicate that, from a purely private perspective, returns to forest conversion are high in Sumatra's peneplains. Because all derived land uses are inferior to natural forest, based on global environmental concerns (e.g. C stocks and biodiversity conservation), ASB research in Indonesia has shown that land-use changes involve tradeoffs between these environmental concerns and the objectives of poverty alleviation and national development. If there is no action on these tradeoffs – by identifying workable options either to change incentives for conversion or to restrict access to the remaining natural forests – these rainforests will continue to disappear.

This research also provides evidence that land-use alternatives differ significantly in their ability to substitute for the global environmental services provided by natural forests. So, although forest conversion has the largest negative effect on these environmental services, the alternative land uses matter too. Carbon stocks are similar for long-rotation tree-based systems, which are superior to all other land uses by this criterion except for natural forests themselves. Similarly, alternative land uses also differ significantly in their potential for biodiversity conservation, ranging between the extremes of smallholders' complex, multistrata agroforestry systems (agroforests) and large-scale plantation monoculture. While there may be a tradeoff between potential profitability and biodiversity in tree-based production systems, this requires further verification. There may be little or no tradeoff between policy-makers' objectives and those of smallholder households, since the potential profitability of some tree-based alternatives for smallholders appears to be comparable to that of large-scale estates; however, this also requires further verification. There are also important institutional questions that must be addressed to enable widespread adoption of profitable alternatives by smallholders.

To obtain estimates of regional or global impacts directly from measures like those in Table 12.2, it is necessary to assume independence, and hence additivity, across space. This assumption is reasonable for some measures (e.g. C stocks), but it is only a rough approximation for others. Among these measures, biodiversity is the most sensitive to scaling issues. While the agronomic sustainability measure used here concerns only on-site, field-level effects, the extent and spatial arrangement of land-use alternatives also produce environmental externalities (e.g. siltation, smoke, fire and floods). One of the key challenges of future research is to be able to assess these phenomena at the landscape level. Ultimately, instead of a single land-use system or technology, the most attractive way to achieve the multiple objectives is likely to come from combinations of complementary land-use practices within a varied landscape. This landscape-level analysis is not feasible now. The land-use-specific analysis presented here is a necessary precursor to that work.

Acknowledgements

The Alternatives to Slash-and-Burn (ASB) Programme has received financial support from the Global Environment Facility (GEF) under United Nations Development Programme (UNDP) sponsorship and from DANIDA. Additional sources of funding for ASB work in Indonesia include the Asian Development Bank (ADB), the Australian Centre for International Agricultural Research (ACIAR), the Ford Foundation, the Government of Indonesia, the Government of Japan and the US Agency for International Development (USAID). The authors thank Arild Angelsen and the editors for their helpful comments.

Notes

1 For further details on the Alternatives to Slash-and-Burn project, see Chapters 1, 11 and 13 of this volume. Detailed results of ASB's Indonesia work are presented in Tomich *et al.* (1998a).

2 These small family groups do not contribute to deforestation, so they have not been emphasized in the ASB research project.

3 Time-averaged C is annual average C stock over the production cycle (standardized at 25 years) and can be interpreted as the C stock of the land use at steady state (see Gockowski *et al.*, Chapter 11 of this volume, Appendix 11.1). Although timing is not considered, this measure is an indicator of net release or sequestration of C resulting from land-use change. Methane and nitrous oxide were measured for the same land-use systems studied for C stocks. Pronounced seasonality was discovered in greenhouse-gas emissions, so additional measurements will be necessary to derive reliable estimates of annual fluxes. However, these fluxes from living systems are a tiny fraction of the C release resulting from forest conversion (Tomich *et al.*, 1998a).

4 Below-ground biodiversity assessments focus on organisms that influence agronomic sustainability. There appears to be less variation among land uses in below-ground biodiversity compared with above-ground biodiversity, which supports our reliance on above-ground biodiversity as an overall indicator of richness.

5 Most of the indicators could be measured relatively easily, often using data collected as part of the surveys of biodiversity, C stocks and greenhouse-gas emissions or other aspects of the analysis of the land uses (for details, see Tomich *et al.*, 1998a). Indicators for weeds, pests and diseases, however, are based on expert opinion rather than field data. The overall ratings in Table 12.2 are also based on expert judgement, with '1' indicating

'no major problems'; '0.5' meaning 'problems that farmers can manage'; and '0' meaning 'problems beyond what farmers can manage'.

6 By most assessments of economic fundamentals (e.g. purchasing-power parity), the Indonesian *rupiah* was not greatly overvalued at that time. The consensus was that the overvaluation of the *rupiah* relative to the dollar may have been 10–15% in June 1997. Some expert analysts even expected the *rupiah* to appreciate if it were floated in 1997 (McLeod, 1997). During the economic crisis of August 1997 to 1998, the Indonesian *rupiah* was undervalued, by any economic measure.

7 Although local land abundance with household labour scarcity has prevailed historically and certainly continues in the ASB sites in Brazil and Cameroon, this fundamental relationship seems to be shifting in Sumatra. Nevertheless, it still is reasonable to believe that local land abundance and household labour scarcity continue in the forest margins, at least from the point of view of smallholder households in central Sumatra. This is supported by the result that returns to labour for rubber agroforests, the predominant smallholder land use, are almost identical to the wage rate (Table 12.2). This implies that no 'rent' accrues to land under the dominant system and is consistent with land abundance.

8 One could argue that the estimated NPV of logging activities of over Rp 2.1 million ha^{-1} – about US$875 in mid-1997 – should be added to the social profitability for all the other activities, at least for large-scale estates, which can often market the timber felled as a by-product of land clearing. All of the forest-derived land uses (rubber, oil-palm, cassava and even upland rice) started out with the felling of forest timber. There is also substantial (but as yet unquantified) timber felling in conjunction with NTFP extraction. Thus, in many cases, it would be appropriate to add the value of the harvested wood to the profitability of each activity. However, this was not done in Table 12.2, because the one-off value of timber extracted as a by-product of land clearing often exceeds the value of the other land uses. Although technically correct to do so, adding the value of timber would simply obscure differences in profitability among the other land uses. This problem is linked to one in conservation: if regulations can be circumvented, forest conversion is privately profitable simply for the value of timber, regardless of the subsequent land use. Of course, for the social profitability calculations, timber values would have to be balanced against the losses stemming from ecological and other environmental functions of natural forests.

9 The initial study of rubber agroforests planted with seedlings ('jungle rubber') was supplemented with data from another ongoing International Centre for Research in Agroforestry (ICRAF) study (Suyanto *et al.*, 1998b). Subsequently, additional data were added from an ICRAF/CIRAD project in Jambi, analysing rubber agroforests planted with higher-yielding PB 260 clones (E. Penot, 1998, Montpellier, France, personal communication). Since smallholder rubber monoculture is rare in Sumatra outside government projects, the data for rubber monoculture are based on a specific project in Jambi Province. The estimates for rubber agroforests planted with clones and for rubber monoculture may not be widely representative of smallholder experiences.

10 At an exchange rate of Rp 2400 per US dollar, the rate prevailing in July 1997, just prior to the economic crisis.

13 Intensifying Small-scale Agriculture in the Western Brazilian Amazon: Issues, Implications and Implementation

STEPHEN A. VOSTI,[1] JULIE WITCOVER,[1] CHANTAL LINE CARPENTIER,[2] SAMUEL JOSÉ MAGALHÃES DE OLIVEIRA[3] AND JAIR CARVALHO DOS SANTOS[4]

[1]Department of Agricultural and Resource Economics, University of California, Davis, California, USA; [2]Commission for Environmental Cooperation (CEC), Montreal, Canada; [3]Empresa Brasileira de Pesquisa Agropecuária (Embrapa), Rondônia, Brazil; [4]Empresa Brasileira de Pesquisa Agropecuária (Embrapa), Acre, Brazil

with Eufran do Amaral,[a] Tâmara Claudia de Araújo Gomes,[a] Henrique José Borges de Araujo,[a] Evaldo Muñoz Braz,[a] Jailton Carneiro,[a] Carlos Castilla, Divonzil Gonçalves Cordeiro,[a] Idésio Luis Franke,[a] Andy Gillison,[b] Ângelo Mansur Mendes,[a] Marcus Vinicio Neves d'Oliveira,[a] Cheryl Palm,[c] Vanda Rodrigues,[a] Luís M. Rossi,[a] Claudenor Pinho de Sá,[a] Abadio Hermes Vieira,[a] Judson Ferreira Valentim,[a] Stephan Weise[d] and Paul Woomer[d]

Introduction

Intensifying small-scale agriculture at the margins of tropical moist forests is increasingly prescribed as a method for ensuring environmental sustainability, boosting these areas' contributions to national gross domestic product (GDP) and relieving the poverty of inhabitants (NRC, 1993; Soares, 1997; Vosti and Reardon, 1997). However, precise definitions of agricultural intensification are not generally offered, and the extent to which any form of intensification fits existing adoptability constraints of small-scale farmers is not addressed. Even if some forms of intensification do fit into existing systems, whether they will be necessary or sufficient to address the key development issues of agricultural growth, poverty alleviation and environmental preservation is not generally discussed. More importantly, the potential tradeoffs among environmental,

[a]Empresa Brasileira de Pesquisa Agropecuária (Embrapa), Brazil; [b]Centre for International Forestry Research (CIFOR), Australia; [c]Tropical Soil Biology and Fertility Programme (TSBF), Kenya; [d]International Institute of Tropical Agriculture (IITA).

(eds D.R. Lee and C.B. Barrett)

growth and poverty-alleviation objectives that small-scale agricultural intensification may bring (in the likely absence of 'win–win–win' situations) are only rarely addressed (Tomich *et al.*, 1998b). If some forms of intensification are found to effectively address one or more of these objectives, technology, policy and/or institutional changes necessary to promote such intensification are not generally identified.

This chapter presents a framework for filling in some of these gaps, as applied to the socioeconomic and agroecological context of Brazil. The study area is the Brazilian site of the Alternatives to Slash-and-Burn Agriculture (ASB) Programme, located in a swathe of territory between central Acre (AC) and southern Rondônia (RO), two states in the western Brazilian Amazon (Avila, 1994). In the chapter, we first describe the area's land-use systems and suggest entry points for agricultural intensification strategies which are balanced with concern for the environment. The subsequent section explores the relative scarcity of land and labour – the key factors of production at the study site – and the resulting implications for technology adoption and sustainable intensification. The following section presents evaluations of pairs of land-use system alternatives – traditional systems versus proposed, more intensive alternatives – regarding adoptability, agronomic sustainability and broader environmental impacts, as well as tradeoffs between intensification and adoptability, plant biodiversity and carbon stocks. Finally, we discuss our conclusions and policy alternatives and the constraints involved in promoting successful intensification.

Towards a Working Definition of Agricultural Intensification

Land uses characterizing the region were drawn from samples of approximately 150 households in two government-sponsored colonization projects in the benchmark site – Pedro Peixoto (AC) and Theobroma (RO) – which were surveyed in 1993/94 and 1995/96. The trends characterizing these households are illustrative of broader trends in the region. In 1993/94, although primary forest covered the majority of the farm area (62%), pasture dominated the open areas, accounting for 20% of total land. The most prevalent food crops were (and remain) rice, maize, beans and manioc, which together accounted for 7% of the land area. Cultivated perennial tree crops accounted for 4% of the total land area and included coffee and, to a lesser extent, bananas and cacao (Vosti and Witcover, 1996; Witcover *et al.*, 1996; Vosti *et al.*, 1998a).

By 1996, pasture dominated to an even greater extent, reaching 28% of the farm area at the expense of virtually all other land covers, including forest (Vosti *et al.*, 1998a). In other words, agriculture extensified. Can the intensification of agriculture – including its forest and livestock components – modify this trend, and with what consequences for environmental sustainability, growth and poverty alleviation? Some characteristics of small-scale agriculture in the region help explain why it continues to encroach upon forested areas, and provide clues to mechanisms that might counter this trend.

Land-use trajectory from forest[1]

Smallholder agriculture in the forest margins of western Brazil generally begins from completely or partly forested lots, and proceeds by slash-and-burn agriculture rather than shifting cultivation (Fujisaka *et al.*, 1996). May, the onset of the dry season, marks the start of the agricultural year. Labour demand (especially for adult males) peaks early, in the forest-felling months of May to July. In August or September, before the start of the rains, farmers burn deforested areas and plant them to annuals, perennials or pasture. The rainy season begins in October and intensifies in January and February, tapering off in April. After the first-season harvest in December and January, a second crop (beans) can follow, planted in March and harvested in June.

Without external nutrient sources, a given plot of land is typically constrained to proceed through a trajectory of uses over time. Slash-and-burn agriculture, in effect, exchanges the relatively efficient, continuous nutrient cycle of the forest ecosystem for a one-time transfer of forest-stored biomass to the soil. In the first year after burning, the increased availability of soil nutrients from the biomass and a reduction in soil acidity (thanks to the basic properties of ash) help ensure good yields. But crop uptake of nutrients at planting time and the release of litter (crop residues and root decay) at harvest favour a net loss of nutrients from the site. Depending on the soil type and crop planted, nutrient deficiencies appear in the second or third year of consecutive planting after burning if no fertilizer is added (Palm *et al.*, 1996). Yields can drop off steeply, especially for some annuals. This drives the farmer to place the plot of land into an agricultural activity for which the nutrient-depleted soil is more agronomically adapted (although productivity still falls over time in the absence of external inputs) or to let the land go into fallow to increase its nutrient level and reduce compaction.[2]

Figure 13.1 presents a schematic of this process with the primary trajectories of land uses observed in the region, beginning with forest. Each type of land cover (each box in the figure) may have resulted from a number of historical land-use pathways since the initial clearing of the forest. These pathways are jointly termed a common 'meta' land use identified by current land cover. Along the way, land grows more susceptible to weed invasion; if not attended to, this can progress to the point where secondary forest regrowth dominates other cover. In forest-margin areas, natural biological processes in the weed and pest environments combine with nutrient outflows such that continued production from a given piece of land requires external inputs, increased labour and careful choice of land cover, or some mix of these (Ruthenberg, 1980). As seen in Fig. 13.1, current meta land-use patterns, which trace out an implicit path of environmental degradation, coexist with deforestation, which continues at the average rate of about 4.7 ha every 2 years (out of an average farm size of 83 ha).

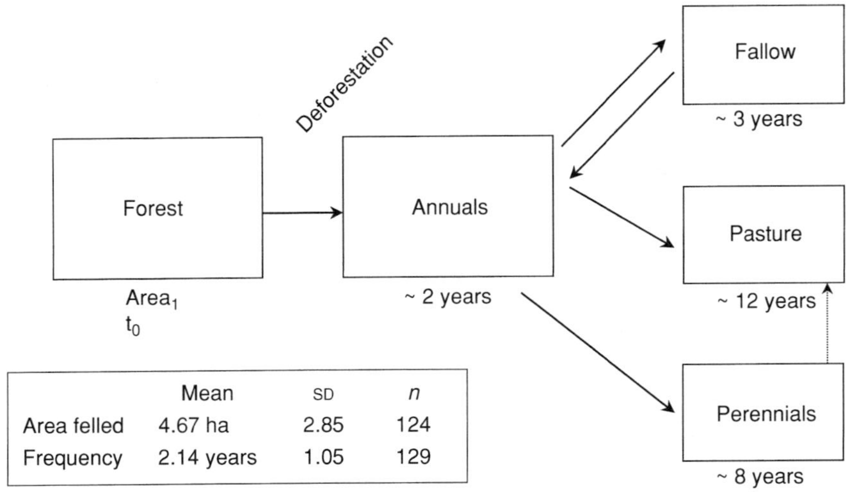

	Mean	SD	*n*
Area felled	4.67 ha	2.85	124
Frequency	2.14 years	1.05	129

Fig. 13.1. Observed land-use trajectories. Number of years noted below each land-use box indicates time continuously in a given land use (not time elapsed from t_0). (From Embrapa/IFPRI surveys.) SD, standard deviation.

Can agricultural intensification increase output and slow degradation?

Given this context, what does it mean to intensify small-scale agriculture in the western Brazilian Amazon, and what mechanisms or entry points exist for doing so? Intensification can mean enhancing the productivity of a given meta land use at a given point in time, extending its productive life, or both. Intensification can also mean generating new methods for converting from one use to another (creating new arrows between the boxes in Fig. 13.1); some of these uses might 'climb' back up the implied degradation trajectory, unless an irreversible threshold is crossed. Both avenues for intensification involve increasing productivity over time.

How does concern for the environment enter into such an approach to intensification? The literature linking agricultural intensification and environmental sustainability is extensive and fast-growing, and identifies the former as a key element in the latter.[3] Definitions for developing-country contexts stress that technologies, while often involving additional inputs (capital, labour or purchased), need to complement biological and environmental processes 'that determine crop productivity and other aspects of agroecosystem characteristics' (NRC, 1993, pp. 72–73).[4] In the forest margins, this includes environmental services performed by standing forests. Concern about the environment can be incorporated into intensification strategies, focusing on activities for existing land uses or land-use conversion. This assumes that the extent to which land-use systems match forest-margin ecosystem characteristics – and how they affect the incentives to clear remaining primary forests – can be gauged.

This chapter examines the incentives to deforest based on relative factor endowments and then focuses on assessments of activity-specific intensification strategies. This leaves aside regional land-use issues, such as which forested areas, if any, should be converted to agriculture and which cleared areas should be considered for what types of intensification. A number of indicators are adopted, consistent with the ASB analytical framework.[5] More intensified systems have higher productivity, measured by the net present value of the system with respect to land and labour, assessed over a 20-year time horizon, which carries a given piece of land through a defined sequence of uses, beginning with forest. 'Agriculture' is broadly defined to include production activities in each meta land use, including the forest. Agroecosystem suitability has one component related to agronomic sustainability and another related to the environmental services provided by the forest. The first is measured by identifying the extent to which biological conditions or processes prevent systems from being replicated over future periods. Environmental services are measured by indicators of the carbon stocks and above-ground plant biodiversity that are associated with each system.[6]

The Context for Agricultural Intensification: Key Ratios

If farmers want to boost output (and profits) and have the option of doing so by expanding the area under production, under what conditions might they forgo this option and instead concentrate on increasing the productivity of their land already in agriculture? In the forest margins, where agriculture encroaches upon forest, labour is scarce and land is abundant; any recipe for sustainable agricultural intensification must confront this question and be compatible with incentives that reflect factor endowments.

Table 13.1 (central columns) presents 1993–1994 factor endowment data for the colonization projects of Pedro Peixoto (AC) and Theobroma (RO). These are compared with state-level data (first two columns) from 1996 and median data from two clusters of households identified by cluster analysis,[7] which demonstrate considerable within-project variation (last two columns). The first set of rows speaks to the propensity to intensify by comparing the area still in forest with the area cleared (in percentage terms), where the use of the cleared area is detailed in terms of the land-use trajectory outlined above. The extent of forest clearing (e.g. ongoing extensification) in one year, 1994, is also described.

At no level had the forest been absolutely exhausted. In fact, even though no set threshold for land scarcity at which the incentive to preserve forest is known *a priori*, the situations shown here appear to fall well on the 'land-abundant' side of any such threshold. Land scarcity was, in no instance, severe enough to provide evidence that incentives changed so as to reduce pressure on the forest. On the contrary, the degree of land scarcity appears to have had little appreciable effect on how much forest was cut in a given year, suggesting that land scarcity here was not yet the predominant factor determining the degree of forest felling. Still, one site, the longer-colonized project of Theobroma, was

Table 13.1. Key ratios for study areas: state, project and household levels (from Instituto Brasileiro de Geografia e Estatística; *Folha de São Paulo*, 6/22/98; Instituto Nacional de Pesquisa Espacial (*www.inpe.br* as cited in Faminow *et al*., 1999); unpublished Embrapa/IFPRI/ASB survey data).

	State		Project		HHs (within PPxto)	
	Acre	Rondônia	PPxto	Theobroma	A	B
Land cover						
% in forest	91.0	79.6	66.7	53.0	72	73
% open (total)	9.0	20.4	33.3	47.0	28	27
% in annuals (of total)	0.7	0.7	5.0	6.5	5	5
% in fallow/secondary forest	0.4	–	5.7	7.9	8	5
% in perennials	0.1	1.1	1.1	6.3	2	1
% in pasture	5.9	12.3	20.6	27.2	13	16
Deforestation						
Area felled (km^2 in 1994)	1,242	4,097	1.31	1.51	0.015	0.028
% of forest felled in 1994	0.89	2.16	4.0	6.2	4	5
% of total area felled in 1994	0.81	1.72	2.7	3.3	2	3
Productivity[a]						
Value of output per area (R$ km^{-2})	9,141	17,190	3,768	4,316	3,652	3,084
Value of output per open area (R$ km^{-2})	–	–	10,090	9,132	13,277	10,228
Value of output per capita (R$ $head^{-1}$)	2,894	3,331	617	596	378	569
Value of output per AE[b] (R$ AE^{-1})	–	–	1,036	1,023	767	1,128
Population						
Pop. growth (% $year^{-1}$, 1991–1996)	3	2	–	–	–	–
% pop. urban (1996)	65	62	–	–	–	–
Population density (pop. km^{-2})						
Pop. per land	3	5	6	7	10	5
AE per land[b]	–	–	3	4	6	3
AE per cleared area[b]	–	–	10	9	21	12

[a]The value of output measure is not directly comparable across levels in these ratios. The state-level statistic is GDP in US$ per area for 1996; project and household-level statistics are value of total output (VTO) from 1993/94 on-farm production activities in 1996 currency (R$), except in the case of VTO/open area, where the numerator excludes extractive activities. In 1996, 1 R ≈ US$1.

[b]Adult equivalent (AE) measures agricultural labour, weighting (individuals as follows: 1 = males 15–49 years; 0.5 = adolescents and females 15–49 years; 0.6 = males 50–64 years; and 0.3 = females 50–64 years. Dependents under 5 and over 65 were not included. State-level data were not available to calculate a comparable ratio of agricultural labour to land. State-level statistics are from the 1996 Agricultural Census; project-level statistics are based on 1993–1994 surveys.

Ppxto, Pedro Peixoto; HHS, households.

relatively more land-scarce, having proportionally less forest and more perennials and pasture, with the latter accounting for most of the cleared area.

The third set of rows in Table 13.1 presents indicators of intensification, expressed as value of output per unit of input.[8] The project where forested land was scarcer, Theobroma, showed higher land productivity than did Pedro Peixoto, which is not surprising, since the productivity of cleared area exceeds that of forest (Witcover *et al.*, 1996).[9] Given this disparity in productivity between open and forested area, separating the value of output derived from each side of the forest margin can clarify the intensity of land use and can highlight the livelihood consequences for farmers of deforesting or not. Looking at the value of output from open areas only, that in Pedro Peixoto exceeded that in Theobroma, perhaps reflecting the fact that less cleared land in the former had marched down the 'degradation trajectory'. In Pedro Peixoto, better-situated farmers generated a higher value of output per hectare.[10] When measured in terms of labour productivity, though, the projects rated nearly the same. More remote farms (with less on-farm labour) got more out of each unit of labour than their well-situated counterparts.

Aggregate statistics, then, suggest little link between land abundance/scarcity and the rate of deforestation, and point to higher value of activities on cleared than forested areas (with mild evidence of lower land productivity in less land-abundant areas, perhaps due to degradation). At the same time, evidence points to wide differentials in value of output in terms of labour, with households where labour is the scarcest displaying the highest productivity figures. Aggregate statistics mask this trend. Since technologies that stand a chance of adoption must have labour requirements in line with labour availability, labour scarcity must be considered in strategies for sustainable agricultural intensification. All too often, this aspect is overlooked in strategies which, focusing on international concerns of forest retention rather than local concerns of income growth and poverty alleviation, emphasize land-saving almost exclusively. The next set of indicators quantifies labour scarcity across and within projects.

Simple human/land ratios show a low population density in the colonization projects, which, when expressed in terms of labour available (second row of the population-density ratios), exhibit even less manpower per land unit (three or four adult equivalents per square kilometre). Farms with better market access exhibit a higher concentration of labour available on-farm by a factor of nearly two, evaluated across all measures of population density, including labour available per unit of cleared land. This measure may have, at least in the short run, the greatest practical implications for the 'fit' of particular agricultural intensification strategies with current practices, given that most labour is expended on cleared land (and on clearing land). In the longer term, the preceding row (AE/land) may be more important as a limit to human/land availability in the absence of effective policies to regulate how much private land can be cleared. Together, the last two rows of ratios define a range of labour/land ratios relevant for agricultural intensification.

Meta land-use practices in the context of key ratios

Current practices already demonstrate that farmers are striving to achieve more output per unit of land. For each type of land cover, there are various indications of possible intensification. Fallow land is cut after a mean fallow period of 2 years, even though, on average, farmers say that 4 years are needed to restore nutrient content. Roughly 75% of sample households intercropped or multiple-cropped land in annuals or perennials (alone or in combination), slightly more so in the project with less forest remaining (Theobroma). Farmers also expressed interest in extending pasture life through a range of options from replanting to planting legume-based systems, perhaps showing a propensity to follow the pattern of farmers living in the longer-settled eastern Amazon basin, where the intensification of pasture/livestock systems has already begun (Mattos and Uhl, 1994). Room for boosting productivity using available technologies exists – external input use, while low, is present and growing (Witcover *et al.*, 1996), with 20% of farmers roughly matching experiment-station yield results.[11]

These qualitative signs of potential intensification for each meta land-use cover are no substitute for actual measurements of inputs and outputs, which can determine precise productivity consequences expressed as returns to land and labour. Nor do they illuminate small-scale farmers' capacity – in terms of labour and capital available, plus market and institutional support – to adopt systems that hold more promise. And these indicators give no suggestion of what impact land-use changes may have on particular environmental services and how more promising technologies might do better. We now turn to a framework that includes just such measures.

Evaluation of Land-use Systems vis-à-vis Agricultural Intensification and the Environment

The ASB matrix framework is used here to evaluate alternative land-use systems based on their ability to address international environmental concerns, agronomic sustainability issues and farmer adoption concerns. The framework presented in Table 13.2 differs from others presented previously (e.g. NRC, 1993; Tomich *et al.*, 1998b) in that it: (i) specifies land-use system trajectories, including technology, land area and the time line associated with each system (matrix rows); (ii) defines indicators corresponding to interests of various stakeholders (matrix columns); (iii) presents measurements for each land-use system selected (matrix cells); and (iv) specifies the socioeconomic and geographical setting for the analysis.

For discussions of agricultural intensification, this matrix can provide a useful first characterization of potential systems and make transparent some consequences of different technologies for forests and farmers. The systems identified here (column 1) are compared for pairs of meta land-use trajectories (all beginning with forest) broadly representing the traditional systems currently prevalent in the landscape, evaluated against other systems capable

Table 13.2. Evaluation of land-use systems (biophysical and socioeconomic indicators) – Brazil sites. Prices are based on 1996 averages and expressed in December 1996 R$ (US$1 = R$1.04). (Data derived from Vosti *et al.*, 2000b; and the following ASB Phase II Working Group Reports: Gillison, 1999; Gillison and Liswanti, 1999; Palm *et al.*, 1999; Muñoz Braz *et al.*, 2000; Weise, 2000a, b.)

	Global environmental concerns – biophysical indicators		Global and smallholder concerns: agronomic sustainability[a]			Adoptability concerns – socioeconomic indicators					
						Profitability					
										Institutional requirements[d]	
Land-use systems	Carbon: above-ground t ha^{-1} (time-averaged)	Biodiversity: above-ground (plants) species : modi	Soil structure	Nutrient export	Crop protection	Returns to land (private prices) (R$ ha^{-1})	Returns to labour (private prices) (R$ per person-day)	Labour requirements[b] – (person-day ha^{-1} year^{-1})	Food security[c] – entitlement path (operational phase)	Market	Non-market
Forest (AC)	148	1.82	0	0	0	−2	1	1	n/a	o	–
Managed forestry (AC)	~148	n/m	0	0	0	416	20	1.22	$	inp, lb, k, o	N, R, S
Coffee/ *bandarra* (RO)	56	1.29	−0.5	−0.5	−0.5	1955	13	**27**	$	inp, o, lb, k	n
Coffee/ rubber (RO)	56	1.1	−0.5	−0.5	−0.5	872	9	**59**	$	inp, o, lb, k	n
Traditional pasture (AC)	3	1.45	0 to −1	−0.5	−0.5 to −1	2	7	11	$+ consumption	inp, o	–
Improved pasture (AC)	3	n/m	0 to −1	−0.5	−0.5 to −1	710	22	13	$+ consumption	inp, lb, k	N
Annual/ fallow (AC)	7	1.96	0 to −0.5	0 to −0.5	−0.5 to −1	−17	6	23	$+ consumption	lb	le
Improved fallow (AC)	~3–6	n/m	0 to −0.5	0 to −0.5	−0.5 to −1	2056	17	**21**	$+ consumption	LB	n

[a]0 indicates no difficulty, −0.5 indicates some difficulty, −1 indicates major difficulty.
[b]A bold figure indicates competition for labour with other agricultural activities (including deforestation).
[c]'Consumption' and '$' reflect, respectively, whether the technology generates food for own consumption or income that can be used to buy food.
[d]Letters indicate institutional constraints to, or impacts of, adoption (upper case indicates a serious problem; lower case indicates a more minor problem): inp, input markets; o, output markets; lb, labour markets; k, capital markets; n, information; r, regulatory issues; le, local environment constraint/problem; s, social cooperation required for adoption.
n/m, not measured; n/a, not applicable.

of potentially improving performance. These systems are compared using indicators on alternative development outcomes of interest: growth, poverty alleviation and environmental sustainability.[12] The exception to this treatment is perennials, for which the prevalent monoculture is a system still under evaluation; in this case, the results of two alternative intercropped systems are presented. These measures allow assessment of whether each technological change: (i) leads to agricultural intensification (using value-based productivity measures); (ii) is adoptable by smallholders (taking into account household-livelihood security constraints and the broader institutional context); and (iii) complements biophysical processes (measured in terms of agronomic sustainability and impacts on particular ecological services associated with humid tropical forests).

Land-use systems under evaluation

As noted above, all systems summarized in Table 13.2 start from forest clearing and follow a described trajectory over a 20-year time horizon, given a particular socioeconomic and geographical setting (Vosti *et al.*, 1998b). They represent both traditional systems and proposed technological advances.[13] Most of the experimental systems were selected as potential candidates for sustainable intensification because they involve an upward shift over the life of the system in returns to labour and land relative to the traditional system, and include elements thought to enhance ecosystem compatibility. The proposed systems do not, as yet, appear broadly on the landscape, however, and one of the 'traditional' systems – a long-term annual–fallow cycle of shifting cultivation – has vanished.[14]

As outlined in the previous section, factors affecting the adoptability of each system – relative profitability and whether the labour, capital and land available to farmers are sufficient to begin or to maintain the system – must be examined in greater detail. This must be done bearing in mind that intensification strategies falling within the reach of an average smallholder with reasonable market access may lie outside the means of a sizable number of others less well situated.

Each system in Table 13.2 was evaluated using comparable value-based productivity measures of intensification plus quantitative and qualitative assessments of environmental impacts. The first two columns assess biophysical indicators of international interest: carbon storage capacity and above-ground plant biodiversity.[15] Criteria that are presumed to be linked to the productivity and longevity of the system – soil structure, nutrient exports and crop protection – measure agronomic sustainability and are of immediate importance to smallholders and local decision-makers, but are also of interest to the international community. Economic returns to land and labour gauge relative productivity, while revealing systems more profitable for farmers.[16] Adoptability issues beyond financial attractiveness and agronomic soundness are measured by total labour requirements (noting seasonal bottlenecks), institutional constraints to adoption (market- and non-market-related) and food security, this last indicating a typical household's increased exposure to food insecurity because of additional production or price risk compared with the traditional system (see Vosti *et al.*, 1998b).

Land-use systems – intensified and traditional systems compared

Table 13.2 confirms that the proposed systems are more intensive than the traditional ones, as evidenced by the higher returns to land and labour for the former. Taking pairwise comparisons, beginning with the first two rows, managed forestry dramatically improves the returns to land and labour compared with simple Brazil-nut extraction (forest), for example. High start-up costs (mostly large capital investments for establishment) and substantial maintenance requirements (a 12-fold increase in labour use over the life of the system) may, however, place the system out of reach for many smallholders.[17] Managed forestry does not disturb the forest ecosystem and can be replicated indefinitely without marked changes. Also from the table, an improved pasture/livestock system, although requiring initial capital outlays and slightly more labour, pays off in significantly higher returns to land and labour than the traditional system. However, it can damage soil structure and provides little ground cover, perhaps resulting in productivity losses over a longer time horizon. This system involves no greater loss of sequestered carbon than traditional pasture, but a massive decline from that found in standing forests (Faminow *et al.*, 1997a; Carpentier *et al.*, 2000). Comparisons of annual/fallow systems (rows 7 and 8) show similar results for the intensified system in terms of significantly higher returns, although labour requirements actually decrease under the improved system. Doubts about system 'longevity' also persist, however. Slightly more carbon is sequestered by the traditional system, but again the substantial loss is from forest conversion. Finally, of the coffee–tree intercropped systems (rows 3 and 4), the one involving *bandarra* (a fast-growing timber species) is the more intensive and requires less labour. However, it may strain farmers' budgets in the start-up phase. The systems are similar regarding environmental impacts, agronomic longevity and carbon and biodiversity scores.

Agricultural intensification measures: relevance to farmers' concerns

This chapter suggested earlier that, in a relatively labour-scarce environment, returns to labour would outweigh returns to land in farmers' decisions to adopt. Table 13.2 presents both profitability measures for comparison. Systems at or below the average rural daily wage for unskilled labour, approximately R$6.50 (1 Brazilian *real* was equal to approximately US$1 in 1996), would probably not be attractive to farmers, although imperfections in the labour market, seasonality of labour demand and heterogeneity of labour within the household make this a less than firm rule.[18] Indeed, the annual/fallow system no longer practised yields slightly lower returns than working for wages. Traditional pasture/

livestock production systems, the most prevalent in the study area, yield slightly better returns than wage labour. All of the intensified systems yield even more, with the higher of the two coffee-based systems (coffee/*bandarra*) generating about twice the wage rate. Improved pasture/livestock and managed forestry systems bring in returns nearly three times more than the traditional livestock system.

Analysis of returns to land shows that all intensified systems appear relatively attractive and the two traditional systems are below average. Farmers who are more interested in returns to labour than to land would probably select improved pasture/livestock systems, while those more concerned with per hectare asset values (including improvements in current production systems) might prefer systems scoring high on both counts, such as managed forest, improved fallow and coffee/*bandarra*. But factor returns are not the only issue in adoption; to achieve those returns requires certain quantities of important factors needed at specific times during production cycles. This can mean land of an appropriate agronomic profile, but, more importantly, labour (due to its relative and absolute scarcity) and, for systems requiring purchased inputs, capital.

Adoptability of intensive agricultural systems

Labour requirements

A system with high returns to labour may simply be out of reach of many small farmers in the area, given the current labour scarcity and imperfectly functioning markets. The coffee/rubber system demands the most labour by far, nearly 60 person-days ha^{-1} $year^{-1}$ (Table 13.2). At the other end of the spectrum lies the low-level and more intensive forest extraction system in Acre, which requires only about 1 person-day ha^{-1} $year^{-1}$ to manage. The system currently forming the end of the land-use trajectory, traditional pasture, requires the least labour of any system other than the forest systems, approximately 11 person-days ha^{-1} annually; its intensified version, improved pasture, needs just slightly more than this. Clustered at 1.5–2.0 times the labour requirements of these systems are two other intensified systems, coffee/*bandarra* and improved fallow, as well as the vanished shifting cultivation (annual/fallow) system.

Figure 13.2 illustrates which systems might fail farmer adoption criteria, due to excessive labour needs. The data from Table 13.2 for system labour requirements are plotted on the horizontal axis in relation to reference lines marking observed household labour availability for each project,[19] and against data measuring returns to labour (vertical axis). The horizontal dashed line indicates the average wage as a reference point.

Since households in Theobroma (RO) are smaller and older than their counterparts in Pedro Peixoto (AC) and have, on average, about a third less available household labour (22 person-days per cleared hectare per year vs. 38 in Acre), they may not have the family labour necessary to adopt systems feasible in the 'younger' colonization project in Acre. One of the proposed systems that performs well relative to the prevailing wage – the coffee/*bandarra* system – is

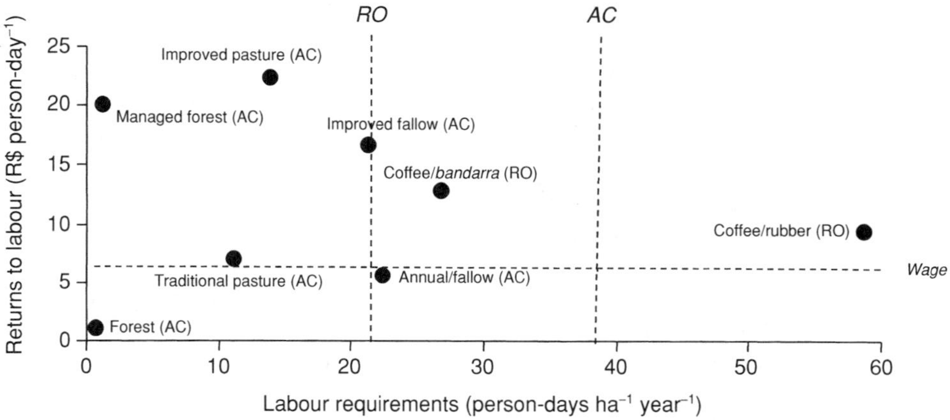

Fig. 13.2. Returns to labour versus labour requirements. The horizontal 'Wage' line represents the daily wage for hired labour; vertical lines represent average person-days available per cleared hectare per year in AC and RO. All prices are in December 1996 R$ (US$1 = R$1.04). Labour requirements (person-days ha^{-1} year^{-1}) are based on total requirements over the system period. (From IFPRI, Embrapa and ASB data.) AC, Pedro Peixoto, Acre; RO, Theobroma, Rondônia.

feasible in Acre but not in Rondônia; the same is true for the defunct annual/fallow system. The other coffee-based system (coffee/rubber), which is less attractive in terms of returns to labour, rests far beyond a typical household's ability to manage without hired labour. Yet the overall picture may favour the adoption of intensified systems. Those systems with the highest returns to labour (improved pasture and managed forestry, followed by improved fallow) fall within the limits imposed by on-farm labour availability – a combination seen in the upper-left quadrant of Fig. 13.2 – although not necessarily within the limits posed by capital and credit constraints. This is due to large start-up costs, which require several years to achieving positive cash flow and credit limitations for managed forestry, coffee/*bandarra* and improved pasture systems.

Institutional requirements and food security

The final three columns of Table 13.2 identify other factors that condition profitability or that might otherwise expose farmers to increased production or price risk, perhaps threatening their livelihoods or food security. Substantial institutional obstacles face farmers who might try to establish intensified systems and achieve the financial returns reported; their degree and type vary widely from one system to another. Imperfections in the labour market have already been discussed, and are considered a factor in the adoption in all intensified systems, particularly the improved fallow. Beyond this, though, practically all intensified systems include reliance on other markets plagued by imperfections as well, including the capital market and markets for specific inputs and outputs. The exception is precisely the system most dependent on labour – improved fallow.

Non-market institutions can also impede or facilitate adoption. The regulatory environment, for instance, can be friendly (or not), and the knowledge needed to apply the technology may be a small extension of farmers' practices or may necessitate broad new understanding. Although all the systems were deemed to face some non-market institutional obstacles, the number and severity of those faced by managed forestry – an attractive system in terms of returns to labour and labour needs – stood out. Managed forestry requires expertise about tree species and felling techniques not readily available in the area. It also demands a level of social cooperation in order to achieve a sustainable production scale and navigation of regulations limiting or monitoring extraction in forest reserves. Improved pasture and livestock management also calls for knowledge of new techniques and seeds, but these can be adopted piecemeal, with a focus on aspects more similar to traditional livestock practices. The non-intensified counterpart of managed forestry, low-level forest extraction, a system that has been in place for many years, has the fewest institutional obstacles. The ability to overcome many of these institutional obstacles is presumably the most restricted for precisely those farmers who are most at risk of food insecurity and who have fewer resources (financial, know-how and time) at their disposal. These households especially may need to take stock, prior to adoption, of the limitations posed by any particular system's reliance on markets to meet food needs.

Tradeoffs between farmers' concerns and environmental concerns

If farmers were to intensify via these improved systems, would their search for profitability in general, and higher returns to labour in particular, come at high environmental cost? From available evidence in the study area, agricultural intensification (using the returns-to-labour measure most suggestive of adoptability) appears to be at odds at least with plant biodiversity.[20] Systems harbouring the most biodiversity (the fallow phase of the annual/fallow cycle, and forests) do not include any of the intensive systems (see Fig. 13.3). They yield lower returns to labour than simply participating in the (imperfect) hired-labour market (see wage reference line in Fig. 13.3), and thus are unlikely to have a long-term future unless returns to labour can be increased. What is more, the systems highest in returns to labour for which biodiversity measures were made, the coffee-based systems, have the lowest biodiversity. The traditional pasture system (where weed invasions can mean higher biodiversity) scored better than the perennial systems on this measure. Thus, while replacing traditional pastures with coffee-based systems, as is happening in some parts of Rondônia, will help farmers' incomes, plant biodiversity is likely to suffer.

Available evidence shows that the tradeoff between agricultural intensification (returns to labour) and carbon stocks (Fig. 13.4) is even more stark in some ways. Forests are by far the best way to store carbon, but extracting Brazil nuts from them yields less per person-day than manual labour (the wage reference line). Managed forestry, if able to overcome institutional constraints, looks promising as an intensive system that sequesters a fair amount of carbon. The most attractive system along the returns-to-labour spectrum, however, improved pasture/livestock, is among the worst ways of storing above-ground

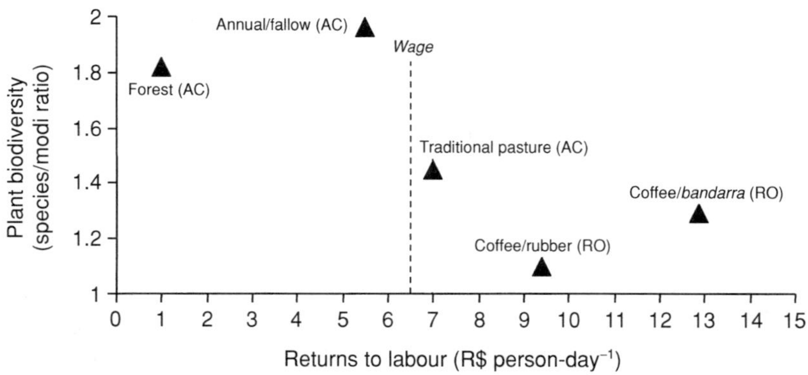

Fig. 13.3. Plant biodiversity vs. returns to labour. The vertical 'Wage' line equals the average daily wage for hired labour during the study period. All prices are in December 1996 R$ (US$1 = R$1.04). Species/modi ratio measurements are for the stable system land cover; for the annual/fallow system, the measurement is for the fallow phase. (From IFPRI, Embrapa, CIFOR and ASB data.) AC, Pedro Peixoto, Acre; RO, Theobroma, Rondônia.

carbon. The coffee-based systems occupy intermediate positions. Moving from coffee/rubber to the coffee/*bandarra* system (the more attractive in terms of labour requirements) improves returns to labour without sacrificing carbon stocks.

Summary and Conclusions

This chapter has examined the potential for, and the economic and environmental consequences of, intensification of selected land-use systems in the western Brazilian Amazon, concentrating on the key development objectives of agricultural growth, poverty alleviation and environmental sustainability. The research uncovered no obvious 'win–win–win' land-use systems that meet all three objectives; none the less, land uses within each meta category chosen for their potential to 'win' on one or more criteria proved promising. Trends in the study area may point the way towards how technologies can be adjusted to improve their impacts.

Some spontaneous agricultural intensification is occurring in the western Amazon, and other types of intensification – generally capital- and knowledge-intensive – have been proposed, some driven by farmer interest. Farmers have spontaneously intensified within meta land-use systems and have moved from one system to another, thereby extending the effective life of particular activities and systems, but not indefinitely so (except possibly coffee-based systems). Although some policy-makers and researchers have focused on land-saving intensification technologies in the hope that this would save the forests, indications are that factors other than land scarcity drive land-use trends in the area.

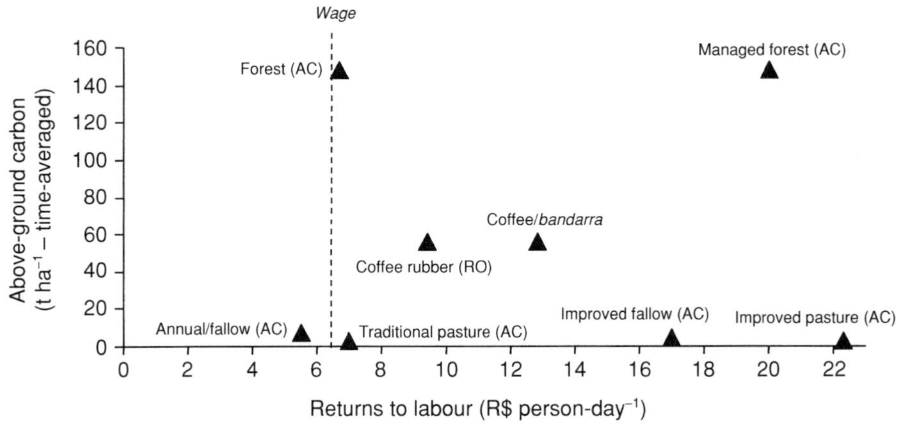

Fig. 13.4. Carbon vs. returns to labour. All prices are in December 1996 R$, US$1 = R$1.04. Evaluations for AC, RO systems use prices and parameters from Pedro Peixoto, Acre and Theobroma, Rondônia, respectively. Returns do not take into account known difficulties in marketing. The vertical 'Wage' line represents the daily wage for hired labour during the study period. (From IFPRI, Embrapa, TSBF and ASB data.)

Deforestation has continued alongside spontaneous intensification, but the forest has not been exhausted. Labour scarcity – absolute and relative – may play a larger role in land-use decisions, and thus in driving intensification and deforestation. A broad array of market failures in input, output and credit markets, as well as other institutional obstacles, also influence farmers' abilities to adopt technologies and thus significantly affect land-use outcomes.

Yet development concerns do not end with farmer adoption of intensified systems. The recognition of effects on environmental sustainability, growth and poverty alleviation must be part and parcel of intensification efforts. One aspect of this is the pressure on standing forests; since labour returns from traditional extractive forest activities remain far below those from any alternative land-use system examined, the prospect of continued deforestation looms large under most types of intensification alternatives. Because of the region's labour scarcity, labour-saving systems are attractive to farmers, but labour intensity may be needed to slow forest clearing. Thus, successful intensification strategies that purport to do more than raise incomes (that is, reduce incentives to deforest) must walk a line between absorbing labour – focusing on unskilled labour, perhaps in peak months when deforestation occurs – and yielding a good return, while remaining within farmers' means (in terms of labour and capital).

That said, within the technologies examined here, some did better than others, not only in regard to farmers' well-being, but also in providing ecological services, such as carbon sequestration and plant biodiversity. And, despite the myriad obstacles to intensification, policy changes can play a role in overcoming those obstacles and in making systems more compatible with stated objectives.

What types of agricultural intensification should be promoted?

Each proposed intensified system is shown to offer some benefits either to the farmer or to the environment, but none comes without some tradeoffs or obstacles to adoption. Managed forestry (although still experimental) holds great promise environmentally, in terms of meeting agronomic sustainability goals and international concerns and in terms of income generation. But it remains unsuitable for many of the smallholders who face depleted forest reserves or who are too poor to make needed upfront investments. For perennials, the coffee/*bandarra* system may compete closely with more intensive pure stands of coffee in terms of profitability (if start-up costs can be covered) and will almost certainly sequester more carbon. For pasture, an improved system involving changes in both pasture and livestock management can dramatically boost incomes, but establishment costs are high and the environment will suffer much more than it would under tree-based systems. An annual/fallow cycle using improved fallow could prove viable, given its higher profitability. While carbon gains would be negligible, plant biodiversity would benefit.

Despite these apparent differences, the intensified systems evaluated here also show some commonalities, particularly regarding adoption. They all increased returns to land and labour (with respect to traditional-system comparisons) and, except for managed forests, involved no major shifts in either the array of non-market institutional obstacles they confront or their primary paths of food entitlement (e.g. mechanisms for providing food for households). They did, on the other hand, entail greater labour and capital inputs (with the exception of the improved fallow system) and heightened dependence on those markets. From the farmers' perspective, they all offered some benefits, but their adoption also presented some obstacles not easily overcome. Barriers to capital, and perhaps labour, availability might be eased if more attention were paid to technologies that could be adopted piecemeal or easily adapted by farmers given their current knowledge, thus accommodating elements of risk important in the frontier environment (Faminow *et al.*, 1997b, 1999).

If minimizing ecosystem disruption is also a goal, the agronomic sustainability of any proposed system, as well as its effects on factors such as carbon sequestration and plant biodiversity, should be monitored and, where possible, improved. But systems that approximate some aspect of the forest at some point of their cycle and which score better on environmental indicators may face the greatest uncertainties regarding price risk and institutional change. Indeed, the intensified version of the most prevalent system, traditional pasture, does not score well on environmental indicators. Here, too, easing uncertainties that discourage farmers from taking the land-use trajectory less frequently chosen must be a focus. This could be supplemented by efforts to convert existing pasture to more intensive systems, perhaps with some environmentally friendly components.

As already noted, strategies to take pressure off the forest will require more attention. To be most effective, a strategy should work both sides of the forest margin, increasing the intensity of use of cleared land while either pushing up the costs of deforesting or improving the value of standing forest (e.g. via sustainable harvesting of timber products from privately held forests). Intensification of meta land uses, then, should be considered as part of policy packages, not in isolation.

Principal obstacles to agricultural intensification

This chapter has outlined some obstacles to smallholder adoption of intensified agricultural systems that are strictly economic in nature. Other non-economic factors, such as biophysical and risk-related obstacles stemming from the impoverished conditions facing small-scale farmers, as well as institutional constraints, also affect economic outcomes. Enough is known about these systems and farmers' concerns to help farmers overcome some of these obstacles, but more research is needed on how to overcome others, as well as on how adoption and non-adoption affect the environment. In particular, more attention needs to be paid to the adoptability and environmental effects of non-market elements and how changing market conditions alter economic and environmental decisions and outcomes. It is worth noting that the specific economic and non-economic obstacles to intensification described here can be expected to arise in groups. Biophysical limitations of the generally poor, acid soils of the region are also known to pose fundamental constraints, with steep yield declines under continuous cropping of some annual systems. But solutions, such as the combination of chemical and organic fertilizers to maintain and restore soil fertility, are not yet entirely clear. How to bring those solutions within the reach of farmers remains a concern.

Production, health and price risks are known to affect land-use decisions in the area, but how they differentially influence the adoption of particular systems and particular adopters – specifically, those most vulnerable to food insecurity – is rarely obvious. Nor is it clear how this may change over time as the frontier develops. During the initial stages of colonization, when food-security problems loom large, making land-use decisions that require complete dependence on purchased foods may be too risky unless complemented by other choices allowing home consumption. In today's context, it is often overlooked that farmers may have risen above abject poverty but may still be unable to meet high establishment costs; they may, in effect, be 'investment-poor' (Reardon and Vosti, 1995, 1997a). Piecemeal or partial investment in technological change may be their only options. Foreseeing this and designing technologies amenable to such a process and in line with capital market imperfections (e.g. credit systems that will enable farmers to wait out the several years needed to garner positive cash flow) may influence both adoption and environmental consequences (Faminow *et al.*, 1997b; Vosti *et al.*, 1998c).

Institutionally, all systems face barriers posed by the region's notoriously weak and seasonally imperfect markets. Perhaps even more daunting, each system faces an almost unique set of non-market-related institutional and organizational constraints. Slow and partial flow of technical information on system management and the dependence of systems on social cooperation in frontier settings not known for such cohesion may further complicate adoption.

The generally low levels of education (30% of household heads surveyed could not read or write) and the large distances separating producers and consumers both contribute to and follow from these obstacles.

When taken together, this set of obstacles may undermine agricultural intensification efforts and threaten to place intensification out of the reach of many small-scale farmers. Regional development involving the expansion of more intensive land-use systems could help eliminate the obstacles to adoption tied to underdeveloped markets, but, if efforts are not broad-based, intensification may be limited to better-off farmers, while poverty and displacement will continue to threaten poorer smallholders. If efforts are focused on intensification alone, they could, as already seen, heighten pressures on the forest.

Policy and technology options for overcoming obstacles to intensification

Despite these obstacles, investing in the most promising land-use systems is still possible and worthwhile, if done with the awareness that intensification *per se* of a given meta land use may have unintended consequences; that is, in the absence of other measures, policies can establish a profit motive for accelerated deforestation. Policy has a role in promoting the establishment (where necessary), expansion and improved efficiency of institutions related to promising land-use systems, but must work in concert with other actors, such as non-governmental organizations (NGOs) and private firms.

For example, policy changes can improve incentives for the formation of farmers' groups capable of managing small-scale managed forestry systems that do not now exist. Almost all intensive systems require rural credit; this credit will not flow from commercial sources, so policy action is needed. Reducing transportation costs will also be key to intensifying agriculture, so, again, policy action is required. Finally, direct subsidies may be needed to promote some systems; chemical fertilizers for establishing some systems are an example. To reiterate, policies that boost incentives to maintain forests must complement these efforts.

Targets of policy action also need to be made more explicit. This chapter has focused on the intensification of small-scale farming, but the measures described here may also have consequences for land use by other rural actors (large farm enterprises, extractivists, etc.). Policy changes need to take these groups into account. For example, intensifying large farmers' holdings could displace smallholders and yield indeterminate effects on poverty. Thus, the environmental and growth consequences of policy targeting must be considered.

Technological improvements are also needed. Harvesting techniques for managed forestry may have to become more sustainable. As already noted, modifications that facilitate technology adoption without compromising profitability are needed. This means striking the right balance in terms of labour requirements and productivity in this labour-scarce region.

Other analytical tools remain to be developed in order to address sustainability, growth and poverty issues effectively. Ways to boost smallholder systems' returns by extending their effective life have not been fully explored or exploited, but these systems cannot be refined without more knowledge about tropical soils' interactions with different cropping systems over the long term. More importantly, while the comparative static analysis presented above highlights key issues and can provide some policy guidance, the scope of the problem calls for tools that take a whole-farm view and consider the intertemporal nature of land-use decisions, as well as the competition for land, labour and capital on farms and for nutrients among alternative land uses. Prototype methods are available (see Barbier, 1998; Barbier and Bergeron, 1999; Carpentier *et al.*, 2000), but more comprehensive efforts must be mounted. In short, we must look beyond individual parcels of land moving through given trajectories of use and move to more complex evaluations of environmental impact, in order to assess whether combinations of actions ease or halt incentives to deforest.

Acknowledgements

This chapter benefited greatly from comments provided by Tom Tomich, Jim Gockowski, Erick Fernandes, participants in the American Agricultural Economics Association (AAEA) conference on agricultural intensification and especially this volume's editors. Financial support for this research was provided by the Global Environmental Facility with United Nations Development Programme (UNDP) sponsorship and the Danish Ministry of Foreign Affairs (DANIDA) (via the ASB Agriculture Research Programme of the Consultative Group on International Agricultural Research (CGIAR), the Inter-American Development Bank, the Empresa Brasileira de Pesquisa Agropecuária (Embrapa) and the Government of Japan.

Notes

1 This section borrows heavily from Carpentier *et al.* (2000).

2 Some agroforestry and pasture systems, properly managed, come closer to replicating the forest system, and should thus use nutrients more efficiently and be more sustainable than annual cropping in tropical conditions (Palm *et al.*, 1996).

3 This literature often highlights the role of technological change in sustainable intensification and examines methods to embed environmental benefits (and losses) within productivity measures of land-use intensity (see, for example, Ehui and Spencer (1993) and Harrington (1994) for methods that use total factor productivity measures).

4 Assuming low levels of purchased inputs and, thus, reliance on the productivity response of the natural resource base over time has led to confounded meanings of 'intensity' in some discussions: a biophysical meaning, concerned primarily with the export of nutrients (such as Ruthenberg's (1980) characterization of fallow-based systems by the prevalence of annual cropping ('R value') over time); and a socioeconomic meaning focused on labour requirements (Ruthenberg (1980) also uses labour requirements and returns as intensity indicators).

5 Additional information on these indicators and other data collected as a part of this evaluation framework appear in Vosti *et al.* (1998b) and the ASB Phase II Working Group Reports: Gillison, 1999; Gillison and Liswanti, 1999; Muñoz Braz *et al.*, 2000; Palm *et al.*, 1999; Weise, 2000a,b).

6 The treatment of intensification employed here thus provides specifics on time period, spatial scale, scope of 'agriculture' and precise technology under evaluation, and distinguishes between biophysical and socioeconomic inputs and impacts.

7 State-wide figures show how one project differs from state-wide realities. Household groups within projects having better and worse access to markets were determined via cluster analysis, based on characteristics taken as exogenous to small-scale agriculturalists: soil type, farm size, distance to market, adult male labour available and time since land was first settled. Two clusters ('A' and 'B') covered most sample farms (fairly evenly split between them). 'A' cluster farms were: more accessible to roads, particularly in the wet season (an average of 3 h to market versus 6 h for cluster 'B'); smaller in area (60 ha versus 90 ha for cluster 'B' farms); and had more labour available on farm (3.2 adult equivalents (AE) of labour versus 2.4 for cluster 'B'). Farmers in different clusters varied in their ability to undertake intensification and possibly faced differing risk levels, which affect overall livelihood security.

8 Low external input use in the area makes total value a good proxy for potential net value, although actual gains could fall short of this (due to farmer choices not to sell, or marketing difficulties) (Vosti *et al.*, 1998a).

9 This is part of the reason for the less-forested Rondônia's land productivity being greater than Acre's. Note that each project's land productivity fell below that of its respective state (which included GDP from urban areas).

10 Table 13.1 reports cluster medians for value-based measures as more representative than means, since the more remote cluster had a highly skewed distribution, making average land productivity higher.

11 For instance, sample households produced an average of 969 t ha^{-1} and 1241 t ha^{-1} of rice in Pedro Peixoto and Theobroma, respectively, in the agricultural year 1993/4 (Embrapa/IFPRI/ASB survey, unpublished results).

12 Use of a similar framework to that applied in other ASB benchmark sites facilitates cross-country comparisons of land-use systems.

13 Land-use systems (including scale of evaluation, management practices and expected outputs) are described in Muñoz Braz *et al.* (2000) and involve a 20-year sequence of land uses from forest, beginning with 2 years of annual cropping.

14 Improved fallow systems and coffee-based agroforestry systems are in the experimental stage (with some experiments in farmers' fields), while managed forestry is in the initial stages of experimentation. The swidden agriculture cycle was constructed based on current practices for a single annual/fallow cycle, with fallow length adjusted to allow repetition of the system over 20 years, and to show why such systems are no longer viable (and to compare them with improved fallow).

15 For carbon indicators, managed forest and improved fallow (AC) systems were not measured directly; approximations were based on expert opinion. For biodiversity indicators, measurements were taken in the operational phase, with the exception of fallow systems, where measurement reflects the fallow part of the cycle.

16 The ASB framework uses social prices in profitability calculations; private prices are used in the calculations presented here, due to lack of social price data. Although prices for most commodities track national trends, and prices here are taken as exogenous, the framework easily allows for re-evaluation of returns under other price schemes.

17 Managed forestry permits low-impact per hectare annual extraction of limited amounts of timber from selected tree species, a rate judged by forestry experts to be sustainable with rotational practices on subplots equal to 10% of total forest area in any given year (Muñoz Braz *et al.*, 2000).

18 For example, Brazil-nut extraction, at returns near R$0, does occur. The activity peaks during a trough in labour demand, which may lower household opportunity costs of labour off the farm to near zero. Moreover, children, with a lower opportunity cost of labour than the prevailing wage, also engage in this activity, so extraction might be observed even if the labour market worked perfectly.

19 Household labour availability is mean person-days available for economic activities (on or off the farm) per hectare of cleared area per year, averaged over the life of the

system, adjusting for the gender and age characteristics of the average household and for leisure patterns prevalent among small-scale farmers (for details, see Carpentier *et al.*, 2000). Labour requirements may differ widely between establishment and operational phases for some systems, or from one month to the next, affecting adoptability (for details, see Muñoz Braz *et al.*, 2000).

20 Measured in terms of species/modi ratio (see Gockowski *et al.*, Chapter 11 of this volume). No 'weights' favour some species (e.g. those derived from forest) over others in the assessment. 'Degradation' from the farmers' point of view (in terms of output that can be derived from a given area) lends some systems their biodiversity (see Gillison, 1999). Biodiversity measures for managed forestry, improved pasture and improved fallow are not currently available.

14 Integrated Bioeconomic Land-use Models: an Analysis of Policy Issues in the Atlantic Zone of Costa Rica

ROBERT A. SCHIPPER,[1] HANS G.P. JANSEN,[2] BAS A.M. BOUMAN,[3] HUIB HENGSDIJK,[4] ANDRÉ NIEUWENHUYSE[5] AND FERNANDO SÁENZ[6]

[1]Development Economics Group, Department of Social Sciences, Wageningen University, Wageningen, The Netherlands; [2]Agricultural Economics Research Institute (LEI – Wageningen UR), The Hague, The Netherlands; [3]International Rice Research Institute (IRRI), Makati City, the Philippines; [4]Laboratory of Theoretical Production Ecology, Department of Crop Science, Wageningen University, Wageningen, The Netherlands; [5]Proyecto ZONISIG, La Paz, Bolivia; [6]Centro Internacional de Política Económica, Universidad Nacional Autónoma, Heredia, Costa Rica

Introduction

The debate about development of the agricultural sector has often centred around the question of how to achieve adequate food security while simultaneously providing sufficient income for food producers (Timmer *et al.*, 1983). Recently, two other concerns have entered the debate – achieving sustainability and environmental protection (Spiertz *et al.*, 1994; Kuyvenhoven *et al.*, 1995). Agricultural-sector development is strongly related to land use. The way land is used has obvious implications for farm income and the various dimensions of sustainability and environmental impact. The environmental effects of agricultural production can range from pollution and nutrient losses to negative externalities resulting from the use of biocides (which include all types of pesticides and herbicides) and trace-gas emissions (Bouman *et al.*, 1998c). Sustainability, on the other hand, is closely associated with maintaining soil nutrient balances.

In most developing countries, methodologies that are capable of simultaneously addressing the various dimensions of agricultural development are lacking, thus seriously compromising informed decision-making by policy-makers. Particularly important are tradeoffs among economic, sustainability and environmental objectives (Crissman *et al.*, 1998). The main challenge in the development of such methodologies is the integration of biophysical and socioeconomic information. In this chapter, we present such a methodology, sustainable options for land use (SOLUS), developed by the Research Program on Sustainability in Agriculture (REPOSA) in Costa Rica.[1] The SOLUS methodology operates at the level of relatively large regions within a country, is interdisciplinary by construction and is able to simulate the effects of different events or policies, including assessing the tradeoffs between different objectives regarding land use. These characteristics are important for a comprehensive analysis of possible policies.

In subsequent sections, we shall explain the SOLUS methodology with emphasis on its economic aspects. The effects of a number of policy scenarios relevant to the agrarian situation in the northern part of the Atlantic Zone (AZ) of Costa Rica will be analysed in terms of aggregate land use and associated economic and environmental indicators.

A Methodology for Land-use Analysis

Land-use analysis aims at assessing present and future land uses in a specific area. It involves a description and diagnosis of the present land-use and farming systems, followed by an exploration of land-use options for the future. As such, land-use analysis forms a part of more general procedures for land-use planning (Fresco *et al.*, 1992; FAO, 1993b; Schipper, 1996). One approach to land-use analysis is the land evaluation and farming systems analysis (LEFSA) sequence developed by Fresco *et al.* (1992). This approach emphasizes the importance of analysing land use at different levels (farm, subregion, region and nation) and highlights the roles of land evaluation and farming systems analysis as two complementary tools for land-use planning.

Using concepts from the LEFSA sequence and tools from systems analysis, the SOLUS methodology was developed (Bouman *et al.*, 1998c) as a general methodology for land-use analysis (Fig. 14.1). The core of the methodology

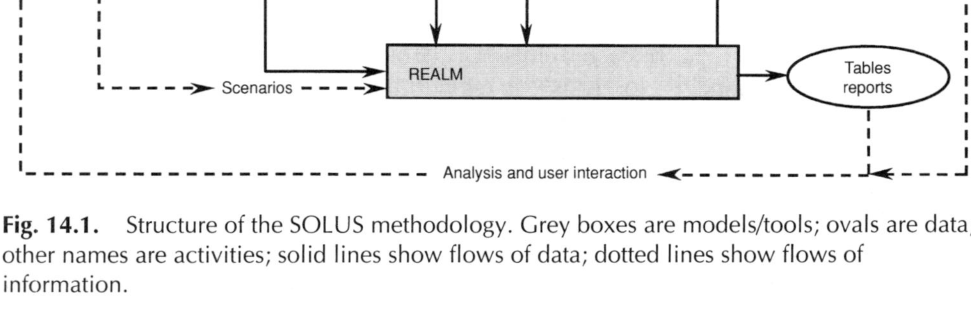

Fig. 14.1. Structure of the SOLUS methodology. Grey boxes are models/tools; ovals are data; other names are activities; solid lines show flows of data; dotted lines show flows of information.

consists of a linear programming model, two technical coefficient generators (TCGs)[2] and a geographical information system (GIS). The linear programming model – regional economic and agricultural land-use model (REALM) – selects land-use systems by optimizing towards a specific goal, for instance, the maximization of economic surplus from the agricultural sector. Optimizations are performed under restrictions, which may be absolute (e.g. no more land can be used than is available in the area) or normative (e.g. threshold boundaries may be imposed on certain sustainability parameters). The maximization of a specific objective function under a set of coherent restrictions is called a 'scenario'. Tradeoffs between economic and sustainability objectives are quantified by running the linear programming model for different scenarios and/or by generating alternative land-use systems, as generated by the TCGs. GIS plays an important role in archiving and manipulating georeferenced input data and in presenting the results spatially.

The REALM linear programming model selects, per subregion and per land unit, the optimal combination of land-use systems, herds and feed supplements by maximizing regional economic surplus. Economic surplus is defined as the sum of producer and consumer surplus. Additional details regarding REALM are presented in Appendix 14.1. A concise mathematical formulation of REALM, as well as the complete model written in GAMS (Brooke *et al.*, 1992), can be found in Schipper *et al.* (1998, 2000).

In the present application of the SOLUS methodology, several thousand actual and alternative land-use systems are included, all relevant to the study area (see Appendix 14.1). Major land-utilization alternatives included nine crop and forest uses and five pasture types. For each land-use system, technical coefficients were generated with the land-use technical coefficient generator (LUCTOR) and the pasture and animal system technical coefficient generator (PASTOR); these included labour requirements; costs of inputs; yields; soil nutrient balances for nitrogen (N), phosphorus (P) and potassium (K); N-denitrification losses; N-leaching losses; N-volatization losses; biocide use;[3] and a biocide index.[4] The technical coefficients are either annual averages (for labour use, soil nutrient balance, N losses, biocide use and the biocide index) or annuities of the present value over the lifetime of the land-use system (yield, input costs and labour use). All systems describe specific quantitative combinations of physical inputs and outputs, representing fixed input–output technologies.

The study area is the northern part of the AZ of Costa Rica. This region is located in the Caribbean lowlands of Costa Rica (Fig. 14.2) and has a humid tropical climate with a mean annual rainfall of 3500–5500 mm. All months of the year have a precipitation surplus. The elevation varies from sea level to 400 m above sea level. The total surface of the study area is 447,000 ha, of which 340,000 ha are suitable for agriculture. In this latter area, three major soil groups were distinguished, based on criteria of fertility and drainage (Nieuwenhuyse, 1996). In this same area, 61,000 ha are protected for nature conservation (including 12,000 ha of national parks) or have a 'semi-protected' status (indigenous reserves, forest reserves, protected areas and wetlands). This leaves an estimated 279,000 ha available for agriculture. Current land use (in the total study area) is dominated by natural forests (about 191,000 ha), cattle ranching (about

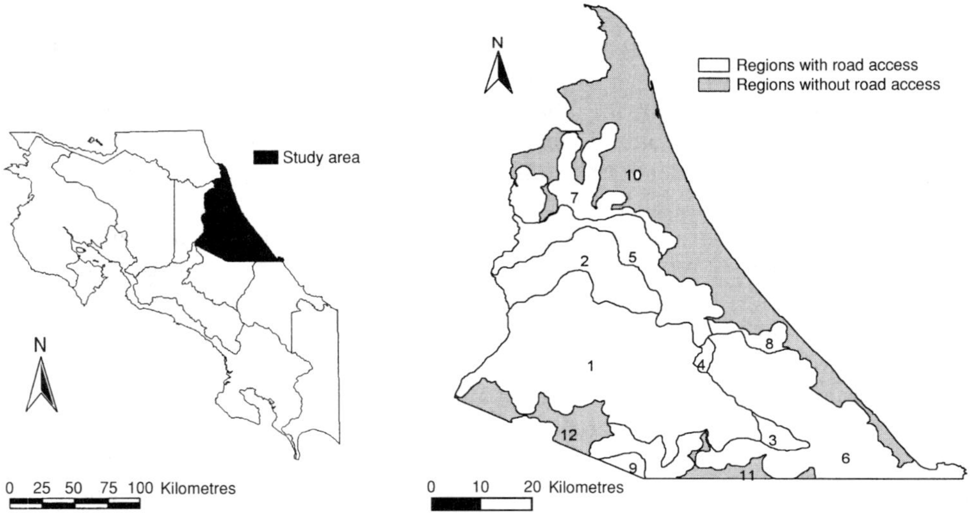

Fig. 14.2. Map of northern part of the Atlantic Zone of Costa Rica (a) with 12 subregions (b).

200,000 ha) and banana plantations (33,000 ha). Secondary crops, which total about 23,000 ha, include plantain, palm heart, root and tuber crops, maize, papaya, pineapple and ornamental plants.

Tropical rainforests, once covering the whole area, are now largely restricted to wetlands, inaccessible mountain areas and semi-protected and protected areas. Colonization, which started at the end of the last century, has accelerated over the past 30 years. At the same time, rapid structural transformations were and are taking place in the ecological, agricultural and socioeconomic conditions of the area – in part, as a response to various structural adjustment programmes. A considerable area has been planted to bananas, in view of its high profitability, both for the owners (multinationals and national companies) and for the country, through the generation of foreign exchange, tax revenues, wage income and employment. However, some of the forest areas that have been cleared should have been protected because of the fragility of their ecosystems.

In certain areas in the AZ, there is a difference between actual and potential land use. Some soils are underutilized; others are damaged as a result of inadequate management. Apart from the introduction of a number of crops that are attractive to both small and large farmers (e.g. palm heart), some crops have been introduced that are unsuitable for the conditions facing parts of the region (e.g. maize). Still other promising crops are not successfully cultivated. Moreover, farmers are frequently burdened by a lack of appropriate technology and knowledge, insufficient credit, lengthy distances to market and unstable prices and marketing conditions (e.g. those for root and tuber crops). There is a rapid turnover of land ownership, with land prices increasing in anticipation of future uses or speculative values. Government legislation lags behind, due to the rapid changes that are taking place, and policy-makers typically have insufficient knowledge for adequate analysis of present and possible future land uses.

Under these circumstances, it makes sense to evaluate future land-use options and their consequences for farm incomes, employment and the environment. Assessing the tradeoffs and outcomes involved in different land-use decisions can inform the development of future policy measures.

Main Economic Aspects of the Optimization Model

In the earlier subregional models developed by REPOSA (see note 1) for the Neguev settlement (Schipper, 1996) and Guácimo county (Jansen and Stoorvogel, 1998), fixed prices were assumed. The size of the study area considered here, considered to be a 'region' in the terminology of land-use analysis, as explained above, requires a different approach with regard to the treatment of prices. Three key issues are at stake: (i) location within the AZ; (ii) downward-sloping demand functions; and (iii) upward-sloping labour-supply functions.

Location within the Atlantic zone

Prices of outputs depend on geographical location within the AZ, due both to variation in distance to markets and to the quality of roads. Likewise, wages in the different subregions are related to the travel costs between the subregions. However, prices for agricultural inputs (e.g. seed, fertilizer and biocides) are assumed to be the same in each subregion, as research by REPOSA revealed only minor geographical differences in input prices.

Geographical variation in product prices is simulated in REALM by dividing the AZ into 12 subregions (Fig. 14.2), each with its own specific transport costs (based on distance and quality of roads) to the most relevant market (depending on the product and final destination). These transport costs are calculated on the basis of a regression model estimated by Hoekstra (1995), similar to that of Jansen and Stoorvogel (1998). The subregions are the result of a GIS overlay of three zonification maps based on equal transport costs (Bouman *et al.*, 1998b).

Downward-sloping demand functions

For a number of products, the AZ share of national supply is considerable, and thus the prices of these products are likely to be influenced by regional supplies. This applies to bananas, palm heart, plantain and animals (meat), all incorporated in the model. For some products (e.g. palm heart and bananas), the supply from the AZ is even a considerable part of the world supply. In these cases, product prices become endogenous, determined by demand and supply. Research in Mexico (Duloy and Norton, 1973, 1983) and Brazil (Kutcher and Scandizzo, 1981) resulted in a methodology capable of incorporating variable prices in linear programming models (Hazell and Norton, 1986). Downward-sloping demand functions, based on econometrically estimated demand price elasticities, are linearized around an observed base quantity and price. Celis (1989) used the same technique in an agricultural sector model for Costa Rica.

In the earlier subregional models with exogenous prices, the producer surplus was maximized. In contrast, in the linear programming model for the AZ region, the area below the demand function for each product is calculated at different prices. These areas less the costs of production represent the sum of the producer and consumer surpluses at different price–quantity combinations. The model selects those price–quantity combinations for all products that, taken together, maximize the sum of producer and consumer surpluses.

The above approach applies to a country as a whole. In the case of the AZ model, the situation is more complicated in that one should take into account not only the supply originating from the AZ, but also the potential supply from other regions in Costa Rica, as well as, in the case of export products, supplies from other countries. Under these conditions, the demand function facing the producers within the AZ is different from the national demand function. It can be shown (Hazell and Norton, 1986) that the regional demand elasticity, η_r, can be expressed as follows:

$$\eta_r = \eta \frac{1}{K} - \sigma_{nr} \frac{1-K}{K} \qquad (1)$$

where η represents the national (or, where appropriate, world) demand elasticity, K is the AZ share of national production (or world production) and σ_{nr} is the supply elasticity of regions (or countries) other than the AZ.

Upward-sloping labour supply function

The labour available for agriculture and its remuneration have a considerable influence on production possibilities and land use. Schipper (1996) demonstrates that labour constraints have an important impact on land-use decisions. In the original Neguev and Guácimo models, both labour supply and wages were fixed, which can lead to undesirable results, in that labour is not truly fixed (Schipper, 1996; Jansen and Stoorvogel, 1998).

In order to relax the assumptions of fixed wages and labour supply, two types of labour are distinguished: labour belonging to the agricultural sector within the AZ, and labour from outside this sector and/or region. With regard to the first type, it is assumed that in each subregion there is a certain amount of labour working in agriculture at a fixed wage (the 'agricultural labour force').[5] This labour can also work in other subregions, in which case transaction costs are taken into account. Of course, in reality, labour allocation within the agricultural sector is more complicated. Farmers can work full-time or part-time on their own farm, on other farms or plantations or outside agriculture; also, they can easily switch between different jobs. Even assuming uniform labour quality, these sectors pay different wages and involve different transactions costs for individual workers. As the present model does not distinguish between different farm types, it is not possible to take all of these details into account. The number of people working in agriculture (expressed either in person-years or in person-days) is assumed to be stable over the years. This justifies – for purposes of the present model – the use of a labour pool from which the agriculture sector can draw its labour needs (up to its upper limit).

In regard to the second type of labour, it is assumed that the agricultural sector can attract labour from outside the sector, the quantity of which depends on the wage the sector is willing to pay. Consequently, the model contains a linear (upward-sloping) labour supply function. How much 'outside' labour is supplied depends on labour response along this function.[6] It is assumed, then, that the total labour supply function (that is, the sum of the agricultural and the non-agricultural labour force) is a kinked curve with a vertex. Up to a well-defined level of supply (the currently available agricultural labour force), the wage is fixed at the present market wage; thereafter, the function is linear and upward-sloping.

As with agricultural products, the market for agricultural labour in the AZ is only a part of the national labour market. Therefore, the national labour supply elasticity has to be adjusted before it can be applied at the regional level. Thus, labour's reaction in sectors or regions other than the AZ agricultural sector – caused by an increased labour supply to the AZ agricultural sector and leading to increased wages – has to be taken into account. Analogous to the situation for product markets (see above), the following relation can be shown to exist (assuming no obstacles to labour mobility other than the previously mentioned transaction costs):

$$\varepsilon_r = \varepsilon \frac{1}{M} - \theta_{nr} \frac{1-M}{M} \tag{2}$$

where ε_r is the labour supply elasticity for sector/region r (here, the agricultural sector in the AZ), ε is the national labour-supply elasticity, θ_{nr} is the labour-demand elasticity in the remainder of the economy and M is the share of labour in r in the national labour market.[7]

The result of using this method is that the agricultural sector in the AZ can use more labour than the estimated agricultural labour force, albeit at (slowly) increasing wages. In this way the AZ agricultural sector competes for labour with other economic sectors and regions. These assumptions allow for the calculation of a labour supply function at different labour supply/wage points. Such an approach makes it possible to linearize the resulting labour supply function in a way similar to the linearization of product demand functions (Hazell and Norton, 1986). Following Hazell and Norton's (1986) treatment of endogenous input prices, labour is valued at the average wage, rather than at the marginal wage (Bell *et al.*, 1982).

Results of Policy Simulations

Policy issues

Regional land-use policies ideally should be based on an evaluation of the socioeconomic and environmental implications of both actual and potential technological options for land use. This has been explicitly recognized by the Costa Rican government (SEPSA, 1997).

Although it is acknowledged that, ultimately, land-use decisions are made at the farm level, this is often not modelled explicitly in the type of methodologies of which SOLUS is an example. At the farm level, factors such as the knowledge level of the farmer, an emphasis on short-term income maximization (and thus, relative neglect of long-term resource productivity considerations), farmers' neglect of negative externalities and issues surrounding property rights are important for understanding the effects of policy. However, these factors are not directly taken into account in SOLUS. This limits the type of scenarios and policies that can be evaluated. For example, improved agricultural extension and markets (leading to better information upon which producers can base their input demand and output supply behaviour) have been identified as promising strategies for improving smallholder farming (Jansen and van Tilburg, 1996; SEPSA, 1997). While these effects cannot be fully addressed with SOLUS, it is capable of exploring, at the regional level, the scope and possible effects of changing policy incentives. Examples are environmental taxes and subsidies to induce certain desired land uses, taking simultaneously into account criteria related to income, sustainability and the environment. The consequences of other policy measures, such as prohibiting the use of certain toxic and persistent biocides, or the creation of national parks or buffer zones (thus effectively restricting agricultural land use), can be simulated as well. In addition, methodologies like SOLUS can analyse the consequences regarding land use of likely changes in key model parameters, such as an expected increase in real wages or decreasing interest rates.

In this context, we employ the SOLUS methodology to evaluate two policy scenarios related to the issues mentioned above and emphasizing tradeoffs between different policy objectives: (i) taxing biocides to reduce environmental contamination; and (ii) maintaining natural forests. In addition, the effects of expected increases in real wage levels are analysed, reflecting expected continued increases in per capita incomes. The effects of each policy are examined by comparing the results of each scenario with the results of a 'base run' of the model, which is used as a benchmark. Before presenting these analyses, we compare land-use patterns that would exist if only the present technologies in crops, pastures and animal husbandry systems were available (the 'present technology' run), with a situation in which alternative, improved technologies are also available (e.g. the 'base run'). In this way, the effects of improved technologies on income, employment, land use and environmental indicators can be assessed.

A last note may be appropriate with regard to the robustness and sensitivity of the results to variations in parameter values. The 'base run' results have been tested extensively by estimation for many different values of parameters for yields, current input costs, wages, demand and supply elasticities, base-year quantities, and prices and market shares. The overall results regarding land use (shares of crops and pastures, type and size of animal husbandry practices), income and employment remain highly similar from run to run. Expected changes, such as higher crop prices, do induce more cultivation, but these increases are controlled by the limited demands for the crop.

Land use under present and improved technologies

Technological progress – essentially producing more with the same or fewer resources (land, labour, capital) – has an important effect on economic surplus, employment, land use and environmental indicators (Table 14.1). Economic surplus (i.e. the value of the objective function[8]) increases by 21%, comparing the present technology and the improved technology scenario (i.e. the base run). Overall land productivity increases by the same percentage (the area used remains constant), while labour productivity increases by 27% (less labour used to produce a higher surplus). Employment decreases by 4%. The quantity of biocide active ingredients (BIOA) decreases by 33%, while the biocide indicator (BIOI) decreases by 16%.[9] Thus, both environmental indicators used in this study are more favourable in the base run, with improved technology, than in the present technology run. It can be concluded that improved technologies result in a 'win–win' situation, that is, a larger economic surplus and less environmental contamination.

This general conclusion is not surprising. The alternative technologies are more productive and more environmentally friendly than the present

Table 14.1. Assessing impacts of technological change: present technology versus base run (improved) technology.

		Scenarios	
	Units	Present technology	Base – improved technology
Objective function	10^6 US$	220.5	267.6
Labour use	10^6 days	9.0	8.7
BIOA[a]	10^6 kg	2.9	1.9
BIOI[b]	10^6	100.0	84.1
Crops	10^3 ha	76.6	61.2
Pastures	10^3 ha	174.4	189.8
natural	10^3 ha	174.4	150.6
grass–legume	10^3 ha	0	39.2
Animals, breeding	Animal units × 10^3	197.7	252.5
Animals, fattening	Animal units × 10^3	115.5	138.3
Crops			
Banana – area	10^3 ha	42.7	31.6
Banana – production	10^3 t	1851.7	2064.2
Pineapple – area	10^3 ha	2.8	2.3
Pineapple – production	10^3 t	196.8	194.3
Palm heart – area	10^3 ha	11.2	7.7
Palm heart – production	10^3 t	60.1	82.2
Plantain – area	10^3 ha	1.9	1.9
Plantain – production	10^3 t	37.1	35.3
Cassava – area	10^3 ha	17.7	17.1
Cassava – production	10^3 t	90.2	87.2

[a]Amount (kg) of active ingredient in all biocides.
[b]Index of environmental effects of all biocides together.

technologies. They were generated based on both theoretical agronomic results and on observed current practices used on better farms and plantations. Obviously, in practice it is more difficult to realize these potential gains. The results of the optimization model are exaggerated because of its linear structure and the absence of rigidities that do exist in reality. None the less, these limitations apply to both scenarios, so the comparison between the two is still valuable.

The present model is not well equipped to analyse distributional aspects of economic surplus increases, which are important in view of poverty-alleviation goals. However, some speculation on this score is possible. First, given that demands for most products produced in the AZ are price-inelastic, relatively large increases in supply can only be sold at lower prices. Therefore, consumer surplus generally increases more than producer surplus, or may even increase at the expense of producer surplus. With regard to the latter measure, its distribution between different types of farms depends on the type of crop. For example, banana cultivation requires a minimum area of about 100 ha for economic viability, so its commercial cultivation is virtually impossible for small farmers. On the other hand, palm heart can be cultivated by small farmers if processing plants are not too far away and are willing to buy farmers' output. Comparing the results regarding individual crops between the present technology and the base run demonstrates signficant gains in productivity and production for bananas and palm heart, while the results regarding pineapple, plantain and cassava show only small declines in productivity (Table 14.1). So, in this instance, both large plantations and smallholders might gain. Regarding the distribution of producer surplus between owners and workers, in the base run, employment is lower and labour productivity higher than in the situation with only present technology available. If wages do not increase and no alternative employment is available, such a situation is not desirable for labour. However, higher labour productivity can be assumed to eventually lead to higher wages in a growing economy with employment opportunities also existing outside the agricultural sector.

In the following sections, comparisons to assess the effects of policy or economic changes are made with the 'base run', assuming the availability of improved technologies.

Taxing biocides

The regulation and control of agricultural input use has been identified as an important policy option for the Costa Rican government to reduce certain negative externalities of agricultural production (SEPSA, 1997). The structural changes of the agricultural sector over the past decade have clear links to increasing trends in biocide use. Biocide policies have traditionally consisted of restrictive legislative measures, without considering the potential effects of economic instruments (Agne, 1996). It can be expected that taxing an input which is currently not taxed but has clear negative externalities (e.g. environmental contamination and human health damage caused by excessive biocide use (Jansen *et al.*, 1998)) will lead to diminished use of this input. Such a tax can be implemented in a variety of ways and at different levels. In this policy scenario, we distinguish between a flat tax and a progressive tax. The latter is proportional to the environmental damage caused by a specific biocide. Such damage, in turn, is related both to the toxicity of the biocide and to its persistence in the environment. Both of these aspects are taken into account in the BIOI (for a precise definition, see Jansen *et al.*, 1997; Bouman *et al.*, 1998c).

Taxing all biocides at a uniform rate of 100% leads to a reduction in the use of biocides in terms of BIOA, of 13% relative to the base scenario, while the BIOI measure is reduced by only 4% (Table 14.2). However, the economic surplus (the objective function value) is reduced by nearly 19%. Thus, a relatively modest environmental gain is obtained at high economic cost. In contrast, a progressive tax regime where different tax rates are applied to three categories of biocides depending on their degree of toxicity (i.e. slightly, medium and very toxic) results in a much larger reduction of BIOI, while at the same time preserving more of the economic surplus. For example, applying taxes of 20%, 50% and 200%, respectively, to the categories of slightly, medium and very toxic biocides (Table 14.2, tax system A) leads to a reduction in economic surplus of only 4%, while reducing BIOI by over 80%. When tax rates are reduced to 10%, 30% and 150%, respectively, for the three categories of biocides (tax system C), economic surplus decreases by just 2% with the same environmental improvement. On the other hand, limiting the tax on biocides in the very toxic category to 100% (tax system B) does not result in a significant environmental gain.

It can be concluded that, even though a flat tax results in the highest reduction in the absolute quantities of biocides applied, such a tax is relatively ineffective when it comes to protecting the environment. In addition, the flat tax is not efficient in the sense that it leads to a large decrease in economic surplus. In contrast, a biocide tax differentiated relative to the degree of toxicity of each biocide can reduce environmental damage to a substantial degree at relatively low economic cost. A comparable conclusion is reached in Jansen *et al.* (1997) for Guácimo county in Costa Rica. It should be kept in mind that rather large

Table 14.2. Effects of different types and levels of a tax on biocide use.

				Tax system		
Type of biocide		Base	Flat tax	A	B	C
Slightly toxic		0%	100%	20%	20%	10%
Medium toxic		0%	100%	50%	50%	30%
Very toxic		0%	100%	200%	100%	150%
	Units	Absolute value	% change	% change	% change	% change
Economic surplus	10^6 US$	267.6	−18.7	−4.3	−4.1	−2.2
BIOA[a]	10^6 kg	1.7	−13.1	−3.9	−3.2	−3.8
BIOI[b]	10^6	84.1	−4.0	−81.9	−1.5	−81.9

[a]Amount (kg) of active ingredient of all biocides.
[b]Index of environmental effects of all biocides together.

differences in tax rates were evaluated; for greater policy precision, a more refined analysis would be warranted, involving estimation of optimal tax levels.

Conservation of natural forests

Agricultural policy in Costa Rica has put increasing emphasis on environmental protection (SEPSA, 1997). Consequently, the identification of efficient instruments to realize protection of natural resources at minimal economic cost has become increasingly important. Currently, the Costa Rican government has a policy to stimulate landowners to maintain part of their property under natural forest. In return for not cutting down these forests, a landowner can obtain a subsidy of about US\$40 ha^{-1} $year^{-1}$.[10]

To analyse the effect of such a subsidization policy on regional land use, premiums were allocated to the 'natural forest' land utilization type. Natural forests can be exploited in a sustainable way, yielding about 0.6 m^3 of wood ha^{-1} $year^{-1}$, with an annual return of about US\$16 ha^{-1}. The linear programming model was run with all the available land in the AZ (340,000 ha) suitable for agricultural use (crops, pastures and forests) allowed to revert to agriculture, including the protected areas (e.g. national parks) and semi-protected areas (e.g. buffer zones, forest reserves). In the base year of the model (1996), a subsidy of US\$111[11] ha^{-1} of forest $year^{-1}$ is not sufficient to induce landowners to maintain natural forests (Table 14.3). On the other hand, subsidies of US\$122 and US\$133 ha^{-1} $year^{-1}$ would lead to forest areas of about 120,000 ha and 200,000 ha, respectively. This is compared with about 84,000 ha currently in primary or secondary forest in the area suitable for agriculture. In the case of these subsidies, a large part of the land used for pastures in the base scenario is converted to natural forests, while the cropped area remains virtually constant.

This analysis to determine the optimal level of a natural-forest subsidy can be extended by analysing land use at the margin from a different perspective. Suppose there does not exist a subsidy for natural forests. In the base scenario,

Table 14.3. Effects of subsidies for non-timber products in maintaining or creating 'natural' forests.

			All land, with natural forest			
Soil type	Units	Available land	No non-timber value to forest	US\$111 ha^{-1} subsidy	US\$122 ha^{-1} subsidy	US\$133 ha^{-1} subsidy
SFW	10^3 ha	118.4	0	0	0	0
SFP	10^3 ha	136.0	0	0	62.6	122.4
SIW	10^3 ha	85.7	0	0	56.4	77.1
Subtotal	10^3 ha	340.1	0	0	118.9	199.5
Economic surplus	10^6 US\$		275.8	275.8	276.1	277.9

SFW, soil fertile, well drained; SFP, soil fertile, poorly drained; SIW, soil infertile, well drained.

only land outside the protected and semi-protected areas is considered (278,900 ha). Extending the potential availability of agricultural land to include the semi-protected areas only (48,900 ha) and then both semi-protected and protected areas (61,200 ha) gives an indication of the incremental increase in economic surplus if these areas could be used for agricultural purposes. Table 14.4 shows that the increments in economic surplus are not substantial. Extending the base-case area into the semi-protected areas (17.5% more land) leads to a marginal increase (2.5%) in economic surplus. Average yearly returns per hectare decrease from US$960 to US$837. The incremental economic surplus attributable to the semi-protected areas is only US$134 ha^{-1}. Similarly, extending this scenario to include the land in protected areas (another 3.8% of land available) results in only an additional 0.6% increase in economic surplus. Average returns per hectare drop to US$811 $year^{-1}$, as the incremental returns on protected area lands decline to US$131 ha^{-1}. A comparison of land-use patterns in each of the above cases reveals that all additional land used in agriculture is used for pasture. This is a consequence of the limited demand for crop products at sufficiently remunerative prices (incorporated in the model through downward-sloping product-demand functions). The incremental increases in economic surplus gained by extending the area in pasture are consistent with the earlier analysis regarding the minimum effective subsidy for natural-forest preservation.

Increasing wages

In the previous section, it was explained that wages are endogenous to the SOLUS model. That is to say, the model formulation assumes that the agricultural sector of the AZ is able to attract labour from other sectors and regions with increasing wages. However, other influences on wages are not incorporated in the model. Growing demands for labour in other sectors due to economic expansion, income growth leading to increased demand for leisure and demographic changes in altering labour supply are all examples of forces that may change real wages. International development banks expect real gross national product (GNP) growth in Costa Rica to reach between 4.5% and 5% $year^{-1}$ over the next 5 years. Given the expected population growth of 2% $year^{-1}$, this translates into a 2.5–3% annual increase in real per capita GNP, similar to the average per capita real GNP growth of 2.8% realized between 1985 and 1995 (World Bank, 1997). Assuming a continuation of such rates of growth in the future (i.e. beyond the year 2003), it is likely that real wages will increase concurrently, independent of any endogenous effects resulting from agricultural expansion.

To simulate the potential impact of real-wage increases on land use, three scenarios were evaluated, assuming total real-wage increases of 50%, 75% and 100%, respectively. A 75% real-wage increase can be expected if past trends continue (i.e. a 2.8% increase per year over 20 years). The remaining two scenarios were evaluated to explore the sensitivity of the model to alternative increases in real wages. A real wage increase of 75% results in less cropland and

Table 14.4. Using semi-protected and protected areas for agriculture.

Scenario	Objective function (10^6 US$)	Increase in objective function (10^6 US$)	% Change objective function (%)	Area (10^3 ha)	Area increase (10^3 ha)	% Change in area (%)	Average returns (US$ ha^{-1})	Incremental returns (US$ ha^{-1})
Base	267.4			278.9			960	
Base and semi-protected land	274.0	6.6	2.5	327.6	48.7	17.5	837	134
All land	275.8	1.7	0.6	340.1	12.5	3.8	811	131

Table 14.5. Real-wage increases: effects on economic surplus, employment and land use.

Scenario	Objective function (10^6 US$)	Labour AZ (10^6 days)	Labour income (10^6 US$)	Crops (10^3 ha)	Pastures (10^3 ha)	Animals – breeding (animal units × 10^3)	Animals – fattening (animal units × 10^3)
Base[a]	267.6	8.7	76.6	61.2	189.8	252.5	138.4
Wage + 50%	232.2	7.6	100.7	54.9	196.1	233.0	128.2
Wage + 75%	215.8	7.3	112.3	43.7	207.3	250.3	137.1
Wage + 100%	200.2	7.0	124.1	42.0	209.0	252.7	138.4

[a]Wage rate in base run is US$8.84 day^{-1}.

more land under pasture (Table 14.5). Wage increases of 50% and 100% result in similar types of land-use changes. The principal reason for these results is that crops use relatively more labour than pastures; this is also indicated by a decrease in the number of animal units per hectare, which drops from 2.1 to 1.9, making animal husbandry less labour-intensive. Thus, labour-intensive activities are replaced by labour-extensive ones. However, this result depends on technology development. The different wage-increase scenarios assume that technology is kept constant. However, it may be that the development of new labour-saving crop technologies keep up with real wage increases. Further, as the demand for the agricultural products concerned (bananas, plantain, palm heart, cassava and pineapple) is likely to continue to grow in the future, whether or not cropland is converted into pastures in the AZ will also depend on real-wage increases in competing countries.

Thus, *ceteris paribus*, labour use in the agricultural sector of the AZ diminishes with increasing real wage rates. This, of course, is not surprising. Moreover, in a growing economy, it can be expected that labour will be increasingly employed in non-agricultural sectors. Another consequence of increasing real wages is a decrease in economic surplus (Table 14.5), as wages constitute a cost component in the model. On the other hand, wages represent income to labourers. Thus, the sum of the economic surplus and total labour income (number of labour-days times wages) better reflects the economic gains resulting from agricultural activities in the AZ to Costa Rican society as a whole. In the base scenario, this sum is US$344 million, while, with a 75% real-wage increase, these gains total US$328 million. Thus, as the economic surplus created in the agricultural sector in the AZ decreases by 19% (from US$268 million to US$216 million) as a result of the 75% wage increase, the sum of economic surplus and labour income decreases by only 5%.

Conclusions

Land-use policies at either national or regional levels ideally should be based on a simultaneous evaluation of the socioeconomic and environmental implications of both actual and potential technological options for land use. The REALM model within the SOLUS methodology provides a useful tool for such an evaluation under the particular conditions of the AZ region at the close of the 1990s. Policy decisions aiming at improved land use in the medium and long term can be analysed with such an instrument.

The impacts of technological change, the desire to reduce biocide-related pollution to avoid negative effects on human health and the environment, and the goal of fostering natural forests (to maintain ecological balance, avoid global warming and encourage tourism) are all important issues that affect economic and agricultural policies relevant to the AZ. Another important issue is what kind of effects on land use can be expected from continuous real wage increases as a consequence of general economic development. REALM has been shown to be a useful instrument for the assessment of each of these issues and, as such, can usefully inform policy-making in the region and country.

The following conclusions can be drawn, based on the policy simulations described above:

1. Development of new land-use options through technological change can bring simultaneous economic and environmental gains, thus providing a 'win–win' situation.
2. A progressive tax on biocides related to their toxicity and persistence, aiming at a reduced use, proved to be more efficient than a flat tax.
3. Subsidizing natural-forest management – which implies that society at large recognizes that the value of natural forests goes beyond potential timber revenues – could help maintain existing natural-forest areas or create new ones, though such subsidies might be rather substantial (i.e. higher than the current subsidy provided in anticipation of future carbon bonds). At the same time, extending the agricultural area (crops and pastures) into existing protected and semi-protected areas brings only marginal economic benefits, mainly because these areas would probably be converted to pastures.
4. Higher wages as a consequence of overall economic development would lead to a reduction in cropland and an expansion of pastures. Given constant technology, labour-intensive activities are substituted by labour-extensive ones. Further, if real-wage increases in Costa Rica exceed wage increases in competing countries, Costa Rica may gradually lose its competitive advantage for crops like bananas, plantain, palm heart, cassava and pineapple. However, development of labour-saving techniques in crops might well moderate or prevent the shift to animal husbandry.

A general evaluation of the different scenarios presented in this chapter leads to the conclusion that new technology of the right kind could generate a complementarity between production, income and environmental outcomes. However, policies to reduce negative environmental consequences – other than policies to stimulate appropriate technological development – often come at the expense of farmers' incomes. In connection with the first conclusion, it should be noted that research, development and extension of appropriate technology remain important for generating such possible complementarities between economic welfare and the environment. This implies that investment in these activities – after appraising their likely costs and benefits – remains necessary, too. In regard to the second conclusion, it is important to recognize that, although an improved environment may come at the expense of income generation, this may be socially acceptable, as both are components of broader societal well-being. The size of the tradeoff between income and environmental benefits can be reduced by carefully targeting policies. *Ex ante* analysis of such policies can help in the search for the right policies at the right time and place. REALM, as part of the general SOLUS methodology, has been shown to be a suitable tool for policy analysis related to land-use choices and, as such, is capable of better informing policy-makers of the likely consequences of alternative policies.

Notes

1 REPOSA is a joint project of Wageningen University (WAU) of the Netherlands, the Ministry of Agriculture and Livestock (MAG) in Costa Rica, and the Tropical Agricultural Research and Higher Education Center (CATIE) in Costa Rica. The SOLUS methodology evolved from the *uso sostenible de tierras en el desarrollo* (USTED) – 'sustainable land use in development' – methodology (Bouman *et al.*, 1999b). USTED was developed initially at the level of a farm household settlement (Schipper *et al.*, 1995; Stoorvogel *et al.*, 1995) and was gradually scaled up via the county level (Jansen *et al.*, 1997) to the regional level (Bouman *et al.*, 1998c).

2 The land-use crop technical coefficient generator (LUCTOR) (Hengsdijk *et al.*, 1998) and the pasture and animal system technical coefficient generator (PASTOR) (Bouman *et al.*, 1998a).

3 Expressed as the amount (kg) of active ingredients in all biocides, abbreviated as BIOA.

4 The biocide indicator (BIOI) is an index of the environmental effects of all biocides together, taking into account the amount of active ingredients, toxicity and persistence in the environment.

5 The agricultural labour force is estimated for each district on the basis of the agricultural labour force in 1984 (DGEC, 1987), taking into account the population in 1996, the population growth between 1984 and 1996 (annual registration of births and deaths (DGEC, 1997a)) and the estimated migration to each county, based on the assumption that migration rates between 1979 and 1984 still applied between 1984 and 1996. The outcomes were compared with more recent survey information for the AZ as a whole (DGEC, 1997b) and were deemed to be not unreasonable. Subsequently, labour-force estimations per district were distributed over the 12 subregions on the basis of population density estimates, using a GIS-based procedure.

6 In this case, transaction costs are taken into account as well; however, these transaction costs are generally higher than those for labour already working in the AZ agricultural sector.

7 In the current version of REALM, at the national level, a wage-labour supply elasticity of 0.2 has been assumed, which is not out of line with other studies (Bosworth *et al.*, 1996). Regarding the labour-demand elasticity in the remainder of the economy, –0.5 would be a plausible approximation (Bosworth *et al.*, 1996). Using equation (2) and in view of a labour share of 0.05, an estimated ε_r of 13.5 for the labour-supply elasticity in the AZ agricultural sector would thus be a reasonable approximation. Such an elasticity implies a very gently upward-sloping labour supply function, in which case more labour would be required than is presently available (i.e. after the horizontal part of the labour supply function is passed).

8 All values in Costa Rican colones (¢) were converted to US dollars at the average 1994–1996 exchange rate of US$1 = ¢ 181.

9 The different rates of change in the two biocide indicators are due to several factors: their difference in measures (BIOA is measured in terms of kilograms of active ingredients, while BIOI takes into account both toxicity and persistence), the quantities and types used and the technologies used for each crop.

10 The subsidy was introduced in 1997 and consisted of ¢ 10,000 ha^{-1} of forest year^{-1}, initially for a period of 5 years. Taking into account an obligatory first-year cost of ¢ 3000 for an officially approved forest management plan, this means an annuity of US$40 ha^{-1} year^{-1} at the 1997 exchange rate of US$1 = ¢ 232. The subsidy is provided in anticipation of future carbon bonds.

11 ¢ 20,000 converted at the average 1994–1996 exchange rate of US$1 = ¢ 181.

Appendix 14.1: Notes on the REALM Model

As a module of the SOLUS methodology, REALM is an optimization model that maximizes the sum of the producer and consumer surpluses. More precisely, it maximizes the sum of the areas below demand functions plus the value of exports, minus product market transaction costs, minus production costs. The latter consist of input costs (i.e. costs of fertilizers and pesticides, annualized costs of capital items, such as machinery, corrals, etc.) and labour costs (area below the labour supply function and transaction costs).

Several thousand actual and alternative land-use systems are included in the model as activities. These consist of 14 land-utilization types: nine crops (black beans, cassava, maize, palm heart, pineapple, plantain, melina and teak plantations and sustainably managed natural forests) and five pastures (three fertilized improved grasslands, a grass–legume mixture and a mixture of natural(ized) grasses), each with a number of different technology options, combined with three major land units identified in the northern part of the AZ of Costa Rica. These major land units were subdivided into two mechanization classes. Actual land-use systems are derived from descriptions of current practices by farmers in the AZ, while alternative systems were generated using the target-oriented approach with a zero soil nutrient-balance restriction. For alternative systems, different technology levels were generated by combining levels of fertilizer use, crop protection, substitution between manual weeding and herbicide use and, for pastures, stocking rates. Moreover, two herd types were distinguished (beef breeding and beef fattening), with, for each, four animal growth rates and five feed supplements.

Beyond constraints concerning land units, constraints are formulated regarding labour supply and the demand for agricultural products, and regarding equilibria between breeding and fattening activities.

All calculations regarding the modelling of product markets are executed in GAMS. The parameters used for each product are a base price and quantity in 1996, a price demand elasticity, the 1996 share of the supply from the region in the national (or international) supply (under the base situation) and a price supply elasticity of the remaining regions (i.e. regions outside the model), or countries in the case of export products. Price demand elasticities used are taken from Geurts *et al.* (1997) and van der Valk (1998). In the absence of estimated data, price supply elasticities were assumed to be 0.7, as many studies suggest that supply elasticities between 0.4 and 1.0 are not unreasonable (Sadoulet and de Janvry, 1995; Mamingi, 1997).

15 Land Tenure and the Management of Land and Trees: a Comparative Study of Asia and Africa

KEIJIRO OTSUKA[1] AND FRANK PLACE[2]

[1]*International Food Policy Research Institute (IFPRI), Washington, DC, USA, and Tokyo Metropolitan University, Tokyo, Japan;* [2]*International Centre for Research in Agroforestry (ICRAF), Nairobi, Kenya*

Introduction

Massive deforestation has been taking place in the Third World, resulting in environmental deterioration and a greater scarcity of forest resources, such as firewood, fodder and timber. Cleared land is generally used for cultivation to feed growing populations, which suggests that the most fundamental cause of deforestation is the expansion of cropland. In some areas, however, reforestation has followed deforestation, due to the formation of forest-user groups that have introduced strict management rules for forest resources. In other areas, the growth of agroforestry, in which commercial trees are intercropped with annual crops when the trees are young, has also been significant. This is a socially desirable development, as agroforestry is typically more sustainable and profitable than shifting cultivation where increasing population pressure exists on limited land resources. Yet, to our knowledge, no broad-based empirical study has previously been carried out to identify, systematically and quantitatively, the factors affecting the use of land for forests, agroforests and cropland, as well as management efficiency at the community and household levels.

The use and allocation of forestland, agroforests and cropland are governed by property-rights institutions and land-tenure systems, ranging from communal ownership of land to state ownership, common property and private ownership. Forests are often state-owned and these forests are seriously denuded in many areas, due to the absence of protection. At the same time, people in these areas commonly have no incentive to plant trees, as they possess no property rights on state land. Similarly, communal ownership of forestland in customary land areas frequently leads to destruction of forests, because forest conversion confers strong individual land rights under customary land law and, in turn, provides strong incentives to clear forests. Over time, however, the individualization[1] of communal-tenure institutions proceeds, which tends to provide incentives to plant valuable trees on the converted land. Under the common-property regime, forest management performance varies considerably from case to case, depending on the importance of forest resources for livelihood and subsistence farming and the nature of major forest products – all factors that affect the incentives to manage forest resources collectively.

A seven-country comparative study was organized – using a consistent and quantifiable framework – with a view to identifying the causes of choices of land-tenure institutions and their consequences for the use and management efficiency of land and trees. This chapter attempts to summarize the major findings of this comparative study.[2] The case studies were conducted in Sumatra in Indonesia, northern Vietnam, the Hill and Inner Tarai regions of Nepal, central Japan, western Ghana, east-central Uganda and Malawi. In general, two sets of surveys were conducted at each site: an extensive survey of large-scale areas to collect community-level information, and an intensive survey of households in selected communities to collect detailed data on actual forest resource management practices and agricultural production. The former was designed to identify the determinants of land-tenure institutions and the consequences of land-tenure choice for changes in land use; the latter was intended to assess the efficiency of managing land and trees by individual farmers (Otsuka and Place, 1998). The intention of the project was to draw generalizable conclusions through a comparative study of Asian and African countries, where cultural, natural and policy environments are vastly different.

Four major issues were addressed in this project. First, how do land-tenure institutions affect the pace of deforestation? Secondly, under what conditions is the common-property system a viable and efficient institutional arrangement? Thirdly, how do land-tenure institutions affect the incidence of planting commercial trees and their management efficiency? And, finally, how do land-tenure institutions affect the efficiency of annual crop farming? The major conclusions on each topic from the case studies are discussed below.

Land Tenure in Study Sites

At the outset, a few key concepts pertaining to land-tenure institutions must be clarified. We propose to classify these institutions into the following four broad categories: (i) communal property; (ii) common property; (iii) state property; and (iv) private property.[3] The difference between the first two is not always clear. Under the communal-ownership regime, virgin forests and uncultivated woodlands are owned communally and controlled by the village chief; exclusive use rights to cultivated land are assigned to individual members of the community and ownership rights are held by the extended family. The ownership of cultivated land, however, has evolved towards more individualized ownership over time. This system is alternatively called the 'customary' land-tenure institution. Land-tenure institutions that are observed in Sumatra, Ghana, Uganda and Malawi all fall in this category.

On the other hand, common property is defined by the joint ownership and use of forest by a group of community members. This joint ownership and use is rationalized by the non-excludable nature of forest resources, which makes it difficult for individual community members to effectively protect properties under private ownership. Common property can be 'open-access', if the community or group of forest users do not manage the forest or enforce rules of use or if the resources are truly non-excludable. There are also cases in which the use of common-property forests is strictly regulated by community members, as in the case of selected forests in the Hill region of Nepal. Uncultivated portions of communally owned land can also be regarded as common property, as they are owned by a group of people and are used jointly for hunting and extraction of trees and minor forest products. It is our observation, however, that these areas are characterized by open access for community members almost without exception. In these areas, forest resources are typically depleted and forest areas decline due to population growth. While we seldom observed the case in which private property is characterized by open access, we found that state-owned forests are generally open-access.

There are similarities and dissimilarities among the seven study sites (Table 15.1). Both the Ghana and Sumatra sites are characterized by the communal-ownership systems. Moreover, both sites have a comparative advantage in agroforestry over pure food production under shifting cultivation, due to the hilly or mountainous topography, where annual crops cannot be grown sustainably without large investments. In western Sumatra, large areas of primary forest still exist in the Kerinci Seblat National Park, even though some portions have been converted to irrigated paddy-fields, crop fields (under slash-and-burn farming systems) and fields of commercial trees, such as rubber, coffee and cinnamon. In western Ghana, primary forests have largely disappeared and have been replaced either by crops under shifting cultivation or by cocoa.

While Malawi is also characterized by communal ownership, mixed food-crop and tree systems are less profitable than food-cropping systems, compared

Table 15.1. Characterization of study sites.

Country	Land-tenure system	Major agroforest and forest products	Topography
Ghana	Communal	Cocoa	Hilly
Sumatra (Indonesia)	Communal	Rubber, cinnamon and coffee	Hilly/mountainous
Malawi	Communal/ common property	None/minor forest products	Flat/hilly
Uganda	Communal/private/state	Coffee/charcoal	Hilly/flat
Vietnam	State/private	Timber and fruit	Mountainous
Nepal – Hill	Common property	Firewood, fodder, grass and leaf litter	Mountainous
Nepal – Inner Tarai	Common property (collective and centralized management)	Timber	Hilly
Post-war Japan	Common property (collective and individual management)	Timber	Mountainous

with Ghana and Sumatra, since many areas are characterized by flat topography. There are also communally owned forests on hilly portions of Malawi, but they are largely open-access, except for a few small village forest areas strictly protected by village organizations and the forestry department. Most community woodlands have been converted to crop fields in Malawi (with the exception of the sparsely populated north). In the fourth study area, Uganda, the sites consist of communal, private and state-owned areas, in which coffee is grown in the hilly and humid areas north of Lake Victoria and charcoal is a major product of woodlands in the remaining areas, which are generally flat and dry. As in Malawi, woodlands have been degraded and converted to crops in most areas.

In northern Vietnam, natural forests in mountainous areas have been seriously denuded, due to the expansion of food cultivation up to the end of the 1980s, when Vietnam launched major economic reform programmes, including the privatization of land tenure. In Vietnam, the village areas typically consist of flat lowlands, mostly occupied by paddy land, with forestland confined to mountainous areas. Although the whole of Vietnam is state-owned, use rights to forestland for a 50-year period have been distributed to individual farmers in selected areas in recent years. This policy has stimulated the regeneration of forests through forest protection and the planting of timber, fruit and other perennial trees through the spontaneous decisions of farmers. Firewood and other minor forest products are amply available from remaining forests and wooded areas.

In contrast, minor forest products are extremely scarce in the Hill region of Nepal, where community forests are seriously degraded. In these regions, as in the mountainous areas of pre-War Japan (McKean, 1992), common-property systems with strict management rules for resource extraction have emerged in selected locations. This suggests that the common-property regime may work effectively if major forest resources consist of minor forest products, which are costly to protect. In other words, the basis for the establishment of common property seems to rest on the non-excludability of resources. In the Inner Tarai region of Nepal, which is flatter and has better access to markets, timber is a major forest product. Compared with minor forest products, timber is less costly to protect, because it is bulky and difficult to haul without being discovered by other members of the community (even though illegal grazing in the same areas needs to be regulated). Timber production, moreover, requires not only protection, but also the management of trees through such silvicultural activities as weeding, pruning and thinning. Thus, trees for timber will not be managed well without providing proper management incentives. It is therefore instructive to find in the Inner Tarai region complex forest management systems. There is collective management, in which all community members are supposed to participate in protection and management activities and to receive, in principle, equal benefits. There is also centralized management, in which community members at large participate in forest protection and receive certain benefits, but user-group committees employ hired labour for the management of trees. In the final study site, Japan over the post-War era, both collective and individualized management systems have coexisted, even though forestland is commonly owned by the community.

We expect that the comparative analyses of land-tenure institutions and the characteristics of dominant natural resources will shed new light on the conditions under which different land-tenure institutions and farming systems emerge and function efficiently.

Land Tenure and Deforestation

Communal ownership

It is our observation that uncultivated forest areas under communal ownership are generally loosely regulated, to the extent that the conservation of forest resources is conditioned on low population pressure. The conversion of forestland is most closely described as open-access, at least for members of the community. We repeatedly encountered cases in Ghana, Uganda, Malawi and Sumatra where, although the village chief is a custodian of the communal forests, he commonly allows villagers to clear the forests for cultivation. In Malawi, this permission often extended to non-villagers as well. We were also told that, in some villages in Ghana, village members report the clearance of communal forests to the chief after clearance is completed, in order to 'register' their ownership. Nowhere in our study sites did we encounter cases in which the village chief strictly controlled the conversion of forestlands to cultivated lands. It appears that the responsibilities of the village chief are generally to approve the clearance of forests and woodlands and to record the changes in land ownership, so as to avoid possible land disputes in the future.

While it is difficult to prove rigorously that primary forest is typically open-access, there is no question that forest areas have increasingly been cleared, with population growth, under communal ownership. As shown in Table 15.2, the proportion of agricultural land increased substantially over the last few decades at the expense of forests and woodlands in Uganda and Malawi, according to the analysis of aerial photographs at the community level (Place and Otsuka, 2000a,b). At present, there remain only small areas of forests and woodlands in these countries. In Uganda, it was statistically demonstrated that the speed of expansion in agricultural land was greater under communal ownership

Table 15.2. Changes in land use in Uganda, Malawi and Vietnam (% of total).

	Uganda[a]		Malawi[b]		Vietnam[c]		
Land use	1960	1995	1971	1995	1978	1987	1994
Agriculture	57	70	52	68	48	69	76
Forest and woodland	32	20	34	19	52	31	24
Others	11	10	14	14	–	–	–

[a]Average of 64 parishes in east-central region.
[b]Average of 57 enumeration areas throughout the country.
[c]Average of formerly forest areas with slope greater than 24° in 56 communes in two northern provinces.

than under private ownership, suggesting that the protection of woodland has been less strictly enforced under communal ownership (Place and Otsuka, 2000b). This finding supports our argument that forestland and woodland are typically open-access under communal ownership, unlike the private ownership system, in which owners compare the costs and benefits of land conversion. In these two countries, the major factor accounting for the decrease in forest and woodland area is population pressure, measured by population density and population growth. In neither of these countries, however, is there any evidence that population pressure decreased tree cover within agricultural or non-agricultural land.[4] These findings strongly indicate that cropland expansion, rather than the extraction of trees and other natural resources, is a major cause of deforestation.

At the household level of analysis, we found that the age of the head of household is the most critical variable explaining the area of acquired forestland in Ghana and Sumatra (Otsuka *et al.*, 1998a; Suyanto and Otsuka, 1998). This seems to reflect the fact that cultivable primary forestlands have largely disappeared in recent years, due to population growth and the clearance of these forests on a 'first-come first-served' basis. In Sumatra, it was also found that the number of male family members who engage in the clearance of forest areas was significantly related to the amount of land acquired through forest clearance.

An important factor affecting deforestation is the conventional rule of communal-land tenure that the effort of clearing the forest is rewarded by the granting of strong individual land rights. Those who have cleared village forestland in Ghana are granted the right to plant trees and to rent out and pawn land (Otsuka *et al.*, 1998a). In Sumatra, those who clear forests are granted almost complete private land rights, including the right to sell, until cleared land is put into fallow (Otsuka *et al.*, 1998b). This accrual of strong individual land rights explains, at least in part, why the conversion of woodland has been faster under communal ownership than under private ownership in Uganda (Place and Otsuka, 2000b). To conclude, there is no built-in mechanism under communal ownership to protect forest areas and forest resources. Thus, forests have been widely converted to agricultural land to feed growing numbers of households. In order to halt this trend, either food production from existing agricultural land must increase, the number of children must decrease or the value and tradability of forest products must increase. Increases in food production will reduce food prices, which will discourage deforestation. It is not clear, however, whether the dissemination of profitable crops and appropriate technologies for marginal lands leads to further deforestation or not (Kaimowitz and Angelsen, 1998; see also Angelsen and Kaimowitz, Chapter 6 of this volume). While the enhanced profitability of cultivation increases the incentive to clear forestland, increased labour demand from existing farmland may reduce the labour supply available for forest clearance.

Other ownership systems

Woodland in Malawi, called *miombo* woodland, may be considered as common property, in which community members collect firewood, poles and other non-timber products and graze cattle. According to our informal interviews with village chiefs, collection of only dead and dry branches is allowed, and no sale of firewood in the market is allowed under the existing management rules. If only dead and dry branches are collected from communal woodlands, sustainable management is assured as long as grazing is regulated so as to avoid damage to young seedlings. In fact, however, woodlands have rapidly degraded and disappeared due to excessive tree cutting, especially for drying tobacco in the central region of Malawi.

In Vietnam and Nepal, forests and woodlands are owned by the state. In northern Vietnam, the proportion of bare land or denuded forestland, which is widely used for shifting cultivation, increased by 21% from 1978 to 1987, due to food shortages and the resulting conversion of forests to cropland (Table 15.2). It is our observation that forest areas have essentially been open-access with no regulation by the state. Since long-term 50-year leases were granted to individual farmers in the early 1990s, not only has the pace of deforestation declined but also the planting of timber and fruit-trees and the regeneration of forests through protective activities have been initiated by farmers. In the perception of farmers, a 50-year lease means almost complete privatization. This has resulted in recovery of forests in recent years in many areas (Tachibana *et al.*, 1998).

It is important to observe that the management of forests by user groups or under the common-property regime did not develop spontaneously in Vietnam, even though communal irrigation systems have been maintained spontaneously for long periods and other collective activities have been carried out under the commune system. As we shall argue below, community management of forests is effective when key forest products are minor products, such as firewood, grasses and fodder. In the case of these products, protection is costly and the intensive care of trees (e.g. through weeding and pruning) is not required. Thus, scale advantages associated with community management in the protection of minor forest resources outweigh the disadvantages of community management in providing work incentives to take care of trees. In the mountainous sites of Vietnam, these minor forest products are amply available, even though massive deforestation has taken place.

In Nepal, forestland was nationalized in 1957 and widespread deforestation followed. Whether deforestation would have been largely prevented if forestland had not been nationalized is a difficult question to answer. Yet, it is true that deforestation was followed by reforestation with the establishment of forest-user groups, particularly formal groups that have received formal use rights to forests from the state (Upadhyaya and Otsuka, 1998; Otsuka and Tachibana, 2000). This suggests that formal user-group management may be more efficient than informal user-group management, because the lack of formal use rights under the latter system fails to provide proper management incentives. Moreover, it appears that the active and effective informal user groups, which were established voluntarily in earlier years, tend to have chosen to become formal user groups. Therefore, the significant impacts of formal user-group management may reflect self-selection rather than the incentive-enhancing effect of establishing land-use rights for formal user groups. In other words, there is a possibility that the management by informal user groups would have been efficient even if they had not been converted into formal user groups. Although we cannot fully settle this issue, our conjecture, based on informal interviews, is that both the incentive-enhancing effect and the self-selection effect were at work. Unique to the case of the Hill region of Nepal is the absence of expansion of agricultural land. In fact, deforestation primarily took the form of a reduction in forest tree density. As will be explained later, this was due to the critical importance of minor forest products for the livelihoods of subsistence farmers, which prevented further encroachment on the forestland.

One may wonder why deforestation has taken place so widely in developing countries under state ownership, in contrast to the experience of industrial nations, in which state jurisdiction has often resulted in a good record of forest conservation. We maintain that the main reason is, in developing countries, strong pressure to clear forestland for cultivation and, secondarily, the non-sustainable extraction of trees for fuel and other purposes. Rapid population growth, coupled with stagnant growth in urban employment and in revenues from food crops grown on existing cropland, has widely exerted formidable pressures on forestlands. Insufficient supplies of alternative sources of energy for firewood have also contributed to the degradation of forest resources, but this effect has been mitigated, at least in some locations, by the planting of trees on farms.

To sum up, our case studies demonstrate that forests and woodlands are not well protected under the state and the communal-ownership systems. In fact, primary forests have largely disappeared in areas owned by the state and controlled by local communities under communal-ownership regimes. It must also be pointed out that the common-property regime is not always effective in preserving forests and woodlands, judging from the experience in Malawi.

Management of Common-property Forests

Whether the management of forests and other natural resources under the common-property regime is an efficient institutional arrangement has been widely debated in the literature. Yet the empirical evidence is deplorably weak. The issue is of practical importance, as tree-planting projects supported by international organizations and aid agencies are often designed as social forestry or community forestry projects, with the principle of equal participation of all community members and the equal sharing of benefits. The same set of principles is usually adopted in voluntarily established community forest management. This comparative research project analysed the management efficiency of common-property forests in the Hill and Inner Tarai regions of Nepal and the mountainous regions of Japan, in addition to the case of Malawi mentioned earlier.

In Nepal, community management of forests under formal user-group management has been successful in reducing the extraction of firewood and other resources, preventing the cutting of green branches and felling of trees and preventing cattle grazing (Upadhyaya and Otsuka, 1998). On the other hand, informal user-group management – which generally stipulates fewer rules and enforces them less strictly – has had no appreciable impact on forest resources, even compared with the absence of user groups or open-access conditions. This may be due to the fact that informal user groups were newly formed primarily to fulfil one of the obligations for acquiring official use rights of forestland. There is the observed trend that management rules have been strengthened as the scarcity of forest and tree resources has increased over time (Tachibana *et al.*, 1998). It is costly to organize collective action, set out management rules and enforce those rules. Thus, forest user groups tend to be formed when the scarcity of forest resources reaches a threshold level sufficient to warrant the initiation of user-group management (i.e. when the values from protection rise sufficiently).

There are at least two fundamental factors that explain the success of common-property forest management in the Hill region of Nepal. First, the success of subsistence farming and livelihoods depend critically on the availability of forest and tree resources, such as firewood, grasses and leaf fodder (Thapa *et al.*, 1998). The transportation network is poorly developed and access to markets and alternative sources of energy and soil nutrients, which can be substituted for firewood and compost, is very poor in most communities in the Hill region. Furthermore, the large and continuous application of compost is essential to sustain upland farming in terraced fields on steep slopes. Thus, the degradation of community forests immediately and significantly jeopardizes the livelihood of farm populations. This implies that there are strong incentives for farmers to protect and manage forest resources. In contrast, *miombo* woodlands in Malawi are located in relatively flat areas and the access to markets is much more favourable. This seems partly to explain why the management of *miombo* woodlands has been much less successful. In fact, the management of common-property forests has never been successful in the flat areas of contemporary Nepal (i.e. Tarai), as well as in pre-War Japan.

Secondly, it is important to recognize that the key forest resources successfully protected by communities are minor forest products, for which the costs of protection relative to benefits would be extremely high if they were owned individually. It is less costly to protect these resources communally by hiring a selected number of guards or adopting a rotational system of patrolling among community members, than by hiring guards individually. In other words, the advantage of community management rests on the economies of scale realized in protection activities. This hypothesis is consistent with the finding from the case study of young timber plantations in the Inner Tarai region of Nepal, where the cost of protection per unit of area is much larger under private than under community land ownership (Sakurai *et al.*, 1998).

The cost of protection of forest resources by private owners is much less, however, if the main forest product is mature timber, because it is not feasible to fell big trees, haul them from forests and process them without being noticed by village people. There is no question that it is less costly to protect timber than such minor forest products as firewood and grasses. It is also important to realize that timber production, to be successful, requires silvicultural activities, such as weeding, pruning, thinning and singling. As in the case of collective farming in socialist economies, each farmer producing timber has little incentive to carry out the necessary forest management activities as long as the benefits are shared more or less equally among community members.

The issue of efficient forest management led us to examine the case of community forest management in post-War Japan, where forests have been converted from miscellaneous broad-leaved trees to timber forests. In Japan, community forests were rigorously managed before the Second World War, in order to extract firewood, charcoal and grasses for feeding horses, which were used to produce compost and to plough crop fields. The demand for firewood and charcoal declined drastically after the War, due to the increasing availability of kerosene, as did the demand for grasses, due to the increased supply of chemical fertilizers and tractors. Thus, the need for protecting community forests was lessened considerably and the management of forestlands was focused on the efficiency of managing trees. Some communities divided forest areas into smaller units and allocated them to user-group members for individual management. We found through regression analyses using the data generated from aerial photographs that timber trees have increased more rapidly and the number of trees per hectare has declined more sharply due to thinning on individually managed portions of community forests, compared with those managed communally (Kijima *et al.*, 2000). The emerging conclusion from this case study is that the individualized management system appears to be more efficient than community management in the case of timber forests.

This conclusion, however, does not necessarily hold in developing countries, where, unlike in post-War Japan, the grazing of cattle and goats is a major threat to sustainable forest management. In fact, one of the reasons for the higher cost of protecting timber plantations under private ownership in Nepal's Inner Tarai is the cost of keeping cattle and goats away. They eat young seedlings and step on them. In order to protect timber forests, collective management with regulated use of forest resources and mutual supervision is more cost-effective. In order to carry out management activities, however, an individualized system is more efficient. In our view, the best management system is, therefore, a combination of the two, in which protection is carried out through participation of all community members and management activities are carried out under more individualized management incentives. Such cases are found in the Inner Tarai, where forest user-group committees employ hired labour for the management of timber trees, with a view to making profits from the sale of timber. This system has proved to be superior to conventional community management in the case of timber forests (Sakurai *et al.*, 1998). It is worth stressing that the possibility of selling harvested timber in the market is a prerequisite for the efficient management of timber forests. Our findings further indicate that social forestry projects often fail to produce valuable timber trees because of the lack of incentives to manage trees, even though trees may be protected in project sites.

Communal Land Tenure and Agroforestry Management

As we have argued, the single major cause for deforestation in the study sites is the expansion of farming areas to grow food and other crops. Crop cultivators belong mainly to the poorer segments of society, who have already cleared and occupied large forest areas. This land is often marginal in terms of agricultural potential. Unless decent work opportunities are made available, it would be practically impossible to relocate inhabitants to restore the forest conditions. To replace the tree-production function of forests, it is socially more efficient to plant commercial trees, such as rubber, cocoa, coffee and cinnamon on former forestland currently used for shifting cultivation or to plant useful trees, such as nitrogen-fixing species, in the midst of crop fields.[5] In fact, as population pressure increases, the fallow period under shifting cultivation becomes shorter, resulting in declining soil fertility and crop yields. Commercial tree farming, one form of agroforestry, and mixed tree/crop farming are more sustainable and often more efficient than pure cropping systems in these marginal areas.

It is widely believed, however, that, because of weak individual land rights, trees are not planted and managed under communal ownership, in which the extended family controls the use rights on cultivated land (e.g. Besley, 1995). If this is indeed the case, it will be difficult to disseminate agroforestry in marginal areas, where agroforestry has an obvious comparative advantage over food cultivation. If, however, communal tenure institutions do provide sufficient incentives to plant and manage trees, the incidence of poverty in marginal areas can be reduced by enhancing the efficiency of land use in such areas. Furthermore, the establishment of agroforesty systems on sloping lands will help to reduce soil erosion and contribute to the partial restoration of tree biomass and biodiversity.

In none of the study sites did we obtain any strong evidence to support the validity of the popular argument that communal ownership or ownership by the extended family in customary land areas deters the development of agroforestry. Commercial trees have been planted under the communal ownership system as widely and actively as under the private ownership system in Sumatra, Uganda and Ghana (Otsuka *et al.*, 1998a,b, 1999; Place and Otsuka, 1998a; Suyanto and Otsuka, 1998). This occurs because, as in the case of forest clearance, the effort to plant trees is rewarded by strengthened individual land rights.

In matrilineal communities of Sumatra, cultivated land was traditionally owned by lineage members, typically consisting of three generations descended from the same grandmother. Gradually, joint ownership by sisters became common. At present, however, the ownership of agroforestry plots has been more individualized and these plots are more likely to be bequeathed from mothers to individual families of their daughters and even to those of their sons (Otsuka *et al.*, 1998b). Such systems may be called ‘single-family’ ownership systems. As shown in Table 15.3, single-family ownership accounts for 40–60% of the areas under agroforestry. It is also clear from this table that private ownership acquired through land-market purchase and forest clearance is also very common. An important observation is that land rights are stronger under

Table 15.3. Distribution of area under different land-tenure institutions and the index of land rights on agroforestry plots in Sumatra.

	Lineage ownership	Joint-family ownership	Single-family ownership	Private ownership: Purchase	Private ownership: Forest clearance
Distribution of area (%)					
High region	3	5	42	10	37
Middle region	5	2	62	14	19
Low region	0	3	46	12	39
Index of land rights[a]					
High region	0.0	0.6	1.6–2.0[b]	3.1	
Middle region	0.8	0.9	1.9–2.9	3.8	
Low region	0.0	1.0	1.9–2.8	3.8	

[a]The following four rights are considered: rights to rent out under share tenancy, rent out under leasehold tenancy, pawn and sell. Numbers refer to the average number of rights (ranging from 0 to 4) that can be exercised without prior approval of family and/or lineage members.
[b]The first number refers to the case of single-family ownership by daughters; the second number corresponds to the case of single-family ownership by daughters and sons.

single-family ownership than under collective ownership, and the transformation from the latter to the former has been facilitated by tree planting. Strengthened land rights under single-family ownership have also promoted the development of land markets.

In indigenous villages in Ghana, so-called uterine matrilineal inheritance has been practised, in which land is bequeathed from the deceased man to his nephew along the mother's family line. While those who clear forestland are granted strong land rights, the cleared land eventually becomes the property of extended families, and is temporarily allocated to those family members in need of land or is bequeathed in accordance with the traditional rules of inheritance. As is demonstrated in Table 15.4, individual land rights on these family-owned plots in indigenous villages are very weak. However, the individual rights of inherited land in migrant villages are much stronger, since patrilineal inheritance has been practised in these villages. This individualization of rights to land in the patrilineal society has been facilitated by the fact that only a small number of family members, typically a father and his sons, are involved in the inheritance decision. If land-tenure security affects land use, which our evidence suggests, the inheritance system is an important factor. It is important to observe that the importance of allocated and inherited family land is relatively low. Instead, cocoa fields are often transferred to spouse and children as *inter vivos* gifts with the permission of members of the extended family (Otsuka *et al.*, 1998a). This transfer to spouse and children represents a reward for the work effort involved in planting and managing cocoa trees.

Given the positive and significant effect of tree planting on individual land-use rights, it is no wonder that sufficiently strong incentives to plant

Table 15.4. Distribution of area under different land-tenure institutions and the index of land rights on cocoa plots in western Ghana.

	Temporarily allocated family land	Inherited family land	Acquired village forestland	Gift	Others
Distribution of area (%)					
Indigenous villages	22	13	19	33	13
Migrant villages	9	18	22	26	26
Index of land rights[a]					
Indigenous villages	0.3	1.1[b]	3.0	4.9	–
Migrant villages	0.2	3.3[c]	3.9	5.3	–

[a]The following six rights are considered: rights to plant trees, rent out, pawn, bequeath, give and sell. Numbers refer to the average number of rights (ranging between 0 and 6) that can be exercised without prior approval of family members or the village chief.
[b]The number pertains to the practice among matrilineal Akan people.
[c]The number pertains to the practice among patrilineal non-Akan people.

commercial trees exist under the communal ownership system. Once trees are planted, the land ownership system is converted to *de facto* private ownership, and the management efficiency of commercial tree fields under the communal system would be expected to be generally comparable to other ownership systems. Indeed, in our study sites, regression analysis revealed no differences in profits per unit of land area across different land-tenure institutions for cinnamon and rubber (Suyanto *et al.*, 1998a,b), coffee (Place and Otsuka, 1998a) or cocoa (Otsuka *et al.*, 1998a). In other words, the communal system evolves towards an individualized system and does not impede the development of agroforestry.

Agroforestry is a far more intensive farming system than food-crop farming under shifting cultivation. Table 15.5 illustrates this point by comparing labour use and residual profit (i.e. revenue minus both actual and imputed costs of non-land inputs) or net revenue (i.e. revenue minus actual cash costs) per hectare between rubber agroforestry and upland rice cultivation in Sumatra (Suyanto *et al.*, 1998b) and cocoa agroforestry and pure food-crop cultivation in Ghana (Otsuka *et al.*, 1999). Note that, although fallow periods are 6–7 years in upland rice production in Sumatra and 7–8 years in food production in Ghana, the data in Table 15.5 pertain to fields actually under cultivation at the time of the study. Thus, although labour inputs look higher under shifting cultivation than under agroforestry, the average labour inputs over time under the former are approximately only one-third of those in the latter case. The estimated residual profit is much higher in agroforestry compared with upland rice in Sumatra, whereas the net revenue during the production year is only slightly higher in agroforestry than under food cultivation in Ghana.[6] There is no question that these agroforestry systems are more profitable than shifting cultivation, given the increasing scarcity of land, which is reflected in shortened fallow periods. This finding is consistent with the evolution of customary land-tenure

Table 15.5. Comparison of labour use and profit/net revenue per hectare between agroforestry plots and food-crop plots under shifting cultivation in Sumatra and Ghana.

	Labour-days per hectare[a]	Profit/net revenue per hectare[b]
Sumatra		(1000 rupiahs)
Rubber[c]	101.2	217
Upland rice	172.7	4
Ghana		(1000 cedis)
Cocoa[c]	68.0	384
Miscellaneous food crops	158.0	494

[a]Assuming 7 h of work day^{-1}.
[b]Residual profit, deducting imputed family labour costs and cash input costs from gross revenue, for Sumatra; net revenue, deducting only cash costs from gross revenue, for Ghana.
[c]For comparison, the averages of fields with trees 11–15 years of age are shown for both Sumatra and Ghana.

institutions towards more individualized systems in both countries, which stimulates the development of agroforestry. Finally, it should be emphasized that agroforestry is sustainable, as is illustrated by the fact that, beginning with the fourth year (in the case of cocoa) to the eighth year (in the case of rubber), tree crops yield positive returns for well over 20 years.

It is important to point out that the institutional rule of granting strong individual land rights on fields planted to trees has been established in areas where agroforestry has a comparative advantage and hence is more profitable than other crops. Since most areas of Malawi are characterized by flat topography with relatively well-developed road networks, many types of agroforestry systems may not have a comparative advantage over maize and tobacco production. In fact, chemical fertilizers are more readily available and the marketing of output is less costly in most areas of Malawi than in remote mountainous areas in other countries which have less access to markets. In such production environments as Malawi, we observed that strong individual land rights do not often follow tree planting (Place and Otsuka, 1998b, 2000a).[7] Further, tree planting is more common in patrilineal areas than in matrilineal areas, reflecting higher initial tenure security for males.

The major policy implication is that, given the existence of strong incentives to manage agroforestry plots on sloping lands under communal ownership, it makes sense to develop and disseminate profitable agroforestry systems through such means as the development of improved species of commercial trees and the improvement of social infrastructure and marketing systems. The enhancement of profitability will be an effective way to promote the widespread adoption of agroforestry and to strengthen individual land rights. In this way, the high incidence of poverty in marginal areas can be reduced by improving the efficiency of land use. It must also be clearly understood that, while the development of roads may accelerate deforestation by enhancing the profitability of timber harvesting, it will also accelerate the development of agroforestry where primary forests have already been cleared, as in Uganda (Place and Otsuka, 1998a).

Land Tenure and Cropland Management

Land-tenure security affects the expected future benefits accruing to those who invest in land improvement, including tree planting. Therefore, it affects long-term but not short-term management incentives. Thus, land-tenure institutions do not have any impact on the production efficiency of paddy-fields in Sumatra, which do not require much investment (Suyanto *et al.*, 1998a). Nor did they have any effects on the efficiency of maize farming in Malawi, for which we did not observe any difference in profits between patrilineal and matrilineal inheritance systems, despite much stronger land rights for males under the former system (Place and Otsuka, 1998b).

We did observe, however, some differences in the management efficiency of annual crop production under different land-tenure institutions. First, farmers subject to patrilineal inheritance have introduced profitable tobacco farming more quickly and more widely after abolition of the policy prohibiting tobacco production by smallholders in Malawi (Place and Otsuka, 1998b). Because tobacco was a relatively new crop, and one where organizational links are important, investment in the acquisition of relevant new farming knowledge in production, financing and marketing was required for tobacco production. The adoption of new cropping technology does not confer strong individual land rights and hence those who are subject to tenure insecurity tend to adopt the new crop less actively.

Secondly, we found that cropland owned and allocated by the extended family under communal ownership is less frequently fallowed than land owned privately in Uganda (Place and Otsuka, 1998a) and in Ghana (Otsuka *et al.*, 1999). Land-use rights are fairly well established under communal ownership, as long as land is used for cultivation. Once it is fallowed, however, individual rights are substantially weakened and extended-family members may attempt to claim such land. Thus, land that is frequently fallowed under communal ownership is subject to weaker tenure security, which forces farmers to continue to cultivate the land to secure use rights.

While customary land-tenure institutions are not significantly inefficient for the management of land in trees due to the tenure security-enhancing effect of tree planting, they are likely to be inefficient in the management of cropland under shifting cultivation. According to the accumulated empirical evidence from sub-Saharan Africa, however, land-tenure institutions do not appear to significantly affect the productivity of sedentary farming (Place and Hazell, 1993). This is because tenure security does not have an appreciable influence on investment decisions or long-term productivity in areas where agricultural investment opportunities are few and of a short-term nature. In other areas, a plausible hypothesis seems to be that, like tree planting, investment in land improvement – e.g. terracing and destumping – strengthens one's land rights in cases where such investments are highly profitable.

Concluding Remarks

Although it may be hazardous to draw strong policy implications from the selected case studies covered by this research, it is worth emphasizing that highly consistent findings were generated regarding the nature and efficiency of state, communal, private and common-property systems across diverse sites in Asia and Africa. Our comparative research has identified that the most fundamental reason for deforestation is the expansion of croplands. In other words, the crux of the forestry problem is the food problem. Accordingly, in order to prevent further deforestation, increasing food supplies from existing farms is essential in order to reduce food prices – that is, if decreasing population growth is an infeasible option in the short run. In the longer run, adequate employment opportunities outside the farming sector must be created and coupled with sufficient investment in the human capital of the farm population.

The most inappropriate land ownership system is generally state ownership. The governments of developing countries do not usually possess the capacity to manage large forest areas, which are continually sought for conversion into cropland, as well as for agroforestry. The weakness of state ownership and the superiority of more individualized ownership is evidenced by the strengthened forest management effort in Vietnam, where the use rights for state forests have been transferred to individual farmers. Except where needed for the protection of the biodiversity of flora and fauna, the ownership of forestland by the state should be abolished.

It must be recognized that the common-property forest regime is effective where key forest resources are minor forest products, whereas high-value tree production (e.g. timber) is less amenable to common management. Thus, the incentive systems in social forestry projects need to be redesigned. In particular, the system of equal benefit sharing should be replaced by systems that provide appropriate incentives for individual farmers to manage high-value trees. The element of community management, however, should be maintained for the protection of trees. It is also important to provide profit incentives to grow and manage timber by promoting the marketing of harvested trees.

Given that farmers in customary land-tenure areas have strong incentives to plant and grow trees for commercial purposes, the effort to develop profitable agroforestry systems should be strengthened. The production of commercial tree crops is often more advantageous than food production under shifting-cultivation systems in marginal areas. Moreover, agroforestry is more sustainable than shifting-cultivation under conditions of increasing population pressure. The establishment of profitable agroforestry systems will contribute significantly to the reduction of poverty by enhancing the efficiency of farming in marginal areas, since farmers in these areas are particularly poor. In addition, agroforestry can contribute to the prevention of soil erosion and to the creation of tree biomass. Thus, the development of agroforestry is expected to be conducive to both efficiency and equity when evaluated from both private and social viewpoints.

Notes

1 The term 'individualization', or alternatively 'privatization', is defined as any change that enhances the property rights of individuals.

2 This chapter summarizes the findings of papers presented at the Workshop on Land Tenure and the Management of Land and Trees, held at Tokyo Metropolitan University in July 1998, as well as other unpublished project papers.

3 Open-access or non-property is treated as a possible outcome under any of the four tenure categories.

4 Trees are grown and managed not only on non-agricultural land but also on agricultural land in these countries. According to our data, the average tree cover on agricultural land amounted to 28% in Uganda and 2% in Malawi. Some trees – e.g. coffee, fruit and those trees with nitrogen-fixing capacity – are intentionally planted as a part of agroforestry systems, whereas other trees grow naturally and are retained for collection of firewood and other tree products.

5 Forests will still be valued for ecosystem functions that cannot be fully replicated in agricultural settings.

6 Although we showed data for tree fields of 11–15 years of age in Table 15.5, the same qualitative conclusions hold for any mature tree-crop fields.

7 For example, males do not increase their rights to residence and control of land by planting trees in their wife's village under the matrilineal system. Also, rights of sale remain non-existent and unresponsive to tree planting and other investments.

16 Sustainable Agriculture and Natural Resource Management in India's Semi-arid Tropics

JOHN KERR,[1] GANESH PANGARE,[2] VASUDHA LOKUR PANGARE[3] AND P.J. GEORGE[4]

[1]Department of Resource Development, Michigan State University, East Lansing, Michigan, USA; [2]Indian Network for Participatory Irrigation Management, Central Soil and Materials Research Station, New Delhi, India; [3]OIKOS, New Delhi, India; [4]Farming Systems International, Cochin, Kerala, India

Background

India's semi-arid tropics (SAT) are home to hundreds of millions of poor rural people. This region, which covers interior areas in the southern half of the country, has limited irrigation potential and low, erratic rainfall. It has been largely bypassed by the Green Revolution, which led to dramatic production and yield gains in more favourable environments of the country (Walker and Ryan, 1990). The Indian SAT also suffers from a highly degraded natural resource base, which is characterized by soil erosion, deforestation, pasture degradation and groundwater depletion.

In recent years, watershed development has been promoted as a means of raising agricultural productivity, conserving natural resources and alleviating poverty. While the watershed approach offers the potential for complementary gains in meeting all three of these objectives, in practice it presents challenges, due to conflicts between technical optimality and socioeconomic feasibility. This chapter presents research findings showing that some watershed projects have succeeded in promoting natural resource conservation and higher agricultural productivity by reconciling these conflicts through a focus on village-level social organization. Other, more technocratic, projects, which ignored issues of social organization, have had less success. The Indian government is currently attempting to expand successful approaches to a national scale, but various constraints remain. One of these is the need for better diffusion of project benefits to landless people, who are the poorest members of the community. In fact, as will be explained below, ensuring an equitable distribution of benefits may prove critical to achieving production and environmental objectives in watershed development.

The literature on watershed development in India is growing rapidly, but most of it is confined to qualitative descriptions of success stories. Some of these contain excellent insights into the social processes that contribute to successful watershed development, but there is little frank discussion of less successful projects. The few quantitative studies available tend to be based on a small number of heavily supervised projects, with no information about long-term impacts. Benefits after the first year or two are typically assumed but not documented and, not surprisingly, cost–benefit findings are almost always favourable. Many such studies are summarized in the *Indian Journal of Agricultural Economics* (1991). At the same time, relatively few projects have been subject to detailed evaluation, and there are good reasons to suspect that many of them have had little impact (Kerr and Sanghi, 1992). The more recent literature analyses the new generation of projects that combine technical development and social organization (Farrington *et al.*, 1999; Hinchcliffe *et al.*, 1999). Most of this literature focuses on project approaches (especially participation) that can contribute to project success, but with less rigour in measuring impact.

With this background, the International Food Policy Research Institute (IFPRI) and the Indian Council of Agricultural Research (ICAR) jointly initiated a detailed study of agricultural productivity, natural resource management and poverty alleviation under a wide range of watershed projects in the Indian SAT. The study explicitly examined the effects of non-project factors, such as infrastructure, access to markets, social institutions and agroecological conditions, not only in controlling for the effects of these factors on project success but also in searching for other policy-relevant determinants of improved natural resource management and economic development. This chapter presents some findings of the study; more detailed information is available in Kerr *et al.* (1998b).

Watershed Development Projects

A watershed (or catchment) is a geographical area that drains to a common point, which makes it an attractive planning unit for technical efforts to conserve soil and maximize the utilization of surface and subsurface water for crop production. In watershed management projects, mechanical or vegetative structures are installed across gullies and rills and along contour lines, and areas are earmarked for particular land uses, based on their production capability and susceptibility to degradation. Cultivable areas are put under crops according to strict principles of contour-based cultivation. Erosion-prone, less favourable lands are put under perennial vegetation. This approach aims to optimize moisture retention and reduce soil erosion, thus maximizing productivity and minimizing land degradation. Improved moisture management increases the productivity of improved seeds, organic matter and fertilizer, so conservation and productivity-enhancing measures are complementary. (Chemical inputs in low-potential rain-fed areas are commonly below polluting levels and will remain so for the foreseeable future.)

Excess surface runoff water is harvested in irrigation tanks (catchment ponds), while subsurface runoff recharges groundwater aquifers, so adopting conservation measures in the upper watershed has a positive impact on productivity in the lower watershed. Reducing erosion in the upper reaches of the watershed also helps to reduce sedimentation of irrigation tanks in the lower reaches. The watershed approach enables planners to internalize such externalities and other linkages among agriculture and related activities by accounting for all types of land uses in all locations and seasons. This systems-based approach is what distinguishes watershed management from earlier plot-based approaches to soil and water management.

Constraints to implementing watershed development

Although a watershed is a convenient hydrological unit for planning and implementing a technical programme, it is not a useful socioeconomic unit. Watersheds contain administrative and property boundaries, lands that fall under different property regimes and farmers whose actions may affect each other's interests. Socioeconomic boundaries do not normally match biophysical ones, and this can create major obstacles to successful implementation.

Tradeoffs between equity and productivity/sustainability objectives

Because watershed development is a land-based activity, the principal beneficiaries are owners or users of the land that the project improves. This raises concerns about tradeoffs between productivity and sustainability, on the one hand, and poverty alleviation on the other. For example, a typical microwatershed in semi-arid India contains the best farmland in the lower reaches, much of it irrigated by groundwater or surface water from a tank. The wealthiest farmers typically hold land in this part of the microwatershed. In the middle reaches, the agricultural plots may be less fertile and of greater slope, and they are less likely to be irrigated. The upper reaches may contain steep, uncultivable land suitable only for pasture or forestry. These areas are utilized mainly by the poorest people in the village, who graze their herds and gather products from common-property pastures and forests because they have no land of their own. Common lands, in general, are extremely important to poor people's livelihoods in India's semi-arid areas (Jodha, 1997).

When a watershed project is introduced, soil conservation measures are installed throughout the microwatershed, and restrictions are placed on access to the upper reaches. Eliminating grazing and tree harvesting enables gradual revegetation of the upper reaches, which increases water infiltration and percolation. This, in turn, raises irrigation potential in the lower reaches. As a result, the people who utilize the upper watershed bear the brunt of the costs of watershed development, which benefits mainly the wealthiest people in the lower watershed. Not surprisingly, those whose interests are threatened by a watershed project commonly refuse to go along with it, actively undermining its efforts. Herders, for example, may refuse to abide by grazing bans, trespassing on the common lands if they are able to do so.

On the other hand, landless people can potentially benefit from watershed projects in three ways. First, in India, virtually every watershed project doubles as an employment project, so local landless people may enjoy increased wage income while the project is being implemented. Secondly, if the project is successful in expanding irrigated area and rain-fed agricultural productivity, labour demand may rise in the long term. Thirdly, if the project succeeds in regenerating uncultivated common lands, poor people, who depend on them for fodder, fuel and other products, may also benefit directly from watershed development after a lag of about 3–7 years, depending on agroclimatic conditions.

The objectives of raising agricultural productivity, protecting the environment and alleviating poverty may be highly interdependent, since ignoring the interests of the landless may lead them to undermine project activities. Alternatively, productivity and environment objectives may still be achieved under a highly inequitable system whereby poorer people are given no voice and their actions tightly controlled. This situation may pertain to villages where the landless are a small, politically weak minority.

Alternative approaches to watershed development

The first watershed projects in India's SAT were strictly technocratic in their orientation. They were organized around large watersheds that covered multiple villages in whole or part. Project staff built small dams and contour earthen bunds, or barriers, to capture water and conserve soil, and they planted trees and pasture grasses on common lands. Local people were expected to maintain these assets, but experience quickly showed that they did not do so automatically (Vaidyanathan, 1991). By the late 1980s, disappointment at the outcome of early projects gave rise to calls for 'people's participation'. While virtually everyone agrees that participation is a good idea, different people define it in different ways. Two extremes help to characterize the experience to date with participatory watershed management.

One extreme is based on the view that people will accept watershed technology once they are made aware of its benefits; this requires a mechanism for project officials to explain to watershed inhabitants how the various recommended practices operate and why it is important to adopt and maintain them. In order to take people's involvement a step further in such projects, local committees are established to mobilize labourers for moving earth and planting vegetation and to facilitate communication within the village to improve the management of common lands.

The opposite extreme is based on the view that people know best how to take care of their land and simply need outside assistance to help organize them to resolve conflicts, solve problems and gain access to resources, including funds and social services. Project staff are more likely to be trained in social work than in agriculture, and the village rather than the watershed is the primary project unit. Watershed projects that emerge from this approach are typically planned and implemented in full participation with local people. They usually involve 'social fencing'[1] to protect common lands, indigenous soil and water conservation technologies and water harvesting structures built by engineers.

Until about 1990, all Indian government projects operated close to the first extreme while most non-governmental organization (NGO) projects operated closer to the second. Since then, however, there has been a gradual movement toward the middle ground, beginning with a few of the best NGOs, which combined high-quality technical input with a strong emphasis on social organization. These projects used the village as the unit of implementation, working in small microwatersheds within villages and then aggregating upward. In the early 1990s, some collaborative government/non-government programmes followed suit and, in late 1994, the Ministry of Rural Development (MORD) adopted new guidelines along the same lines. After a slow start, it has begun to make a substantial investment in truly participatory watershed development, with an annual budget of over US$600 million.

Analytical Approach

This study examines watershed project performance as defined by various indicators of agricultural productivity, natural resource management and human welfare. Performance indicators were collected at the village, plot and household levels. The village, as opposed to the watershed, was selected as the primary sampling unit, for three reasons: (i) most projects studied operate at the village level; (ii) the village provides a standard unit for collecting comparable data across locations and over time; and (iii) this approach facilitates analysing the impact of projects on people, rather than watersheds *per se*. Most of the treated watersheds fall entirely within a single village, but some transcend village boundaries.

The data come from a survey of 86 villages in the Indian states of Andhra Pradesh and Maharashtra. The villages are covered by a variety of project approaches and include control villages with no project. Quantitative data collected at the village, plot and household levels provided the basis for analysing the determinants of changes in pre- and post-project conditions. Open-ended discussions provided further qualitative information on the impact of projects on people from various interest groups.

Projects covered under the study

All categories of projects operating in the two states are covered by the research. These include the following types of projects:

- *Ministry of Agriculture (MOA)*: projects that focus primarily on technical aspects of developing rain-fed agriculture. These include the National Watershed Development Project for Rainfed Areas (NWDPRA), ICAR's model watershed projects and the World Bank-funded Pilot Project for Watershed Development in Rainfed Areas.

- *MORD*: engineering-oriented projects that focus on water harvesting through construction of percolation tanks, contour bunds and other structures. These fall under the Maharashtra Department of Soil and Water Conservation projects (Jal Sandharan) and the Drought-Prone Area Project (DPAP).[2]
- *NGOs*: projects that typically place greater emphasis on social organization and less on technology, relative to government-led programmes. Individual NGO programmes vary in their emphasis and professionalism, but they are grouped together none the less.
- *NGO–government collaboration*: projects jointly managed by government and NGOs (e.g. the Indo-German Watershed Development Project (IGWDP), Adarsh Gaon Yojana (AGY)), which combine the technical approach of government projects with the NGOs' orientation toward social organization. These projects are found in Maharashtra but not Andhra Pradesh.
- *Control*: villages with no project.

This research was originally designed to examine only completed projects where the staff had withdrawn. However, despite the large literature on watershed development in India, the number of projects in which work has actually been completed is quite small, so the intended approach was not feasible. Instead, the study covers mainly well-established projects, with a few that have been completed.

Selection criteria under each project

How project personnel select participating villages can have a major impact on what a project can achieve, since various conditioning factors have strong influences on people's incentives to invest in land improvement. In addition, differences in selection criteria complicate the statistical analysis of the determinants of higher agricultural productivity and improved natural resource management (this issue is discussed below). This section summarizes project policies regarding village selection; actual data relating project categories to village characteristics are analysed below.

Whereas all projects focus on relatively unfavourable areas with low rainfall and low irrigated area, NGOs favour the most remote villages with the poorest, most marginalized people. On the other hand, the NWDPRA intentionally selects easily accessible villages. This reflects the orientation of these projects towards planning and supervision by people located outside the village, with the optimistic view that visibility will lead to dissemination of practices introduced by the project (GOI, 1991). More subtly, the approach also leads to an apparently unintentional bias in the selection of project sites towards more densely populated areas with better access to transport and markets. Since these conditions may be especially favourable for the promotion of rain-fed agriculture, these villages may be predisposed to high agricultural productivity. In this case, it may be difficult to assess whether watershed success is determined by project activities or by pre-existing village characteristics.

NGOs and NGO/government collaborative projects (the AGY and the IGWDP) attach great importance to pre-existing social institutions in the villages, while government projects ignore them. The most revealing factor here is that the AGY and the IGWDP, along with some Maharashtra NGOs, require that participating villages undertake community voluntary labour (*shramdan*) on a regular basis and ban grazing and tree cutting on common lands. The *shramdan* effort, which is devoted to some issue of common concern in the village, is intended to foster a spirit of self-sufficiency and mutual dependence. It may also be an indicator of a village's propensity for collective action, which will be necessary for the successful protection of common lands. If the premise that social organization is important for watershed development is correct, these villages are self-selected for success, because villages unwilling to undertake *shramdan* will not join the project.

Another important factor is that many projects take advantage of work done by earlier projects. Virtually all projects in Maharashtra are located in villages already treated with soil and water conservation investments in the 1980s by the state government's Comprehensive Watershed Development Project (COWDEP). For NGOs, this makes sense, as their work in social organization is complementary to earlier technical inputs. In addition, the start-up phase of the IGWDP, which is covered by this study, was restricted to well-established NGOs, which were already familiar with the community in which they initiated the project, and many ongoing activities were simply brought under the flag of the IGWDP. These facts about pre-project history are important to keep in mind for two reasons. First, project impacts must be jointly attributed to both the new project and the old COWDEP. Secondly, impact is likely to come more slowly when these projects expand to areas not previously covered by an earlier project.

Data

Evaluating watershed projects requires baseline and monitoring data for the comparison of pre- and post-project conditions, but unfortunately no such information was available. As a result, the quantitative analysis is based on secondary data, which were available for both the pre-project period (1987) and the present (1997), primary data on current conditions, based on interviews and visual assessments, and primary data on past conditions, based on recall by local inhabitants.

A major element of the research was the development and collection of data on various performance indicators, such as natural resource conservation, agricultural productivity and equitable distribution of project benefits. These data were collected through direct observation, group discussions and government records. Quantitative data were also collected on the background characteristics of the projects, villages, households and plots covered under the study. Some of the village-level information came from public sources, but most was collected from group and individual interviews in each village. In addition, qualitative data were collected regarding the natural resources people use to earn their livelihoods, the social institutions that govern access to those resources and any changes in resource access resulting either from changes in their quantity or changes in social institutions. This information was collected in open-ended discussions with members of specific interest groups in each village, such as farmers with irrigated land, farmers with rain-fed land, landless people, herders and women.

Villages included in the study were selected randomly and stratified by project category. A full set of quantitative and qualitative data at the village, household and plot levels was collected in 13 villages in Maharashtra and 16 villages in Andhra Pradesh, for a total of 29 villages. Village-level data were collected in an additional 57 villages in Maharashtra. The village-level analysis is confined to the 70 Maharashtra villages, while the plot-level analysis covers the 29 villages from both states, where more detailed data were collected.

Econometric Model

The basic conceptual framework for this study is represented by the following model:

$$Y = f(W, V, H, P)$$

$$W = f(V)$$

where:

Y = outcome (from performance indicators)

W = village's watershed project category

V = village-level characteristics

H = household-level characteristics (omitted from the village-level analysis)

P = plot-level characteristics (omitted from the village-level analysis)

The key analytical feature of this model is the endogeneity of W, the project category. This results from biases in the way the villages in each project were selected. In short, if each project employs different criteria for selecting villages, it is possible that differences in performance can result more from differences in initial, pre-project conditions than from the work undertaken by the watershed project. Differences in selection criteria for each project may be based on both observed and unobserved village characteristics. Unobserved differences cause standard econometric approaches to yield biased coefficients.

Econometric methods of controlling for the problem of selection bias are well established in the analysis of only two categories (such as a single treatment and a control), but the current study covers five project categories, making the problem much more complex. The econometric model is still being developed, so this chapter presents preliminary findings based on two types of regression models, one with and one without correction for selection bias.

The uncorrected model is presented in two specifications. In model 1, all five project categories identified above are included; in model 2, these are consolidated to only three categories: purely government projects (MOA and MORD), projects with input from NGOs (NGO and NGO–government collaboration) and control villages with no project. (The project categories are represented by dummy variables; the 'control' category is omitted.)

A second set of models uses the treatment-effects approach (Greene, 1990) in order to correct for selection bias.[3] These models can only contain two project categories, so they are run separately under two alternate specifications. In model 3, the project dummy variable is equal to 1 if any kind of project (MOA, MORD, NGO or NGO–government collaboration) operates in the village and is equal to zero for the control villages. In model 4, the project dummy variable is equal to 1 for villages with an NGO or NGO–government collaborative project and is zero for villages that have either an MOA or MORD project or no project.[4]

Econometric analysis provides insights into causal relationships between various biophysical and socioeconomic factors and the project objectives of increased productivity, better natural resource management and poverty alleviation. However, econometrics is by no means the only valid analytical tool for examining these relationships, and the discussion of selection bias shows that it has important limitations. Another shortcoming is that the econometric database, as described below, is limited by the absence of systematic baseline and evaluation data. As a result, this study uses descriptive and qualitative analysis in addition to econometric analysis to provide a broader perspective of the relationships under study. Due to space constraints, this chapter only presents the econometric analysis; further descriptive and qualitative findings are available in Kerr *et al.* (1998b).

Variables Used in the Analysis

The analysis presented here contains three dependent variables – project category, soil conservation and net returns to cultivation – along with a variety of explanatory variables. Soil conservation is an indicator of natural resource conservation measured at the village level, while net returns to cultivation are an indicator of agricultural productivity measured at the plot level. Quantitative analysis is not applied to questions of poverty alleviation, which is discussed on the basis of qualitative data below. Further information on performance indicators is available in Kerr *et al.* (1998b).

Dependent variables

Project category

A categorical variable represents each of the five project categories; it is used as the dependent variable in a multinomial logit model to analyse the determinants of how villages are selected in different project categories.[5]

Soil conservation

Trained investigators conducted a village-wide transect to assess the extent of soil erosion and measures to arrest it. One transect followed the village's main drainage line and a second crossed the fields along a route perpendicular to the drainage line, covering a representative sample of different types of land in the village – irrigated, rain-fed and uncultivated.[6] Investigators selected the transect routes based on discussions with villagers, published soil maps and maps made by villagers in participatory mapping exercises.

Investigators divided the transect into segments that were uniform in land use, land capability classification, erosion status and extent of conservation investment. Any time any of these features changed, a new segment began. Based on visual assessment of each segment, investigators assigned scores for erosion and conservation, indicating high (3), medium (2) or low (1); the overall village score is the average score of all the segments, weighted by their length. Accordingly, the soil conservation variable used here is a village-level score; the analysis is accomplished with a Tobit model, which accounts for the fact that the score is bounded at 1 and 3.

Soil erosion in the transect refers to visible rills and gullies. Soil conservation measures refer to any management practice that contributes to soil conservation; these include terraces, field bunds, land levelling and protection of perennial vegetation, for example. Inhabitants of the study villages accompanied the field investigators to help ensure that they did not overlook local land management practices, which they might not have recognized. The soil conservation findings are presented here, but not those for soil erosion.

Returns to cultivation

Returns to cultivation are an obvious plot-level indicator of agricultural productivity. The data cover gross returns per hectare and both cash costs and imputed costs of household resources and labour. Investigators collected this information in interviews with farmers for the year immediately prior to the interview. The analysis covers only rain-fed crops since irrigation is the most important determinant of potential returns and because developing rain-fed agriculture is a critical objective of the watershed projects. The estimation model employs ordinary least-squares (OLS) linear regression.

Explanatory variables

This section lists explanatory variables for each model.

Determinants of project category

The factors determining a village's inclusion in a given project represent conditions prevailing in 1987, before the projects began. Altitude range (the difference between the highest and lowest points, in metres) is important, since many projects focus on areas with high potential for water harvesting. Infrastructure variables include the distance to *taluka*[7] headquarters, the population density in 1990 (which is positively correlated with most indicators of infrastructure development), percentage of area irrigated, adequacy of drinking-water availability, distance to market, distance to the nearest bus-stop and distance to the nearest public health centre. Other infrastructure variables are omitted, due to high correlation with those included. One such variable is the existence of a former COWDEP project in the village; it is omitted from the analysis because it perfectly predicts the existence of a current project, making multinomial logit estimation infeasible. Explanatory variables representing social conditions and social institutions are included, representing whether the village practised *shramdan*, the number of communal groups and the male literacy rate. Male literacy was used instead of overall literacy or female literacy, because the latter were highly correlated with some of the infrastructure variables.

Determinants of soil conservation

Variables explaining soil conservation status in 1997 include agroclimatic characteristics, project inputs, economic factors and social organization. Some of these, such as agroclimatic characteristics and caste structure, are fixed over time, while others, such as project inputs, may change over time. For the latter, 1987 values are used, since soil erosion and conservation are long-term processes; 1997 values would not be the correct explanatory variables for conservation measures that took place prior to 1997.

Agroclimatic variables include the village's altitude range (difference between the highest and lowest points, in metres), which is reflected in the transect route, mean annual rainfall (measured at the *taluka* level) and the respective shares of the transect line under irrigation, rain-fed agriculture and pasture. Farmers are well known to take the best care of irrigated land, while pastureland is often unmanaged and subject to open access (Kerr and Sanghi, 1992).

Social institutions and characteristics include a dummy variable indicating whether *shramdan* is practised in the village, a dummy variable indicating the presence or absence of a strong leader (determined subjectively by the investigators) and the number of different communal (caste and religious) groups in the village. As mentioned above, *shramdan* is an indicator of social organization and collective action, which may contribute to better land management, particularly on common lands, whereas greater communal diversity may increase the coordination costs of working together. The expected impact of a strong leader is ambiguous, dependent on the interests of the leader (Wade, 1988).

Economic factors influencing soil conservation include infrastructure, such as the presence or absence of a paved road, distance in kilometres to the nearest bus-stop, distance in kilometres to the *taluka* headquarters (where markets and other services are located), population density (inhabitants per square kilometre), and the percentage of people in the village who earn most of their income from a source other than cultivation, livestock or agricultural labour. Population density, infrastructure and access to markets can increase the pressure on natural resources, but they can also raise the returns to better land management. Off-farm income also has an ambiguous effect; it can help finance land improvement or it can lead people to focus their interests elsewhere, making them less willing to participate in social action to develop the village's natural resources. Finally, as discussed above, project inputs are represented by dummy variables for each project category. Variables are available for the percentage of each village covered by the project and the number of years of project activity, but these were highly correlated with the project dummy variables and did not vary much across project category, so they are omitted.

Determinants of returns to cultivation

Variables for the plot-level analysis of returns to cultivation include plot-, household- and village-level characteristics. Values for 1997 are used for variables that change over time, since cultivation took place in 1997. The plot characteristics include area, land capability classification (which incorporates both slope and soil fertility), the rank of the plot within the farmer's overall holding and the number of seasons the plot is cultivated each year. Irrigation is probably the single most important determinant of net returns; it is likely to dwarf the effects of all other factors. For that reason, irrigated plots are omitted from this analysis for the time being, leaving a sample of 246 plots from 29 villages. Household characteristics include the farmer's total landholding size, percentage of income from off-farm sources, number of household workers and the cost of past soil and water conservation investments. Village-level characteristics are included, as in previous analyses, but altitude range and the percentage of people in the village working off-farm are excluded.

Although this model is estimated using data for both states, a state-level dummy variable is not used because it is so highly correlated with many other explanatory variables. Government policies affecting agriculture do not differ greatly between the two states, but there may be other state-level differences that are not accounted for in this model.

Results of Econometric Analysis

Determinants of project category in Maharashtra

The multinomial logit analysis supports some of the expectations about project selection but not others. For example, with control villages as the base category, the analysis (Table 16.1) shows that all projects have a greater range in altitude between the highest and lowest point in the village compared with control villages; this difference is significant for all except the NGO–government collaborative projects. This difference is to be expected, since hilly areas are most suited for water harvesting. Project villages are also more likely than control villages to exhibit communal diversity; this difference is significant for the NWDPRA and DPAP villages. NWDPRA villages are likely to be more densely populated and NGO villages less densely populated than control villages, but this difference is not significant. NGO villages are significantly further from markets than control villages, and their literacy rate is lower. They are also more likely to practise *shramdan*, but the difference is not significant. NGO–government projects are significantly more likely to practise *shramdan* and to be located further from the nearest public health office.

Table 16.1. Village characteristics determining selection by different projects[a]: multinomial logit regressions (standard errors in parentheses).

	Project category			
Variable	NWDPRA	DPAP/Jal Sandharan	NGO[b]	NGO/government collaboration
Paved road in 1987	0.031	0.510	0.371	0.365
	(1.372)	(1.143)	(1.391)	(1.389)
Distance to *taluka* headquarters	0.014	−0.045	−0.037	0.002
	(0.063)	(0.048)	(0.046)	(0.040)
Population density in 1990 (persons km^{-2})	0.005	−0.019*	−0.009	−0.001
	(0.012)	(0.010)	(0.012)	(0.007)
Distance to regulated market in 1987 (km)	0.084	0.007	0.176*	−0.016
	(0.094)	(0.083)	(0.093)	(0.106)
Whether *shramdan* was practised in 1987	−1.232	−1.196	1.780	3.925**
	(1.33)	(1.108)	(1.313)	(1.856)
Percentage of area irrigated in 1987	1.708	−1.546	1.10	4.97
	(4.353)	(3.911)	(4.59)	(5.62)
Number of communal groups in the village	0.617**	0.550**	0.373	0.137
	(0.253)	(0.225)	(0.235)	(0.229)
Altitude range (m)	0.039**	0.029*	0.038**	0.010
	(0.016)	(0.016)	(0.017)	(0.023)
Male literacy rate in 1987	−6.695	−9.265	−10.71*	−8.02
	(7.87)	(6.4)	(6.18)	(6.05)
Whether the village had sufficient drinking-water in 1987	1.265	−1.052	0.397	0.144
	(1.397)	(1.106)	(1.155)	(1.227)
Distance to nearest bus-stop in 1987 (km)	0.225	−0.294	0.151	0.015
	(0.407)	(0.3)	(0.283)	(0.266)
Distance to nearest public health centre in 1987	0.010	0.115	0.108	0.194**
	(0.143)	(0.101)	(0.104)	(0.098)

[a]Reference category is control (no project); variables reflect values in the pre-project period; the model is not corrected for choice-based sampling (i.e. sample stratified on the dependent variable).
[b]NGO projects.
*, **, statistical significance at the 10% and 5% levels, respectively.

Analysing the data further, using the NWDPRA villages as the base instead of the control villages, provides additional insights concerning differences between projects in different categories (as opposed to differences between project villages and control villages). In this case (results not shown), the NWDPRA villages demonstrate a significantly higher population density (compared with the DPAP villages) and are significantly less likely to practise *shramdan* than the NGO or NGO–government collaborative villages. They are also likely to have more communal diversity than the villages in the NGO–government projects and they are likely to be closer to a public health office than villages in any other project. On the whole, these findings confirm that NGO-collaboration projects demonstrate more evidence of collective action than others, while NWDPRA villages are located closer to towns and have better infrastructure than those under other projects. Control villages are also favourably located.

Soil conservation at the village level

Table 16.2 presents the results of the analysis for soil conservation scores in the transects. The model 1 results show that NGO and NGO–government collaborative projects are characterized by significant positive coefficients, indicating that they contribute to good performance in soil conservation. This contrasts sharply with the insignificant coefficient for government projects under the MOA and MORD. In model 2, the projects are aggregated to three categories and the finding is the same. Other variables with significant coefficients in both models 1 and 2 include paved roads, which contribute to improved soil and water conservation status, and greater communal diversity, which has the opposite effect. By far the largest and most statistically significant coefficients are those for share of the transect line under different land uses. A larger share of irrigated land improves soil and water conservation status, while a larger share of pastureland reduces soil conservation; this finding is consistent with other evidence described above. The coefficient on the variable representing NGO and NGO–government collaborative projects also has a high positive coefficient. In comparison with the insignificant coefficient for government projects, this suggests that the NGOs' greater attention to social organization has a high pay-off in stimulating conservation investment.

The treatment-effects model (model 3) contains almost identical findings. The project dummy variable represents all projects and is not significant, but, in model 4, the project dummy variable represents only NGO projects and it is large and highly significant. All other significant variables are the same as in models 1 and 2.

In the results given in Table 16.2, the variable indicating the presence of *shramdan* in 1987 was excluded from the models 1, 2 and 4 because of the high correlation between the practice of *shramdan* and the presence of an NGO project. Combined with the significant, positive coefficient for *shramdan* in model 3, there is reason to suspect that positive coefficients for NGO and collaborative NGO–government projects may be driven more by the practice of *shramdan* than by the activities of the project. However, when models 1 and 2 were respecified to include the *shramdan* variable, the coefficients of the NGO and NGO–government collaboration dummy variables remained significantly positive, while the *shramdan* variable was insignificant. Further work is needed to determine the optimal model specification in order to account for the effects of both project categories and *shramdan*.

The negative effect of greater communal diversity in all four models is consistent with expectations; this suggests that watershed projects can more easily achieve success in more homogeneous villages (but does not provide any clues about how to proceed in more diverse villages). The positive and significant coefficient of the 'roads' variable in all four models suggests that the benefits of access to the outside world may outweigh the associated costs of increased pressure. None of the other infrastructure variables – distance to the *taluka* headquarters, distance to a bus-stop – are significant in any of the models.[8] This suggests that any effects of infrastructure on soil conservation are dwarfed by project inputs and land-use status, which is not surprising.

To summarize the findings regarding the determinants of success in soil conservation efforts, projects may benefit from greater emphasis on social organization and by focusing their work on rain-fed land rather than irrigated lands, which farmers manage well in the absence of projects. Promoting soil conservation on common lands poses particular problems which cannot be fully addressed here. One option is to attempt to strengthen common-property resource management institutions; this would require supportive changes in government policies to create conditions conducive to collective action (Jodha, 1997).

Returns to cultivation

Table 16.3 shows the findings regarding plot-level returns to cultivation.[9] In this case, the model is not corrected for selection bias. However, the fact that there was no difference in the findings between corrected and uncorrected models in the village-level analysis suggests that this is probably not a major concern.[10]

The table shows that NGO and NGO–government collaborative projects have a significantly positive contribution to both net returns and gross revenues. Projects under the MORD also have a significant positive return, and those under the MOA have a positive but insignificant coefficient. The finding that MOA projects have the least demonstrable impact on net returns per hectare is disappointing, because these projects devote the most attention to rain-fed plots. The positive contribution of NGOs is interesting, because NGOs place less emphasis on technical assistance than government projects. In fact, some NGOs have no staff with technical skills. On the other hand, many NGOs lobby government agencies to provide better services in the villages in which they work, and this may bring additional technical expertise. In addition, they help farmers work together to buy inputs or outputs at better prices, contributing to increased profit margins. The finding of higher profits is supported by further results (not presented here) that farmers in NGO villages use more fertilizer and improved varieties than farmers in other villages.

The coefficient on high-quality land is positive and significant, as expected, as is the value of soil and water conservation investments made prior to 1987. However, the coefficient of the latter is very small, suggesting a low return to such investment.[11] The value of more recent conservation investments has a statistically insignificant return; this may be because the full benefits of soil and water conservation investments are realized with a lag. This may also reflect the fact that some of the more recent investments made by the watershed projects are not always maintained by the farmer (Kerr *et al.*, 1999). The number of seasons cultivated has a positive effect on net returns, as expected. Rainfall also

Table 16.2. Determinants of soil-conservation transect score under alternate models[a]: Tobit regressions[b] (standard errors in parentheses).

Variable	Model 1 Tobit model not corrected for selection bias (five project categories)	Model 2 Tobit model not corrected for selection bias (three project categories)	Model 3 Treatment-effects model, all projects versus control	Model 4 Treatment-effects model, NGO projects compared with government projects and control
NGO project operates in the village	26.0***			
	(9.2)			
NGO/government collaborative	26.5***			
project	(8.4)			
Ministry of Agriculture project	8.0			
	(9.8)			
Ministry of Rural Development	1.9			
project	(8.2)			
NGO or NGO/government		26.02***		37.52***
collaborative project		(7.45)		(0.05)
Government project (MOA or		3.94		
MORD)		(7.56)		
Any project (government or NGO			23.23	
or NGO/government)			(20.72)	
Altitude range (m)	0.05	0.05	0.04	0.06
	(0.07)	(0.07)	(0.07)	(0.07)
Distance to bus-stop in 1987 (km)	1.8	1.88	2.41	14.66
	(1.6)	(1.56)	(1.67)	(16.57)
Paved road in 1987	14.1**	13.94**	15.84**	13.42**
	(6.5)	(6.45)	(6.81)	(6.79)
Number of communal groups	−2.1**	−2.03**	−2.48***	−1.84**
	(0.09)	(0.90)	(1.05)	(0.90)

Population density (persons km^{-2})	0.01	0.02	0.04	0.03
	(0.05)	(0.05)	(0.06)	(0.05)
% of households working primarily	–0.30	–0.31	–0.41	–0.33
off-farm	(0.27)	(0.26)	(0.27)	(0.26)
Mean annual rainfall at *taluka* town	–0.02	–0.02	–0.03	–0.03
(mm $year^{-1}$)	(0.04)	(0.04)	(0.04)	(0.04)
Share of transect line that is	36.8**	35.25***	37.89***	33.92**
irrigated	(14.0)	(13.41)	(14.03)	(13.38)
Share of transect line that is	–73.0***	–73.59***	–62.98***	–77.53***
uncultivated	(17.6)	(–17.65)	(17.99)	(17.76)
Strong leader in the village	–5.6	–05.08	–1.22	–9.56
	(7.3)	(–73.03)	(7.43)	(8.05)
Distance to *taluka* headquarters	–0.04	–0.04	–0.03	–0.13
	(0.26)	(0.26)	(0.27)	(0.27)
Shramdan (voluntary community			14.19**	
labour)[c]			(6.19)	

[a]Model is not corrected for choice-based sampling.
[b]Coefficients and standard errors are multiplied by 100 for easier reading.
[c]*Shramdan* is excluded when NGO project category is included as an explanatory variable, because they are highly correlated. Of villages covered by these projects, 85% practise *shramdan*.
*, **, *** denote statistical significance at the 10%, 5% and 1% level, respectively.

Table 16.3. Determinants of net returns from the plot-level survey.[a]

Variable	OLS linear regression coefficients (standard errors)
NGO project operates in the village	3030*** (1107)
NGO/government collaborative project	2393* (1402)
Ministry of Agriculture project	1041 (1032)
Ministry of Rural Development project	2124* (1169)
Distance to bus-stop in 1997 (km)	103 (231)
Paved road in 1997	−1575* (832)
Distance to *taluka* headquarters (km)	−36 (39)
Mean annual rainfall at *taluka* HQ (mm $year^{-1}$)	8*** (3)
Farmer's total landholding (ha)	11 (24)
% of farmer's income that comes from off-farm work	−7 (13)
Number of workers in farm household	−7 (92)
Farmer is high-caste	11 (749)
Number of years schooling of best-educated household member	45 (64)
Number of seasons per year the plot is cultivated	3139*** (751)
Plot ranks highly within the farmer's holding	742 (640)
Plot is of land-capability classification 2	3080*** (1139)
Area of the plot (ha)	−131 (213)
Value of land-improvement investments made before 1987 (Rs ha^{-1})	0.028* (0.015)
Value of land-improvement investments made after 1987 (Rs ha^{-1})	−0.0001 (0.03)

[a]OLS model is not corrected for selection bias or choice-based sampling.
*, **, ***, statistical significance at the 10%, 5% and 1% level, respectively.

has a positive coefficient. This makes sense in the dry areas covered by the study, but it may also capture other regional variations. One unexpected finding concerns the significant, negative coefficient of the variable indicating that the village lies on a paved road. Roads should improve access to markets, thus reducing the costs of buying inputs and selling crops; thus this result is difficult to explain.

In summary, the analysis of net returns to cultivation, an indicator of agricultural productivity, shows again the importance of social organization as a component of watershed projects. Helping farmers gain better access to information, markets and government services may benefit them more than transferring specific technologies.

Equity Impact of Watershed Projects: Findings from Qualitative Data

As mentioned above, qualitative data focused largely on project impacts on specific interest groups in each village. The main interest here is in landless people, who are the poorest people in rural areas and who depend disproportionately on the uncultivated lands that watershed projects try to protect.

Landless people indicated overwhelmingly that they had benefited from labour employment provided directly by the project. In fact, employment was the most commonly cited project benefit among all respondents. However, landless people rarely reported any other benefits, and they expressed concern that employment opportunities would cease when projects ended. In some villages, respondents said they thought that employment had risen permanently, due to an increase in irrigated area stimulated by the project, but this was not common.[12]

Livestock herders in many villages complained that they had suffered from loss of access to their traditional grazing lands, which were sealed off under the projects to promote regeneration. All of these projects had provided employment opportunities to the herders, but many of them said it was not enough to compensate for their losses. This problem commonly arose in Maharashtra, where landless, low-caste people are a small minority in most villages and the decision to close the common lands was based on a majority-rule vote. In some villages, herders said that they had been promised that access restrictions would be temporary while vegetation was allowed to regenerate. However, they complained that regeneration had already taken place, but the common lands remained off limits to them. Ironically, such inequities are more likely to be a problem where projects succeed in addressing productivity and environmental objectives. In other places, herders were able to ignore grazing restrictions, protecting their immediate livelihoods but undermining project objectives.

A few NGOs, particularly in Andhra Pradesh, have worked to overcome this kind of problem by trying to build the interests of different groups into the project design at the outset. For example, in some projects, landless people are granted fishing rights to the bodies of water protected by soil conservation and revegetation. Projects may encourage farmers without irrigation to dig group-owned wells, so that they have an interest in promoting groundwater recharge. Outside the study area, in the well-known Sukhomajri and Pani Panchayat projects, landless people even own irrigation water rights, which they can utilize by leasing in farmland or, in the case of Sukhomajri, selling to other farmers. And, in several Andhra Pradesh villages not covered by any kind of project, shepherds lease cultivated land and manage it as pasture. Such an arrangement could be made within a watershed project, as it would give the shepherds an incentive to manage those lands more productively. A wide assortment of such arrangements can be devised to spread the benefits of watershed development and, as a consequence, increase its chances of success.

Discussion and Conclusions

The findings of this analysis show that many of the participatory watershed projects managed by NGOs, or in collaboration between NGOs and governments, have made significant contributions to agricultural productivity, natural resource conservation and poverty alleviation in the study areas. More technocratic, top-down government projects, on the other hand, have fared less well. The key factor in projects with NGO involvement appears to be their attention to issues of social organization, not just technology transfer.

Social organization matters because a watershed's environmental processes often conflict with human-defined boundaries, so that human intervention in the watershed ecosystem may create externalities and divide costs and benefits unevenly among its inhabitants. Mobilizing local people to work collectively, resolve conflicts and create institutional mechanisms to give them common economic interests, can help overcome some of the social complications of working at the watershed level and can help realize a synergy among the objectives of natural resource conservation, agricultural productivity and poverty alleviation. When projects transfer technology without regard to social organization, conflicts among users limit the scope for coordinated watershed management and its associated benefits. In particular, uneven distribution of benefits and costs typically leads to one of two outcomes. The first of these is conservation and productivity with inequity, which occurs if impoverished landless people lose access to the common lands so important to their livelihoods. The alternative is a failure to achieve any benefits whatsoever, which happens if landless people are able to undermine efforts to protect the commons, thus causing loss of vegetation and soil erosion, which in turn reduce water infiltration and its associated downstream benefits.

A second reason for the important role of social organization in watershed projects concerns the strong role of markets and government services in the rural economy. Projects that organize people to access existing markets and services can help them obtain higher prices and better support from government offices.

Additional findings not presented in this chapter support the overall conclusion that projects perform better when they focus on social organization (Kerr *et al.*, 1998b). For example, analysis related to indicators such as productivity of common lands, changes in irrigated area and changes in access to employment all point to the benefits of promoting social organization. In fact, for many performance indicators, the technocratic government projects did not perform any better than control villages with no project. Overall, the multiple sources of information available in this analysis offer broadly consistent findings; all support the conclusion that the more participatory projects have had greater impact.

The most successful projects have several common features. They are truly participatory, in that project staff consider local people as equal partners who help design and implement the work, help pay for it and help manage the funds. Project funds flow only after villagers prove that they can work collectively; otherwise social organization may be superficial and will not be sustained after project funds and staff are withdrawn. Field staff are encouraged to take the initiative and participate with farmers, and are given more authority to make project decisions. Since successful watershed management depends on organizing communities to work together, in the most successful projects, the village is the primary project unit rather than the watershed, which would be the logical unit in a purely technical programme. In addition, some of these programmes work only in villages with social conditions conducive to collective action for successful watershed development. Screening villages in this way reflects good financial management in a country with 0.5 million villages and a finite watershed budget.

Collaborative projects between NGOs and government agencies have performed particularly well, and this appears to bode well for efforts to expand participatory approaches to a larger scale. However, it is important to acknowledge that the NGO–government collaborative projects analysed in this study have benefited from favourable treatment, which cannot be extended on a large scale. For example, all of the study villages had been the site of previous watershed projects (as had almost all other projects in Maharashtra) and, in most of them, an experienced NGO had already been active in the village for several years. Moreover, as these were high-profile projects, subject to frequent visits from high-ranking government officials, project staff worked particularly hard and development funds were allocated on a priority basis. Such special treatment will not be possible as these projects continue to expand, so it is premature to draw conclusions about the potential for scaling up based on the findings presented here. However, these comments are not meant to detract from the good performance of these projects; resources should be allocated to experiment further with NGO–government collaborative projects and other efforts to introduce more participatory approaches to government-funded projects.

Generating widespread benefits from watershed projects will require that large government projects do, in fact, scale up the successes achieved so far in small projects. Attaining this goal will require devoting greater attention to poverty-alleviation objectives, in addition to those of raising agricultural productivity and protecting the environment. The poverty-alleviation effects of watershed development depend on the impact on landless people, who stand to gain wage employment in the short term and perhaps the long term, but who lose access to common lands for several years in the medium term. In villages where the commons are important to poor people's livelihoods, they may undermine project objectives in order to protect their interests, thus reducing or even eliminating productivity and environmental gains. In this case, equity is a prerequisite to productivity and sustainability. The current approach to solving this problem is to 'buy' poor people's cooperation through employment benefits under the project. A superior alternative would be to find ways to give landless people a direct self-interest in project success. As suggested above, this could be achieved by granting landless people rights to water and other resources generated by the project. Sharing project benefits would both raise the benefits to poor people and increase the likelihood of success by minimizing conflicts in different groups' objectives.

Acknowledgements

The authors express their thanks to Peter Hazell, Derek Byerlee, John Pender, Anna Hazare, G.B. Singh, Dayanatha Jha and numerous investigators and respondents who contributed to this study.

Notes

1 'Social fencing' means that social institutions rather than physical barriers are used to protect the common lands.

2 In 1995, the DPAP guidelines were restructured under radical new, participatory guidelines, but only pre-reform DPAP projects are included in this research.

3 The treatment-effects method is similar to the two-step Heckman correction, but for uncensored data, in which the dependent variable is observed for all cases. The first step is a probit model, which predicts the probability that a village will fall into one project category or the other. It also calculates an adjustment factor, which is included as an explanatory variable in the second-stage equation and is an ordinary least-squares (OLS) regression to predict the outcome variable. The treatment-effects model uses OLS instead of Tobit for the analysis of soil-conservation status. Although Tobit is a better specification than the OLS, there was virtually no difference between the two in the model that does not correct for selection bias.

4 A subsequent version of this analysis incorporates methods to control for selection bias when the dependent variable is categorical (Kerr *et al.*, 2001).

5 The model is not corrected for choice-based sampling, i.e. the fact that the sample is stratified on the dependent variable.

6 The straight-line design of the field transect oversamples plots close to the centre of the village relative to those at the periphery, which are more likely to be hilly and degraded.

7 A *taluka* is a subdistrict administrative unit. In the study area, a *taluka* contains 80–150 villages.

8 The data include many infrastructure variables such as distance to a regulated market, distance to an industrial unit, percentage of houses in the village that are electrified, etc. These were tested in other specifications of the model but were never significant.

9 The model was also run with gross returns instead of net returns as the dependent variable, and the results were roughly the same. Net returns are the more relevant number, but they may suffer from inaccuracies because imputed values of family resources may be overestimated.

10 Kerr *et al.* (2001) adapt this model to correct for selection bias. Results suggest that the effects of this correction are small.

11 Both the dependent variable and the explanatory variable for soil conservation investment are measured in terms of Rs ha^{-1}. The coefficient is significant only at the 10% level.

12 Findings not presented here showed no difference in increased irrigated area between control villages and those under any of the projects.

17 Soil Fertility, Small-farm Intensification and the Environment in Africa

PEDRO A. SANCHEZ, BASHIR JAMA, AMADOU I. NIANG AND CHERYL A. PALM

International Centre for Research in Agroforestry (ICRAF), Nairobi, Kenya

Introduction

The continued threat to the world's land resources is exacerbated by the protracted poverty and food-security crisis in sub-Saharan Africa (henceforth, Africa). Per capita food production continues to decrease, in contrast to increases in other parts of the developing world. Africa also ranks first in human population growth rates (2.9% year^{-1}), in poverty – with half its population living on per capita incomes of less than US$1 day^{-1}, in the number of malnourished children and in the proportion of arable land that is degraded (Cleaver and Schreiber, 1994; Badiane and Delgado, 1995; World Bank, 1995c). The vast majority of the poor in Africa live in rural areas (World Bank, 1990).

It is widely recognized that sustained policies that favour smallholder rural development can help address widespread hunger, malnutrition, poverty and environmental deterioration. Agriculture – broadly including food crops, tree crops, livestock, forestry and fisheries – is very much the engine of economic growth in Africa (Brown and Haddad, 1994). Agriculture contributes the most towards meeting the countries' basic needs for human survival and employment, and also catalyses the development of agroindustry. Mining, tourism and other economic sectors contribute significantly to growth but are not the engine that agriculture is, because such sectors do not address the basic needs of the majority of the population, nor do they provide sufficient employment opportunities for the expanding rural labour force. African countries that are relatively better off, such as South Africa and Zimbabwe, have made agricultural development a high priority, in spite of their abundant mineral resources and a strong tourism sector.

The main reasons for Africa's position at the bottom of the development scale relate to policy and structural factors (Badiane and Delgado, 1995), rather than an inferior biophysical endowment. In fact, Africa compares favourably with other tropical regions in terms of natural resources, particularly climate and soils (Sanchez and Leakey, 1997; Smithson and Sanchez, 1999), despite pronouncements to the contrary (Sachs, 1997). However, Africa lags behind other developing regions in governance and in basic infrastructure.

Yet good governance, macroeconomic policies and infrastructure development are necessary but not sufficient conditions for poverty alleviation, food security and improved natural resource management (Reardon and Vosti, 1997a). To achieve these goals fully, efforts directed towards the smallholder farming sector need to focus on two key areas: (i) reversal of soil fertility depletion as the fundamental biophysical constraint to food security; and (ii) intensification and diversification of smallholder farming by producing high-value products as the key to poverty reduction in rural areas (Sanchez and Leakey, 1997). The first of these goals is now beginning to receive attention, but the second – the subject of this book – is yet to come to the forefront in the development community. Since replenishing soil fertility is a prerequisite for farm intensification, we deal with both in this chaper, with an emphasis on experiences in eastern and southern Africa.

Soil Fertility Depletion: the Biophysical Root Cause

Soil fertility depletion in smallholder farms is increasingly being recognized as the fundamental biophysical root cause responsible for declining per capita food production in Africa (World Bank, 1995c,d, 1996c; IFPRI, 1996; Sanchez *et al.*, 1996, 1997a,b; Pieri, 1998). By fundamental root cause, we mean that no matter how effectively other constraints are remedied, per capita food production in Africa will continue to decrease unless soil fertility depletion is effectively addressed.

During the 1960s, the fundamental root cause of declining per capita food production in Asia was the lack of short-statured, high-yielding varieties of rice and wheat. Asian food security was only effectively addressed with the advent of improved germ plasm. Then other key aspects that had previously been largely ineffective (enabling government policies, irrigation, seed production, fertilizer use, pest management, research and extension services) came into play in support of the spread of the new varieties. The need for soil fertility replenishment in Africa now is analogous to the need for 'Green Revolution-type' germ plasm in Asia three decades ago, a belief that is supported by two of the 'fathers' of the Green Revolution, Norman Borlaug and M.S. Swaminathan (Borlaug and Dowswell, 1994; M.S. Swaminathan, personal communication, 1998, Nairobi, Kenya).

Socioeconomic externalities

The negative effects of soil nutrient depletion extend beyond the farm to community, regional and national levels. Especially during drought years, food shortages and famines are exacerbated. Soil nutrient depletion lowers the returns to agricultural investment, which reduces incomes at the community level (Delgado *et al.*, 1994). Several studies from both Africa and Asia indicate that reduced community income and employment effects are substantial because of the large multiplier effects of agricultural growth (Hazell and Hojjati, 1995). Lower overall production also results in higher food prices, increased government expenditures on health, more famine relief and reduced government revenue due to lower tax collection on agricultural goods. Poor soil fertility has also rendered much of the agricultural research on improved crop germ plasm ineffective; new varieties of hybrids, regardless of their positive traits, simply will not yield well in nutrient-depleted soils.

Perhaps the most insidious negative social externality of soil fertility depletion is its link to lower employment and increased poverty. As long as agricultural returns are limited by nutrient depletion, farm employment and spillover non-farm employment opportunities will remain low, sustaining severe poverty. The Kenyan Minister of Agriculture, Musaila Mudavadi, after discussing these aspects, recently noted, 'Now I understand why my people [in western Kenya] are getting poorer' (personal communication, 1999, Maseno, Kenya).

But these externalities are not confined to rural communities, as poverty often pushes individuals and households into urban areas. The influx of rural migrants puts a greater strain on the limited urban infrastructure; unemployment, crime and political unrest sometimes follow (Homer-Dixon *et al.*, 1993).

Environmental externalities

Soil fertility depletion also exacerbates a host of environmental problems. These include increased sediment loading of water bodies, increased greenhouse-gas emissions and decreases in biodiversity and watershed stability. Increased soil erosion, particularly in steep areas, causes sedimentation and siltation of reservoirs and of coastal areas and, in some cases, leads to the eutrophication of rivers and lakes and increased loading of nutrients into coastal areas. There is evidence of siltation and eutrophication occurring in some African rivers and lakes (Melack and MacIntyre, 1992), including Lake Victoria, where erosion from surrounding lands provides nutrients that exacerbate the current water-hyacinth epidemic.

The loss of topsoil organic carbon (C) associated with soil nutrient depletion results in additional carbon dioxide (CO_2) emissions to the atmosphere from decreasing soil and plant C stocks. We have estimated that 27 Gt (10^{15} g) of C are emitted annually to the atmosphere as CO_2 from cultivated land in Africa, largely due to the depletion of soil organic C (Sanchez *et al.*, 1997b). This amounts to about 21% of the C annually emitted from land degradation at the global scale (Lal *et al.*, 1995). Carbon loss, however, is a reversible process in most soils.

Soil fertility depletion is likely to decrease above- and below-ground plant and animal biodiversity, due to increased encroachment on forests and woodlands in response to the search for new fertile land (Sanchez, 1995). There are no quantifiable estimates available, but biodiversity losses appear to be particularly relevant in the *miombo* woodlands of southern Africa, in the rainforest remnants in the Great Lakes region and on the eastern coast of Madagascar. The role of soil fertility depletion in below-ground biodiversity – responsible for basic ecosystem functions – is largely unknown.

Lastly, deforestation resulting from the search for the few remaining pockets of fertile land in densely populated areas often results in an almost total removal of trees from the landscape. This is seen, for example, in parts of Ethiopia. Lack of tree protection in the upper parts of watersheds severely affects hydrological functioning.

The following paragraphs describe the central concepts and processes of soil nutrient depletion and the major options available for overcoming this constraint.

Key concepts: nutrient capital stocks, flows, inputs and cycling

There is an exact congruence between the concepts of capital stocks and service flows in economics and those of nutrient pools and fluxes in soil science (Izac, 1997). Nutrient capital stocks can be defined as the amounts of nitrogen (N), phosphorus (P) and other essential elements in the soil that become available to plants over a time-scale of 5–10 years (Sanchez and Palm, 1996). Nitrogen capital stocks consist of the more dynamic pools of soil organic matter, while P capital stocks include such soil organic-matter pools and the 'fixed' P at the surface of clay particles. Nutrient capital may be expressed as kilograms per hectare of N or P within the rooting depth of plants (Sanchez *et al.*, 1997b).

The transfer of nutrients from the capital stocks to the soil solution, where they are taken up by plants, includes the flows illustrated as arrows in Figs 17.1 and 17.2 at the field scale of smallholder farms in Africa. There are also service flows related to soil nutrient capital beyond the field scale. Examples are on-farm crop, fodder and fuelwood production (valued by individual farmers), food security and soil conservation (valued by national societies and farmers), poverty alleviation, enhanced carbon sequestration and biodiversity conservation (valued by the global society and future generations).

It is important to distinguish between nutrient inputs and nutrient cycling (Sanchez and Palm, 1996). At the field scale, nutrient inputs are additions from outside the field, such as N fixed from the air by legumes or the addition of mineral fertilizers. Cattle manure is an input if the manure is produced from forage grown outside the field.

Nutrient cycling, on the other hand, refers to the transfer of nutrients already in the field from one compartment to another. Examples include the return of maize stover back to the soil, cattle manure and urine deposited by animals while grazing crop residues, and the transfer of nutrients from trees to the soil through prunings, leaf drop or root decomposition in agroforestry systems. Enhancing nutrient cycling is extremely important, but nutrient-depleted soils need inputs from outside the field.

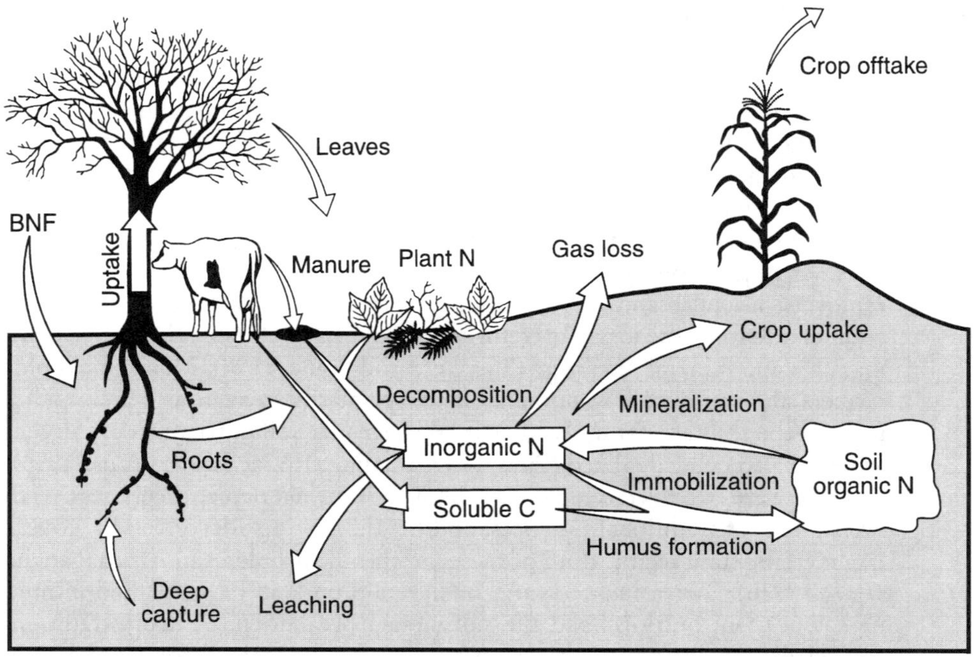

Fig. 17.1. Main features of the nitrogen cycle using agroforestry in smallholder farms of Africa. Soil organic N is the capital stock, while arrows illustrate the inputs, outputs and flows. BNF, biological nutrient fixation. (From Sanchez *et al.*, 1997b.)

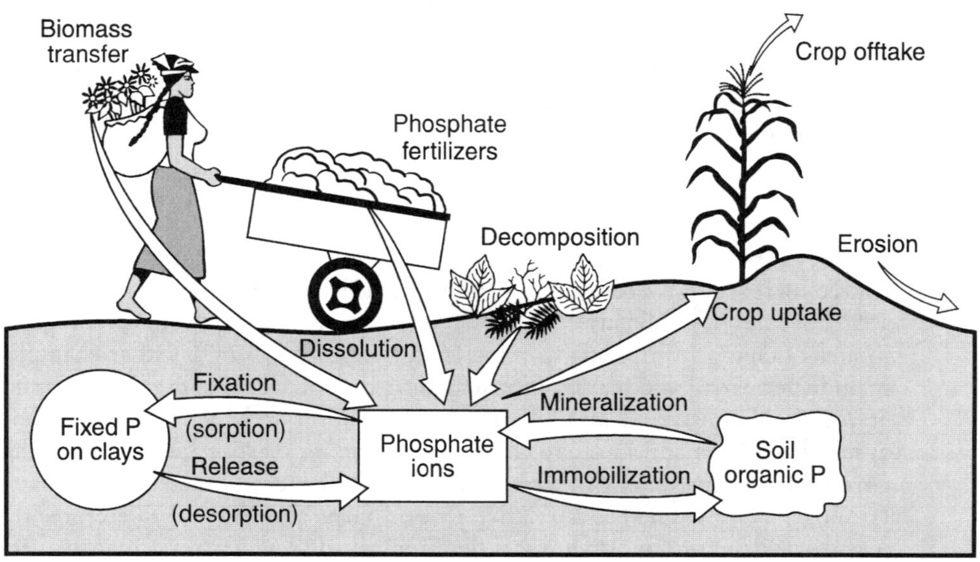

Fig. 17.2. Main features of the phosphorus cycle using agroforestry and phosphate fertilizers in smallholder farms of Africa. Soil organic P and fixed P on clays are the capital stocks, while arrows illustrate the inputs, outputs and flows. (From Sanchez *et al.*, 1997b.)

The nutrient depletion process: causes, magnitude and consequences

Everywhere in the world, people have first settled in high-potential areas with fertile soils, adequate rainfall and mild temperatures (Sanchez and Buol, 1975). The Lake Victoria basin in East Africa is one example. This area now supports one of the densest rural populations in the world – 500–1200 inhabitants km^{-2} (Hoekstra and Corbett, 1995).

Such settlements were originally supported by the high fertility of the soils. As populations grew, nutrients were gradually depleted by crop harvest removals, leaching and soil erosion, as farmers were unable to sufficiently compensate for these losses by returning nutrients to the soil via crop residues, manures and mineral fertilizers (Shepherd and Soule, 1998). Now, smallholder farmers also cultivate low-potential lands, primarily in semi-arid areas, where most of the soils are naturally infertile. The smaller soil nutrient stocks in these areas are also being depleted (Pieri, 1989; Smaling, 1993a; Sanders *et al.*, 1996).

There are two overarching reasons for the nutrient depletion process: (i) the breakdown of traditional practices; and (ii) the low priority given by governments to the rural sector. Both of these are chronic problems in Africa (Sanchez *et al.*, 1997b). Increasing pressures on agricultural land have resulted in much higher nutrient outflows and the subsequent breakdown of many traditional soil fertility maintenance strategies, such as fallowing land for several years, mixed crop–livestock farming and opening new lands. Such strategies have not been replaced by an effective fertilizer supply and distribution system (Sanders *et al.*, 1996). Due in large part to policy and institutional constraints, African agriculture did not intensify as quickly as in other parts of the world following increased population pressure (Cleaver and Schreiber, 1994). Nevertheless, African farmers repeatedly outperform the weather; current crop yields over time fluctuate considerably less than indices of rainfall (Dommen, 1988). Continued population pressure has, however, reduced farm sizes in many parts of Africa to the point where farmers can only provide an adequate living for their families if the land is farmed very intensively and if there is off-farm income.

In terms of the second problem – the low priority rural areas have been given by governments – this is reflected in many sectoral pricing and marketing policies. First, most African governments have used agriculture as a principal source of revenue by restricting producer prices and taxing exports (Cleaver, 1993). Secondly, public investments have been focused away from rural regions in areas ranging from roads to research (Pardey *et al.*, 1995). Other examples include poor road and market infrastructure, lack of access to credit and inputs at reasonable cost, and lack of timely information and ineffective extension systems (Badiane and Delgado, 1995; Tomich *et al.*, 1995a). The result of this chronic underinvestment in agriculture and rural areas is that the returns to farming have remained low relative to non-farm alternatives. Therefore, the incentive structure is such that few improved soil fertility management technologies have been widely accepted by African smallholder farmers (Conway and Barbier, 1990). Soil fertility depletion, therefore, is largely a consequence of socioeconomic constraints and policy distortions.

Nutrient depletion rates are field-specific; they depend on the way each particular field has been managed over decades. This process results in a mosaic of degrees of nutrient mining in the broader landscape. At the national scale, there are areas that have not suffered much from nutrient depletion, whether because of the low intensity of use or from the use of fertilizers for export crops. Still, nutrient-depleted smallholder farms in Africa are much more common than those where this constraint is not a major problem.

Nutrient balance studies initiated by Smaling (1993a) show that enormous losses of N, P and potassium (K) have taken place in about 100 Mha of cultivated land in Africa during the last 30 years (Table 17.1). In contrast, commercial farms in North America and Europe have averaged larger but positive nutrient balances during the last 30 years on over four times the amount of cultivated land of Africa. (This has, however, often resulted in groundwater and stream pollution.) Nutrient mining in Africa, therefore, contrasts sharply with nutrient build-ups in temperate regions.

Africa is now suffering estimated net losses of 4.4 Mt of N, 0.5 Mt of P and 3 Mt of K every year from its cultivated land. These rates are several times higher than Africa's annual fertilizer consumption, which are included as inputs in these calculations (Sanchez *et al.*, 1997b). Unfortunately, there are no comparable estimates of nutrient balances at the continental scale in Latin America and Asia, but since fertilizer use in these regions is an order of magnitude higher than in Africa, the problem is probably much less widespread.

Because the soil resource has not kept its productive capability over time, African farmers have witnessed low and declining yields. For example, a long-term trial in Kabete, Kenya, indicates that a fertile, red soil lost about 1 t ha^{-1} of soil organic N and 100 kg ha^{-1} of soil organic P in 18 years of continuous maize–bean rotation in the absence of nutrient inputs. Maize grain yields decreased from 3 to 1 t ha^{-1} during this period (Qureshi, 1987, 1991; Bekunda *et al.*, 1997).

In addition to continuing decreases in crop productivity, a marked decline in food security results from soil fertility depletion in Africa. In the densely populated districts of western Kenya, it is common for villages to experience from 3- to 6-month-long periods of hunger, after their maize supplies are exhausted and before the next crop is harvested, unless they have off-farm income to buy maize (Niang *et al.*, 1996). This is also quite common in Malawi and Zambia (Kwesiga *et al.*, 1998; Mann, 1998).

Nutrient depletion also produces negative side-effects on and off the farm. On-farm effects include less fodder for cattle, less fuelwood for cooking and less crop residue and cattle manure for recycling nutrients. The result is often

Table 17.1. Net nutrient balances in cultivated lands of sub-Saharan Africa and Europe and North America (from Sanchez *et al.*, 1996).

Region	Nitrogen (kg N ha^{-1} in 30 years)	Phosphorus (kg P ha^{-1} in 30 years)	Potassium (kg K ha^{-1} in 30 years)
Africa	−700	−75	−450
Europe and North America	+2000	+700	+1000

increased runoff and erosion losses, because there is less plant cover to protect the soils. A build-up of the parasitic weed striga (*Striga hermonthica*) takes place in N-depleted soils of East Africa, severely decreasing yields of maize and sorghum.

Responses to soil fertility depletion

Traditional approaches to soil fertility management range from recurring fertilizer applications to low external-input agriculture based on organic sources of nutrients. Although both extremes work well in specific circumstances, they pose major limitations for most smallholder farmers in Africa.

Recurring fertilizer applications

Fertilizer use is the obvious way to overcome soil fertility depletion. Most smallholder farmers in Africa appreciate the value of fertilizers, but they are seldom able to apply them at the recommended rates and at the appropriate time, because of high costs, lack of credit, delivery delays and low and variable returns (Heisey and Mwangi, 1997). These constraints are largely due to the deficient road and market infrastructure typical of most African countries, which has impeded the development of efficient markets (Donovan, 1996). As a result, the price of fertilizers in rural areas of Africa is usually at least twice the international price (Bumb and Baanante, 1996).

In the past, many African counties subsidized fertilizers. However, the removal of fertilizer subsidies by most African governments as part of the structural adjustment programmes in the last decade has often tripled or quadrupled fertilizer prices in relation to crop prices (Holden and Shanmugarathan, 1994; Bumb and Baanante, 1996). Furthermore, since fertilizer recommendations are normally formulated to cover broad areas with diverse soils, farmers also lack information about the best fertilizer to use for their particular fields and cropping practices, making the crop response to fertilizers more erratic and less profitable.

These policy and information constraints can certainly be overcome, thereby resulting in increased food security and reducing poverty. An example of a promising fertilizer-based approach is the Sasakawa Global 2000 project in Ethiopia, where many policy distortions have been overcome by government intervention at the highest level (Quiñones *et al.*, 1997). However, if recurring fertilizer applications are aimed at only replacing the nutrients that are removed by crops, the lost soil nutrient capital will not be replenished. Recurring P fertilizer applications at rates above replacement (and accounting for P fixation), coupled with effective sources of C inputs, are needed for soil fertility replenishment.

Organic farming

The exclusive use of organic inputs as nutrient sources has been advocated as a logical alternative to expensive mineral fertilizers in Africa (Reijntjes *et al.*, 1992). The main advantages of this approach are the replacement of scarce or non-existent capital for labour and the fact that cattle manures or green manures contain all essential nutrients plus C, the source of energy for the soil biota, which regulates nutrient cycling.

One of the main arguments against the use of organic inputs is their low nutrient concentration. Animal manures and plant materials contain from 1% to 4% N (on a dry-weight basis), while mineral fertilizers contain from 20% to 46% N and are already dry. To provide the 100 kg N generally needed to produce a 4 t ha^{-1} maize grain crop, it would be necessary to transport 217 kg of urea or 20,000 kg of leaf biomass or manure (with 80% moisture and 2.5% N concentration). Organic inputs are also poor suppliers of P because of their low concentrations (Palm, 1995; Palm *et al.*, 1997). Considering the transport costs of the high amounts of nutrients needed just for replacing crop nutrient requirements, the pure organic alternative is seldom an option for fertility replenishment.

It takes soil fertility to grow organic inputs, whether it is livestock manure, litter fall or plant biomass for transfer, composts or green manures. In nutrient-depleted soils, it is difficult to grow enough forage to feed cattle to produce significant quantities of manure with adequate nutrient content. This is why African farms severely depleted of nutrients have few or no cattle. The nutrient content of cattle manure varies widely with soils and fodder availability in Africa, reflecting the nutrient status of the soils; it is often extremely low in N and P (Murwira *et al.*, 1995; Probert *et al.*, 1995). African farmers do not need low-input, low-output systems, which perpetuate food insecurity and poverty.

Soil Fertility Replenishment

Given the limitations of the conventional pure organic and pure fertilizer approaches to increasing soil fertility, it is time for a more robust approach that provides fresh alternatives and increases the options available to farmers and policy-makers (Buresh *et al.*, 1997a). We focus here on ways to replenish the two main limiting nutrients, N and P. The strategies are quite different. Phosphorus replenishment strategies are mainly mineral fertilizer-based with biological supplementation, while N replenishment strategies are mainly biological with mineral fertilizer supplementation (Sanchez *et al.*, 1997b). Our experiences in eastern and southern Africa suggest that the following basic concepts should be considered in new approaches.

Combine organic and mineral sources of nutrients

The plant does not care whether the N and P ions it absorbs from the soil come from organic or mineral sources of nutrients. Organic inputs, however, also provide C, the energy source for microorganisms, which enhances nutrient cycling and forms new soil organic matter. Mineral fertilizers do not add C. Nutrient combinations may have additive effects from the two types of inputs (Palm *et al.*, 1997) and provide tradeoffs between capital and labour costs.

Capture most of the nitrogen from the air by biological nitrogen fixation in the same fields where crops are to be grown, largely through short-term leguminous fallows (Niang *et al.*, 1996, 1997; Giller *et al.*, 1997; Kwesiga *et al.*, 1998; Rao *et al.*, 1998; Maroko *et al.*, 1999)

Such fallows are grown during the dry season, when no crops are in the field, or during extremely unreliable short rainy seasons near the equator. Leguminous fallows can capture from the air 100–200 kg N ha^{-1}, which is then transferred to the soil by litter fall and made available to crops when the leaves decompose. These amounts are respectable N inputs by developed-country farmer standards. The responses to such practices in Zambia are summarized in Table 17.2. A 2-year *Sesbania sesban* fallow produced sufficient N for three subsequent maize crops, doubling maize yields over a 5-year period, in spite of missing two crops while the fallow was growing. It also tripled the net returns to labour and increased the net present value of returns to land from US$4 to US$150 ha^{-1}, at a discount rate of 20% $year^{-1}$ (Rao *et al.*, 1998).

Mineral N fertilizer, at the recommended rate of 112 kg N ha^{-1} $year^{-1}$, was even more profitable (Rao *et al.*, 1998), but its use would require new credit schemes or subsidies (Mann, 1998). Our view is that mineral N fertilizer should be used strategically, as top-dressing applications when needed, while relying on biological N fixation and subsoil nitrate capture by the tree fallows as the principal sources of N (Sanchez *et al.*, 1997b).

Table 17.2. Maize yields following 1- and 2-year *Sesbania sesban* fallows in eastern Zambia (from ICRAF, 1996).

	Maize yield (t ha^{-1})					
Treatment	1990/91	1991/92	1992/93	1993/94	1994/95	Cumulative total yield
Continuous maize production (without fertilizer)	2.72	1.07	1.86	1.03	0.94	7.62
2-year fallow	F	F	4.67	2.84	2.28	9.79
1-year fallow	2.72	F	2.89	1.44	1.15	8.20
2-year fallow	2.72	F	F	3.36	2.30	8.38
1-year fallow	2.72	1.07	F	1.87	1.16	6.82
2-year fallow	2.72	1.07	F	F	3.06	6.85
1-year fallow	2.72	1.07	1.86	F	1.75	7.40
Continuous maize production (with fertilizer)	7.15	1.95	7.68	3.83	3.75	24.36
SED	0.615	0.239	0.293	0.446	0.289	

F, period under fallow (1 or 2 years) in phased-entry design (fallow periods phased in over time to compensate for lower than average yields, usually due to drought); SED, standard error of the difference in means of maize yields.

Recapitalize the soil's phosphorus stocks, where needed, by investment applications of mineral phosphorus fertilizers (Buresh *et al.*, 1997b)

Cheaper, indigenous forms of P, such as locally available rock phosphates of medium to high reactivity, can be used where available (Sanchez *et al.*, 1997b). The residual effects of mineral P applications will last for at least 5 years. A variety of options exist, including a one-time large application or a gradual build-up of P capital through annual applications. Logistics costs, credit schemes and soil properties will be the main factors involved in choosing among these options.

Supplement mineral phosphorus with organic inputs, particularly those that enhance phosphorus cycling as they decompose in the soil

A good example of this is the biomass transfer of *Tithonia diversifolia*, a common hedge and roadside shrub in eastern and southern Africa. *Tithonia* has high concentrations of N, P and K in its biomass, decomposes very rapidly in the soil, and produces sharp increases in grain yields (Gachengo, 1996; ICRAF, 1996; Niang *et al.*, 1997; Nyasimi *et al.*, 1997; Palm *et al.*, 1997; Rao *et al.*, 1998). The processes are not thoroughly understood, but seem to be related to the straight nutrient effects, particularly N, P, and K fertilization (ICRAF, 1998), a slight reduction in the soils' P fixation (Nziguheba *et al.*, 1998) and higher microbial C, N and P (Nziguheba *et al.*, 1998). Given the large additions of soluble C and nutrients to the soil when *Tithonia* leaves decompose, we speculate that these processes may enhance nutrient cycling and therefore the conversion of mineral forms of P into organic ones.

The combined options of two sources of P fertilizers, with and without *Tithonia* biomass transfers, are shown in Fig. 17.3 (B. Jama, unpublished data). The data represent the means of two rainy-season maize crops from three on-farm experiments in western Kenya. A total of 1.8 t ha^{-1} of dry *Tithonia* biomass was transferred from nearby hedges, which provided 60 kg N ha^{-1}, 6 kg P ha^{-1} and 60 kg K ha^{-1}. The equivalent amount of N was applied as urea, so *Tithonia* is also compared with mineral fertilizers as far as N is concerned. The first aspect to be noted in Fig. 17.3 is the doubling of maize yields with *Tithonia* alone, compared with urea, with no P fertilizer additions. Doubling or tripling of maize yields with *Tithonia* biomass transfers of about 2–5 t ha^{-1} of dry matter without mineral fertilizers has been recorded in many other on-farm trials conducted by non-governmental organizations (NGOs) and farmers in the region (Nyasimi *et al.*, 1997). A large part of this effect may be due to the addition of P from *Tithonia* in soils marginal in K.

Phosphate additions were done either gradually (50 kg P ha^{-1} $year^{-1}$) or at a one-time investment rate of 250 kg P ha^{-1} applied to the first crop, which we call the recapitalization rate. At the end of 2 years, the recapitalization rates produced 30% more maize than the incremental approach, which at this time added only 100 kg P ha^{-1}. The equivalent comparison will be tested in the fifth year where both strategies will total 250 kg P ha^{-1}. The selected recapitalization rate was a researcher's 'best bet'.

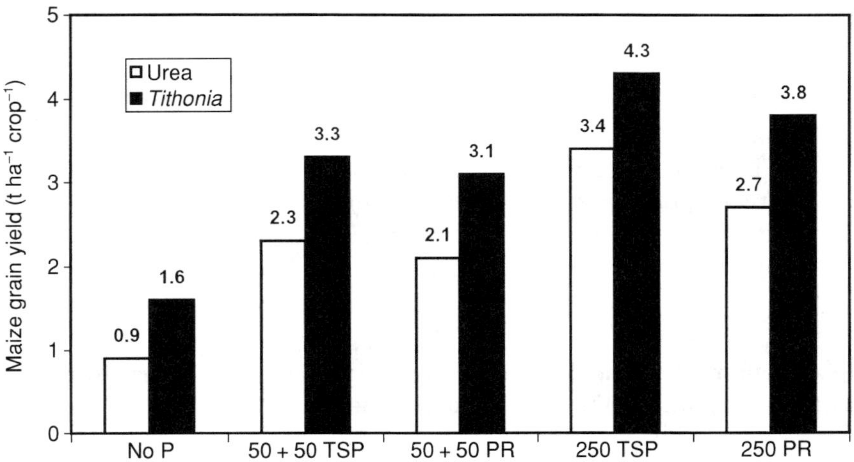

Fig. 17.3. Effects of 60 kg K ha^{-1} and N sources on maize grain yields in western Kenya. TSP, triple superphosphate; PR, Minjingo rock phosphate; average of two consecutive crops; N and K inputs applied only to the first crop. (From B. Jama, unpublished data.)

Without *Tithonia* additions, the high-reactivity Minjingo rock phosphate was 69% as effective as triple superphosphate. But, when *Tithonia* was added, the agronomic effectiveness of the rock phosphate increased to 91%. Considering the lower cost of rock phosphate in western Kenya, the recapitalization rate using rock phosphate with *Tithonia* biomass transfers – two resources widely available in East Africa – seems to be the 'best bet' at this point. By the end of 2 years, the overall response to mineral P applications generated increased maize yields of 138% (from 1.3 to 3.1 t ha^{-1} $crop^{-1}$), and the overall response to *Tithonia* versus the equivalent amount of N as urea resulted in increased maize yields of 39% (from 2.3 to 3.2 t ha^{-1} $crop^{-1}$). About 4000 farmers are currently trying these techniques, along with improved fallows (Niang *et al.*, 1998).

Maximize the recycling of potassium and other nutrients with the use of leguminous tree fallows and the return of crop residues to the fields where they were grown

Both fallows and crop residues return all nutrients to the soil, including the C needed by soil microorganisms to enhance nutrient cycling. Potassium is marginally deficient at current crop-yield levels in western Kenya, but, when yields are increased by fertility replenishment techniques, K becomes a clear limiting factor. In the long run, unless crop residues are effectively returned to the soil, there will be a need for K fertilizers beyond the amounts recycled by fallows. In the short run, the use of fallows and biomass transfer will largely eliminate the cost and logistical nightmares of applying mineral K fertilizers.

Have soil conservation structures in place to ensure that replenished soils do not erode and become a source of pollution

For example, hedgerow barriers – contour strips of trees and shrubs – may be used to control soil erosion effectively, even on steep slopes (Kiepe and Rao, 1994; Njoroge and Rao, 1994; Garrity, 1995, 1996; Kiepe, 1995). Data from the International Board for Soil Research and Management (IBSRAM) Sloping Lands Network trials in six countries have confirmed that, with hedgerow systems, annual soil loss is typically reduced 70–99% (Sajjapongse and Syers, 1995). Contour strips can also serve as sources of nutrients for biomass transfer or as cattle fodder (Niang *et al.*, 1998). In the latter case, many of the nutrients are cycled though the ruminant animals as manure.

After the fertility is restored, take advantage of the existing improved crop germ plasm, integrated pest management and other sound agronomic and post-harvest practices

A large arsenal of technologies are usually available from research services, but many of them are simply not profitable in fertility-depleted soils, because low crop yields commonly give negative returns to investments.

Improve policies to facilitate the provision of appropriate fertilizers and planting materials at a reasonable cost and at the right time

These policies include access to microcredit, timely access to markets, better infrastructure and community-based extension services. Policy reforms to secure opportunities for smallholder development and to eliminate policies that discriminate against the smallholder agricultural sector therefore remain a top priority. In addition, studies on the effectiveness of technology dissemination have shown that women tend to be the fastest innovators of new agricultural technologies, as their main concern is to make sure their children get fed properly. Therefore, public investments to expand access to primary education for girls and improve public health services in rural areas can also play an important role in fertility replenishment.

The fact that smallholders – often female farmers – produce most of the food in Africa is frequently considered a major constraint to agricultural development. In contrast, we believe that small-scale farms can be an asset, rather than a liability, when supported by appropriate policies. An example to look to is the agricultural production boom in Asia, which was a product of smallholder farms.

Consider soil fertility replenishment as an investment in natural resource capital with beneficiaries at the household, community, watershed, national and global levels

This provides a different array of financing possibilities (Izac, 1997; Pieri, 1998). Soil fertility replenishment was found to be profitable at the farm level in three contrasting case studies (Table 17.3), but the levels of investment in purchased inputs are beyond the means of most smallholder families. The initial investment required for P replenishment in western Kenya (US\$40 0.3 ha^{-1} $farm^{-1}$) is greater than the annual farm income generated by the existing system and about 10% of the average annual household income of about US\$420 for a low-resource farmer (Shepherd and Soule, 1998). Such households are unlikely to undertake these investments without some form of financial assistance.

In such circumstances, government intervention may be needed to alleviate rural poverty, improve household food security and provide a boost to poor farmers to help lift them out of poverty. We believe that the way forward is a cost-shared initial capital investment to purchase P fertilizer and germ plasm to grow organic inputs and high-value trees, combined with effective microcredit to cover recurring costs, such as N fertilizers and hybrid seeds.

Given the positive social and environmental externalities associated with soil fertility replenishment, an equitable cost-sharing mechanism can be developed and implemented, similar to existing ones in developed countries that deal with positive environmental externalities (Izac, 1997). Cost sharing of the capital investments should be done on the principle that whoever benefits should pay. Subsidies, in our view, have a limited role to play. If P recapitalization is done as a one-time-only investment, some of the 'dark sides' of subsidy use, such as continuity and dependency, are not likely to play major roles (Gladwin *et al.*, 1997).

Table 17.3. Simulated net present values (NPV) for fertility replenishment strategies in typical resource-poor farms of western Kenya, eastern Zambia and central Burkina Faso grown to basic food crops (maize, beans, sorghum, millet) (from Sanchez *et al.*, 1997b).

Region	Technology	NPV (US\$)
Western Kenya	Basal application of 250 kg P ha^{-1} as Minjingu phosphate rock + *Sesbania* fallow on half of the field area each year	250 per 0.2 ha
Western Kenya	Basal application of 250 kg P ha^{-1} as Minjingu phosphate rock + transfer of 1.9 t ha^{-1} $year^{-1}$ of dry *Tithonia* biomass from farm boundaries	350 per 0.2 ha
Eastern Zambia	Rotation of 2-year *Sesbania* fallow followed by 3 years of maize. No mineral fertilizers	588 ha^{-1}
Central Burkina Faso	Basal application of 48–72 kg P ha^{-1} of Kodjari phosphate rock + annual maintenance applications of 12–124 kg P ha^{-1} and urea. No organic inputs.	396 ha^{-1}

Calculations assume a discount rate of 20%.

Intensification and Diversification through High-value Products

President Yoweri Museveni of Uganda, in his opening address to the meeting of the Special Programme for African Agricultural Research (SPAAR) in February, 1996, stated: 'It does not make sense to grow low-value products [maize and beans] at a small scale; instead high-value products should be grown at a small scale, while low-value products should be grown on a large scale.' This strategy may be the most direct way out of poverty.

Some of these high-value crops (e.g. vegetables, oil-seeds) are also key staples for rural and urban African populations. For example, a few farmers who previously lived in cities have moved back to western Kenya and have used their recapitalized soils to grow high-value vegetables, such as *sukuma wiki* (kale), tomatoes and onions. Some farmers have reported increases in their net profits from US\$91 to US\$1665 ha^{-1} $year^{-1}$ (Nyasimi *et al.*, 1997). The extent to which this is a viable solution on a wide scale will depend critically on broad economic development, the expansion of storage and processing opportunities and urban population growth rates.

A further step may be the switch to newly domesticated tree crops that produce high-value products. These 'Cinderella' species – so called because their value has been largely overlooked by science, although appreciated by local people – include indigenous fruit-trees and other plants that provide medicinal products, ornamentals or high-grade timber (Leakey *et al.*, 1996).

One example is *Prunus africana*, a timber tree indigenous to montane regions of Africa. A substance is extracted from its bark to treat prostate gland-related diseases, and this has an annual market value of US\$220 million (Cunningham and Mbenkum, 1993; Simons *et al.*, 1998). Because of the high demand for these trees and the destructiveness of indigenous harvesting methods, *P. africana* is now in the Convention on International Trade in Endangered Species (CITES) Appendix II list of endangered species. With domestication, this tree is now being turned into a crop, as researchers select superior ecotypes and discover ways to harvest the bark sustainably; they may eventually develop extraction industries located in nearby rural areas (Simons *et al.*, 1998). Other examples of intensification through high-value products are the bush mango (*Irvingia gabonensis*), a nutritious fruit from the humid tropics of West Africa (Ladipo *et al.*, 1996), and *Sclerocarya birrea*, from the *miombo* woodlands of southern Africa, the fruits of which are used to make expensive liqueurs.

Techniques are being developed to convert other wild species into domesticated crops in agroforestry systems, including vegetative propagation and clonal selection, which capture genetic diversity (Leakey and Jaenicke, 1995). Domestication involves the formulation of a genetic improvement strategy for agroforestry trees and a strategy for the use of vegetative propagation to capture the additive and non-additive variation of individual trees in tree populations (Simons, 1996). Furthermore, guidelines have been developed for determining the species prioritized by farmers (Franzel *et al.*, 1996; Jaenicke *et al.*, 1996).

Through domestication, these tree crops could become higher-yielding, have higher quality and more uniform products, be more attractive commercially and help diversify diets. Such progress could improve household welfare by providing traditional food and health products, boosting trade, generating income and diversifying farming systems, both biologically and economically, beyond the production of basic food crops. Generally, tree crops have lower labour requirements than basic food crops or vegetables and their exploitation could thus free farm household labour to be used in seeking off-farm income-generating opportunities.

Through these types of production strategies, a new paradigm for smallholder farming in Africa emerges: one which, instead of being based on a limited number of highly domesticated food crops, often grown in monoculture, is based on a much greater diversity of plants, which together produce food and high-value products in agroforestry systems (Leakey and Izac, 1996). It will be important for scientists involved in domestication to work closely with the food and pharmaceutical industries, since the agroforester needs to know that there will be a market for these products, while the industry needs to know that there will be a regular supply of uniform and high-quality products before committing capital to develop these markets (Leakey and Izac, 1996).

High-value trees can fit into specific niches on farms, while leaving open land for growing staple food crops or other profitable crops, such as vegetables. Timber trees can also be grown on farm boundaries, with leguminous fodder trees under them. Similarly, fuelwood trees can be grown on field boundaries or as contour hedges on sloping lands. In such schemes, improved fallows become a crucial part of the crop rotation. On these farms, income is increased and diversified, providing resiliency against weather-related crop failure or price disruptions. Soil erosion is minimized, nutrient cycling maximized and above- and below-ground biodiversity enhanced. The farm thus approximates a functioning ecosystem (Sanchez and Leakey, 1997). The latest definition of agroforestry summarizes this approach as a 'dynamic, ecologically-based, natural resource management system that, through the integration of trees in farm and rangeland, diversifies and sustains smallholder production for increased social, economic and environmental benefits' (Leakey, 1996, pp. 5–7).

Environmental Implications of Soil Fertility Replenishment

Combinations of fertilizers and organic inputs can replenish soil nutrient stocks in smallholder African farms and restore service flows to those approaching their original levels (Izac, 1997). Such restoration is a prerequisite for food security. Shifting part or all of the farm enterprise to high-value vegetables or trees, with value-added processing industries nearby, is probably the first step out of rural poverty for many households. But the question remains: Can the replenishment of soil fertility also provide environmental benefits to farmers, communities and society as a whole? Having considered how to help satisfy the basic human needs for food security and reduced poverty, we now consider the environmental dimensions. Unfortunately, no hard data are available on the global environmental benefits of soil fertility replenishment and tree domestication and the resulting tradeoffs with private benefits, as has been done in the

Alternatives to Slash-and-Burn (ASB) Programme (see Chapters 11–13 of this volume). The following represent hypotheses of what the environmental implications might be.

Decreased soil erosion and sedimentation

With improved soil fertility levels, farmers would have greater incentives to protect their fertile soils from erosion than before fertility replenishment. The use of multipurpose contour strips, which arrest erosion and provide nutrient inputs through biomass transfers, was previously discussed. It is estimated that 40% of the nutrient losses in Africa are due to soil erosion (Smaling, 1993a). A major reduction of this figure due to protected soils and more tree crops on farms might significantly decrease the rate of sedimentation of coastal-area coral reefs and the rate of eutrophication of African rivers and lakes (Melack and MacIntyre, 1992). This positive externality could become a large negative one if soil conservation measures are excluded from fertility replenishment schemes.

Increased carbon sequestration

A recapitalized soil is likely to increase above- and below-ground C sequestration. Nutrient-depleted fields have little biomass C stock; one time-averaged estimate is of the order of 0.3 t C ha^{-1} as above-ground C (Shroeder *et al.*, 1993; Palm *et al.*, 2000). For the same year, with replenished soils producing 6 t per ha^{-1} of a dry-matter maize crop (half grain and half stover) plus 7.5 t ha^{-1} of dry-biomass *Sesbania* fallow, the time-averaged above-ground C content would increase to 1.6 t C ha^{-1}. That is not much.

Soil C stocks have significally decreased with soil nutrient depletion. Long-term experiments on clayey red soils in Kabete, Kenya, show a 28% decrease of soil C, from 36 to 26 t C ha^{-1}, in the topsoil over an 18-year period (Kapkiyai, 1996). Another long-term trial, in Muguga, Kenya, with similar soils, shows a total loss of 91 t C ha^{-1} from 0–120 cm depth with 8 years of continuous cultivation without inputs; about half the loss (48 t C ha^{-1}) occurred in the top 15 cm (Woomer *et al.*, 1997). While all such losses are reversible in principle – as long as clay contents do not change by erosion – there are no hard data to predict the magnitudes of increased C stocks in the soil that result from improved management.

Taking into consideration a complete agroforestry system with niches existing at the farm scale, as previously described, Woomer *et al.* (1997) have estimated increased farm-level C sequestration following 20 years of nutrient recapitalization and tree intensification in the Kakamega District of western Kenya. The results shown in Table 17.4 indicate a doubling of total-system C, an increase in above-ground C from 13 to 64 t C ha^{-1}, and a lesser increase in below-ground C (from 57 to 72 t C ha^{-1}). Total C sequestration may, therefore, be quite considerable with agroforestry systems that involve soil fertility

Table 17.4. A scenario for carbon sequestration at the farm level before and 20 years after nutrient recapitalization: domesticated high-value trees grown as orchards or live fences and contour grass strips for soil conservation (from Woomer *et al.*, 1997).

Farming system carbon pools (t C ha^{-1})	Before	20 years after	Difference
Structures	0.4	1.6	1.2
Crops	2.1	3.2	1.1
Tree orchards	4.0	12.0	8.0
Live fences (shrubs to trees)	6.5	47.0	40.5
Grass contour strips	0.0	0.2	0.2
Total above-ground carbon	13.0	64.0	51.0
Surface soil (0–15 cm)	27.0	36.0	9.0
Subsoil	30.0	36.0	6.0
Total below-ground carbon	57.0	72.0	15.0
Total system carbon	70.0	136.0	66.0

replenishment and intensification, and diversification of farming systems through the use of high-value domesticated trees.

It is not possible to calculate time-averaged C sequestration from the above data. As it stands, Table 17.4 assumes that no trees are being removed from the system for either fuelwood, timber or other purposes, but that crop residues are being taken out of the system. A more appropriate accounting method for C sequestration is estimating the difference in time-averaged C stocks, which takes into account periodic woody biomass harvests. Taking tree removals into consideration allows farmers to earn income from tree products as well as from crops, with the average C sequestered estimated over several years. The ASB Carbon Working Group has used this concept in evaluating the tradeoffs between C sequestration and private profitability (Palm *et al.*, 2000). Currently, the financing mechanisms for C offsets in the Kyoto Protocol are largely based on growing trees forever without any economic benefits (IGBP Terrestrial Carbon Working Group, 1998). Time-averaged C sequestration estimates may help justify the financing of C offsets in agroforestry systems, while allowing farmers to achieve food security and get out of poverty.

Protecting biodiversity

With improved agroforestry systems of the type described in this chapter, above- and below-ground biodiversity can be expected to increase over that in nutrient-depleted systems. To the authors' knowledge, there are, however, no conclusive studies on this subject for smallholder systems in Africa.

People deforest in the search for new lands to replace nutrient-depleted land, and degrade forests to obtain tree products, such as fuelwood, poles, timber, fruits and medicinals. Profitable agroforestry systems, through increasing returns to land in existing agricultural areas, may deflect deforestation on the remaining patches of primary forest at the margins of smallholder-dominated landscapes by easing the 'push' factors motivating migration and new land clearing. In eastern and southern Africa, affected areas include the Kakamega Forest in western Kenya, the Bwindi Impenetrable Forest in Uganda and Rwanda and the remaining patches of rainforest in eastern Madagascar. The latter two regions harbour unique animal biodiversity. The *miombo* woodlands of southern Africa are also being heavily deforested in many areas and could similarly benefit from profitable agroforestry systems.

Land-use intensification caused by demographic pressure is generally associated with environmental degradation, as is the case with soil fertility depletion in Africa. A point is reached, however, where degradation can be alleviated with the incorporation of trees within the farming system. There is emerging evidence from Kenya, Uganda, Burundi, Nigeria and Nepal that the number of trees planted on-farm is rapidly increasing with increasing human population density (Sanchez *et al.*, 1997b; Sanchez *et al.*, unpublished results). Holmgren *et al.* (1994), for example, estimate a strong positive correlation between population density and the standing volume of planted trees on farms in high-potential districts in Kenya. As farm size decreases and most off-farm sources of tree products are exhausted, smallholder farmers plant more trees under a variety of agroforestry systems to ensure access to fuelwood, fodder and other products, thereby increasing the woody biomass land cover of their farms.

This evidence of increased on-farm tree planting comes from the subhumid tropical highlands or 'high-potential areas', as well as from semi-arid or 'low-potential areas' of Africa (Sanchez *et al.*, 1997a). However, all the cases cited in this source occur where farmers have assured land tenure. Where this is not the case, the tree population on farms decreases or vanishes, as seen in much of the central highlands of Ethiopia and also in the matrilineal societies of southern Malawi (ICRAF, 1998). Situations where tenure is assured thus pose major opportunities to diversify tree planting. It seems likely that the future of trees in Africa is on-farm, where increases in tree populations are the highest.

Conclusions

Humanity has a moral imperative to eradicate the last bastion of widespread hunger, malnutrition, poverty and environmental deterioration in the world existing at a continental scale – sub-Saharan Africa. To do so, the main need is for sustained policies that favour smallholder rural development. This chapter focuses on two other key factors: (i) the reversal of soil fertility depletion as the fundamental biophysical constraint to food security; and (ii) the intensification and diversification of smallholder farming by growing plants and trees that produce high-value products, as the key to poverty reduction.

To achieve these objectives, ten main components of a more robust, integrated approach are proposed: (i) combine organic and mineral sources of nutrients; (ii) capture most of the N from the air by biological N fixation in the same fields where crops are to be grown, largely through short-term leguminous fallows; (iii) recapitalize soil P stocks, where needed, by investment applications of mineral P fertilizers, whose residual effects will last for at least 5 years; (iv) supplement mineral P with organic inputs, particularly those that enhance P cycling as they decompose in the soil; (v) maximize the recycling of K and other nutrients with the use of leguminous tree fallows and the return of crop residues to the fields; (vi) have soil conservation structures in place to ensure that replenished soils do not erode; (vii) after soil fertility is restored, take advantage of improved crop germ plasm, integrated pest management and other sound agronomic practices and post-harvest technologies; (viii) improve policies that facilitate the provision of appropriate fertilizers and planting materials at a reasonable cost and at the right time, as well as access to microcredit, timely access to markets, better infrastructure and community-based extension services; (ix) consider soil fertility replenishment as an investment in natural resource capital; and (x) shift to high-value products, such as vegetables or trees, which are in demand in both domestic and international markets. Pilot projects currently involving thousands of farmers in eastern and southern Africa appear to be highly productive and profitable.

Replenishing soil fertility with new, ecologically sound approaches and policies can reverse the downward spiral of food insecurity in Africa. Intensifying farming with high-value crops and trees as a diversified but intensive agroecosystem is one clear way out of poverty. Value-added enterprises for product transformation in towns will further reduce rural poverty. Both the food security and poverty-alleviation measures proposed appear to have highly positive environmental benefits – soil conservation, C sequestration and biodiversity protection. While these benefits have not been quantified in a sufficiently rigorous manner to ascertain the tradeoffs between global environmental and private benefits, the strategies proposed are sufficiently promising for them to play a central part in a renewed commitment to agricultural and rural investment in Africa.

Acknowledgements

The authors are grateful to Thomas Tomich, Steve Franzel, Roland Buresh, Frank Place and Fiona Chandler for their helpful comments on the drafts of this manuscript.

18 Livestock–Environment Interactions under Intensifying Production

STEVEN J. STAAL,[1] SIMEON EHUI[2] AND JON TANNER[1]

[1]*International Livestock Research Institute (ILRI), Nairobi, Kenya;*
[2]*International Livestock Research Institute (ILRI), Addis Ababa, Ethiopia*

Introduction

As human populations continue to grow, the intensification of agricultural production is occurring in order to maintain food consumption and rural income levels in developing counties. The demand for livestock products is generally income-elastic, so that, with increases in real incomes in most regions, the demand for livestock products is growing as well, resulting in greater projected numbers of both ruminant and monogastric livestock.[1] Both trends and, indeed, their interaction will have consequences for rural and urban environments in the future and for the sustainability of land-use systems. The intensification of agricultural land use may degrade soils and alter natural habitats and biodiversity or, in some cases, may lead to excessive levels of inputs, with consequent effects on the environment. Similarly, increased livestock production may lead to animal-waste pollution in some settings, and contribute to greenhouse-gas production. What may not be well recognized, however, is the role that ruminant livestock, in particular, play in assisting the attainment of higher and more sustained levels of agricultural production, often in ways that have net positive environmental impacts. This role is particularly important in the crop–livestock systems on which (sub)tropical agricultural production is increasingly dependent. 'Win–win' situations are clearly possible, especially through the contribution of intensive livestock production to soil fertility, which is relevant for large areas of the world, where soil nutrient deficits and slow nutrient turnover rates remain the primary agricultural constraint.

This chapter begins by describing the relative importance of different livestock production systems and the global trends associated with them. Within each system, the net impact of livestock production on the environment is discussed and case studies of positive livestock and environment interactions are presented. Finally, the possibilities for and implications of further intensification of livestock production are explored in some depth.

Global Livestock Production Systems and Future Trends

Livestock are an important part of agricultural systems nearly everywhere. In Asia, 20% of gross agricultural output is attributed to livestock (30% if draught power and manure are included). In Latin America, this contribution reaches 40% (Vandermissen and Symoens, 1997). In Africa, livestock account for 25% of agricultural output; this rises to 35% with the inclusion of manure and draught. Further, many crop products derive most or all of their use value as inputs into livestock production. If one includes the value of these products, such as feed grains or straw for fodder, these proportions are much higher.

Livestock production systems can be categorized into three broad types: grazing, mixed crop–livestock farming systems and landless or industrial systems. De Haan (1997) defines these as follows:

- Grazing systems: where at least 90% of feed is from pasture. Examples are the traditional pastoral systems in sub-Saharan Africa (SSA) and ranching in South America.
- Mixed crop–livestock farming systems: where at least 10% of feed is obtained from crops and residues from own farm. These cover a wide range of systems, from Wisconsin dairy farms to smallholder mixed crop and livestock farms across the developing world, under a considerable variety of ecozones.
- Industrial or landless systems: where less than 10% of feed is from the production unit, but, feed is purchased or brought in. This mostly describes modern commercial operations, such as large-scale pig or poultry production, but also encompasses small backyard dairy, beef fattening or sheep/goat raising in African cities, and could include landless rural dairy production in India.

The intensification of livestock production generally occurs as a movement from grazing systems to industrial systems. As production intensifies, less land is used per unit of output, with labour and/or capital substituted instead. Intensification often means increased reliance on feed resources imported from outside the production unit. The nature of livestock intensification and its interaction with the environment are discussed in more detail below.

These categories can be used to begin to characterize livestock and environment interactions. The grazing and mixed farming systems are considered closed systems from a nutrient-cycling perspective. In such systems, animal waste is generally used within the system and most costs are contained within, so that there is a direct incentive to manage wastes in an environmentally sustainable manner (de Haan *et al.*, 1997). In contrast, industrial systems are, by their nature, generally 'open', in that, for both inputs and outputs, there is heavy dependence on markets or other forms of resource waste transfer. In developed countries, where some 8 billion t of animal waste are produced per year, the handling and disposal of wastes from industrial livestock production impose constraints on production and usually impose costs on society that are not transferred back to producers. Occasionally, and particularly in developing nations, there is a market for animal manure, which alleviates some of these constraints.

Importance of mixed farming systems

Mixed farming systems encompass 2.5 billion ha of land globally and produce most livestock products. Mixed systems produce 92% of the world's milk, the rest coming from grazing systems. In 1991–1993, mixed systems accounted for an estimated 54% of the world's meat, with 9% produced in grazing systems and 37% in landless systems (de Haan *et al.*, 1997). Mixed systems produce all buffalo meat and 70% of sheep and goat meat. About half of the meat produced in mixed farming systems is produced in developing countries – the rest is from the Organization for Economic Cooperation and Development (OECD), Eastern Europe and former Soviet Union countries. In 1993, about a third of milk production occurred in developing countries, nearly all from mixed systems (Seré and Steinfeld, 1996).

The trend worldwide is increasingly towards more intensive livestock production. Over the last decade, meat production in mixed farming systems grew at a rate of 2% annually, compared with 1% growth in grazing systems, but was still behind the global growth in demand for meat (Blackburn, 1998). This implies that greater intensification will be required to meet demand. Because livestock products tend to represent higher value protein than do crop products, production is closely tied to demand driven by income growth, at least outside subsistence settings. A recent report by Delgado *et al.* (1999) provides systematic projections of expected growth in livestock products by region and describes a coming 'livestock revolution', stemming from increases in demand stimulated by income growth, particularly in Asia. Between 1982 and 1994, meat consumption increased globally at an annual rate of 2.9%, with most of that occurring in developing countries, where growth was 5.4% annually (compared with 1.0% in developed countries). This trend will continue as incomes rise. During the period 1993–2020, it is projected that consumption of beef and pork in developing countries will grow by 2.8% annually, consumption of milk by 3.3% and that of poultry by 3.1% (Delgado *et al.*, 1999). Because livestock products are generally less easily traded internationally than grains, given the need for product transformation to maintain shelf-life, most of the production growth will occur in the same regions where demand growth occurs. As an example, in developing countries, meat imports or exports comprise less than 5% of production by weight, while imports of cereals are substantially higher, at around 12% (FAOSTAT, 1997). As a consequence, the effects of demand growth will be largely determined by its location.

As a result of the dependence on domestic production rather than imports, by 2020, developing countries are projected to produce, on average, 38% more meat and 54% more milk per capita than in the early 1990s. There will be a corresponding increase in the shares of developing countries in the global production of meat and milk, rising from 31% and 25%, respectively, in the early 1980s, to 60% and 52% (Delgado *et al.*, 1999). Much of the expansion in meat production will be of monogastric livestock, such as pigs and poultry, the production of which will create increased demand for high-energy feed, such as cereals. As demand increases and the opportunities for industrial livestock production grow, monogastric production is economically favoured because of higher feed conversion ratios for pigs and poultry.

These increases in livestock production might be expected to drive up grain prices. However, Delgado *et al.* (1999) predict that, even with these large increases, inflation-adjusted prices of animal feeds are expected to fall, although not as rapidly as they have during the past 20 years. They suggest that, even under the conservative scenario of declining livestock productivity, world maize prices would rise by only 21% by 2020, which, in real terms, is only half the prevailing price of the early 1980s. This would be achieved through continued productivity increases in grain production.

The resource costs of livestock production may not be high, as often suggested. Because of the use of roughages, on a global basis, less than 3 kg of grain are required to produce 1 kg of meat of any kind, and less than 1 kg of grain is required per kilogram of milk (CAST, 1999). Globally, animals produce about 1 kg of human food protein for each 1.4 kg of human-edible protein consumed. Further, the biological value of protein in foods from animals is about 1.4 times that of foods from plants. Thus, diverting grains from animal production to direct human consumption would, in the long term, result in little increase in total food protein and would decrease average dietary quality and diversity.

Future increases in livestock demand and production appear certain. However, the nature of the production increase and its effects on rural and urban environments are much less certain. Much of the projected growth is expected to occur in mixed systems, which will retain their dominant role among livestock systems. The share of production from industrial systems will nevertheless grow, as greater intensification is required to meet demand using existing land resources. The key issues, then, are not whether sufficient animal products and cereals will be available, but what impacts increased production and consumption will have on the environment, human health and the incomes of the poor. Clearly, some form of intensification of production will need to occur, since simply increasing livestock numbers will be limited by land resources, which are already stretched in many areas. This intensification and its consequent impacts will depend locally on the technologies available, market access, agroclimatic conditions and the relative values of land, labour and capital.

Livestock and the Environment

In considering the environmental effects of livestock intensification, it is useful to distinguish two main settings where environmental outcomes are different. In what might be termed 'natural' settings – recently preserved forest, grasslands, desert, etc. – the preservation of the natural environment is aimed mainly at sustaining natural ecosystems and biodiversity. In the 'settled environment', where people live and conduct agriculture, environmental preservation is aimed mainly at sustaining human health and comfort and agricultural productivity. The distinction is made simply to help recognize the fact that, while the negative impacts of livestock production occur in both environments, the positive effects accrue mainly in the settled environment. The discussion of the contributions of livestock to the settled environment should be seen in light of this limitation.

Detrimental effects of livestock intensification

Livestock production has frequently been criticized as being inherently harmful to the environment: the more livestock, the worse the environmental effects, measured in a number of different ways. These effects can be through soil, water and air. According to these arguments, livestock trample fragile grasslands and shrub systems in semi-arid areas and thus contribute to desertification; they add measurably to the atmospheric accumulation of greenhouse gases; they pollute aquifers and waterways with excessive animal wastes; and they cause rainforest to be permanently lost.

Although these claims will continue to be made, much evidence now refutes some of these arguments. The Winrock International Institute (1992) suggests that animal agriculture is seldom the initial cause of degradation and does not significantly contribute to desertification, deforestation or erosion. Although livestock may contribute to these effects locally and temporarily, overall trends are apparently determined by biotic conditions, such as climate, land and ecosystem. Studies have shown that the effects of grazing and drought are often confused and that irreversible effects of livestock on vegetation are found mostly around local watering points and settlements (de Haan *et al.*, 1997). In the Sahel, where much of this attention is focused, de Haan *et al.* (1997) further report that arid grazing systems are not generally under threat.[2] Instead, there has been increasing productivity in terms of meat and milk per hectare and per animal head since the early 1980s, although animal-protein consumption per household continues to decline.

Other evidence shows that vegetation dynamics is more linked to climatic variability than to livestock. Due to the extreme climate in semi-arid areas, the majority of the rangeland area is dominated by annual plant species, which offer poor rangeland quality. It is low rangeland quality that leads to a high degree of mobility in livestock systems, rather than livestock mobility affecting the vegetation (Behnke *et al.*, 1993). The mobility of livestock is often criticized as contributing to pasture degradation. But, in a study using satellite imagery in inner Asia, it was found that, while degraded pastures were found in most parts of Chinese and Russian inner Asia, little pasture degradation occurred in the independent state of Mongolia (Sneath, 1998). The author attributes these differences directly to collectivization in the former regions, which reduced or eliminated large-scale seasonal pastoral movements; those districts with the highest degradation were also those with the most limited livestock movement. Fencing led to continuous grazing and trampling of vegetation. These results show that livestock intensification can indeed lead to degradation, but particularly when the structure of that change is driven by inappropriate policies. Traditional grazing systems may be more environmentally robust.

In the settled environment, most of the negative impacts of livestock are a result of surplus animal waste brought about by industrial livestock production, mainly in the developed world, and mostly related to pigs and poultry. In these systems, where pig factories can house well over 1000 animals, waste problems are concentrated and large-scale. Industrial production has been increasing in East Asia and Latin America as well, however, with industrial pig and poultry complexes a serious source of nutrient surpluses in Beijing, Shanghai, New Delhi and parts of Mexico and Brazil (FAO, 1998). Landless cattle production in Eastern Europe is one example of an industrial ruminant system that poses similar problems. Most of the negative environmental impacts result from surpluses of nitrogen (N): pigs and poultry excrete some 65% and 70%, respectively, of their N intake (de Haan *et al.*, 1997). The nitrogen surplus can enter water and soil systems, particularly in the form of nitrates, causing human health problems, algal bloom-and-bust cycles and eutrophication. Nitrogenous gases, mainly ammonia, are also emitted from manure and urine, and can affect plant growth and acidify the soil. In 1993, ammonia from manure contributed some 55% of total acid deposits in the Netherlands, although, in developing countries, industrial pollution is a more significant factor (de Haan *et al.*, 1997). The environmental effects are not all related to animal excreta, however, as water pollution may also result from pesticides used to control animal diseases and other waste disposal from slaughterhouses, dairies and tanneries.

Gases released by livestock include carbon dioxide, methane, ozone and nitrous oxide, which contribute to greenhouse effects and acidification. The livestock contribution to global greenhouse gases is 2.5%, including 20% of methane (Sansoucy, 1995). Nevertheless, paddy-fields alone are a much larger source of methane globally than livestock (de Haan *et al.*, 1997). In spite of growing livestock populations, methane emissions from livestock remain static, due to increases in productivity (leading to lower emissions per animal) and greater numbers of monogastrics, which emit smaller amounts (de Haan *et al.*, 1997).

The impacts of animal wastes on the environment are determined by the manner in which wastes are handled and the nutrient balance in the immediate and larger system, including organic matter. Generally, one finds nutrient-deficit systems in most developing countries and nutrient-surplus situations in developed countries and some parts of East Asia and Latin America (Blackburn, 1998). Nevertheless, regionally, within any country, there are surplus and deficit areas. In Europe, in particular, strict regulations control animal stocking and manure application rates, in an attempt to ameliorate the negative effects of animal-nutrient surpluses. Anaerobic digestion is used in some cases, and large-scale processing and transport of manure have developed in certain regions (de Haan *et al.*, 1997), based on regional surpluses and deficits. As will be discussed below, relative factor values and policies are key to utilizing nutrient surpluses, not just to address negative environmental effects, but also to promote positive ones.

Importance of Livestock for Soil Fertility

The positive contributions of livestock to the environment accrue in the settled environment, especially in mixed systems. As noted above, these are expected to intensify and grow in importance. In such circumstances, soil nutrient conservation or enhancement techniques, particularly fallow, are not sufficient to sustain cropping. Without external inputs, the integration of crops and livestock is the main avenue for intensification. Livestock in mixed systems can diversify risk, use labour more efficiently and add value to crop by-products. Such integration also maintains ecosystem function and health and provides resilience to shocks (Blackburn, 1998).

Soil fertility is generally recognized as the most important constraint to sustaining yields in agriculture, particularly under heavily cropped land. This is particularly true in the developing world. Pests, weeds and soil erosion are also important constraints. Animal-waste nutrients have long played a critical role in alleviating soil fertility constraints, particularly in mixed systems. Animal manure application in agriculture returns nutrients and organic matter to the soil, contributing to the maintenance or restoration of soil fertility in terms of nutrient content, soil structure, water retention and drainage capacity. Whereas pigs and poultry rely on concentrated feeds, ruminant livestock produce mainly from roughage. Thus, ruminant livestock also recycle by-products and wastes from households, crops (stems, husks and weeds), other livestock (pig/poultry manure)[3] and agroindustry (slaughterhouse and fish wastes, oilcake, grain-milling by-products) into higher-value products. Many of these wastes would not otherwise be used. The burning or incorporation into the soil of crop stalks could have harmful environmental effects by releasing carbon dioxide and by creating a N sink, reducing N availability to crops (Steinfeld *et al.*, 1997). Manuring, in addition to directly contributing soil nutrients, promotes the accumulation of organic matter, enhances cation exchange capacity and appears to have beneficial effects on biological processes (increased earthworms and microbes). The high levels of excreted N, which are a problem in industrial production systems, generally become an asset in mixed systems, where nutrient deficits and soil fertility are major constraints. Approximately 100–140 g N per tropical livestock unit (TLU) (1 TLU = a 250-kg animal) is produced daily, 60% of which is in manure and the rest in urine (Smith *et al.*, 1997).

As a result of the ability of ruminant livestock to contribute directly to sustained crop production, generally using inputs that are otherwise not easily used, the role of livestock in the intensification of agriculture is critical. In developing countries, more than half of total fertilizer is provided in the form of manure; this rises to 70% in lower-income countries (Fresco and Steinfeld, 1997). In mixed irrigated (sub)humid systems in Asia, animal wastes provide 40–60% of all inputs to soil fertility (Jansen and de Wit, 1996). Although increased productivity in agriculture will require the use of chemical fertilizers, manuring will remain critical for sustaining soil fertility and will probably grow in importance in intensified smallholder systems. To deal with impending food insecurity in Asia, for example, the integration of livestock and crop production through nutrient recycling is considered by some as the key (Vandermissen and

Symoens, 1997). Soil fertility maintenance is not the only contribution of livestock to agricultural intensification, however. In developing countries, more than half of the arable area (52%) is cultivated with the help of animal draught power (Fresco and Steinfeld, 1997).

Case studies from South Asia and East Africa

Some of the strongest potential avenues for 'win–win' situations, where smallholder farmers and the agricultural environment both benefit from livestock production, occur through the intensification of smallholder mixed crop–livestock farming systems. In such areas, the heavy cultivation of land can only be sustained with large inputs of nutrients and organic soil amendments. To illustrate these positive interactions and to clarify their long-term potential, several specific case studies are discussed here.

Local cattle in wheat-growing areas of Nepal

In a study of the socioeconomic determinants of sustainable crop production in Nepal, Ali (1996) combined indices of long-term sustainability and resource-use inefficiency to create an index of sustainable resource-use efficiency. The Rupandehi district is an intensive wheat-growing area, an area of concern because wheat yields fell by 0.12% annually in the 1980s, suggesting sustainability problems. About half of the wheat plots on 170 sample-farm parcels regularly received an application of manure (4 t ha^{-1} $year^{-1}$), and nitrogen–phosphorus–potassium (NPK) fertilizer was also applied (85 kg ha^{-1} $year^{-1}$). The positive and sustained effects of livestock manure on soil fertility and crop productivity were found to be clear. Marginal productivity analysis indicated that an additional 1 t of farmyard manure increased wheat production by 42.1 kg ha^{-1} on good fields (2% of average yield levels of 1.7 t ha^{-1}). More important, however, are the residual and cumulative effects. Discontinuity in farmyard manure application had a significant impact on productivity. Parcels where farmyard manure was not applied to every crop had about 18% lower productivity than those where manure was always applied, given current levels of fertilizer application. Perhaps the largest threat to sustained productivity in the area is the growing demand for manure as fuel, due to depletion of forest fuelwood sources (Ali, 1996). This study illustrates the direct role that livestock can play in maintaining long-term soil fertility.

Local and improved cattle on maize/bean farms in western Kenya

Shepherd and Soule (1998) carried out a simulation analysis of agroforestry and soil management strategies in the highlands of western Kenya, a zone of high agricultural potential but severe nutrient depletion, with N and P the main limiting nutrients. In recent decades, human populations have grown, to the point that population densities now exceed 1000 km^{-2} in some places, with farm sizes averaging only 0.65 ha. Fallow periods have declined or disappeared, and soil fertility has become the primary constraint to production. The study modelled nutrient flows and financial returns. Wealth ranking was used to separate farms into high-, medium- and low-resource endowment farms. Cattle were kept by high- and medium-resource endowment farms only, and only the former used any inorganic fertilizer. The model results showed that soil carbon (C), N and P balances were positive only in high-resource households, due to fertilizer purchases and cattle manure use. Recycling of N was several times higher in high- and medium-resource households because of recycling through livestock, as were inflows of N, mainly due to purchases of livestock concentrate. Thus, most of the net positive balance in N was due to livestock. This is an example of a 'win–win' scenario: total farm revenue is seven times higher on high-resource compared with low-resource farms, due largely to the presence of dairy cattle or even a single cow, which provides higher returns than food crops. Studies have shown near-universal negative nutrient balances on farms in Africa. In this case, positive nutrient balances are due primarily to the presence and management of livestock manure and the import of inorganic fertilizer. The indirect contribution of livestock through concentrate feeds is also clear.

Dairy cattle on maize/coffee/tea farms in highland Kenya

While the previous study estimated the contribution of manure to soil fertility maintenance, another study examined, in a very similar farming system, the response of farmers to the perception of that contribution. In research on manure management in highland Kenya (Lekasi *et al.*, 1998), a structured survey was carried out in Kiambu and Murang'a districts of Central Province, areas where coffee, tea, dairy and maize are the main agricultural activities. Again, small farms predominate, averaging 1.6 ha in size. Most cattle are Friesian or Ayrshire dairy crosses with local zebu, two to three per farm.

In this area, dairy cattle have been an integral part of the farming system since at least the late 1950s, and cattle manure is critical to sustained multiple cropping. In a ranking by small farmers, milk and manure were ranked about equally in importance. Most farmers actively manage manure, keeping animals in permanent confinement to allow maximum manure and urine collection and adding organic amendments, such as feed refusals and weeds. As a result, most farmers applied more than 10 t ha^{-1} $year^{-1}$ of manure to crops. Some farms reported purchasing manure to make up for on-farm deficits. This intensive reliance on manure results in relatively high values of manure (approximately US$100 t^{-1}) in the local market. On the farms sampled, the value of manure produced was estimated to be some 30% of the value of the milk produced. Comparisons with local inorganic fertilizer prices suggest that 83% of the value of the manure may reflect the organic component, rather than N and P. This may be due to the value that is placed on the physical and residual benefits of organic matter in manure in terms of enhancing soil structure and nutrient uptake. This case clearly shows the strong value placed on the nutrients and organic matter in animal manure, which are responsible for long-term sustained cropping.

Increasing Intensification: Livestock and People

An important indicator of the positive contribution of livestock to agricultural intensification can be seen in the statistical link between human and livestock population densities. Both global and regional detailed studies have firmly established this link. In a global study, Slingenbergh and Wint (1997) found that spatial livestock population densities occur at a rate of approximately double that of the human population density. Approximately the same results were found in SSA and, at a slightly lower level, in Asia. Spatially, therefore, livestock are found where people are, and they increase in density at roughly twice the rate that human populations do. This challenges the perception that livestock occur mainly in less populated areas where people are few and grazing land is abundant. Rather, as agricultural activities intensify with more dense human habitation, the number of livestock increases more than proportionately, seemingly regardless of climatic or ecological factors. This suggests that people keep their livestock close by for convenience of supply, with feed resources brought from other areas, and, more generally, that livestock are an important ingredient globally for the intensification of agricultural systems.

More evidence of the importance of the livestock–people intensification link is provided by a detailed regional study by Wint and Bourn (1994). Looking first at SSA in general, they find that most ruminants in SSA are found in the 500–1500 mm rainfall range, which covers most of the continental land mass. In areas receiving less than 500 mm annually, 70–90% of ruminants are in rural systems; in areas receiving 500–2000 mm of rainfall, 50–60% of ruminant biomass is in rural systems; and, in the wettest areas, it is 10–15%. The remainder are, in each case, in villages or within settlements. They find that a significant proportion of ruminant livestock – more than a third in > 500 mm rainfall areas – are in village settings and may not have been included in previous estimates. Most of these are small ruminants. Regression analysis shows that biomass densities are closely linked to cultivation levels, but are less strongly associated with measures of natural grazing. Thus there is a much greater association of livestock density with the consequences of human activity, as measured by either cultivation percentage or habitation density, than with the extent or distribution of natural grazing. This includes pastorally managed livestock and, in fact, in pastoral areas, density is often negatively correlated with natural grazing. At least in regard to ruminants – the large majority of livestock in Africa – these results clearly undermine the traditional view that livestock are mainly linked to grazing. Rather, livestock are linked to cultivation and the vegetation associated with it.

Wint and Bourn (1994) maintain that, based on mounting evidence, conventional distinctions between pastoralists and arable farmers are becoming less clear-cut. Although specialist producers are likely to persist in many regions, there is a more general tendency for pastoralists to settle and grow crops and for arable farmers to take up animal husbandry. Based on the knowledge of livestock's contribution to soil fertility, the presumption is not that such trends are mining soil nutrients, but rather that they result from the implicitly higher sustainability of systems that more intensively exploit positive livestock, crop and soil interactions.

Figure 18.1 illustrates the potential net impact on the environment of livestock intensification if the above arguments are accepted. On the left side, at low levels of intensification – typically, agropastoral or grazing systems – the net environmental impact may be negative. Although livestock in these systems contribute somewhat to nutrient cycling, the positive effects may be outweighed by the impact of erosion and land degradation under some policy and land-tenure regimes. In the mixed systems that dominate most livestock production and occupy the middle of the range of intensification, we suggest that the effects are increasingly positive, through nutrient cycling leading to improved soil fertility and more sustained cropping. In the high-intensity mixed system, the role of livestock in maintaining agricultural sustainability may not be fully recognized. Finally, the right-hand side of the graph shows the highest level of intensification, the industrial system, where environmental impacts turn strongly negative, through air, soil and water pollution. As discussed further below, where labour costs are relatively low, the utilization of animal wastes is facilitated, so that, at the extreme right of the scale, the environmental costs of industrialization are reduced.

Livestock Production Intensification

The interactions of livestock and the environment have been reviewed, based on experience to date from different development settings. In order to better

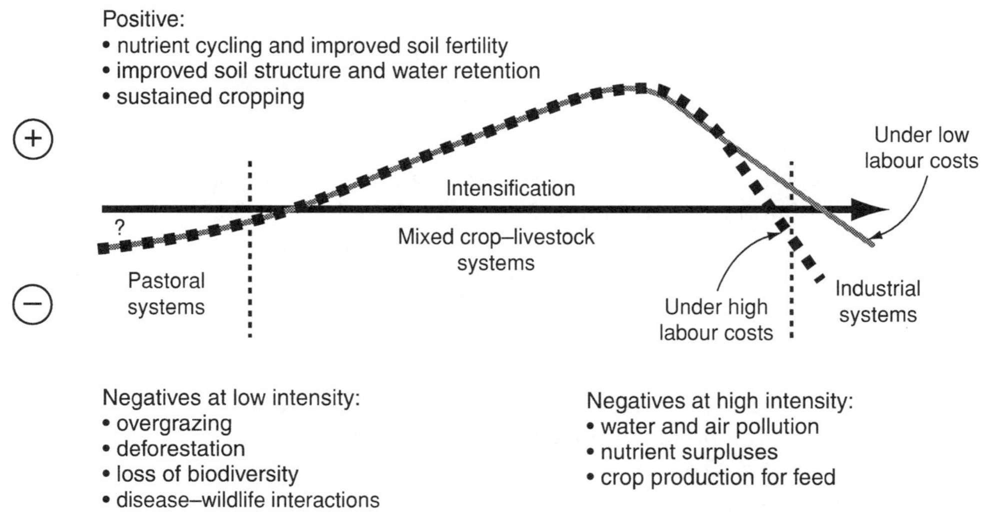

Fig. 18.1. Possible net environmental impact of intensifying livestock production.

understand how these interactions will evolve, it is useful to explore the implications of several principles of livestock intensification.

Principles and measures of livestock intensification

The intensification of livestock production follows the same principles as agricultural intensification in general, but is measured against a different resource base. Human population pressure on relatively fixed land resources is seen as the driving force behind an evolutionary process of agricultural intensification (Boserup, 1965; Binswanger and McIntire, 1987; McIntire *et al.*, 1992). Boserup postulated that population pressure on land brings about agricultural intensification through induced innovation of technologies for increased productivity, including soil conservation. At the level of communities and larger systems, institutions, infrastructure and markets also respond to population pressure to help achieve higher levels of land productivity. Similarly, population pressure eventually induces investments in resource improvements as well, particularly where tenure is secure (Pender, 1998).

In spite of the implications for intensification at different system levels, the degree of agricultural intensification is generally measured simply in terms of productivity per unit of land. Agricultural intensification has generally been defined in terms of the cropping intensity of land (Byerlee, 1990), measured as total cropped area per year as a proportion of total cultivated area per year. This reflects the fundamental constraint on crop agriculture, which is the availability of land, the primary productive resource. Livestock intensification is typically measured in a similar manner, such as milk or meat offtake per unit of land per year, or stocking density.

However, feed, rather than land, is the primary resource in animal agriculture, and feed availability per animal remains the basic constraint. The usual measures of intensification, therefore, may not be adequate to capture the great variation in livestock production systems. To understand better why that is, it is useful to examine the process of livestock intensification, and particularly the manner in which livestock production may or may not be linked to crop production at different levels of intensification.

Within the induced-innovation context, McIntire *et al.* (1992) describe the evolution of crop–livestock interactions as livestock production intensifies as an 'inverted U' pattern of interaction. At low human populations, crop–livestock systems are extensive, with more land per unit of output and few direct crop–livestock interactions. Under these conditions, there is a cost advantage in specializing in crop or animal production and interacting through markets and contracts. Examples include ranching and traditional pastoral systems. Intensification occurs in response to population growth and market forces. As this happens, crop–livestock interactions also increase. Population pressure increases both the net demand for agricultural products and the opportunity cost of land, raising the derived value of crop–livestock interactions. These include straw and other crop residues for feed, and manure and animal traction for crop production. In developing-country settings, market constraints to obtaining inputs result in a cost advantage in providing such inputs directly on-farm. At further stages of intensification, as markets develop, there can be a movement towards specialization again as a result of changing relative values of land, labour and capital. In these cases, fertilizer replaces manure, tractors replace draught animals and concentrate feeds substitute for crop residues. Livestock are thus raised in specialized, industrial-scale settings. Direct crop–livestock interactions diminish, and instead, markets provide the remaining necessary linkages, mainly through feed resources from crops. Critical to this process is an increase in the value of labour relative to capital. The labour required to handle and use bulky animal-waste and crop-residue nutrients gives concentrated feed and fertilizer nutrients, imported from off-farm, a clear cost advantage.

The delinking of crop and livestock production in highly intensive agriculture reinforces the point that the primary constraint to animal agriculture is not land, but rather feed resources. Land becomes a constraint to livestock production only in those systems in which feed resources are derived directly from it, and not imported from outside the system. This raises the recognition of an important distinction between the constraints to crop and animal agriculture: while land resources are fixed in a particular farming system and location, feed resources are not and can be imported.

Any useful measure of livestock intensification should, then, encompass both feed and land-resource constraints. Two principal axes of intensification can be identified, both centred on the primary constraint, feed supply. These are illustrated hypothetically in Fig. 18.2. The first is the standard measure of crop intensification, the increased intensity of land cultivation (crop or fodder). This results in higher plant biomass production and feed availability per animal from

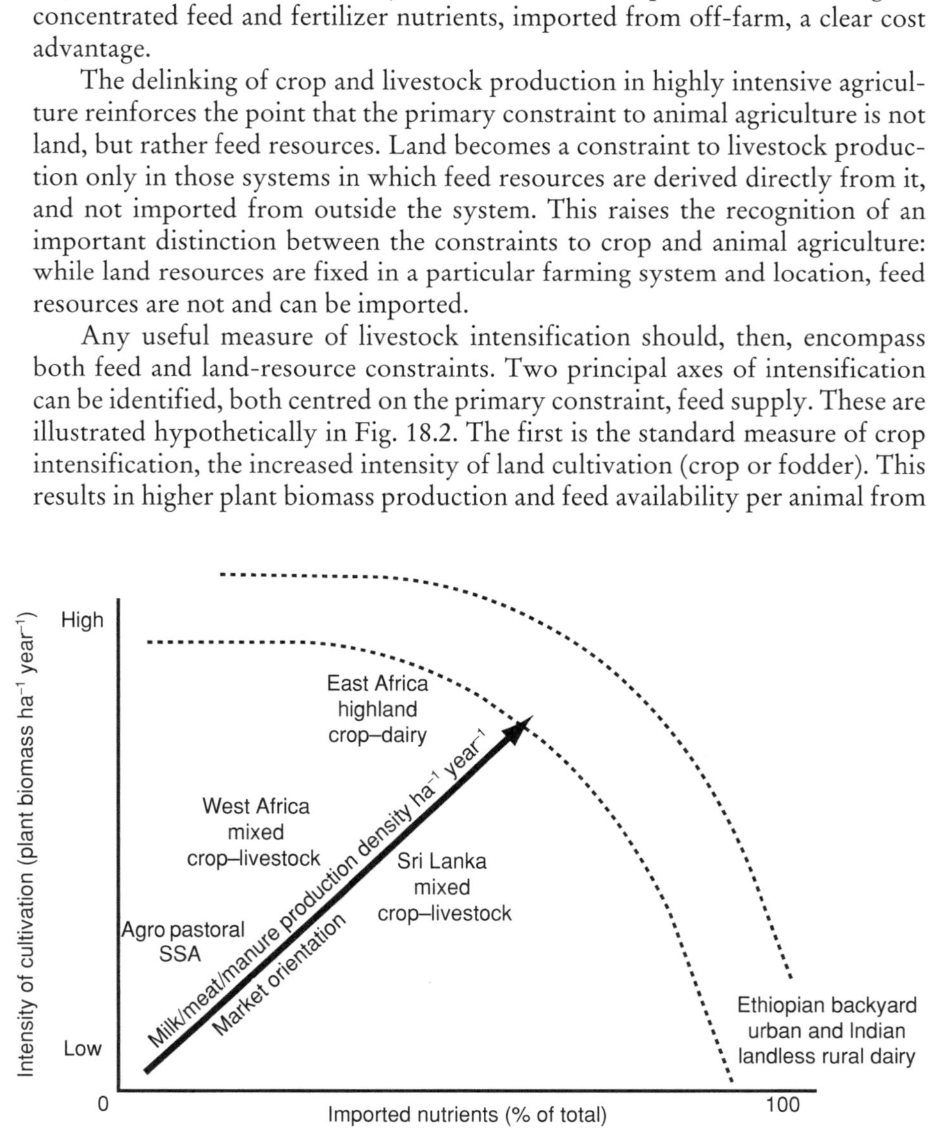

Fig. 18.2. Axes of livestock intensification: examples from smallholder ruminant systems.

land resources within the production unit. This often results in a closer integration of crop and animal agriculture, but may also result in specialized fodder cultivation. A second avenue for intensification is through higher levels of animal feeds imported into the production unit, measured on the horizontal axis of Fig. 18.2 as the proportion of total feed. Although typical of industrial production systems of the West, a high proportion of imported feed is also found in backyard urban agriculture (as in urban SSA) or rural landless livestock systems (as in rural India). Figure 18.2 shows where some specific types of livestock production systems in SSA and South Asia fall within this framework. Increases along either or both of the axes results in higher density of livestock biomass and milk production per hectare, which is represented by a hypothetical isoquant curve.

An empirical example of this framework for analysing livestock intensification can be seen in data gathered in central Kenya from smallholder dairy farmers. Principal-component analysis was used to identify the main underlying axes of variation among 355 dairy farmers across several agroecological zones. Two main principal components emerged, which correspond to farm feed resources used and imported feed resources used, respectively. The results show that the variation among farms along these two axes is relatively independent: even within a given region, farmers can employ either strategy or both (Staal *et al.*, 1999).

Patterns of intensification: local and regional

The primary factors that determine the level of intensification of livestock systems are the agroecological potential of the land (and associated potential technologies), the relative values of domestic factors, such as land, labour and capital, the value of livestock products relative to other potential products and, as a consequence of these, the relative value of imported and own-produced feed resources. These values are naturally dependent on market supply-and-demand conditions, in turn reflecting the level of development and industrialization, disposable income, tastes and preferences for products and economic policies. Based on these, producers choose livestock and crop strategies that best meet their needs for home consumption, income and risk avoidance. For example, in Sri Lanka (Fig. 18.2), where rural labour costs are about double those in Kenya as a proportion of farm-gate milk prices, most dairy producers choose to manage cattle in a semi-intensive manner, relying on a combination of tethered grazing and seasonal cut-and-carry fodder, with some purchased concentrates (Ibrahim *et al.*, 1999). In Kenya, in contrast, the lower relative labour costs often result in labour-intensive dairy production, relying heavily on cultivated cut-and-carry fodder.

Within a given economy, different levels and strategies for livestock intensification occur where there are locational differences in agroecology, domestic factor values and, particularly, market access and infrastructure. In Tanzania, for example, several types of dairy production systems supply the main urban areas. In the northern highlands, far from the urban markets and where market outlets are few, milk prices are low and so farmers purchase few inputs, instead intensively using crop residues and family labour in a closely integrated livestock and crop system. Near Tanzanian cities, on the other hand, milk buyers are nearby and numerous and milk prices are high. Because markets for fodder operate more reliably than those for perishable milk, milk production mostly occurs in backyard urban units, using purchased fodder and concentrates, rather than in peri-urban areas, where the fodder is grown (Omore *et al.*, 1998).

Within a given location, intensification generally occurs when prices or factor values change such that incentives are increased for the use of labour or capital per unit of output. Relative milk prices can rise due to market development or policy reform, relative labour costs can fall with increased population density and relative feed costs can fall with technological innovation. In highland central Kenya, fodder cultivation has increased measurably in the last 10 years, due to rising human population densities, which, unmatched by real economic growth, have led to lower labour/land cost ratios (Staal *et al.*, 1997).

The dynamics of intensification and the effects on crop–livestock interaction remain complex, with the complexity centred on the potential for livestock producers to import the primary resources from outside the farming system. Understanding the potentially 'open' systems[4] of nutrient and waste acquisition and disposal is critical to understanding the effects of livestock development on the environment.

Constraints to industrialization

The industrialization of animal production is characterized by market-based systems where large numbers of animals are kept on relatively small landholdings and fed market-sourced materials. In this way, economies of scale in production, processing and distribution are captured. Such systems are especially common in pig and poultry production, which rely mainly on concentrate feeds and where feed conversion ratios are higher. Ruminant production, which depends more on bulky fodder, is less amenable to industrialization of this type. As a result, only 12% of beef production is from landless (industrial) systems, compared with 39% and 74% of pig and poultry production, respectively (Seré and Steinfeld, 1996).

As suggested previously, more industrial livestock production will be needed to meet rising demand. Such industrialization depends critically, however, on the relative values of domestic resources, particularly labour. The economies of scale of industrial systems are typically captured by substituting capital for labour, such as through mechanized feeding, milking and processing. They also depend on sophisticated output markets, of which processed, high value-added livestock products are an important part. These high-output, high capital-input systems are not generally economically viable, however, where labour costs are relatively low and in markets where the value of manure is positive.

In contrast, in the low labour-value settings of most developing countries, the most efficient livestock production systems tend to be those using low levels of capital (including few animals per farm/household unit and almost no mechanization) and high labour inputs (including family labour), and achieving small output with low levels of value-added before reaching consumers; importantly, they are closely integrated with food-crop production. Such systems also tend to be exhibit strong rural nutrient deficits for N and P, which drives up their value. The close crop–livestock integration allows the capture of the value of manure through the relatively direct return of nutrients to crops. In these circumstances, even production by the landless is often small-scale, based on relatively small household herds. Integration with crops may then occur through the exchange of manure for fodder, as in case of dairy production by the landless in India. This leads to 'area-wide' crop–livestock integration, meaning simply that market or exchange mechanisms transfer manure and crop residues between relatively independent animal and crop production units.

As a consequence of these factors, it has been suggested that there are few economies of scale in livestock production in developing countries, where existing economies lie mainly in processing and distribution (Delgado *et al.*, 1999). It should be noted that, while the real prices of capital and tradable inputs (fuel and fertilizer) or outputs (milk, meat, nutrients) may vary by only factors of 1 or 2, even between rich and poor countries, labour costs, measured by gross domestic product (GDP) per capita, vary by a factor of 30–50 or more. The cost determinants of livestock intensification are thus not simply relative labour/land values, but also labour costs relative to the costs of alternative nutrients (i.e. fertilizer or concentrate feed) and capital. In smallholder systems, labour can be substituted for imported nutrients through better management of manure and crop residues, and can be substituted for capital by hand-milking, herding, etc. Labour can only be substituted for imported nutrients in this way when there is close crop–livestock integration, favouring mixed farming. Smallholders are further relatively advantaged when there are few technological economies of scale under existing factor costs; milking by hand two or 200 cows results in the same unit labour costs. Their advantage arises, then, from their ability to capture the generally lower opportunity costs of family labour over hired labour. These factors combine to favour small-scale mixed farming over industrial large-scale production in many settings where labour costs are low.

The question remains as to how these trends are likely to change in the future. What will happen when relatively land-abundant areas, such as parts of SSA, become relatively more labour-abundant? If general economic development allows a simultaneous increase in the supply of labour and its relative value, there is likely to be a trend towards labour-saving and capital-intensive production. However, we should keep in mind that livestock products are highly income-elastic: demand will grow faster than incomes. The trend towards industrialization will perhaps thus not occur as fast as demand growth, and the factors that favour small-scale and mixed crop–livestock production are likely to maintain their importance for some time to come, particularly in the areas where labour is especially cheap – SSA and South Asia. Not by coincidence, these are two areas where the growth projections in demand for livestock products are among the highest (Delgado *et al.*, 1999).

Policy and Market Approaches to Sustainable Livestock Intensification

As indicated, there is little question that livestock production will increase, that much of that production increase will occur in developing countries and that mixed and industrial production systems will play an increasing role. These processes are already evident. Surveys from the late 1980s, such as in Nigeria (Mohammed, 1990), report significant increases in cattle populations in predominantly crop-producing areas. Large parts of SSA are moving into mixed farming as animal disease constraints are removed. In Asia, a high cropping intensity has been sustained for centuries by crop–livestock integration (Fresco and Steinfeld, 1997). Livestock are currently being displaced from more arid zones (where environmental impacts may be the most uncertain) to subhumid zones, where the environmental impact is more clearly positive. Grazing systems in Latin America, Africa and the Near East are intensifying, which has potential environmental benefits by diverting potential expansion into forest areas (Fresco and Steinfeld, 1997).

Policy interventions and strategies

Given these trends, the uncertainties lie mainly in the extent to which policy-makers can influence them to preserve environmental goals. General policy objectives would appear to be:

1. Preserving and enhancing environmentally sound mobile grazing systems.
2. Encouraging mechanisms for better integration of crops and livestock, including intensifying mixed production, but also market-based integration.
3. Discouraging, where appropriate from a social-value perspective (including environmental costs), the growth in large-scale industrial livestock production.

Appropriate grazing systems may be constrained in the future by rising labour costs, but, where they exist, the key to preserving them may lie in land-use policies. Policies that allow land values to reflect the full economic value for grazing land would help limit the mining of natural and environmental resources. In traditional systems, policies that permit movements of cattle help spread the grazing burden, as well as helping nutrient cycling. Better market and savings institutions in low-income pastoral systems may reduce producer incentives to keep cattle for risk insurance, thereby potentially reducing densities. Reduction of export incentives or trade protection would shift producer returns closer towards socially desirable levels.

Encouraging the intensification of mixed systems will be a key to sustained agricultural production and natural resource preservation (FAO, 1998). Avenues for achieving this include reducing subsidies on inorganic fertilizers, reducing the protection of cereal production, and developing new technologies for handling and treating crop residues, for planted fodders and for small-scale livestock production. The promotion of markets and institutions that serve the needs of smallholder producers for inputs, market outlets and credit and extension services will be important, as will providing for secure land tenure.

Discouraging industrial production, where it is deemed to cause negative net social costs, has typically been approached through regulation, particularly through land-use zoning or regulating the scale of production, animal wastes or pollution. These will need to be complemented by measures that alter the financial incentives for industrial production. Internalizing just the direct environmental costs may, however, increase costs by some 10–15%, not including the less tangible costs of greenhouse-gas production, biodiversity loss, etc. (de Haan *et al.*, 1997). Further, it is not clear that blanket financial incentives will yield an efficient spatial organization and scale of production, which again may require more direct regulation. Other measures may include tradable manure emission quotas, the reduction of subsidies for capital and equipment, increased taxation of some inputs and outputs and, potentially, subsidies for manure marketing. Productivity gains that reduce environmental costs may be available through improved feeding strategies and better manure handling and processing technologies.

New approaches to market integration

Increasing attention is being given to developing market or exchange-based solutions to crop–livestock integration. Sometimes called 'area-wide integration', one can envision a regional mixed farm, where crops are produced in one place and livestock in another, but where animal-waste nutrients and crop residues are transported and exchanged to preserve complementarity. This allows some degree of specialization in production, while capturing the opportunities for both productivity gains and environmental cost reduction of nutrient exchange between animals and crops. An indigenous case of such exchange is found in the large landless dairy system in India, where small-scale landless buffalo milk producers exchange manure for rice straw with nearby farmers.

The success of such integration will naturally depend on the transport costs for manure and crop residues. Due to the weight of moisture and the exorbitant costs of mechanical drying, in some cases the market for manure may only extend for a maximum radius of 10–15 km (FAO, 1998). In other cases, such as use for vegetable production in the Sri Lankan highlands, manure is transported for over 100 km (Ibrahim *et al.*, 1999). The critical variable producing these differences is, again, labour costs, with capital and fuel costs varying much less between countries and regions. Markets for crop residues, such as straw and stover, may also play an important role. While these are already developed in the case of industrial food by-products (e.g. wheat bran), they are less developed for roughages. As conceived on a larger scale in more industrial settings, plans for area-wide integration of crops and livestock typically include control of the location and siting of production. A balance will have to be found between market-led interventions and regulation of scale and location, in order to yield environmental outcomes that do not inordinately compromise economic efficiency.

Changing the paradigm

Livestock's contribution to sustained agricultural intensification is through nutrient provision and turnover, in buffering labour needs and supply and in providing savings for capital requirements. In general, sustained intensification cannot fully take place without the resource-enhancing and stabilizing role of livestock (Fresco and Steinfeld, 1997). Livestock development should be included as a component of most rural development projects since livestock support the livelihood of most rural populations (Vandermissen and Symoens, 1997). The role of markets in integrated systems will necessarily grow if the economies available through specialized production and full nutrient utilization are both to be captured.

The success of market integration of crop–livestock production with regard to nutrient exchange will also depend on changing perceptions towards animal-waste nutrients. Indeed, the concept of 'area-wide integration' suggests an only reluctant deviation from the idea that animal wastes are primarily waste and transport systems are needed to remove them to areas where they can be integrated. It is curious that we do not refer to food markets as 'area-wide integration', even though they also integrate producers' and consumers' needs; we simply call them 'markets', because food is assumed to be a product with intrinsic value. Once we begin to include animal-waste nutrients among the important livestock products – the way we now consider recyclable solid wastes – measures to improve the new animal-product market will improve livestock's contribution to at least the settled environment.

The policy challenge, first, is to recognize those aspects of environmental harm that livestock cause that are not easily managed. Measures to internalize those costs would contribute to enhancing overall social and environmental welfare. Beyond that, the key to capturing the potentially positive contributions of livestock to the environment and natural resource base will depend on promoting the role of integrated crop–livestock production. Smallholder production in mixed systems will remain an important avenue for some time for livestock product supply and for addressing natural resource preservation in the developing countries, where most growth is anticipated. Whether integrated on-farm or through markets, this complementarity will be our means to sustainably intensify livestock production to the levels that future demands will require.

Acknowledgements

The authors gratefully acknowledge Cees de Haan, Nancy McCarthy, Phil Thornton, David Lee and Chris Barrett for comments that substantially improved the accuracy and scope of this chapter.

Notes

1 Ruminants include cattle, buffalo, sheep, goats and camels, while monogastrics include pigs, poultry and equines.
2 Satellite information confirms that the Sahara desert decreased by 15% between 1984 and 1988 following good rains (de Haan *et al.*, 1997).
3 In Kenya, poultry waste is commonly mixed with dairy-cattle feed to contribute protein.
4 'Open' systems may be regarded as those in which a substantial proportion of feed nutrients is imported and/or a substantial proportion of animal wastes is not recycled back into the system.

19 Sustainable versus Unsustainable Agricultural Intensification in Africa: Focus on Policy Reforms and Market Conditions

THOMAS REARDON,[1] CHRISTOPHER B. BARRETT,[2] VALERIE KELLY[1] AND KIMSEYINGA SAVADOGO[3]

[1]Department of Agricultural Economics, Michigan State University, East Lansing, Michigan, USA; [2]Department of Applied Economics and Management, Cornell University, Ithaca, New York, USA; [3]School of Economics, Université de Ouagadougou, Ouagadougou, Burkina Faso

Introduction

African farmers have traditionally pursued shifting cultivation in response to population growth and declining soil fertility. This path of agricultural extensification can still be sustained for some time in land-abundant countries, such as the Congo (Kinshasa), although extensification into pastures, forests and watersheds may lead to potentially catastrophic loss of biological diversity and is therefore an environmental concern. However, in sub-Saharan African (SSA) countries facing rapid population growth and increasing scarcity of cultivable land (per capita), the extensification path is quickly becoming unsustainable or impractical. This scarcity is increasing as the forest, rangeland and wetland margins become exhausted, farmers are often barred from using remaining lands due to the gazetting of parks and protected areas, and soil degradation reduces crop yields and forage growth over time (see Sanchez *et al.*, Chapter 17 of this volume). There are also growing pressures for intensification as population and income growth fuel domestic demand for agricultural products. Farmers respond to these pressures by using more labour and capital per hectare of land, following the theory of induced innovation (Boserup, 1965; Hayami and Ruttan, 1985).

As African farmers are driven to intensify, however, the key emerging issue is what type of intensification they use: sustainable or unsustainable. We contend in this chapter that many African farmers are intensifying in a financially or ecologically unsustainable fashion. For many of SSA's production environments, appropriate technologies for sustainable agricultural intensification (SAI) are available and the crucial problems appear to revolve around thorny questions of institutions, markets and policies that condition the incentives and constraints faced by farm-level decision-makers. Policy researchers, therefore, have a crucial role to play in promoting SAI in Africa. This chapter focuses on two policy research questions: (i) what role market-oriented policy reforms, commonly associated with structural adjustment programmes, have played in diverting African farmers from sustainable intensification; and (ii) what policy actions are needed to put farmers on a more sustainable path. We explain how inappropriate policy reforms and weak markets can lead to either a failure to undertake necessary intensification (versus extensification) or to intensify in ways that are not environmentally sustainable. Appropriate policies and reasonably well-functioning markets, on the other hand, can provide incentives for financially and environmentally sustainable intensification.

The chapter proceeds as follows. In the next section, we discuss the concept of SAI. The following section lays out a simple conceptual framework to link policies and farm intensification choices and discusses in general terms how macroeconomic and sectoral policies have affected farmer incentives for sustainable intensification in Africa. The chapter then presents selected case studies of African farmers following a variety of sustainable and unsustainable intensification paths. A concluding section synthesizes the chapter's main points.

Concepts of Sustainable and Unsustainable Intensification

We define SAI by two criteria: (i) an environmental criterion: the technology protects or enhances the resource base of the farmer and thus maintains or improves land productivity; and (ii) an economic criterion: the technology is financially profitable and meets the farmer's production goals (his/her 'reservation' food and/or cash needs).[1] The sustainability debate in environmental circles has largely focused on the first criterion. But that is insufficient: even if a production technology is environmentally sustainable, if it is unprofitable, farmers will not adopt it and, if it fails to attain their reservation levels for food or cash, farmers will have to push out on to the extensive margin to make up the gap in production, thereby undermining the environmental benefits by harming the fragile margin.[2]

Under present conditions of increasing population density, growing demand for food and fibre and declining soil quality, satisfaction of both of these environmental and financial criteria requires a shift in the use of the three principal factors of production – land, labour and capital. We argue that: (i) capital-led rather than capital-deficient intensification is required (Reardon *et al.*, 1997; Clay *et al.*, 1998); and (ii) capital must shift increasingly from labour-enhancing to land-enhancing technologies.

Prior to the 1980s, land was generally abundant in SSA; hence African farmers needed primarily labour and labour-enhancing technologies to meet food security and cash income needs. Labour-enhancing technologies are primarily those that introduce equipment to speed up ploughing, planting and weeding. Other examples are the introduction of new crops that have peak labour requirements at different times from those of existing crops or the use of herbicides, which reduces demand for weeding labour.[3]

As the population grew in SSA during the 1980s and 1990s, new lands for clearing became scarce and land under cultivation became less productive due to reductions in fallow periods and subsequent soil mining (Smaling, 1993b). This led to a gradual shift in what could be considered 'sustainable' agriculture: farmers needed to start investing in land-enhancing technologies in addition to labour-enhancing technologies. Land-enhancing technologies include those that: (i) provide external sources of nutrients (inorganic fertilizers); (ii) improve soil organic matter and nutrient supplies (animal manures, green manures, crop residues and compost); (iii) reduce soil erosion (terraces, live hedges and tree crops); (iv) increase soil moisture availability (irrigation, rock bunds and all of the above-mentioned techniques that increase soil organic matter); and (v) introduce high-yielding or high-value crops into the cropping system (improved maize seeds or horticultural crops, for example).

Capital-led intensification, with a focus on increasing land productivity, entails a substantial increase in the use of generally expensive, non-labour inputs that enhance soil fertility (e.g. fertilizers) and quasi-fixed capital that conserves soil and soil moisture (e.g. irrigation and soil conservation structures).[4] The labour-enhancing technologies identified above are also capital-intensive, so the need for capital-led intensification is not new. What is new is the urgency of the need for land-enhancing technologies (evidenced by declining yields in SSA despite improved rains during the last decade), the high costs of the capital required to improve land productivity and the greater risks associated with land-enhancing (versus labour-enhancing) capital. If financial capital is not available for investment or price incentives for intensification technologies are not favourable at the point when a production system needs land-enhancing capital investment, farmers will resort to capital-deficient approaches that severely threaten the productive capacity of the natural resource base, presenting great environmental dangers with consequences that can go far beyond the agricultural sector.

The type of 'capital-deficient' intensification that we are most concerned about occurs when a farmer depends inordinately on labour to increase land productivity. Characteristically, farmers following this path will merely add labour – unassisted by improved technologies – to the production process on a given unit of land. This allows them to crop more densely, weed and harvest more intensively, and so on. Over time, this type of intensification depletes soil nutrients and cannot be sustained without shifting toward 'capital-led' investments, such as inorganic fertilizers.

Although 'low-input sustainable agriculture' (LISA) often includes substantial investments in quasi-fixed capital – particularly the type that can be obtained using modest levels of purchased inputs in combination with high levels of labour – it generally fails to meet our criteria for 'sustainable' agriculture, because it does not promote the use of improved seed/fertilizer technologies required to increase land productivity at a pace that can keep up with increased demand for food and fibre and restore soil nutrients. Ruttan (1990) estimates that LISA has the potential to increase food output by only about 1% a year in

Africa, well short of the 3.0–3.5% annual growth in African food production required to meet growing demand. Failure to satisfy food and fibre demand will force a return to extensification on to fragile margins. In addition, capital-deficient intensification uses labour in a way and to an extent that may not be profitable. LISA techniques, such as hand-weeding, recycling organic matter and alley cropping, are labour-intensive. The evidence is mixed on the farm profitability of these practices (Low, 1993), due to the sometimes severe intraseasonal labour shortages common in African agriculture (Byerlee, 1980), as compared with alternative uses of labour off-farm (Reardon and Islam, 1989).

Insufficient use of capital that returns nutrients to the soil, combined with the intensity of land use that characterizes most of the semi-arid and hillside tropics in Africa today, leads to soil mining and degradation. Recent reviews of the soil-science literature show overwhelmingly that inorganic fertilizer is a necessary input for sustainable growth in productivity, even in fragile soils and low-rainfall zones; organic fertilizers simply cannot provide the quantity of nutrients needed to replace those removed by SSA crop production (Weight and Kelly, 1998; see also Sanchez *et al.*, Chapter 17 of this volume). And, when soils degrade due to nutrient depletion, the traditional cycle of shifting cultivation accelerates, limiting the necessary regeneration of forest and range ecologies and undermining the environmental objectives of sustainable intensification.

The capital deficiency of African agricultural intensification is underscored with a few simple facts about inorganic fertilizer use in SSA: 9 kg ha^{-1} in 1995 (Weight and Kelly, 1998), down from 10 kg ha^{-1} in 1993, and compared with 83 kg ha^{-1} in all developing countries in 1993 (Heisey and Mwangi, 1997). The low use of chemical fertilizer is a major reason for worry, both from environmental and food-production perspectives: outside Africa, as much as 75% of crop yield increases since the mid-1960s are directly or indirectly attributable to fertilizer use (Viyas, 1983). Even manure, a key component in most LISA systems, is in short supply in many countries, such as Rwanda, Malawi and Zimbabwe, because of increasing population pressure and the clearing of pasturelands for farming. Tribe (1994) notes that if a LISA food-production strategy, rather than the Green Revolution food-production model, had been pursued in South Asia in the 1960s, 44 Mha that are now under forest would instead be ploughed and cropped. The upshot is that the capital-deficient, unsustainable intensification widespread in Africa today is a major force behind farmland degradation and productivity decline.

In a situation where land constraints increasingly bind and labour/land ratios are rising, one might expect farmers to choose intensification technologies that are as labour-using as possible. Capital-led intensification might, therefore, be a recipe for freeing labour for extensification. This conjecture depends crucially, however, on a 'one-sector' perspective, i.e. that labour is used only in agriculture. In the presence of a multisectoral rural economy, if employment in the non-farm sector pays more than farm labour and/or helps to reduce overall income risk, farmers may eschew labour-using agricultural technologies even in the face of farm labour surpluses. Low (1986) shows this in a case study of hybrid maize adoption in Botswana, where farmers deliberately chose labour-saving technologies in order to free labour for lucrative non-farm work.

Similarly, labour liberated from the farm does not automatically go to bringing more land under the plough. Non-farm work often pays more per day than farm work and provides a means of stabilizing incomes and smoothing consumption through activity diversification (Kelly *et al.*, 1993; Clay *et al.*, 1995). Reardon (1997), in a review of 28 field-survey studies in SSA, found the average share of non-farm income in total income (in cash and in kind) to be fully 45%. Moreover, given weak rural financial systems, the presence of non-farm income can further reinforce capital-led intensification, as these income sources enhance households' capacity to invest in variable inputs and quasi-fixed agricultural capital of the sort emphasized above.[5]

The economic criterion for sustainability (that practices be profitable to the farmer) is perhaps the thorniest issue, and leads to the focus of this chapter on policy issues. Practices can be good for the environment or for meeting household needs, but, if they do not start and continue profitably, the farmer will just ignore them. There is important accumulating evidence on this that needs to have a major impact on future policy-making, technology development and environmental debates.

Market Conditions and Policy Reforms: Mixed Signals

Farmers' choices of agricultural technologies and associated intensification and extensification strategies turn fundamentally on the incentives and constraints they face. This section presents a conceptual model in which policy changes induce changes in market conditions and prevailing price distributions, which, in turn, affect farmers' choices and environmental outcomes.[6] This framework underpins the next section's illustrative case studies.

The basic logic of liberalization is that incentives drive microeconomic behaviour, so removing government-imposed price distortions will lead to greater efficiency and thereby to poverty alleviation and environmental protection. While the logic that incentives affect behaviour is correct, the subsequent expectations too often rely on grossly oversimplified assumptions about the path of transmission from macroeconomic or sectoral shocks to farm-level decisions, including smallholder decisions that affect the natural environment. Policy reforms affect the structure and performance of markets. Government market intervention has generally fallen considerably in Africa over the past decade or so. In some cases, private agents have filled the resulting breach; in some cases, they have not. Market liberalization has been accompanied by state compression, which has often affected public infrastructure provision and services delivery in rural areas, which has further affected market conditions. With rural factor and product markets necessarily responding to macroeconomic and sectoral policy changes, the constraints and incentives (especially price distributions), faced by smallholders have often changed sharply too.

Farmers' production, marketing and technology choices reflect implicit, if not explicit, comparison of expected returns and associated risks across an array of alternative farming technologies and across various farm and non-farm rural sectors. In an environment of imperfect and missing markets, the effective price distributions faced by different decision-makers can vary markedly. As a result, smallholder response becomes quite heterogeneous, since the effective changes to incentives vary widely, depending on households' circumstances. One practical implication is that, where markets are weak, many smallholders rationally opt out of participation in multiple (but usually not all) markets, and their capacity to undertake capital-led intensification falls.

In particular, where asset poverty constrains smallholder liquidity and market imperfections cause complementary variable input (output) prices to be high (low) and volatile, investment in quasi-fixed factors, such as erosion and water control or machinery, which contribute to SAI will be discouraged (Newbery and Stiglitz, 1981; Reardon and Vosti, 1995). Induced changes in crop and technology choice or in input use – including cultivated area, inorganic and organic fertilizer, seeding densities, etc. – can affect the natural environment, thus completing the link from policy changes through markets and farmer microeconomic behaviour to environmental sustainability.

Macroeconomic and sectoral-level policies and incentives for SAI

The main policy changes common in Africa over the past 15 years include macroeconomic and sectoral policy reforms, such as currency devaluation, the liberalization of marketing arrangements (including privatization of government-run parastatals), the removal of fertilizer and seed subsidies and marketing subsidies for crop outputs, and the reduction of farm financial services subsidies. Too often in policy discussions concerning African agriculture, there is a tendency to assume policy effects on output and input prices facing farmers, i.e. on the incentives for SAI. For example, that 'liberalization will raise farm profitability' is a common claim heard in policy debates. Commonly lost amid the *ceteris paribus* assumptions are the complex means by which policies actually affect prevailing price distributions, transactions costs and farmer behaviour. We contend that the effects of the major policy changes associated with structural adjustment have been ambiguous and often disappointing, generating quite mixed farm-level impacts on intensification patterns.

Macroeconomic and related policy reforms (such as currency devaluation and trade or domestic market liberalization) tend to have analytically and empirically indeterminate effects on the incentives facing farmers, either enhancing or reducing net profitability and the relative risk of using crops and technologies required for sustainable intensification. For example, on the one hand, devaluation could raise the output price of an 'intensification crop', such as rice, maize or cotton, more than the increase in input prices. This depends on: (i) the tradability of outputs and inputs; (ii) government 'pass-through' policies, that is, how much of a trade gain they tax away versus passing it on to the farmer; and (iii) private commerce margins. The Malian rice case discussed below is illustrative. Similarly, trade and domestic market liberalization can reduce commercial margins through competition, open up new markets for output and drive down farm-gate input prices, improving profits (Kaufmann and O'Connell, 1991; Savadogo, 1997). Or, devaluation-induced increases in the prices of tradables (increases generally perceived to hurt consumers or farmers) may be counterbalanced by governments that take accompanying measures to reduce these price increases, such as in reducing tariffs on fertilizer in Mali and on rice in Senegal following the 1994 devaluation of the CFA franc (Kante *et al.*, 1995).

On the other hand, devaluation can raise farm-gate prices of imported inputs more than output prices and thus lead to a loss of profitability, as in the Senegal rice case discussed in the next section. Similarly, devaluation and market liberalization can lead to increased isolation of interior markets, raising transport costs, imported input prices and price risk. The limited available evidence suggests that, where state intervention lowered the mean and variance of agricultural product prices, liberalization has increased not only expected prices, but also price variability (Krueger *et al.*, 1988; Barrett, 1997b). The Madagascar rice case reviewed below illustrates this. Price variability can undermine farm investment even where it raises average output prices, because price instability discourages investments in quasi-fixed capital (Reardon *et al.*, 1992; Barrett and Carter, 1999). Price instability also reduces rates of technology adoption, reducing the speed of diffusion of yield-increasing technologies (Kim *et al.*, 1992).

Thus, the effects of macroeconomic and related policy changes on SAI, via output and input price changes, depend importantly on market structure (e.g. concentration of the commercial sector, and entry barriers due to market and physical conditions). The evidence in a variety of African rural areas points to concentration and market entry barriers that tend to produce greater price instability with liberalization and ambiguous effects of currency devaluation on the profitability of intensified input use (Guillaumont, 1992; Barrett, 1997b).

Sectoral policies' impacts on farmers' incentives for SAI

Taken out of the macroeconomic policy-reform context, the effects of most sectoral price policies (taxes, subsidies, price controls) on output or input prices are unambiguous. But, placed in a stabilization context, the effects are uncertain. Sectoral policies can counterbalance macroeconomic reforms, and may even be designed to do so. Moreover, the past generation of policy reforms has tended to emphasize macroeconomic policy and to subordinate sectoral policy, ending sectoral interventions in the interest of achieving fiscal balance, border parity pricing, etc. Sectoral interventions, however, may have important, overlooked 'crowding-in' effects, encouraging private investment in sustainable technologies. We now consider several specific sets of sectoral policies.

Fertilizer/seed policy

African fertilizer use is the lowest in the world and has even decreased over the past decade and a half, that is, over the same period in which fertilizer and seed subsidies and cheap-input financial services programmes have been reduced or eliminated. The effective interest rate for input acquisition rose sharply throughout Africa, as did fertilizer and seed prices. Case study evidence points to a connection between the reduction in fertilizer use and these rising prices for input and financial services (Kelly, 1988; Kelly *et al.*, 1996; Maredia and Howard, 1997; Rusike *et al.*, 1997). Moreover, there is growing evidence that private fertilizer and seed merchants have responded much less than was expected to the liberalization of input markets enacted through the elimination of fertilizer and seed parastatals (Dembélé and Savadogo, 1996; Rukuni, 1996; Rusike *et al.*, 1997). Fertilizer markets in Africa are plagued by a series of fundamental problems, such as risk, seasonal demand, high transport costs, underdeveloped financial services markets and cash-constrained farmers. Small markets add to the problem by limiting economies of scale and product differentiation. Moreover, economies of scale in fertilizer production make domestic production inefficient in most African economies, so domestic fertilizer prices are sensitive to macroeconomic, trade and exchange-rate policies and to volatile international fertilizer prices. African fertilizer subsidies and domestic fertilizer production schemes frequently increased fertilizer consumption in the short run, but they have generally proved ineffective in the long run. Nevertheless, it is becoming clear that private markets in rural Africa cannot at present provide the type of market services needed to stimulate fertilizer demand. Some role for government is inevitable in the short to medium term. Given the considerable costs of delivering fertilizer to farmers on time and the restricted physical availability of fertilizer to most farmers, investment in improved private marketing infrastructure seems to be one of the most promising roles for the state (Ahmed *et al.*, 1989; Rusike *et al.*, 1997).

Financial services policy and capital-led intensification

Before liberalization, delivery of rural financial services was commonly accomplished through vertically integrated parastatals, which controlled all input distribution and output market functions in a subsector. Privatization programmes that eliminated public input and output market interventions not only increased variable input costs for small farmers in many areas, but also raised effective interest rates for rural borrowers, or eliminated their access to seasonal credit altogether. Many private input merchants have found market entry or expansion difficult in the absence of public rural finance schemes (Rusike *et al.*, 1997). While vertically integrated parastatals and private companies (e.g. some of the African cotton schemes) were able to establish functioning credit systems by linking input and the output markets, such integration is rarely feasible, given current institutional and legal arrangements available to newly emerging, undercapitalized private-sector operators. The adverse partial-equilibrium effects of state monopoly or monopsony appear, in hindsight, to have been at least partly mitigated by the favourable general equilibrium effects of loosened rural liquidity constraints, made possible by market concentration and interlinked credit–marketing contracts. The cotton companies of French-speaking West Africa are one of the few surviving examples of this type of vertical coordination, with the success of the systems being ensured, to a large extent, by the effective monopoly in the output market, which results in one of the best agricultural credit reimbursement rates in SSA (i.e. farmers have no other options for selling, so the cotton companies can easily recover the input credit they extend to farmers).

In the wake of reduced rural credit volume, smallholders rely increasingly on cash crop and non-farm earnings, through labour markets or small and medium-scale enterprises, to finance quasi-fixed capital accumulation and to smooth consumption. However, these income sources usually exhibit high Gini coefficients and are distributed regressively – the biggest farmers also earn the most off-farm and from cash crops – so, with the removal of public input and equipment financing, access to alternative (usually self-) financing is often quite concentrated. The effect is concentration of the capacity to follow SAI among only larger operators, leaving the mass of smallholders to either extensify, intensify unsustainably or exit agriculture (Reardon *et al.*, 1994; Clay and Reardon, 1997; Reardon, 1997).

While there is scant empirical evidence of the link between the demise of public financial services and the use of seed and fertilizer, even less is known about how the changes in financial markets have affected physical capital formation. One hypothesis is that effective interest-rate increases undermine such farm investment (Lipton, 1991), but there have been few, if any, studies of this effect. Prices of physical capital items, which are mainly imported in Africa (e.g. animal traction equipment, irrigation pumps, spare parts for vehicles, tractors, ploughs), are driven up by currency devaluation. This translates into higher costs for irrigation schemes, transport and land-conservation investments. We do not know of any studies of the price elasticity of African farm investment, but it is highly likely that the combination of financial-sector retrenchment, contractionary monetary policy and currency devaluation have combined to discourage investment in agrarian quasi-fixed capital. Of course, where such investments had already been made, the capital remained available and seems to have contributed in some places to the maintenance of sustainable intensification practices, as the case-studies of Burkina Faso and Mali in the next section illustrate.

Stimulating rural finance is central to capital-led, sustainable agricultural intensification. While some quasi-fixed capital investment involves considerable commitment of labour – e.g. bunding and terracing – there is usually a complementary commitment of purchased inputs. Moreover, inorganic fertilizer and tools must almost always be purchased. Given considerable seasonality of agricultural incomes (Sahn, 1989), capital-led intensification depends on non-farm earnings, cash-crop earnings or the capacity to engage in intertemporal savings and borrowing. While state-directed rural credit schemes were often fiscally unsustainable and ineffective in serving Africa's most credit-constrained smallholders, there is none the less a strong case to be made for state subsidization of the initial start-up and training costs for self-sustaining rural financial institutions that can mobilize local savings and recirculate them within and across communities as loans (Zeller *et al.*, 1997). Without this, capital-led intensification will probably continue to be beyond the reach of a sizable portion of Africa's smallholders – in particular, the poorest.

Labour and wage policy

In theory, wage policy should have an effect on incentives for SAI: lower rural wages can increase the incentives for using labour to build quasi-fixed capital (anti-erosion structures, for example), but they generally discourage most other types of capital-led intensification (Barrett, 1999). However, in Africa, most wage policy covers only the formal, urban sector. There are few studies mapping the transmission effects of urban wage policy changes on rural workers and hence on induced incentives with respect to technology choice. However, one can hypothesize indirect transmission effects through changed rural-to-urban migration incentives. If the reduction of foreign-aid receipts and shrinkage of the public sector – both of which reduce public-sector employment and wages – contribute to surplus labour in the urban sector and thus lower urban wages, these will discourage rural out-migration and help keep real rural wages low. These factors may also reduce remittances received by smallholder families, and remittance income can be an important source of financing for on-farm capital investment (Collier and Lal, 1984). This may also contribute to a reduction in capital investment in agriculture and, therefore, a rise in labour-led intensification or extensification.

Land policy

Land policy in the past decade has mainly involved titling schemes, gazetting of public areas and some limited land redistribution. The former tends to drive up land prices, spurring intensification and long-term land-improvement investments (Place and Hazell, 1993); land redistribution tends to increase the marginal value product of land use through higher labour/land ratios, as smaller farmers supplant larger farmers (van Zyl *et al.*, 1995; Barrett, 1996; Byiringiro and Reardon, 1996). It must be noted, however, that the impact of land tenure on technology adoption is ambiguous in the case of SSA. Migot-Adholla *et al.* (1991) show that the impact of land-tenure systems is blurred by many other structural factors, such as rural health, education and infrastructure. Land tenure becomes a factor as a country progressively overcomes these constraints.

Finally, the past decade's burst of activity in gazetting lands for protected areas has increased tenurial insecurity for those living in environmentally sensitive areas. If farmers become less certain that the state will not appropriate their land for parks and reserves, they have less of an incentive to invest in the conservation measures required for SAI. The irony is, thus, that pressures for environmental conservation may induce environmental degradation by threatening current operators' control over their land.

Back to the future: projects in lieu of policy?

While public involvement in agriculture is deorganizing with the dismantling of financial services and input parastatals, it is reorganizing as public or non-governmental organization (NGO) projects, which are essentially 'mini-packages' of temporary policies that affect smaller groups. These packages often reproduce a subset of prestructural adjustment policies, including extension services, subsidized 'microfinancial' services, subsidized equipment and input and marketing services, and so forth. These projects are often presented as 'demonstration projects' in areas where diffusion might eventually have a chance.

Good examples include the Sasakawa Global 2000 projects in Ethiopia, Ghana, Tanzania and Mozambique (Putterman, 1995; DE/MAP and DNER/MAP, 1998) and a variety of contract farming schemes (e.g. barley for breweries in the Vakinankaratra of Madagascar, or the case studies in Little and Watts (1994)).

Many of these projects have succeeded in sharply increasing yields on participating farms, but only by circumventing the structural obstacles that often impede adoption of SAI methods. That is, projects have delivered appropriate inputs directly to farmers on a timely basis, often with financial services, obviating potential bottlenecks in commercial distribution systems, and have ensured a market for the output. However, results then often prove non-transferable outside the scheme or unsustainable once the scheme ends. And the schemes themselves are not fiscally sustainable on any significant scale. Such projects do demonstrate that African smallholders can achieve higher-yielding, environmentally sustainable agricultural production. They also implicitly demonstrate how the weak state of rural factor and product markets mutes both incentives to intensify sustainably and the ability of governments and donors to alter those incentives effectively through macroeconomic or sector-level policy. While policy reforms may have been necessary to establish a stable macroeconomic environment, they have generally proved insufficient to remedy the underlying structural problems that induce unsustainable intensification and extensification.

The discussion in this section signals the importance of being able to trace empirically the effects of policy changes on output and input prices in order to provide evidence of a link between policy reforms and changes in the nature of intensification. While this chapter presents a number of hypotheses and some illustrations (below), it is striking how rare empirical studies on these links are. Rather, assumptions and assertions tend to underpin the policy debate. This suggests an important policy research agenda ahead, making use, in the meantime, of whatever evidence is available as to how policy changes are affecting the nature of intensification.

Illustrations

This section presents several illustrative case studies of the effects of policy change and market conditions on the incentives and capacity for farm investments and input use in various African countries, contrasting cases of sustainable and unsustainable intensification. We emphasize especially the role of input and output prices, risk and access to capital – through financial services, non-farm income or cash crops – in the type of agricultural expansion that resulted. The success stories of continued or induced sustainable intensification appear in those cases where necessary investments have been made through projects, in farm-level capital investment or where market proximity and satisfactory infrastructure have enabled markets to function reasonably well. Where state or NGO interventions have resolved structural weaknesses in factor or product markets or have established an agrarian capital base, farmers enjoy incentives and have often pursued SAI. In contrast, many of Africa's poorest smallholders live in relatively remote areas, are poorly served by infrastructure, financial institutions and public services and are faced with poor and volatile terms of trade. In their daily struggle against food insecurity and poverty, the capital-led path to SAI is too often infeasible for these households, frequently leading to a vicious circle of immiseration and environmental degradation.

The centrality of profitability

Farmers exploit crops and/or techniques that prove economically profitable. Profitability entails both the existence of an effective market and a favourable output/input price ratio (Dembele and Savadogo, 1996). Only profitable, commercial agriculture (even if small-scale) induces investment by African farmers in inorganic fertilizers, animal traction, organic-matter use and soil-conservation investments. For example, in Burkina Faso, farmers use 13 times more manure on cotton and maize, both cash crops, than on sorghum and millet, the main subsistence food grains (Savadogo *et al.*, 1998). In Zimbabwe, farmers mainly use improved tillage practices and fertilizers where there are profitable cash crops (Mudimu, 1996). In northern Ghana, fertilizer use is low on average and highly variable over farms, and tends to be applied only to commercial crops (hybrid maize, cotton and rice) and not on subsistence crops (sorghum, millet and cowpea) (al Hassan *et al.*, 1996). In the highland tropics of Tanzania, farmers confine fertilizer and soil-conservation practices to cash crops (Semgalawe, 1998), as they do in Rwanda (Clay *et al.*, 1998) and Kenya (Tiffen *et al.*, 1994). Policy reforms and project-level interventions that render sustainable crops and technologies profitable contribute to environmentally sustainable agricultural intensification.[7]

Onions and rice in Mali: infrastructure investment and policy reform[8]

For years, the Office du Niger (ON) irrigation scheme in Mali was run by a government parastatal, which attempted to control all aspects of construction and maintenance of irrigation infrastructure, crops cultivated and input and output marketing. During the 1980s and 1990s, the government withdrew from many of these activities and reduced import and export restrictions, leaving farmers free to grow crops, purchase inputs and market to whom they wanted. At the same time, the government continued to invest in upgrading the irrigation infrastructure. This enabled farmers in the ON scheme to benefit substantially from the 1994 devaluation of the CFA franc. Both the rice and the onions produced by ON farmers became much more competitive within Mali and the West African region.

Onions, produced during the dry season, are not only much more profitable than the dry-season rice production that the government had formerly been imposing, but also represent a crop in which women are more heavily involved, so there are equity benefits as well. Double-cropping (rice followed by onions) has significantly increased total farm income, the productivity of government investment in infrastructure and the incentives and capacity of farmers to maintain that infrastructure. Improved cash flows leave farmers better able to purchase fertilizer – in 1996/97, 20% did so without relying on financial services – and farm equipment.

Regarding rice, the increased demand stimulated by the CFA franc devaluation led to increases in production, through the adoption of improved practices (transplanting seedlings rather than broadcast seeding, more fertilizer use, more use of hired labour, etc.), and helped farmers realize real increases in net returns per hectare of 8% to 94%, depending on zone and type of production unit. Government price policy (exercised primarily through import duties) also helped. The policy was carefully thought out so that both producer and consumer interests were taken into account. In 1995, rice import taxes were reduced from 46% to 11% in an effort to keep sky-rocketing urban prices under control. This resulted in more imports of low-quality broken rice from Asia demanded by low-income consumers, but did not suppress the producer price of higher-quality domestic rice, which continued to be consumed locally and exported to Côte d'Ivoire. Another important factor was a concerted effort to reform the highly overextended agricultural credit system and to reschedule farmers' debts accumulated in the early 1990s.

Although the ON onion/rice story is an SAI 'success story', there are some potential problems. Overextended farmer credit remains the rule rather than the exception, threatening continued access to fertilizer. In addition, much of the irrigation infrastructure improvement was financed by government loans. The repayment of these loans is costing the government about 300,000 FCFA ha^{-1} $year^{-1}$. This is viewed as a subsidy to ON farmers by the World Bank and there is a movement underway to require farmers to pay these costs. Recent financial analyses suggest that net income per hectare for the principal rice crop barely covers the irrigation subsidy cost for two of the six farm types examined and, for the remaining four, net income is insufficient to cover the subsidy. Achieving further gains in rice yields from such means as greater fertilizer use and triple-cropping, as well as other increases in crop intensification, will have to accommodate an increased share of the infrastructure debt paid by farmers.

Bananas in Rwanda and agroindustrial development

Over the past two decades, there has been a rapid rise in cropping of bananas in Rwanda. Much of the banana crop is processed into banana wine by smallholders. Bananas, as they are grown on the small hillside plots, provide protection against erosion (much more than other food crops), which has become a major concern of Rwandan peasants (Byiringiro and Reardon, 1996). Moreover, the marginal value product of land under bananas (and coffee) is far higher than in alternative uses, and so provides profitable intensification opportunities. Although bananas require a period for initial establishment, food crops can be grown around the young plants, so that even the poor can accept this delay, which is not true of some other cash-crop perennials. Additionally, bananas provide a ready and cheap source of intermediate inputs for banana wine production, a major small-scale agroindustrial activity of the poor. This is an important source of income and one that is far less variable interseasonally than traditional agricultural income. Other non-farm activities have higher entry barriers for the poor, so this is a particularly useful means of income diversification. Although, in the long term, banana wine will probably be supplanted in the diets of urban and even rural consumers, for the medium term, bananas and banana wine are examples of an agroindustrial, cash/food-crop opportunity that meets growing urban demand, and which provides a path for SAI (Kangasniemi, 1998).

Cotton/maize zones in Burkina Faso and Mali

Animal traction equipment programmes, fertilizer and seed subsidies and guaranteed output markets in the Sahel led to expansion of the Guinean savannah zone planted in cotton in Mali and Burkina Faso in the 1970s and 1980s. The results of these programmes – all achieved through vertically coordinated, mixed public/private firms – raised the net profitability of cotton (Savadogo *et al.*, 1998). Relatively large amounts of fertilizer, organic matter and animal traction were used on both cotton and the rotational crop in this setting. Dioné (1989) found that the income generated from Malian cotton cropping increased the ability of farmers to buy inputs for maize production. Through the cotton scheme, farmers also had access to financial services to buy animal traction equipment, which increased the productivity of both cotton and maize production.

Cases of unsustainable intensification and extensification

In most of the cases that follow, policy reforms did not create the adverse conditions that, in turn, sparked unsustainable intensification or extensification; rather, liberalization simply laid bare long-standing structural impediments. Low population densities and high import costs made infrastructure expensive, climatic variability increased output and input market risk, scant cash savings and limited marketed volumes led to thin rural financial markets, and so forth. However, at great expense to the state and to donors, there were widespread efforts to overcome those obstacles through subsidies of various kinds: in output marketing, input provision, financial services, storage and transport. By the mid-1980s, though, state subsidies had proved fiscally unsustainable. In schemes tied to profitable export or domestic markets (e.g. cotton in the Sahel or horticulture in Kenya and Zimbabwe) or to certain politically powerful producer groups, such as large landholders in Zambia and Zimbabwe (Rusike *et al.*, 1997), little was disturbed. There was sometimes even growth, because the capacity and conditions were in place to take advantage of the new incentives wrought by macroeconomic policy reforms. The story was very different, however, for the broader swathe of producers: smallholders producing grains, roots and tubers under rain-fed conditions for domestic markets. The cases of SAI among this strata of smallholders prior to market-oriented policy reforms were essentially through public imitations of private/public cash-crop schemes; when the subsidized elements of these schemes were withdrawn, there were reductions in market coverage for both outputs and inputs, increases in input and financial services prices and greater risk exposure. The result was extensification, as in the Madagascar case, and a shift toward capital-deficient intensification, as in the Zimbabwe, Zambia and Senegal illustrations below.

Rice in Madagascar

Madagascar's economy is dominated by its rice sector. Until economic reforms in the 1980s, the state controlled rice prices, keeping them low and stable, as in most low-income countries (Krueger *et al.*, 1988). With reforms, prices were decontrolled, rose and became more volatile, due in part to sharp exchange-rate devaluation and in part to the weakening state of the private marketing infrastructure and financial markets (Barrett, 1995, 1997a,b). The same factors led to reduced fertilizer use. Fertilizer distribution has long been highly erratic, and most smallholders have faced serious liquidity constraints, which impede fertilizer use. Devaluation, termination of rural finance programmes, more volatile output market conditions and fiscal cutbacks, which effectively ended rural roads maintenance, all created further disincentives to use imported modern inputs like fertilizer. None the less, the increased mean and variance of rice prices induced Malagasy rice producers – most of whom are net rice buyers – to stimulate output by expanding the area in cultivation through the further shortening of fallow periods and extensification into fragile forest margins (Barrett, 1998, 1999). Madagascar's unique ecosystems make the prospective environmental costs of extensification or unsustainable intensification particularly high. Without significant improvements in production technologies or the application of modern inputs, farmers continue to shorten fallow periods, bring new land into cultivation and further degrade already fragile and erosive soils. Economic reforms in the 1980s appear to have exacerbated this unsustainable trajectory.

Maize in Zambia and Zimbabwe

The Zambian and Zimbabwean maize subsectors present interesting cases where pre-structural adjustment policies – notably, massive investment in rural feeder roads and depots to provide reliable market access; subsidized credit, fertilizer and seed inputs; and expanded extension services – created conditions in the early 1980s for smallholders to adopt long-available hybrid maize varieties. This technology was essential to SAI among smallholders in the most environmentally fragile areas of the country, and a boom (a local 'Green Revolution') resulted. However, neither Zambia nor Zimbabwe could afford the public expenditures demanded by these policies and, under pressure from international financial institutions concerned about fiscal deficits, the system was dismantled in the late 1980s and early 1990s. Smallholders in both countries subsequently shifted away from fertilizer use on maize and toward capital-deficient intensification, which threatens soil fertility. Private smallholder output and input markets are slowly emerging, but it is too early to tell how widespread and successful this will be and whether it will render intensification once again financially profitable for the majority of smallholders (Eicher, 1995; Howard and Mungoma, 1997; Rusike *et al.*, 1997).

Groundnuts in Senegal

The Senegalese groundnut basin is a good illustration of a production system evolving from one needing labour-enhancing technologies to one needing land-enhancing technologies. Animal traction equipment and fertilizers were both strongly promoted in the 1960s through government credit and input distribution programmes, but it was animal traction that farmers adopted most enthusiastically. Animal traction, particularly mechanical seeders, permitted farmers to speed up planting and cover more hectares of groundnuts immediately after the first rains. As the land frontier began to diminish and soils began to show signs of loss in productive capacity, farmers became more enthusiastic about fertilizers (Kelly, 1988; Kelly *et al.*, 1996). Unfortunately, groundnuts are not highly responsive to fertilizer, but financial analyses show that fertilizer use was profitable at the subsidized price levels prevailing during the 1960s and 1970s (Tourte *et al.*, 1971).

Then, in the early 1980s, the government-run agricultural credit programme fell apart, due to high levels of debt accumulated during a prolonged period of recurrent droughts. It became virtually impossible for most farmers to obtain fertilizer credit. This was followed in the late 1980s by the complete removal of fertilizer subsidies. The combined impact of these two policies brought fertilizer use on groundnuts to a virtual halt, despite the fact that the sustainability of the agricultural system in the groundnut basin demanded investments in land-enhancing technologies (particularly ones such as inorganic fertilizers, which can replenish soil nutrients). By the late 1980s, extensive groundnut production (adding an extra hectare of groundnuts) was as profitable as using recommended doses of fertilizer on an existing hectare, but less risky because there was no fertilizer debt to reimburse in the case of crop failure (Kelly, 1988). Consequently, extensive production was the predominant response for those farmers who had access to adequate supplies of groundnut seed and land.

From the late 1980s on, however, it has become increasingly difficult to find new land to bring into cultivation and the productive capacity of land already cultivated has continued to decline. Farmers have responded to dropping yields by increasing groundnut seeding densities, to as much as twice the recommended quantities of seed per hectare, but a less risky and more profitable alternative to fertilizer use. Unfortunately, agronomic research suggests that this is not a sustainable practice. Without supplementary fertilizer and organic matter, increased seeding density not only leads to further soil mining but also to reduced seed quality, creating a vicious circle (Gaye and Sène, 1994). While early in the post-devaluation period there was an increase in groundnut area and a small increase in fertilizer use, overall government price incentives for producers have not been sufficient to stimulate a sustained increase in aggregate production and investment. Increased seeding densities were a rational short-run response by producers both before and after the 1994 CFA franc devaluation, but constraints in access to groundnut seed (a major cost of groundnut production), fertilizer, equipment and financial services have sharply limited the post-devaluation response in the groundnut subsector (Diagana *et al.*, 1996; Sène, 1998). In sum, the partial implementation of structural adjustment measures, characterized by government withdrawal from direct participation in credit and input markets but continued intervention in export marketing channels (Freud *et al.*, 1997), appears to have left the groundnut basin no closer to a sustainable production system than it was at the height of government intervention in the sector during the late 1970s.

Summary and Conclusions

Our main contention in this chapter is that SAI in Africa requires capital-led intensification, broadly defined as adequate use of inorganic fertilizer, organic matter and quasi-fixed capital, such as soil-conservation structures and equipment and irrigation. Capital-deficient intensification merely leads to soil mining and a return to extensification, because it meets neither productivity nor sustainability goals. In a highly capital-constrained continent, SAI is clearly a challenge and, at present, most African smallholders appear not to be choosing sustainable paths, leading to the interlinked crises of rural poverty, declining per capita agricultural productivity and environmental degradation. None the less, the vicious circle can be reversed through appropriate investments. Most needed technologies are available already. The key lies in giving African smallholders the capacity and incentives to choose sustainable expansion paths. To date, however, the policy community has focused excessively on macro-level reforms, which have, at best, simply laid bare underlying structural weaknesses in rural markets and, on occasion, have reversed policies that induced sustainable intensification.

As reviewed above, the record of the effects of recent policy reforms on African farmers' incentives and their capacity to undertake investments necessary for SAI is mixed. That mixed record of incentive effects has translated into a mixed record of intensification – including a return to environment-damaging extensification practices in Madagascar, and a shift from sustainable to unsustainable, capital-deficient intensification in Senegal, Zambia and Zimbabwe. In contrast, in the presence of a satisfactory infrastructure and capital base, reasonably stable input and output markets, financial services and agroindustrial opportunities, one finds persistent instances of sustainable intensification – some made even more successful by having been able to take advantage of incentives created by liberalization measures.

The troubling aspect is that the market and infrastructural conditions present in the cases of successful sustainable intensification just serve too few African farmers. They are linked to the fortunate few who have the capacity to produce for the profitable export and urban markets, or to those who happen to fall within well-conceived and generously funded – but inevitably temporary – demonstration projects. The issue, then, is how to reverse the decline in conditions for the broad mass of smallholders producing cereals, tubers and roots under rain-fed conditions for local markets.

To a large degree, this will involve policies that spur private investments, supported by public infrastructure, institutions and goods, to improve the condition of rural factor and product markets. Heavy-handed state interventions in marketing systems have proved to be fiscally unsustainable failures in most of Africa. But necessary state support services for private marketing have also too often been discarded in the course of economic reform programmes. The state needs to steer a middle course between a sole reliance on liberalization and heavy interventionism. The selection of necessary public investments in physical infrastructure and institutional change will need to be made in a country-by-country fashion, supported by cost–benefit analysis, which has only rarely been undertaken since the macroeconomic policy reform period.

Acknowledgements

We are grateful to Arild Angelsen, David Lee, Alex Winter-Nelson and conference participants for comments on an earlier version of this chapter.

Notes

1 We have simply combined sustainability criteria set out by others: in particular, Lynam and Herdt's (1989) definition of sustainability of the resource base as the maintenance of total factor productivity, the combination of ecological sustainability and the meeting of human needs in the Consultative Group for International Agricultural Research/Technical Advisory Committee (CGIAR/TAC, 1988) and the profitability criteria for adopting and maintaining a technology, discussed for example in Feder *et al.* (1985).

2 In this chapter we focus on 'private' (i.e. farm-level) aspects of financial and environmental sustainability. In the long run, however, agricultural intensification will not be truly sustainable unless government policies and programmes to stimulate intensification are 'socially' (i.e. economically) profitable.

3 Labour-enhancing technologies, such as animal traction, may also contribute to increased returns to land (Savadogo *et al.* (1998) on Burkina Faso), but they usually lead to extensive production practices (Kelly (1988) on Senegal; Barrett *et al.* (1982) for Burkina Faso; Sargent *et al.* (1981) for French-speaking West Africa).

4 Note that we use the term 'capital' to refer to both variable and fixed inputs to the production process.

5 The presence of profitable non-farm opportunities can also discourage labour-using investments in quasi-fixed capital if non-farm wages are greater than the expected returns to the labour invested in the quasi-fixed capital, i.e. there are various forces at work here.

6 The heuristic model is drawn from Reardon and Vosti (1992), Reardon *et al.* (1995) and Barrett and Carter (1999).

7 Although the bulk of the evidence suggests that expensive inputs, such as fertilizer, are used primarily on cash crops with relatively good market prospects, there are cases where fertilizers may be used on non-marketed cereal crops (e.g. when other sources of agricultural or non-farm income are available to pay for the fertilizer) and where fertilizers are not used on major cash crops (even when provided to farmers on credit) because the fertilizer response of the cash crop is relatively poor – e.g. certain coffee and groundnut crops. In the latter case, farmers often take the fertilizer on credit and apply it to food crops or sell it to other farmers.

8 This section draws heavily on Mariko *et al.* (1999).

20 Intensive Food Systems in Asia: Can the Degradation Problems be Reversed?

PRABHU L. PINGALI[1] AND MARK W. ROSEGRANT[2]

[1]*Economics Program, International Maize and Wheat Improvement Center (CIMMYT), Mexico City, Mexico;* [2]*Environment and Production Technology Division, International Food Policy Research Institute (IFPRI), Washington, DC, USA*

Introduction

The last three decades or more have witnessed the phenomenal growth of cereal-crop productivity in the developing world, particularly for rice and wheat in Asia. Green Revolution-type growth in cereal-crop productivity resulted from an increase in land productivity and occurred in areas of growing land scarcity and/or areas with high land values as a result of good market infrastructure, and was always associated with strong policy support. High levels of investments in research and infrastructure development, especially irrigation infrastructure, resulted in the rapid intensification of the lowlands. Consequently, the irrigated and the high-rainfall lowland environments became the primary source of food supply for Asia's escalating population.

As we look ahead to the future, the pressure on the irrigated lowlands will continue to be high for extracting food surpluses to meet the rising demand induced both by population and by income growth. Asian wheat demand in 2020 is projected to be around 322 Mt as compared with 205 Mt in 1993 (Rosegrant and Ringler, 1997). The demand for rice, although anticipated to decline in per capita terms, will rise in the aggregate from 309 Mt in 1993 to an estimated 410 Mt. Even with an anticipated expansion in imports into the region, domestic supplies would have to be enhanced substantially in order to meet future food demand. Virtually all future output growth in Asia must come from increased yield per unit of land, since the opportunities for further area expansion are minimal.

Recent signs, however, indicate a slow-down in the productivity growth of the primary cereals, rice and wheat, especially in the intensively cultivated lowlands of Asia. Slackening of infrastructure and research investments and reduced policy support partly explain the sluggish growth. This chapter argues that, in addition to the above factors, degradation of the lowland resource base, due to intensive use over the long term, also contributes to declining productivity growth rates. The root cause of lowland resource degradation is not intensification *per se*, but rather the policy environment, which has encouraged monoculture systems and injudicious input use.

Trade policies, output price policies and input subsidies have all contributed to the unsustainable use of the lowlands. The dual goals of food self-sufficiency and sustainable resource management are often mutually incompatible. Policies designed for achieving food self-sufficiency tend to undervalue goods not traded internationally, especially land and labour resources. As a result, food self-sufficiency in countries with an exhausted land frontier has come at a high ecological and environmental cost. Appropriate policy reform, at both the macroeconomic and the sectoral levels, will go a long way towards arresting and possibly reversing the current degradation trends.

This chapter provides an extensive review of the existing evidence on intensification-induced degradation in Asian rice monoculture systems, as well as rice–wheat systems which are prevalent in the Indo-Gangetic plains of South Asia. Opportunities for arresting and/or reversing the current degradation trends are discussed, along with policy changes conducive to a more sustainable use of the Asian lowlands.

Intensive Lowland Food Systems – Emerging Problems of Productivity Slow-down

Increasing cereal-crop productivity through the application of modern science to agriculture has been most successful in land-scarce economies, particularly in Asia. Partial and total factor productivity studies conducted in Asia attest to the contribution of biological innovations in increasing food production and alleviating food scarcity. The returns to investments in agricultural research and irrigation infrastructure have been the highest in areas of acute land scarcity but with good market infrastructure (Pingali and Heisey, 1996). Rising land values and the rapid adoption of land-augmenting technical change have been the primary factors contributing to productivity growth in much of Asia for both rice and wheat (Byerlee and Moya, 1993; Pingali *et al.*, 1997). By the mid-1990s, 91% of the wheat area in Asia was under modern high-yielding varieties; the corresponding figure for rice was 74% (CIMMYT, 1996; IRRI, 1997).

Intensive rice monoculture systems in South and South-east Asia are the hallmark of the Green Revolution in rice; less well known, although equally important for South Asia, are the rice–wheat systems that emerged at the same time. The rice–wheat systems – a crop of rice followed by a crop of wheat in the same year – cover approximately 12 Mha of the Indo-Gangetic plains, which span an area from Pakistan to Bangladesh. Productivity growth in these two intensive food production systems has had an enormous impact on food supplies and food security in Asia.

Virtually all future output growth in Asia must continue to come from increased yield per unit of land, since the opportunities for further area expansion are minimal – hence the need for sustaining the productivity and profitability of intensive food-production systems. However, there is increasing evidence that the growth in cereal yields has levelled off, especially in the irrigated lowlands of Asia, and there is a danger of future declines in yield growth (Pingali *et al.*, 1997). In the past decade, the growth in aggregate rice output in Asia has declined to 1.5% year^{-1}. Rice-yield growth in Asia also declined sharply in the 1980s, from an annual growth rate of 2.8% in the preceding decade to 1.4% during the period beginning in 1986. Wheat output and yield growth rates exhibit a similar trend, although they tend to be slightly more favourable than the trends for rice (Pingali and Heisey, 1996).

The slow-down in rice and wheat productivity growth in Asia since the 1980s has been caused by: (i) world cereal price-induced factors; and (ii) intensification-induced factors. World cereal prices have been on a declining trend in real terms since 1900 (Mitchell and Ingco, 1995). In the case of rice, declining prices have caused a direct shift of land out of rice and into more profitable cropping alternatives and have slowed the growth in input use and yields. Probably more importantly in the long run, the declining world rice price has caused a slow-down in investment in rice research and irrigation infrastructure (these issues are discussed in detail in Rosegrant and Pingali (1994)). Decreasing investment in research on other cereal crops, such as wheat, is also evident (G. Traxler, D. Byerlee and K.B.L. Jain, unpublished manuscript; Maredia and Byerlee, 1999).

Is there an intensification-induced decline in cereal-crop productivity?

Does intensification of land use, independent of world price effects, lead to a long-term decline in cereal-crop productivity? Pingali *et al.* (1997) argue that the practice of intensive rice monoculture itself contributes to the degradation of the paddy resource base and hence causes declining productivity. Declining productivity trends can be directly associated with the ecological consequences of intensive rice monoculture systems, such as the build-up of salinity and water-logging, declining soil-nutrient status, increased soil toxicities and increased pest build-up, especially soil pests. Many of the above degradation problems are also observed in the irrigated lowlands where wheat is grown after rice in a long-term rotation (Hobbs and Morris, 1996).

Cassman and Pingali (1993) provide evidence using long-term experiment-station data on declining yields and productivity under intensive rice monoculture systems. The essential message from these results is that, under intensive rice monoculture systems, productivity over the long term is difficult to sustain, even with the best scientific management. Similarly, for continuous rice–wheat cropping systems, evidence of long-term productivity declines is beginning to show up at the experiment stations of Bhairahawa, Nepal, and Faizabad and Pantnagar, Uttar Pradesh, India (Hobbs and Morris, 1996).

At the farm level, declining yield trends are usually not observed, since input levels are not held constant over time. However, in areas where intensive rice monoculture has been practised over the past two to three decades, one does observe stagnant yields and/or declining trends in partial factor productivities, especially for fertilizers, and declining trends in total factor productivities (Cassman and Pingali, 1993; Pingali *et al.*, 1997). Similar partial factor productivity trends for the rice–wheat zone are reported by Sidhu and Byerlee (1991) and Hobbs and Morris (1996).

Where intensification is not associated with a change in the inherent productivity of the paddy resource base, declining factor productivity indicates movement along a production function. Where intensification leads to reduced productivity of the resource base, declining factor productivities signify both a shift downward of the production function and a movement along the new production function. The following section discusses possible ecological and environmental factors that could cause a downward shift in the production function and hence a slow-down or, in the extreme, a decline in cereal crop productivity in the lowlands.

Ecological Consequences of Intensification

Intensive rice monoculture, as well as the rice–wheat rotation in the lowlands, results in such production-system changes as: (i) seasonal wet- and dry-crop cycles over the long term; (ii) increased reliance on irrigation and inorganic fertilizers; (iii) asymmetry of planting schedules; and (iv) greater uniformity in the varieties cultivated. In the long run, these changes impose significant environmental costs, due to negative biophysical impacts.

The most common environmental consequences of lowland intensification are: (i) the build-up of salinity and waterlogging; (ii) the depletion/pollution of (ground)water resources; (iii) the formation of a hardpan (subsoil compaction); (iv) changes in soil-nutrient status, nutrient deficiencies and increased incidence of soil toxicities; and (v) increased pest build-up, pest-related yield losses and associated consequences of increased and injudicious pesticide use. A brief description of each of these problems and the possibilities for reversing them are discussed below. At the farm level, long-term changes in the biophysical environment are manifested in terms of declining total factor productivity, profitability and input efficiencies.

Salinity and waterlogging

Intensive use of irrigation water in areas with poor drainage can lead to a rise in the water-table due to the continual recharge of groundwater. In the semi-arid and arid zones this leads to salinity build-up and, in the humid zone, waterlogging occurs. Salinity is induced by an excess of evapotranspiration over rainfall, causing a net upward movement of water through capillary action and the concentration of salts on the soil surface. The groundwater itself need not be saline for salinity to build up; it can occur due to the long-term evaporation of continuously recharged water of low salt content (Moorman and van Breemen, 1978). Postel (1989) estimates that 24% of irrigated land worldwide suffers from salinity problems, with India, China, the USA, Pakistan and the republics of the former Soviet Union being the most affected. In the short term, salinity build-up leads to reduced yields, while in the long term it can lead to the abandonment of croplands (Postel, 1989; Mustafa, 1991; Samad *et al.*, 1992).

Salinity problems are caused by excessive irrigation, poor drainage (especially, seepage from unlined canals), poor irrigation system design and management constraints. For instance, in Pakistan's Sind Province, large areas became saline after the introduction of extensive irrigation, which led to a rise of the water-table from a depth of 20–30 m to 1–2 m within 20 years (Moorman and van Breemen, 1978); other examples from South Asia can be found in Dogra (1986), Abrol (1987), Chambers (1988) and Harrington *et al.* (1992). Dogra (1986) estimates that, in India, nearly 4.5 Mha are affected by salinization and a further 6 Mha by waterlogging.

In higher-rainfall areas, such as in East India, induced salinity build-up is not as much of a problem because the rain flushes out the accumulated salts. However, excessive water use and poor drainage cause problems of waterlogging in this zone. Waterlogged fields have lower productivity levels, because of lower decomposition rates of organic matter, lower nitrogen availability and accumulation of soil toxins. In the case of wheat, low plant populations in some areas can be attributed to waterlogging, especially waterlogging occurring early in the growing season during germination and the emergence stages of wheat. Hobbs *et al.* (1996) report that, in Nepal's Tarai region, waterlogging has reduced yields by 0.5 t ha^{-1}.

Once salinity has set in, it becomes very difficult and expensive to reverse. Salts have to be flushed out of the soil and drained from the area. This process, which can be very expensive, requires large quantities of fresh water and a drainage system to be in place. Often, retiring saline lands from irrigated agriculture may be more cost-effective than trying to fix them. The opportunity costs of fresh water are very high, for both agricultural and non-agricultural purposes; hence, the flushing of salts may not be the socially optimal use of this water. Similarly, where scarce national resources are to be diverted to the construction of drainage systems, the returns are higher for investments in productive systems than in 'dead' systems.

The problem of induced salinity ought to be managed by addressing the causes of the problem – poor system design and management and inefficient use of irrigation water. For systems that are already in place, improving irrigation water use efficiency would lead to a significant slow-down in salinity build-up. An essential component in improving water use efficiency is pricing irrigation water at its 'true' economic cost. New drainage systems must be planned correctly from the start, even if this means that many new systems may not be cost-effective. These issues will be discussed in more detail below.

Groundwater depletion

Development of groundwater resources has been a significant driving force for agricultural intensification in many parts of Asia. The massive expansion of private-sector tube-well irrigation in Bangladesh, India and Pakistan is the most successful example of private-sector irrigation development in Asia. A 'groundwater revolution' in Bangladesh, beginning in the 1980s, was a key stimulant to rapid agricultural growth in the 1980s and early 1990s. Nearly 1.5 Mha of land were newly irrigated after 1980, in significant part from the installation of shallow tube-wells, spurred by the deregulation of tube-well imports (Rogers *et al.*, 1994).

However, just as excess use of unpriced irrigation water can lead to rising water-tables and salinization, it can also lead to falling water-tables in tube-well irrigated areas, with negative environmental and productivity consequences. The problem of the overdrawing of groundwater often occurs because individual pump irrigators have no incentive to optimize long-run extraction rates, since water left in the ground can be captured by other irrigators or potential future irrigators. Groundwater is depleted when pumping rates exceed the natural recharge rate of the aquifer. While mining of both renewable and non-renewable water resources can be an optimal economic strategy, it is clear that groundwater overdrawing is excessive in many intensive agricultural areas in Asia. In parts of the North China Plain, for example, groundwater levels are falling by as much as 1 m year^{-1}, and heavy pumping in portions of the southern Indian state of Tamil Nadu has been estimated to reduce water levels by as much as 25–30 m in a decade (Postel, 1993). Overdrawing groundwater also increases pumping lifts and costs, due to the lowered water-table, causes land to subside (sometimes irreversibly damaging the aquifer) and induces saline intrusion and other degradation of water quality in the aquifer.

Government intervention to prevent depletion of groundwater in the developing world has proved difficult to implement, been subject to corruption and, in many cases, been very costly. The most successful tube-well development has been through small-scale private investment, which is widely dispersed and difficult to monitor. Only when private tube-well imports and markets were deregulated did the small-scale tube-well revolution take off in Bangladesh. An attempt at reregulation through restrictions on well siting slowed growth in tube-well adoption during 1985–1987 (Rogers *et al.*, 1994). Other Asian countries have also been ineffective in managing groundwater. Indonesia and the Philippines have systems of licensing wells, but these have proved difficult to apply in rural areas. India has been ineffective at implementing licensing laws at the state level, where ownership of all water resources resides. Pakistan has no legal system for licensing groundwater withdrawals, and limited attempts to give ownership of underlying aquifers to municipalities have been challenged in the courts. Even China, which applies a strict licensing system, has been unable to avoid the massive overdrawing on the North China Plain (Frederiksen *et al.*, 1993). Governments should move toward incentive-based systems to effectively manage groundwater resources and reduce the negative impacts from overdrawing without imposing unnecessary explicit or implicit taxes on groundwater and stifling the appropriate use of valuable groundwater resources (Rosegrant, 1997).

Changes in soil-nutrient status

The most commonly observed effect of intensive rice monoculture systems, as well as of rice–wheat systems, is a decline in the partial factor productivity of nitrogen fertilizer. (See Hobbs and Morris (1996) for a review of evidence from the rice–wheat systems and Pingali *et al.* (1997) for a review of evidence from the rice monoculture systems.) Work at the International Rice Research Institute (IRRI) (Cassman *et al.*, 1994) indicates that the declining partial factor productivity of nitrogen in rice monoculture systems is due to a decline in the nitrogen-supplying capacity of intensively cultivated wetland soils. Rice–wheat systems could be facing a similar phenomenon. In addition, increased incidence of phosphorus, potassium and micronutrient deficiency has been attributed to a lack of nutrient balance in applied fertilizers (De Datta *et al.*, 1988).

Dynamics of soil nitrogen supply

Fertilized rice and wheat obtain 50–80% of their nitrogen requirement from the soil; unfertilized rice obtains an even larger portion, mainly through the mineralization of organic matter (De Datta, 1981). The nitrogen-supplying capacity of the soil depends on the previous cropping history and residue management, the quantity and quality of soil organic matter and the moisture regime, which affects the composition and activity of the microflora and fauna that govern the decomposition of soil organic matter and crop residues. In continuous cropping of flooded soils with two or sometimes three crops each year, organic matter is conserved or increased even when all straw is removed at harvest. This is due to the large carbon inputs from the aquatic biomass, such as green and blue-green algae, and to a rate of organic-matter decomposition that is slower than that for dry soils. Despite this conservation, the soil nitrogen-supplying capacity decreases, due to chemical changes in the organic matter and the effects of flooded soils on microbial activity (Cassman *et al.*, 1994). The soil's capacity to provide nitrogen for the plant declines with continuous (two to three crops per year) flooded rice cultivation systems. Declining soil-nitrogen supply results in declining factor productivity of chemical nitrogen, since soil nitrogen is a natural substitute for chemical nitrogen.

The magnitude of forgone yields due to declining soil nitrogen supply is estimated by Cassman and Pingali (1993). Using long-term experiment data from the IRRI farm, Cassman and Pingali estimate the decline in yields to be around 30% over a 20-year period, at all nitrogen levels. In other words, the response function for nitrogen shifts downward, due to the declining nitrogen-supplying capacity of the soil. How can these trends in nitrogen productivity be reversed? The primary leverage point in the cropping system is a break in the continuous flooded rice cycle with the growing of a dry season non-rice crop (a crop that does not require standing water). In the southern coastal plains of China, a cropping system consisting of two crops of rice followed by a crop of barley or soybeans, practised over 18 years, has maintained a high and stable crop yield (Li, 1993, as cited in Kundu and Ladha, 1995). The problem of the declining nitrogen-supplying capacity of the soil may also be manifesting itself in the intensive rice–wheat system of South Asia. There is a need to conduct detailed experimental studies similar to those conducted for intensive rice systems.

Macronutrient deficiencies

In addition to nitrogen, phosphorus and potassium are the two other macronutrients demanded by rice and wheat plants. Phosphorus and potassium deficiencies are becoming widespread across Asia in areas not previously considered to be deficient. These deficiencies are directly related to the increase in cropping intensity and the predominance of year-round irrigated production systems. For example, in China, it is estimated that about two-thirds of agricultural land is now deficient in phosphorus, while, in India, nearly half of the districts have been classified as low in available phosphorus (Stone, 1986; Tandon, 1987; Desai and Gandhi, 1989). Desai and Gandhi (1989) note that this is due to the emphasis on nitrogen, rather than a balanced application of all macronutrients required for sustaining soil fertility. The result of this unbalanced application of fertilizers has been a decline in the efficiency of fertilizer use over time (Ahmed, 1985; Stone, 1986; Desai and Gandhi, 1989).

Micronutrient deficiencies and soil toxicities

Perennial flooding of ricelands and continuous rice monoculture, as well as the rice–rice–wheat rotation, lead to increased incidence of micronutrient deficiencies and soil toxicities. Zinc deficiency and iron toxicity are those most commonly observed in the tropics. Waterlogging and salinity build-up aggravate these problems. In Asia, zinc deficiency is regarded as a major limiting factor for wetland rice on about 2 Mha (Ponnamperuma, 1974). In the rice–wheat zone also, zinc deficiency ranks first in importance among the micronutrient deficiencies (I. Ortiz-Monasterio, personal communication, 1998, Mexico). These are mainly soils of low zinc content. Soils that are not initially of low zinc content also show signs of induced zinc deficiency, due to perennial water-saturated conditions and continuous cropping. Drainage, even if temporary, helps alleviate this deficiency by increasing zinc availability (Moormann and van Breemen, 1978; Lopes, 1980).

Most irrigated lowlands do not start off with any soil toxicities; however, toxicities may build up in some soils due to continuous flooding, increased reliance on poor-quality irrigation water and impeded drainage, especially in soils where a hardpan is formed due to alternating wet and dry cycles. Iron toxicity is the most commonly observed soil toxicity due to intensive irrigated crop cultivation.

Once diagnosed at the farm level, micronutrient deficiencies are relatively straightforward to correct. Zinc deficiencies can be corrected by adding zinc, for instance. Diagnosis itself is not easy, though; quite often micronutrient deficiencies are misdiagnosed as pest-related damage. In the case of soil toxicities, farm-level diagnosis is equally complicated and corrective actions are not as straightforward. In both cases, however, the problem ought to be attacked at its root causes. Periodic breaks in rice monoculture systems and rice–rice–wheat systems (two crops of rice followed by a crop of wheat) and improved water-use efficiency go a long way towards reducing the incidence and magnitude of micronutrient deficiency and soil toxicity problems.

Long-term changes in soil physical characteristics

Seasonal cycles of puddling (wet tillage) and drying, over the long term, lead to the formation of hardpans in paddy soils. The hardpan refers to compacted subsoil that is 5–10 cm thick at depths of 10–40 cm from the soil surface. Compared with the surface soil, a plough pan has higher bulk density and fewer medium- to large-sized pores. Its permeability is generally lower than that of the overlying and deeper horizons. The formation of hardpans makes it difficult to grow non-rice crops after rice in a cropping system; for the rice crop, it contributes to impeded root growth and reduced ability to extract nutrients from the subsoil, and leads to the build-up of soil toxicities, due to the perennial waterlogged condition of the soil layer above it.

A striking example of the problem of hardpans is seen in the rice–wheat cropping system in South Asia. The productivity of the wheat crop is affected by the poor establishment of wheat following puddled rice. If the hardpan is broken through, deep tillage and soil structures are improved through the incorporation of organic matter, which also affects the productivity of the subsequent rice crop by reducing the water-holding capacity of the soil (Fujisaka *et al.*, 1994). Thus, intensification has reduced the flexibility of crop choices in the dry season by changing the soil physical structure.

Increasing losses due to pests

The use of purchased inputs for plant protection was unimportant for cereals prior to the mass introduction of modern varieties. Farmers had traditionally relied on host-plant resistance, natural enemies, cultural methods and mechanical methods, such as hand-weeding, to eliminate pests. Agricultural intensification, in general, and continuous cropping of cereals, in particular, have increased the incidence of weed, insect and disease problems (Hobbs and Morris, 1996; Pingali and Gerpacio, 1997).

In the case of rice, relatively minor pests – leaf-folder, case-worm, army worm and cutworm – started to cause noticeable losses in farmers' fields as the area planted to modern varieties increased, hence the rapid increase in insecticide use in intensive rice monoculture systems (Rola and Pingali, 1993). In the case of wheat, insecticide use is not very prevalent and fungicide use has been largely controlled by the development of varieties with resistance to major disease pressures. However, some diseases, such as *Helminthesporium sativum* (spot blotch) are on the rise in intensive wheat-production zones, as well as the rice–wheat zone. Soil-borne diseases are also becoming an increasingly important factor in constraining yield growth in the rice–wheat areas of the Indo-Gangetic plains. On the other hand, the incidence of Karnal bunt, a very important disease problem in wheat, has been reduced with the advent of the rice–wheat system, because the soil-saturated conditions under rice are unfavourable to the disease build-up over crop cycles.

Insect and disease problems that have emerged have been exacerbated by crop management and pesticide-use practices. Injudicious and indiscriminate pesticide application is related to policies that have made these chemicals easily and cheaply accessible. Heong *et al.* (1992) argues that prophylactic pesticide application has led to the disruption of the pest–predator balance and a resurgence of pest populations later in the crop season. Rola and Pingali (1993) have argued that pesticide use has been promoted by policy-makers' misperceptions of pests and the damage they cause. Policy-makers commonly perceive that modern variety use necessarily leads to increased pest-related crop losses and that modern cereal production is therefore not possible without high levels of chemical pest control.

Ecologically safe methods of weed management continue to be a major concern for cereal crop production across Asia. The increasing importance of herbicides in Asian rice production with the shift from transplanting to direct seeding brings with it a concern for changes in weed ecology and the possible emergence of herbicide resistance (Moody, 1994). *Phalaris minor* became the major weed problem with the advent of the rice–wheat cropping system in South Asia (Hobbs and Morris, 1996). The homogeneity of cropping patterns across large areas contributes to the rapid build-up and spread of *Phalaris*. Breaking up the cropping pattern reduces the weed build-up and herbicide resistance problems; however, cropping pattern choices are made on economic grounds rather than sustainability grounds.

The widespread availability of insect- and disease-resistant varieties for the major cereals has reduced the productivity benefits and the profitability of applying insecticides and fungicides (see Pingali and Gerpacio (1997) for a current review of the impact of host-plant resistance for the major cereals and Rola and Pingali (1993) for evidence specific to rice). Even where resistant varieties are used, one could anticipate pest problems, due to a narrowing of genetic diversity on farmers' fields. When many farmers in the same area choose to grow the same high-yielding variety or ones with similar resistance genes, there is a lower level of genetic diversity than would most effectively protect against the emergence and spread of new disease strains (Heisey *et al.*, 1997). However, increasing diversity on farmers' fields is not a simple proposition. The socially optimum level of diversity might differ quite substantially from the private optimum, due to potential yield tradeoffs and the cost of frequent varietal replacement (see Heisey *et al.* (1997) for an excellent case study of wheat-rust management through enhanced genetic diversity in the Punjab of Pakistan).

Intensification Policies – Learning from Past Mistakes

Ironically, the very policies that have encouraged increased food supplies through intensive monoculture systems have also contributed to the declining sustainability of these systems. Intensification policies have operated under two presumptions: (i) that the lowlands are resilient to intensification pressures and that they could sustain productivity growth indefinitely; and (ii) that modern technology has provided a 'silver bullet' solution to food-supply problems of

Asia. Traditional farming systems were sustainable because of lower intensities of cultivation and because they benefited from a stock of farmers' technical knowledge about the crop and paddy resource base, built over millennia. Neither farmers' knowledge nor science was able to predict the changes imposed by intensification and modern technology use on the biophysical resource base. However, by learning from the experience of intensification-induced environmental degradation, degradation of agricultural resources can be brought under control, halted or possibly even reversed through appropriate policy and management reforms. These policy and management reforms are discussed in the following sections.

Water and irrigation policy

Massive investments in irrigation infrastructure were essential for the success of the Green Revolution. However, in retrospect, there were significant design problems in many systems. Drainage investments were deliberately left out of irrigation projects to keep costs down (NAS, 1989). But it was not just the design of irrigation systems that was problematic; prevailing modes of system management, water allocation and water pricing also contributed to the environmental costs of irrigated agriculture. Water has been widely provided at virtually no cost, encouraging its overuse, and administrative control of water allocation has divorced water users from the effective management of the resource.

In order to create incentives for efficient and more environmentally friendly water allocation, water subsidies (and power subsidies for the operation of tube-wells) should be phased out, with more realistic water charges in all sectors. In the longer term, markets in tradable water rights should be established, where feasible. The establishment of secure water rights for water users is an important foundation for creating economic incentives for efficient water allocation. Moreover, responsibility for irrigation-water management should be devolved where possible to autonomous local institutions with user representation and/or joint ownership. Full financial responsibility should be granted, including the right to charge for water and services.

Evidence on user-based allocation, both in traditionally farmer-managed systems and in those systems devolved from public management, indicate substantial benefits from local management when the enabling conditions are supportive (Ostrom, 1992; Meinzen-Dick *et al.*, 1994; Turral, 1995; Vermillion, 1996). This allocation option, nevertheless, requires strong institutions for collective action. The social capital of village institutions in Asia provides a stronger basis for user-based organizations than is found in many other parts of the world. However, even in Asia, turning over the infrastructure and management of systems to local communities has often failed because of flaws in internal structural features or external factors, which affect the viability and sustainability of user associations in managing irrigation systems. Among a wide range of factors affecting the viability of water-management organizations, property rights are critical (Meinzen-Dick *et al.*, 1994). The cohesive force of property rights is important in many aspects of water management, but is especially critical for allocation. User groups cannot make decisions regarding water if they have no *de jure* or *de facto* rights over the resource.

Water rights are also important for incentive-based water allocation. Markets in tradable water rights may have considerable efficiency and other advantages over other allocative mechanisms, thus empowering water users, encouraging investment in water-saving technology and providing incentives for users to consider the full opportunity cost of water, including externality costs. Despite these potential benefits, constraints to incentive-based water-market approaches are significant in most of Asia. The unique physical, technological and economic characteristics of water resources pose special problems for establishing tradable water rights and developing markets for such rights (Young and Haveman, 1985). Multiple reuse of water creates the likelihood of significant externalities imposed on third parties – that is, spillover effects on other people's welfare from water trades, creating further difficulties in enforcing and regulating water trade. Moreover, surface irrigation in much of Asia consists of very large systems serving many small farmers. The development of water markets at the farm level under these conditions will be difficult, because the measurement of deliveries to large numbers of end-users, as well as the initiation of volume-based charges for water use, requires a combination of technology and monitoring efforts that is not cost-effective in many developing countries. One alternative – used in western Hunan Province in China – is the wholesaling of water from the main systems to distributing organizations (Svendsen and Changming, 1990). This in turn requires an effective water-user organization that can collect from the final users.

Because of these complexities, the establishment of markets in tradable water rights will probably be a long-term solution in much of Asia, will be more extensive for groundwater irrigation and will be concentrated (at least initially) around major urban areas in the case of surface-water irrigation systems. But market-type incentives in water allocation can be strengthened through innovative approaches that combine the benefits of water markets and user management. Although appropriate institutions would need to be designed for specific countries and regions, the wholesaling of relatively large blocks of water to user groups or privately run irrigation subunits could establish appropriate incentives for water allocation. The user group would then be responsible for internal allocation of the water and could resell water that was saved through efficient use. The price charged for water under this allocation method would have to be high enough to improve cost recovery for irrigation and to encourage conservation of water. Water prices can also include pollution or effluent charges in order to reduce the incentive to pollute. Principles for groundwater management reform are similar and will require a mix of instruments keyed to local institutional realities, including the establishment of water-user associations, pumping quotas, pumping charges and transferable rights to groundwater.

Although seemingly paradoxical, reduced water use through incentive pricing and user management is likely to be a 'win–win' policy, achieving both a reduction in environmental degradation and an increase in the volume and value of crop production. Efficient water allocation reduces the amount of water required for each unit of crop production, induces productivity-enhancing on-farm investments and causes a shift at the margin to less water-intensive (and often higher-valued) crops. Although the best evidence of positive impacts comes from Chile and California, where strong incentive-based policy reforms have been implemented (Rosegrant, 1995; Gazmuri Schleyer and Rosegrant 1996), the limited empirical evidence from Asia also shows that crop production can increase even as water use declines. When a communal irrigation system in Nepal shifted to a tradable water-share system, the saving of water on existing irrigated area led to an increase in total irrigated area and production (Small and Carruthers, 1991). Informal intersectoral water markets operating in and around the major river basins in Tamil Nadu resulted in both increased income and increased value of crop production for participating farmers (Palanisami, 1994).

Fertilizer and pesticide policy

Government intervention in the cereals market, especially through output price support and input subsidies, long provided farmers incentives for increasing cereal productivity. In addition to highly subsidized irrigation water, Asian farmers benefited from 'cheap' fertilizers, pesticides and credit (see Monke and Pearson (1991) for an example of Indonesian price policy). The net result was that rice monoculture systems, as well as rice–wheat systems, were extremely profitable through the decades of the 1970s and the 1980s, despite a long-term decline in real world rice and wheat prices throughout this period.

Input subsidies that keep input prices low directly affect crop management practices at the farm level; they reduce farmer incentives for improving input use efficiency, which often requires farmer investment in learning about the technology and how best to use it. In addition to inducing increased use, subsidized fertilizer prices have tended to favour the use of nitrogen fertilizers over other nutrients, creating soil fertility imbalances. As discussed below, the reduction and eventual removal of fertilizer price subsidies can substantially improve the efficiency of fertilizer use. Non-price policies are also important, including location-specific research on soil fertility constraints and agronomic practices, improvement in extension services, development of improved fertilizer supply and distribution systems, and development of physical and institutional infrastructure (Desai, 1986, 1988).

To complement the approaches to crop-management improvement in reducing fertilizer-related degradation problems, fertilizer subsidies ought to be removed to eliminate the incentive for unbalanced and excessive use. The financial costs of fertilizer subsidies to government treasuries are high. The true economic costs can be even greater, as subsidies soak up funds that could be used for alternative investments.

The policy scenario for pesticides is similar to that for fertilizers. Pesticide subsidies provided during the early stage of production technology adoption have led to indiscriminate pesticide use, leading to some of the ecological problems discussed above. Policies are often designed based on farmers' and policy-makers' perceptions of pest-related yield losses, anchored around exceptionally high losses during major infestations, even when the probability of such infestation is low (Rola and Pingali, 1993). For various environmental and human health reasons (see Pingali and Roger (1995) for detailed evidence in Philippine rice ecosystems), integrated pest management (IPM) programmes have been vigorously pursued. To make IPM more attractive, pesticides should never be subsidized, since, as in the case of fertilizers, farmers would have no incentive to invest time in acquiring IPM skills. Removing all explicit and implicit subsidies on pesticides is essential to reduce pesticide use on farms.

Price and trade policy and sustainability

Degradation of the intensive food systems in Asia has been caused not only by poor policies at the sector and farm levels, but also by the macroeconomic setting, which has led to unsustainable management practices. Agricultural price and trade policies in many Asian countries have often been both internally inconsistent and costly for long-term diversified growth. On the one hand, general trade and exchange rate policies have penalized agriculture across the board, while, on the other hand, crop-specific interventions, such as output price protection and input subsidies, have attempted to favour individual crops, particularly rice and wheat.

In most Asian countries, the indirect effects of trade and macroeconomic policies have caused overvaluation of the real exchange rate, which, in turn, has lowered effective agricultural prices (Bautista, 1990). In the Philippines, for example, overvaluation of the exchange rate, arising from the protection of the domestic industry, has lowered rice and other agricultural prices by 30% over the past several years (David, 1990). Similarly, policies in Thailand, Sri Lanka and Pakistan induced overvaluation of the real exchange rate by 15–25% during the 1980s (Bautista, 1990).

Historically, many Asian countries compounded these macroeconomic distortions by protecting cereal prices through import restrictions and tariffs, under the rubric of food self-sufficiency. Removal of trade restrictions would reduce the domestic cereal price, improve consumer welfare, release productive resources for other crops with a comparative advantage in production and reduce the pressure for cereal crop-related resource degradation. As David (1990) has shown, there has been gradual, but significant, reform in this area since the mid-1980s, with many countries in Asia reducing the level of rice protection and moving toward a policy of following the long-term world price in setting domestic rice prices. Ironically, the depreciation of currencies in East and South-east Asia during the financial crisis of 1997/98 could provide a significant additional stimulus to the agricultural sectors of several Asian countries.

With the progression toward global integration, the competitiveness of domestic cereal agriculture can only be maintained through dramatic reductions in the cost per unit of production. New technologies designed to significantly reduce the cost per unit of output produced, either through a shift in the yield frontier or through an increase in input efficiencies, would substantially enhance farm-level profitability of cereal crop production systems. Increasing input use efficiency would also contribute significantly to the long-term sustainability of intensive food-crop production and help arrest many of the ecological problems described previously.

Conclusions

While resource-base degradation is increasingly observed in the intensively cultivated lowlands of Asia, intensification *per se* is not the root cause of environmental and ecological damage. Severe environmental degradation in intensified agriculture occurs mainly when incentives are incorrect, due to bad policy or a lack of knowledge of the underlying processes of degradation. As Asian countries liberalize their agricultural sectors and move away from the single-minded pursuit of food self-sufficiency, one can expect positive resource and environmental benefits.

The policies to sustain intensified agriculture while protecting the environment are quite consistent across the issues identified here. In the broadest sense, these are policies to improve the flexibility of resource allocation in agriculture – through the removal of incentive-distorting subsidies and taxes; the establishment of secure property rights; increased investments in research, education and training; improved public infrastructure; better integration of international commodity markets; and a greater inclusion of populations in developing countries into these markets.

The problem of sustaining productivity growth comes about because of inadequate attention to understanding and responding to the physical, biological and ecological consequences of agricultural intensification. The focus of research and policy ought to be on shifting away from a fixation on 'increasing the pile of grain' to a holistic approach to the long-term management of the agricultural resource base. The irrigated lowlands will continue to be the primary source of food supplies for most Asian countries, and it is imperative that long-term solutions to sustaining the productivity of these lands are found. Policy-makers need to look beyond the risk of short-run production shortfalls in order to ensure stable and sustainable long-term food supplies.

Many of the degradation problems observed in the intensively cultivated lowlands of Asia are not irreversible; appropriate policies will provide farmers with the incentives to invest in more sustainable land and crop-management practices. Techniques for improving fertilizer-use efficiency, for example, are available, but will only be viable at the farm level when fertilizer subsidies are removed. The same is the case with the adoption of IPM techniques and more judicious water management.

Meeting future food requirements in Asia requires sustained productivity growth from the irrigated and the high-rainfall lowlands of Asia. Continued high levels of investments in research and infrastructure improvement, as well as institutional and policy reforms, are necessary to meet this goal.

21 Biodiversity and Agricultural Development: the Crucial Institutional Issues

JEFFREY A. MCNEELY

IUCN – The World Conservation Union, Gland, Switzerland

Introduction

Agricultural development is facing many challenges in both temperate and tropical countries. Important issues being addressed in various forums include the growing dominance of transnational corporations in food and agriculture, the multiple challenges of biotechnology, intellectual property rights and equitable access to genetic resources and the social problem of continuing hunger even where agricultural productivity is relatively high. But this chapter will focus on yet another problem, namely how to address the linkages and tradeoffs between promoting agricultural development, alleviating rural poverty and conserving biological diversity. Particular attention will be given to the issue of how areas designated for conserving biodiversity ('protected areas') relate to agricultural development, primarily through helping to maintain the diversity of ecosystems, species, genetic varieties and ecological processes (including the regulation of water flow), which are essential for supporting agriculture.

Ecosystem Services from Protected Areas

Protected areas – those areas devoted to conserving biodiversity or to conservation purposes – now cover over 13 million km^2 or about 9% of the planet's land surface, with 12,754 sites listed in the latest United Nations (UN) List of Protected Areas (IUCN, 1998). This land is thus slightly more than the land area devoted to crops – around 12 million km^2 (WRI, 1998). Many protected areas are justified on the basis of tourism, but the ecological services these areas provide for society are often far more important. A list of some of these services and the functions they carry out is presented in Table 21.1. Particularly important services at the rural-community level include soil regeneration, nutrient cycling, pollination, provision of clean water and maintenance of

(eds D.R. Lee and C.B. Barrett)

the functioning ecosystems that yield harvestable resources. Such benefits are often difficult to quantify, and even local people may take them for granted. Ecological services do not normally appear in corporate or national accounting systems, but they can far outweigh direct values when they are computed. One recent review estimated that coastal ecosystems provide services worth over US\$4000 ha^{-1} $year^{-1}$, while tropical forests are valued at \$3000, wetlands at nearly \$15,000, lakes and rivers at \$8500 and croplands at \$92 (a figure that does not include the value of the crops produced). At the global level, the value of these ecological services was estimated to amount to around \$33 trillion (10^{12}) per year (Costanza *et al.*, 1997; Table 21.2), roughly 50% more than the annual global gross national product (GNP). This figure, presented with considerable bounds of confidence, is not widely accepted by economists on methodological grounds, but has given a useful stimulus to thinking about the value of ecosystem services. While virtually all ecosystems provide at least some of the

Table 21.1. Ecosystem services and functions (adapted from Costanza *et al.*, 1997).

Ecosystem service	Ecosystem functions
Climate regulation	Regulate global temperature, precipitation and other biologically mediated climatic processes at global and local levels
Disturbance regulation	Provide storm protection, flood control, drought recovery and other responses to environmental variability mainly controlled by vegetation structure
Water regulation	Regulate hydrological flows
Water-supply	Store and retain water
Soil formation	Support soil formation processes
Erosion control	Retain soil within an ecosystem
Nutrient cycling	Store, internally cycle, process and acquire nutrients
Pollination	Support pollinators for reproduction of plant populations
Biological control	Enable trophic–dynamic regulation of populations
Food production	Convert a portion of gross primary production into food
Raw materials	Convert a portion of gross primary production into raw materials
Genetic resources	Produce unique biological materials and products

Table 21.2. The economic value of ecosystem services (from Costanza *et al.*, 1997).

Ecosystem	Area (Mha)	Value (\$ ha^{-1} $year^{-1}$)	Global value (\$ trillion (10^{12}) $year^{-1}$)
Open ocean	33,200	252	8.4
Coastal	3,102	4,052	12.6
Tropical forest	1,900	2,007	3.8
Other forests	2,955	302	0.9
Grasslands	3,898	232	0.9
Wetlands	330	14,785	4.9
Lakes and rivers	200	8,498	1.7
Cropland	1,400	92	0.1
Total annual worth of biosphere services			33.3

listed services, protected areas where biologically diverse ecosystems remain intact are likely to be particularly valuable (e.g. Hooper and Vitousek, 1997; Tilman *et al.*, 1997).

Among the most important ecosystem services for agriculture are those related to water. As an example of the economic costs of poorly managed watersheds, about 880 Mt of agricultural soils are deposited into reservoirs and aquatic systems in the USA each year, reducing their flood-control benefits, increasing operating costs of water-treatment facilities, clogging waterways and shortening the lives of dams, not to mention the costs of the loss of prime agricultural soil (Pimentel *et al.*, 1995). The annual damage to water-storage facilities from sediments carried by water erosion in the USA amounts to an estimated \$841 billion (10^9) per year, with another \$683 billion in damage to navigable waterways, \$2 billion in damage to recreational facilities and \$1 billion for other in-stream uses. Watersheds whose functions are stabilized by protected areas could greatly reduce such damage and provide significant economic benefits.

The importance of watershed protection is illustrated by Nepal's 114-m-high Kulekhani hydroelectric dam, which produces 60 MW of hydroelectricity. The 125 km^2 watershed has been heavily used to produce food, fodder, fuelwood, fibre and shelter, leading to significant erosion in the watershed and sedimentation of the reservoir. Due to this rapid sedimentation, the volume of the reservoir has been reduced by some 66% in the past 13 years and it is projected that the reservoir will be completely filled within the following 7 years; thus the proposed 50-year life of the dam has been reduced to just 20 years (Ministry of Forest and Soil Conservation, 1994).

In some countries, the watershed protection benefits of protected areas have been linked directly to agricultural development. In the lower Mekong basin, for example, the general approach is to use the most productive lowlands more intensively through irrigation, thereby reducing agricultural pressure on the more marginal land of the hills. The hills would then revert to their most productive long-term uses, including forestry (both timber and non-timber products), watershed protection and wildlife conservation. Irrigation and hydroelectric dams would be built to meet the energy and water requirements of the region, reducing demand on both firewood and fossil fuels and permitting basic changes in land-use patterns, which would be of benefit to the protected areas. Part of the plan is a series of protected areas located in the forested uplands, which are unsuitable for permanent agriculture but which form watersheds vital to the productivity of the lowland rice-growing areas and also reduce floods and siltation rates, increase usable water yields and decrease water waste (McNeely, 1975).

Despite the creation of these important values, markets do not work very well for watershed management. While it is feasible to charge for forest products, charges for reduction in sedimentation and enhanced flood control are politically very unpopular and essentially impossible. These 'public-good' benefits accrue to all downstream residents and do not directly generate a stream of revenues; even so, they do prevent costs incurred elsewhere in the watershed and their value should help inform decision-making.

Watershed protection has helped to justify many valuable reserves, which might otherwise not have been established, so at least some governments have recognized the linkages (McNeely, 1987). In Indonesia, the Dumoga-Bone National Park was established with a loan of US$1.2 million from the World Bank, justified on the basis of the protection the park provided for a major irrigation project in the lowlands below. And, in Sri Lanka, the Maduru Oya National Park protects the watersheds of four reservoirs, including the major reservoir for the entire Mahaweli Development Project region. Engineers expect the protection of the watershed provided by the national park to effectively double the life of the reservoir. The challenge is to ensure that a reasonable share of such benefits actually reaches the local communities, rather than flowing primarily to distant cities and towns.

Agriculture, Biodiversity and Protected Areas

With their great species diversity, protected areas are potential sources of agricultural crops (especially for agroforestry), new genetic materials for commercial crops, plants useful for regenerating rangelands that have been overgrazed and biological controls. In addition to the benefits to agriculture provided by the high quality and relatively predictable supplies of water produced by many protected areas, more direct benefits to the agricultural sector are also significant. With more appropriate management of at least some protected areas, both farmers and the agricultural industry could become significant partners in managing these areas. A few examples illustrate the importance to agriculture of maintaining species diversity:

- Arizona's Rock Coral Canyon Reserve, covering 2000 ha, is owned by the US Department of Agriculture. One of just a handful of places in North America where wild varieties of chilli peppers grow naturally, it is the focus of the first proposed government-sponsored *in situ* conservation plan for wild native crops. The project is run by a non-governmental organization (NGO), Native Seeds/SEARCH, which aims to preserve and exploit some of the wild crops in the region. Apart from peppers, four other important wild varieties of native crops grow in the reserve: tepary beans, cotton, squashes and the tequila plant. These have traditionally been gathered by the local Tohono O'Odham people. As recently as 70 years ago, this group cultivated 4000 ha of farms without having to pump groundwater, an impossible dream for most farms in Arizona today.
- Biological control makes use of indigenous and introduced natural enemies to control pest populations. According to one estimate, the pool of plant-feeding organisms, predators, parasites and pathogens from which agricultural scientists draw biological control agents constitutes over 50% of the species on earth. Diversity is depressed in agroecosystems by cropping practices that depress important natural enemies, such as fungi and ants. Natural vegetation around crops is important for conserving the diversity of natural enemies, and possibly more so in tropical than in temperate

systems. Thus, protected-area buffer zones may be especially important as sources of biological control agents, particularly when they are part of dynamic, evolving systems. Demand for new biological pesticides and exotic natural enemies to control introduced pests is likely to increase, and source areas for these natural enemies need to be conserved, as many of them are threatened by habitat destruction (Waage, 1991).

- May (1993) reports that more than 500 species of insects and mites that are pests in crops or orchards have evolved resistance, often to a wide range of different pesticides. He points out that, overall, more than one-third of all agricultural production is still lost to pests, the same proportion as a century ago. New genetically engineered crops that contain insecticidal toxins will soon be available commercially, but the utility of such transgenic crops will be short-lived if insects quickly adapt to the toxins. But the time taken for resistance to appear can be significantly lengthened if fresh supplies of susceptible individuals keep appearing, each generation, in treated regions. May suggests that these susceptible individuals could come from untreated refugia – protected areas – deliberately established to preserve susceptibility. The gene flow from such refugia can greatly lengthen the time before resistance becomes a problem, and thus extend the commercial life of the genetically engineered crops developed by the agricultural industry.

Many protected areas contain wild relatives of domestic plants. Because they are still evolving in nature, these wild relatives survive droughts and floods, extreme heat and cold and numerous other natural hazards, including attacks from the pests and diseases that damage their related crops. Thus many of their genetic traits are valuable to agriculture today. In the Asia–Pacific region, for example, wild species of particular importance include the relatives of rice, millet, beans, bananas, coconuts, sugar cane, mangoes, oranges, yams, tea and many others (Vavilov, 1997). Many protected areas contain important genetic resources; as just one example, the Chatkal Mountain Biosphere Reserve in Kyrgyzstan conserves important wild relatives of walnuts, apples, pears and prunes.

India has established a 'gene sanctuary' in the Garo Hills for wild relatives of citrus and further sanctuaries are planned for banana, sugar cane, rice and mango (Hoyt, 1992). These protected areas are especially important, because they enable the wild relatives to continue co-evolving with their pests. The basic principle behind a genetic reserve is to conserve sufficient genetic diversity to enable the species to continue evolving, with sufficient research capacity available to enable the resources to be utilized. Relatively few protected areas have been well surveyed for their wild plant relatives and in even fewer is there the ongoing research capacity to utilize such resources.

While many seed companies might claim that they have sufficient genetic materials in their own collections, the future is uncertain and new threats to crop productivity are a regular feature of modern agriculture. This is one of the reasons why society continues to make a substantial (though probably inadequate) investment in agricultural research, at both national and international levels. For example, the Consultative Group on International Agricultural

Research spent US$325 million in 1996, with a total staff of over 10,000 people (CGIAR, 1997). The work of the International Plant Genetic Resources Institute (IPGRI) is especially relevant to wild relatives, but virtually all of the crop-based international agricultural research centres also study wild relatives, as do many national agricultural research programmes.

Building increased research capacity is justified, because wild relatives of domestic plants have made many critical contributions to human welfare, providing disease resistance, vigour, nutritional content and other desirable characteristics (Prescott-Allen and Prescott-Allen, 1984). Swanson (1997) indicates that the commercial seed and plant-breeding industry requires an injection of around 8% of 'new' genetic material into the system each year, thus totally renewing the stock of germ plasm in use over a period of about 10–15 years. On average, about 6.5% of all genetic research undertaken in the agricultural industry is focused on germ plasm from wild species and landraces (Swanson, 1997). For the USA alone, the contributions of wild genetic resources amount to an estimated \$340 million year^{-1} (Prescott-Allen and Prescott-Allen, 1986).

In conclusion, protected areas provide natural (and thus inexpensive) storage of genetic variety in an evolving state, and therefore may become ever more valuable as pools of genetic material. Much more needs to be done to clarify the role of protected areas, such as serving as *in situ* gene banks in support of agricultural development, and to design and implement management programmes that will mobilize the benefits to agricultural research and development from protected areas.

Farmers and Protected Areas

The stakeholders with the most direct dependence on protected-area resources are the people living in communities in and around protected areas. In most parts of the world, rural villagers strongly believe that they have historical rights to the land and resources that governments have declared 'protected' in the national interest (e.g. Dang, 1991; Vandergeest, 1996). In India, for example, at least 3 million people live inside protected areas and several million more live in areas immediately adjacent to protected areas; all have been historically dependent on the protected areas for various resources and it would be difficult, if not impossible, to find adequate alternatives for them outside the protected-area system (Kothari *et al.*, 1989). Similarly, in South America, about 86% of the national parks have people living within the boundaries (Amend and Amend, 1995). Thus, for the reason of dependency alone, more inclusive conservation policies are imperative.

In general, modern protected areas are a social and cultural response to the overexploitation of natural resources. The ideal promoted by some conservationists is to preserve a sample of intact undisturbed ecosystems. However, no areas of 'empty', 'pristine' or 'undisturbed' land exist that are free of human influence and historical claims, although most governments have declared all forests, watercourses, lakes and oceans to be government property. Further, many local communities have already established their own protected areas and resource-management practices over long periods of time. These include measures such as sacred groves, forbidden practices, hunting seasons, taboos and various other ways of asserting community interests above those of the individual. While these traditional conservation measures do not serve all the functions of modern protected areas, because national-level concerns may not be addressed, they have the great advantage of strong local support.

The conflict between the ideal of 'undisturbed nature' and the reality of long-term human occupation of the land (e.g. Spencer, 1966; Poffenberger, 1990; Turner, 1990) has led to the wide recognition that conservation cannot succeed unless it is linked to economic opportunities and investments aimed at the rural communities which might otherwise threaten the viability of protected areas through their activities in pursuit of livelihood. For example, the Convention on Biological Diversity calls for governments:

- to protect and encourage customary use of biological resources that are compatible with conservation or sustainable use (Article 10c);
- to support local populations to develop and implement remedial action where biodiversity has been reduced (Article 10d);
- to promote environmentally sound and sustainable development in areas adjacent to protected areas with a view to furthering protection of these areas (Article 8e);
- to ensure that an equitable share of benefits from conservation and sustainable use flow back to local communities (Article 8j);
- to promote the exchange of traditional and indigenous knowledge (Article 17.2).

It is up to each government to determine how these objectives are to be implemented at the national level. Moreover, much work remains to be done, both at the policy level and in particular sites, to generate 'case law' on the kinds of development that are appropriate, how the benefits can be distributed, who monitors the distribution of benefits to ensure equitable sharing, and so forth.

Numerous factors can contribute to a productive partnership between protected-area managers and local communities. Perhaps most importantly, when local people are the primary decision-makers in and beneficiaries of these partnerships, they can reasonably be expected to institute their own conservation measures or to support those initiated by government. Numerous examples cited from various parts of the world (e.g. UNEP, 1988; Stone, 1991; West and Brechin, 1991; Birckhead *et al.*, 1992; Wells and Brandon, 1992; Kemf, 1993; Western and Wright, 1994; Kothari *et al.*, 1996) support the general conclusion that earning the support of local communities means giving them a real stake in the success of a well-managed protected area.

Effective protected-area management also typically involves restrictions on using some resources, involving an opportunity cost incurred by those who had previously used the resources. But the notion that protected areas are designed primarily to benefit the nation or the globe has tended to obscure the reality that most protected areas provide numerous benefits for local communities, especially when the sites are consciously managed to provide such benefits. These benefits can include:

- ecosystem services, such as watershed protection, protection against storm surges (in the coastal zone) and protection against landslides or avalanches in mountain areas;
- protection of sacred sites or traditional lands against alternative uses by outside interests;
- economic benefits from tourism development, including markets for agricultural surpluses and handicrafts;
- meeting daily survival needs, such as medicinal plants, other non-timber forest products, firewood and fish, if necessary through the exclusion of outsiders;
- secondary benefits from infrastructure development, such as access roads, communication and health-care facilities;
- returns for outside use of traditional knowledge;
- various kinds of amenity values.

In some countries, increased awareness about the value of protected areas has driven land speculation, with land developers buying up rights to use lands bordering protected areas, leading to the destabilization of some buffer-zone communities as villagers begin to sell their land, often at prices too tempting to reject. At least one danger of this development is that villagers who fail to use their proceeds wisely may face new economic hardships, which could force them back into forest encroachment.

It is important to recognize that the increasing attention given to local communities does not necessarily imply that local communities are the major threat to protected areas. In fact, in most countries, the major threats to protected areas come from outside influences, such as government-supported timber or mining concessions, road-building activities, subsidies to agricultural development or mechanized fishing, dam construction, expanding populations and accompanying resettlement programmes, air and water pollution and (in the longer term) climate change. Most of these problems need to be addressed as part of regional planning and central government policy, rather than protected-area management. On the other hand, protected-area managers are regularly called upon to address conflicts with people living in and around the protected area (Lewis, 1996), so the increasing attention being paid to local communities is a reflection of a real need at the protected-area management level as well.

Many of the conflicts between resource managers and local communities are played out in 'buffer-zone' areas. Buffer zones are areas adjacent to protected areas where the use of resources is managed in ways that are compatible with the objectives of the protected area (Sayer, 1991). They are critical parts of modern protected-area management, permitting forms of resource exploitation that cannot be permitted in protected areas but which are essential to meeting the needs of local communities. Buffer zones have two broad functions. First, they extend the area of protected natural habitat into a larger management area to permit plants and animals to survive outside the reserve. This 'extension buffering' can include land used for selective logging or hunting. Secondly, they provide products of value to local people, enabling them to harvest and use these resources ('socio-buffer function'). For buffer zones to be effective, the communities dependent on these areas must perceive that their interests are being addressed, requiring a high level of joint planning and cooperation. Appropriate activities in buffer zones can include:

- community and farm forestry, with indigenous species of trees planted adjacent to protected areas, providing a protective belt and a source of income for villagers;
- tourism development, providing income for local people while reducing tourism infrastructure within the protected area;
- reservoirs, which can provide a clear boundary and natural barrier, as well as sources of dry-season water for wildlife, and which can help reinforce the watershed-protection value of the protected area;
- hunting of permitted animals, especially those that can be agricultural pests, such as pigs or deer;
- nurseries for native species, which help provide a cash income from a regular supply of propagated ornamental or medicinal plants, which might otherwise be illegally harvested from the protected area; and
- research stations, which address subjects of interest to both the protected area and the surrounding communities.

It is normally inappropriate to have commercial farming and fruit-tree orchards as buffer zones, as these might prove a source of conflict between the farmers and wildlife from the protected area. In any case, the uses permitted in buffer zones need to be clearly specified in the site management plan, which should be prepared with the active participation of all stakeholders and then clearly communicated to all parties.

While community involvement in protected areas or the surrounding buffer zones is widely seen as essential, it is no panacea. First, many of the problems are of fairly recent origin, as a result of expanding populations, immigration and increased consumption levels, so traditional community-based solutions may not be effective in these new circumstances. Secondly, local communities are not noted for their peaceful relations with neighbouring communities; in fact, the history of village boundaries tends to be a history of conflicts, so building supporting networks requiring intervillage cooperation is not easy. And, thirdly, the fact that local communities are often well-adapted to their local environmental conditions does not automatically mean that they are going to make wise decisions. Deciding how to invest scarce resources in assets that mature over several decades (such as forest trees) or which are highly mobile (such as migratory species of waterfowl) is a sophisticated task, and clearly some individuals or communities will be able to organize themselves more effectively and make better decisions than others. Any community faces a challenging set of problems when it tries to govern and manage complex multispecies, multiproduct resource systems, whose benefits mature at varying rates and are under pressure by competing groups of humans at every step (Ostrom, 1998).

The best general approach to this complex of problems appears to be greater commitment from resource management agencies to working with communities, improved community resource management programmes, effective enforcement of agreed-upon regulations, continuing research and monitoring and long-term commitments by both conservation and development NGOs (Wood *et al.*, 1995).

Community Involvement in Management of Protected Areas

Given the great diversity of farming communities, a vast spectrum of potential approaches to involving communities in protected-area management is available. In many cases, it will be a matter of 'learning by doing', building on existing measures, such as legal ownership of land, customary tenure rights of local communities, legislative frameworks, and so forth. Further, local communities are not the only stakeholders, and finding appropriate ways of involving communities in protected-area management often involves negotiations with other stakeholders as well, including the private sector, NGOs and research organizations. However, rural people are widely demanding a greater voice in decision-making, seeking greater control over the resources upon which their welfare depends. In some cases, conservation can become a strong means of enabling communities to re-establish their identities and protect their cultural distinctiveness (Kemf, 1993). Further, the investment by local communities in their own environment is far higher than that of all external actors put together (Murphree, 1994), so their long-term commitment to sustainability is usually much greater. This section will focus on how local communities can become involved in protected-area management and the problems in doing so.

Involving local communities in protected-area management

The involvement of local people in managing protected areas has clearly worked in at least some cases. In Nepal's Sagarmatha National Park, initial hostility to the park by the local Sherpa people was converted into strong support through economic incentives, such as employment in tourism-related activities, preferential employment as park staff, registration of land to establish tenure rights, restoration and protection of religious structures inside the park, the return of forest management to the village and community development activities clearly linked to the park (Norbu, 1985).

In Indonesia's Irian Jaya province, the Arfak Mountains Nature Conservation Area exemplifies the modern approach, actively involving local people in the preparation of the management plan, marking and maintaining the boundary and then benefiting from economic incentives designed to support the area. The protected area and its surrounding buffer zone were divided into 16 management areas, each of which was assigned to a committee of influential local people to manage in accordance with tribal customs and community decisions, including marking of the boundary. At the same time, the government protected-area management authorities were erecting concrete markers along another part of the boundary, but without village consultation and following maps that included some village lands and gardens within the reserve. As a result, the government's markers have since been removed or ignored, while the western boundary marked by the local villagers is still respected. This approach has been effective, because the boundary falls under multiple jurisdictions and allows rapid identification of violators as either landholders or outsiders, with different regulations applying to the two broad groups. The committees have government-sanctioned powers to enforce government regulations, with graduated penalties if required. This effort has shown that building local support from the outset of protected-area establishment can reduce the costs of boundary demarcation and enforcement of regulations (Craven and de Fretes, 1987; MacKinnon, 1997).

Despite these and other promising examples, far more needs to be done to build support from local communities for protected areas. This will require a challenging combination of incentives and disincentives, economic benefits and law enforcement, education and awareness, employment in the protected area and employment opportunities outside, enhanced land tenure and control of new immigration (especially where the buffer zones around protected areas are targeted for special development assistance). The key is to find the balance among the competing demands, and this will usually require a site-specific solution. The policy guidelines in Table 21.3 suggest a basis for seeking such solutions.

In some countries which otherwise might favour the involvement of local communities, policy distortions favour extractive uses of land for agriculture, livestock, mining or logging over non-extractive uses, such as tourism (Wells, 1997). It is often advisable, once the policy decision has been taken to return greater responsibility to local communities, to hand over management responsibilities in a phased manner, as management capacity is developed. In other cases, it may be more appropriate for local communities to be involved in only some parts of the overall management enterprise – for example, providing some of the services in tourism, producing handicrafts or food, contributing ideas to management plans or discouraging poaching or encroachment by outsiders.

Problems with involving local communities in protected areas

Involving local communities in protected-area management will inevitably encounter some difficulties (Borrini-Feyerabend, 1997). These include the following:

- Some rural communities have no interest in protected areas and fear the forest, so they do not want to be involved in management and have little to contribute in any case, while other communities may not have the time and resources available to invest in the protected area, even if they have the interest to do so.
- Joint management can become a means whereby the already powerful concentrate their power still further and may undermine some of the community controls that previously served to protect the environment.
- Some cultures and groups may find the concept of participation to be alien, hesitating to express views and interests that may be different from those of

Table 21.3. Policy guidelines for involving local communities in protected-area management (adapted from Kothari *et al.*, 1997).

- Identify the local communities and other groups and individuals that have a stake in the protected area, and assess the power relationships of the various interest groups to determine patterns of resource use. On the basis of this assessment, enable local residents to derive benefits from the protected area in proportion to their investment in the area and its conservation objectives
- Build sensitivity towards the inequities within and between communities and make special attempts to empower the underprivileged, including women
- Ensure that the benefits of the protected area to the local community are equal to or greater than the potential benefits from other uses of the protected area (i.e. develop means of compensating local stakeholders for their opportunity costs). This may require economic incentives provided by other stakeholders with an interest in the area (e.g. the tourism industry)
- Specify the functions, powers, rights and responsibilities of local communities in relation to the protected area; acknowledge skills and educational and cultural gaps that might exist, and plan for incremental devolution of responsibilities, along with training
- Where the local people are empowered to protect and utilize resources from protected areas, also raise their awareness of broader environmental issues through the implementation of conservation education programmes
- Develop institutional structures at local and wider levels to facilitate community participation in various protected-area management issues. Provide legislative and policy support to build a strong foundation for such arrangements
- Build an appropriate knowledge base regarding the use and conservation of local resources, especially in the buffer zone
- Develop appropriate attitudes of protected-area staff toward local people, replacing the traditional police role with a more cooperative and collaborative role; develop confidence among the local people that the resources belong to them and will not be alienated after the site has been fully protected
- Select the right person to lead the local-level management committee. Many real leaders may not hold any political position, so select the leader through a democratic means rather than through nomination by the protected-area managers

their neighbours; on the other hand, some governments may discourage local participation or empowerment, especially when it threatens their own authority or is seen as encouragement to opposition groups.

- Participatory approaches require commitment over time, requiring patience from all those involved, even though the problems may be urgent.
- Given the investment of time and energy required by protected-area managers to work with local villagers, they may be forced to neglect other duties, such as resource management within the protected area.

Another significant danger of the greater involvement of local communities in protected-area management is that various destructive forces are fully capable of turning this good idea to their own destructive ends. Local communities are not necessarily paragons of virtue. In India, for example, fish contractors in Pench National Park (Madhaya Pradesh), real-estate agencies in Borivali

National Park (Maharahasdra), industrialists in Narayan Sarovar Sanctuary (Gujurat), timber merchants in various protected areas in north-east India and others have shown that they are quite capable of using local villagers as a front for their vested interests to exploit protected-area resources for their own gains (Kothari *et al.*, 1997). The challenge is to be able to sift out these detrimental interests from the local people who genuinely depend for survival on the natural resources of the protected areas, and empower the latter to be able to live with dignity and be instrumental in alienating destructive forces within and outside protected areas.

In short, involving local agricultural communities in protected-area management has both costs and benefits. The balance between these will vary with the site and over time, but, generally speaking, the investment required to build effective partnerships with local communities will pay long-term dividends to the nation and in terms of meeting biodiversity conservation goals.

Costs of protected areas to surrounding communities

The preceding discussion has focused on the benefits of protected areas to the surrounding communities, but protected areas can also involve costs to them. In some cases, these costs can be very significant, but clearly it is important to recognize their existence as a basis for the full consideration of the benefits and costs accruing to farmers living in surrounding lands. Calculating benefit–cost ratios, even in very rough terms, can help to focus appropriate development programmes or forms of compensation.

Potential costs that need to be considered can include the following:

- crop-raiding by herbivores, including deer, elephants, pigs, rhinos and so forth. A single elephant can cause havoc in a maize field that a farmer has carefully tended until the crop is about to be harvested, losing virtually the entire season's work in a single night.
- Increased predation on both livestock and people. Predators, particularly large cats, can have a profound influence on livestock such as sheep, goats and cattle. Smaller predators may prey on chickens, ducks or even pets, and large animals supported by protected areas – elephants, buffalo, rhinos, cats – kill hundreds of rural people every year.
- Transmission of diseases between wild plants or animals and domestic species – and even humans – is a problem in some parts of the world, where protected areas can serve as reservoirs for various kinds of pathogens and pests, ranging from viruses to various kinds of insects.
- Opportunity costs, where the local populations no longer have the option of converting lands contained within the protected areas for agricultural purposes.
- The termination of conditional access, where local populations may be denied access to their traditional sources of fuelwood, water sources or grazing lands (although protected-area managers today are well aware of this problem and are showing more flexibility on such issues).

- Competition with tourists is a problem in some areas. While tourism brings undeniable economic benefits, it also involves economic costs, particularly in remote areas, which may already be in food deficit. In such cases, the demand from tourists for high-quality protein may either drive the price of meat beyond the purchasing power of the local people, or seduce local populations into selling their livestock, rather than consuming it themselves, thereby suffering a protein shortage.

The above is just a partial list, but indicates that any use of resources, including those for conservation purposes, includes some degree of tradeoffs and may change the local distribution of costs and benefits. In seeking to design programmes that will benefit both protected areas and local communities, careful consideration needs to be given to the costs incurred versus the benefits derived and to the incidence of both.

Integrated Conservation and Development Projects

Given that protected areas generate both benefits and costs for rural people, the major challenge becomes designing development activities that maximize the benefits and minimize the costs of conservation. One important way to provide benefits from protected areas to local people is through integrated conservation and development projects (ICDPs). These are efforts to reconcile conservation and community interests through promoting social and economic development among communities adjacent to the boundaries of protected areas, using a bioregional approach that links the protected area to the surrounding lands, often through the mechanism of buffer zones, as reviewed above (Wells and Brandon, 1992). Effective ICDPs require skills beyond those of traditional protected-area managers, involving agriculture, commerce, economics, sociology, anthropology, law and public policy. Many of these skills may be found more readily in civil society than in protected-area management agencies.

The advantage of the ICDP approach is that it treats each protected area and its associated buffer zone as a single planning unit, with the aim of conserving biodiversity by reconciling the management of the protected area's conservation values with the social and economic needs of the local population. ICDPs typically feature activities to involve and elicit support from local communities in designing management programmes, allowing the harvesting of certain natural resources within at least parts of protected areas, while developing environmentally sustainable alternatives to offset possible losses of economic benefits from harvest and other use restrictions in the protected area (in other words, compensating local communities for their opportunity costs).

ICDPs are especially appropriate where multiple but interrelated development problems are facing rural areas with significant conservation values, which require a multisectoral approach to solve them simultaneously. Successfully implemented, such projects can achieve the desired synergistic effect of addressing cross-cutting issues – such as poverty alleviation and environmental protection – which could not be adequately resolved with a single-sector approach.

ICDP approaches can also help to support the move toward decentralization of resource management functions that characterizes most governments today, but they will not solve all problems. ICDPs have sometimes been criticized for not addressing the most significant problems facing protected areas (for example, major infrastructure developments, such as road construction, mining, logging concessions, and so forth), instead providing limited funds for a few villagers in buffer zones and ignoring the need to ensure better enforcement of laws and regulations and the incorporation of protected areas into regional planning. While some have concluded that ICDPs centred on protected areas and directed to local populations are not able by themselves to play more than a modest role in mitigating the powerful forces causing environmental degradation (Wells and Brandon, 1992), it is often better to take a step-by-step approach and aim for incremental gains rather than grand solutions. Moreover, no better approach than ICDPs has yet been found.

Many ICDPs have had an overambitious project scope and complex implementation arrangements involving multiple implementing agencies, so too many resources have been spent on administration and organizational issues. Perhaps a larger problem with many ICDPs is that they are extremely expensive, involving massive subsidized changes in land use, with little assurance that the new land uses will be sustained after the project subsidies expire or are withdrawn. They may also tend to convince governments that conservation of protected areas is of necessity a very expensive process. In fact, more modest approaches, which build on local interests and capacities, may be far more sustainable, if less spectacular.

A final problem of ICDPs is that their success, involving the improvement of living standards, may stimulate immigration, thereby jeopardizing the stability and sustainability of the activities encouraged in the buffer zone. At least part of this problem is that most ICDPs operate in open-access systems, and lack of demographic monitoring may obscure the risk of immigration and thus prevent ameliorative efforts. One option is to favour existing inhabitants over newcomers in ICDPs, and to seek ways to actively discourage development activities that stimulate immigration in the buffer zones, often requiring an appropriate legal framework, accompanied by appropriate enforcement.

Designing successful ICDPs

Good project design includes at least the following components (Barber *et al.*, 1995):

- clear objectives and implementation strategies.
- Manageable size, usually limited to a protected area and its surrounding lands or a large water-catchment area.
- Agreement on appropriate property rights.
- Decentralized implementation arrangements, which are flexible to site-specific conditions or changes during project implementation.
- A strong coordination agency, with authority over the implementing agencies and the ability to provide effective incentives for conservation.

- Proven technologies to be introduced, or those that are not too risky for adopters.
- Effective institutional strengthening measures introduced early in project implementation.
- Adequate field management support, including provisions for effective law enforcement, routine monitoring, feedback and problem-solving.
- Project organization integrated with the existing local government institutions, budgetary planning and fund-disbursement procedures.

The success of ICDPs depends critically on appropriate integrated conservation and development policies and institutional initiatives. Barber *et al.* (1995) suggest four changes in protected-area policy that could facilitate ICDP efforts, which in turn should lead to improvements in protected-area management policy and practice:

- Acknowledge the reality of human use and occupation of protected areas and the resources they contain.
- Design policies and zoning practices to minimize the impact of human use and secure local livelihoods and enlist local residents in controlling access and resource exploitation within the protected area.
- Develop new policies on protected-area boundary establishment and zoning.
- Clarify the structure and mandate of the government agency charged with managing protected areas, leading to possible sharing of some responsibilities with other stakeholders, based on clearly defined roles and responsibilities of a joint management arrangement.

While good project design and sound science are necessary, they are not sufficient for project success, which is more dependent on building and maintaining support from a broad range of stakeholders. A key factor in the success of ICDPs is the quality of the project leader: one who can deal effectively and simultaneously with a multitude of technical, social, political and financial challenges and is able to adapt the project to changing conditions and feedback from early implementation efforts. It is difficult to avoid the conclusion that ICDPs are more a matter of art than of science.

Conclusions

Modern society has brought expanding populations, global markets and new pressures on land and resources. Protected areas are an essential element of the strategies of modern societies to ensure that resources are used sustainably and biodiversity is conserved for present and future generations. Protected areas provide a wide range of economic, social, cultural, recreational, scientific and spiritual values. They provide considerable economic benefits, ranging from tourism development to carbon sequestration and watershed protection. It is now widely accepted that local agricultural communities have a legitimate right to participate in at least some aspects of protected-area management. Indeed, because of a range of economic factors, the supply of biologically rich protected areas is doomed to be suboptimal without a concerted effort by the world community to make conservation an attractive option to the rural people who have had practical jurisdiction over protected-area resources for many generations (McNeely and Guruswamy, 1998).

A key factor in successful protected-area management strategies is the stability of rural communities, implying that governments need to pay particular attention when contemplating major efforts at relocating people from one part of the countryside to another. Those people who have developed long-term relationships with particular settings and have developed knowledge on how to manage the resources contained within those ecosystems are likely to have a very different relationship with the land and its resources than are new immigrants who have no particular linkage to local resources and who may receive considerable subsidies from outside. Given the dynamism of development in many parts of the world, communities in and around protected areas often include both indigenous peoples who have a long history in the region and immigrants who have arrived much more recently. The new arrivals are frequently responsible for more destructive land-use practices than are the long-term residents; but, of course, new technologies and new markets can be expected to change the behaviour of local villagers, regardless of their traditional conservation practices.

At a minimum, local communities need to be deeply involved in buffer-zone development activities and should be consulted on any decisions that affect them. In many cases, giving local people preferential treatment in terms of employment within the protected area, providing economic incentives to establish income-generating activities in the buffer zone and ensuring an appropriate flow of benefits from the protected areas to the surrounding lands can help to build a positive relationship between protected areas and local communities.

In other cases, it might be most sensible to return the full management responsibility for at least the buffer-zone areas to the local community, leading to community-owned forests that serve at least some of the functions of protected areas; these may be managed under forest stewardship contracts between government agencies and local communities. Successful project interventions in buffer zones need to address community priorities, providing incentives such as material benefits in a way that builds long-term partnerships rather than dependency. Successful projects add diversity to the development options available to the local community and enable the community to build self-reliance. While economic incentives are important to compensate local communities for the opportunity costs of resource use, such compensation should be in the form of improving access to suitable productive resources, such as better agricultural land or technology. This can provide some degree of continuity between prior economic modes of production and new opportunities for economic improvement, and is far better than simply providing cash compensation, which is often quickly dissipated.

One useful mechanism for putting this vision into practice is through ICDPs, which seek to reconcile conservation and community interests through promoting social and economic development among communities in and around protected areas. Such projects need to be carefully designed to ensure that the interests of the various stakeholders are well represented. It is critical that the effort involves a clear identification of the problems facing the protected area, so that the proposed measures specifically address the problems identified. In virtually all cases, it will be found that the most significant threats to protected areas originate from government policies affecting land ownership, foreign trade, agriculture, forestry and so forth. ICDPs can be effective only when they are part of an overall strategy, which includes the development of appropriate policies that will enable the ICDPs to function.

Detailed knowledge of the people whose lives are affected by the establishment and management of protected areas is at least as important to protected-area managers as information about the plant and animal species to be conserved. The cultural, socioeconomic and demographic characteristics of local people – including the age and gender divisions of labour – form the basis for measures to promote the sustainable use of natural resources, alleviate poverty, develop appropriate agricultural practices, improve the quality of human life and create support for protected areas.

22 Moving Beyond Integrated Conservation and Development Projects (ICDPs) to Achieve Biodiversity Conservation

KATRINA BRANDON

Center for Applied Biodiversity Sciences, Conservation International, Washington, DC, USA

Introduction

In the past two decades, biodiversity conservation has gone from being a specialized concern among scientists to drawing worldwide attention at events such as the 1992 United Nations Conference on Environment and Development. While a broad, global consensus that biodiversity should be conserved has emerged, there is in fact little agreement on exactly what biodiversity means or how it can best be preserved (Sanderson and Redford, 1997). This chapter summarizes the challenges that are inherent in site-based biodiversity conservation, with a particular focus on rural development issues.

The chapter is composed of four distinct but interrelated sections, which move progressively from addressing field-based projects aimed at preventing biodiversity loss to the role of agricultural intensification in mitigating these losses at local, regional and national levels. The first section provides an overview of integrated conservation and development projects (ICDPs) – field-based projects designed to stop biodiversity loss in protected areas. The second section summarizes what a decade of experience has shown and identifies a set of assumptions that underlie many ICDPs and other efforts to conserve biodiversity. Many of these may be relevant for a variety of projects promoting agricultural intensification as a way of increasing farmers' incomes and reversing patterns of resource overuse and misuse. The third section summarizes selected findings from *Parks in Peril: People, Politics, and Protected Areas* (Brandon *et al.*, 1998) that have implications for agricultural intensification efforts. The final section outlines future directions for agricultural intensification efforts that can promote biodiversity conservation.

Integrated Conservation and Development Projects

Much of the development theory popular in developing countries during the late 1970s and early 1980s focused on the legacy of colonialism, continuing imperialism and ensuing dependency. Coupled with this theoretical base was a wariness of transferring 'developed-country' patterns of development, which included macrostructural views of economies, the importance of large-scale development projects and solutions achieved through macroeconomic reforms. In contrast to these large-scale and heavy-handed initiatives were 'softer' and more populist conceptions of development, which sought to correct development biases through promoting appropriate and small-scale technologies, self-reliant and sustainable agriculture, local empowerment, popular participation, democratization and devolution of power. At the same time as these concepts were being promoted within the development community, there was recognition among conservationists that conservation initiatives had to be integrated with rural development. Many of these alternative views of development were uncritically adopted by non-governmental organizations (NGOs) involved in conservation work from those organizations focused on development objectives. This transfer spurred the design and implementation of programmes that emphasized more populist notions of development and were the forerunners of what are now called 'sustainable development projects' (Gradwohl and Greenberg, 1988; Reid *et al.*, 1988).

Within conservation organizations, these initiatives are known as integrated conservation and development projects. The philosophy underlying ICDPs – that 'protected areas in developing countries will survive only in so far as they address human concerns' – was highlighted at the 1982 World Parks Congress in Bali (Western and Pearl, 1989, p. 134). At that Congress, suggestions on how to support communities living adjacent to parks were made, including local participation, education, revenue sharing, development activities and opening park resources to local use. In response, a variety of organizations began implementing projects linking biodiversity conservation in parks with local social and economic development.

In 1989, the World Bank, the World Wildlife Fund (WWF) and the US Agency for International Development (USAID) sponsored a study to identify the strategies that were being pursued by these projects and the extent to which these investments represented cost-effective, sustainable or replicable approaches to protected-area management (Wells and Brandon, 1992). This study, known as *People and Parks*, introduced the ICDP label for this portfolio of initiatives, based on a review of 23 projects at 18 sites in 14 countries. In contrast with natural resource management projects, where the objective is to change management regimes across an entire area, all ICDPs have 'core' protected areas[1] in which uses are restricted. In addition to this protected core, ICDPs include components designed to promote socioeconomic development outside the core areas and to provide local people with sources of income compatible with park management objectives. The three major strategies that were evident in virtually all ICDPs were: (i) enhanced park management and/or creating buffer zones around protected areas; (ii) providing compensation or substitution for local people for lost access to resources; and (iii) encouraging local social and economic development.

Activities to accomplish park management goals typically included boundary marking, developing park management plans, research, trail maintenance and improved enforcement. Buffer zones, conceptualized as a protective band of land encircling the park, were created to meet both conservation and development objectives. From the conservation side, buffer zones were viewed as extending the ecosystem services and range for wildlife. From the development standpoint, they allowed local people to continue to engage in the sustainable harvest of resources.

Compensation and substitution strategies, parts of many ICDPs, were intended to reduce the economic burden on those people who would otherwise have few alternative means of livelihood beyond continued exploitation of the park's flora and fauna. There are three key ways that this strategy has been implemented: by compensating local people for the economic losses which they have suffered as a result of park establishment; by providing substitutes for specific resources to which access has been denied; and by providing alternative sources of income to purchase substitutes. Projects also provide for cash payments to communities and goods and services (e.g. wells, schools and health clinics) as direct compensation for economic losses. Substitutes are often targeted at specific resource uses, such as providing wood lots outside a park if fuel wood is needed or introducing pig-raising as a substitute for bush meat. Alternative sources of income are provided to purchase substitutes for commodities such as construction materials, medicinal plants or certain fruits, which may not be readily available outside the park.

Promoting local social and economic development adjacent to protected area boundaries has been the most common ICDP strategy. Under this strategy, poverty mitigation and community development activities are promoted to break the destructive patterns of resource use. Projects have generally tried to meet the most evident local needs with project components designed to improve natural resource management outside protected areas, i.e. agricultural development, agroforestry and forestry, wildlife utilization, irrigation and water management, and soil enhancement and erosion control. Many projects have also improved market access, generated employment, promoted ecotourism and/or provided community services, such as health clinics, schools and sanitation works for local communities.

Findings and Critiques of ICDP Approaches

The *People and Parks* study provided the first progress report on how ICDP initiatives were faring. All of the projects in the study represented 'models' of successful projects, had social or economic development components linked to protected areas and had been operational for at least 3 years. Despite the tremendous diversity among project strategies and site-specific conditions, similarities emerged in the overall findings on the factors that affected project performance. Many of these findings were similar to evaluations of rural development projects in general (World Bank, 1988a).[2] Apart from implementation issues, three significant problems with ICDPs were identified. First, ICDPs represented a major investment with no certainty of a return. Based on experiences with rural development, ICDPs were presumed to need long-term (e.g. 15–20 years) funding and time horizons. Secondly, because of the complexity of ICDPs and the time needed to establish connections between conservation and development, they were found to be poor choices in areas with immediate problems, e.g. rapid declines in biodiversity or high levels of social conflict. Finally, ICDPs were found to require both policy-level and project-level components. The *People and Parks* study concluded that, given these findings, large donors (e.g. the World Bank), who both had the money and wielded the political influence to see reforms implemented, should be partners in ICDP implementation. The study cautioned that there would be 'a huge learning curve in developing and implementing ICDPs . . . It is not unwise to question whether they should even be attempted' (Wells and Brandon, 1992, p. 567).

Recent reviews of ICDPs have found that, although there has been an evolution in the scale of projects and how they are conceptualized, most of the projects are still hampered by implementation problems similar to those encountered when implementing rural development projects (Sanjayan *et al.*, 1997; Brandon *et al.*, 1998; Larson *et al.*, 1998). It is obvious that there are certain elements which are key to successful implementation: clear objectives, good stakeholder consultation, collection of baseline data, monitoring and evaluation systems and awareness of local knowledge and practices. It is not surprising, then, that these elements should form part of ICDPs. What is surprising, however, is that projects are only now beginning to broadly implement these components. Apart from implementation problems, there are problems with the underlying ICDP approach, and project success is elusive even in the best of circumstances. But the contexts where ICDP initiatives are implemented do not represent the best of circumstances. These initiatives combine some of the most difficult elements of conservation with the most difficult elements of rural development. While ICDPs are still viewed as 'conservation projects', they are in fact large-scale social interventions in complicated settings.

Another problem with the ICDP approach is the lack of nomenclature or any standardization of what ICDPs are. Must they be based around a core protected area? Is biodiversity conservation the primary objective? Can conservation and development objectives both be equivalent project goals? The initial definition of an ICDP – as having some core protected area – is now largely ignored. According to a recent World Bank report, 'ICDPs have become all things to all people and a number of labels have been developed to meet the particular objectives of the project' (Sanjayan *et al.*, 1997, p. 1). They have been variously referred to as integrated conservation and development (ICAD) and community-based conservation projects. What is now called an ICDP is often any conservation project that deals with people, including more general efforts to promote improved natural resource management activities. The lack of consensus about what these different project types really are, when they should be used and their 'bottom-line' objectives is an indicator of a general lack of rigour in developing projects to link conservation and rural development. This has translated into both poorly designed projects and difficulty in generating lessons from one kind of project and effectively applying them to another. Unfortunately, a discussion of lessons learned is hampered by the clustering and overlapping of these different types of projects, which, one way or another, have all fallen under the rubric of ICDPs.

Evaluations of the activities broadly labelled as ICDPs suggest that few of today's field-based initiatives are living up to their proclaimed potential, even though they are widely supported by organizations such as the World Bank, the Global Environmental Facility, bilateral agencies and international conservation organizations (Kiss, 1990; West and Brechin, 1991; Wells and Brandon, 1992; Western and Wright, 1994; Church and Brandon, 1995; Barrett and Arcese, 1995; Gibson and Marks, 1995; Biodiversity Support Program, 1996; Sanjayan *et al.*, 1997; Brandon *et al.*, 1998). The ICDP approach has become extremely commonplace as perhaps the primary strategy for biodiversity conservation. As Larsen *et al.* (1998, p. 1) state, 'In the mid-1980s, ICDPs were a radical divergence from the norm; today, such projects receive over half of WWF's funding.' A cursory review of World Bank projects with biodiversity components suggests that one-quarter to one-third are ICDPs (World Bank, 1998a).[3] This is noteworthy because ICDPs are now viewed as the main strategy to implement biodiversity conservation.

One key reason for the frequent lack of success of ICDPs is that flawed assumptions often underlie project design. Not surprisingly, many of these assumptions are rooted in the populist slogans and rhetoric that were imported from development NGOs into conservation strategies (Brandon, 1997).[4] A summary of seven of these broad assumptions and a brief critique of each of them is presented in Table 22.1.

Unfortunately, these flawed assumptions frequently serve as the intellectual foundation for the vast majority of conservation activities being developed – whether these initiatives are in parks or forest reserves, ICDPs or locally managed reserves. While some of these assumptions may be accurate at some sites some of the time, evidence is increasingly showing that many of these assumptions do not hold much of the time and that fewer 'win–win' outcomes are possible than might be imagined. The fact that conservation agendas are actively being designed around these assumptions is likely to lead to numerous and significant failures in biodiversity conservation. Perhaps the two most relevant of these assumptions for this volume are that: (i) economic incentives can be readily defined; and (ii) poverty mitigation and economic development will lead to biodiversity conservation. An analysis of the difficulties when projects are designed around these two often-related assumptions provides some insight into the challenges of each. When several of these assumptions are taken together as the basis for project implementation, it is not surprising that objectives are not met.

Regarding the first assumption, using economic incentives and disincentives to achieve biodiversity conservation has been promoted in *Economics and Biological Diversity* as 'the most effective measures for converting overexploitation to sustainable use of biological resources' (McNeely, 1988, pp. vii, ix). For example, one African project trying to 'transform rural hunters into conservationists' by hiring local villagers and encouraging sustainable use of

Table 22.1. Key assumptions underlying ICDPs and their critiques (adapted from Brandon, 1997).

Assumptions	Critiques
Method	
Biodiversity conservation can best be accomplished through field-based activities, such as establishing parks and reserves	A wide variety of policies affect biodiversity conservation in developing countries; evidence shows that projects are unlikely to succeed if policy issues are not dealt with. Furthermore, policy reforms may in many cases do more to conserve biodiversity than site-based projects
Use	
Sustainable use is possible under a variety of management regimes, ranging from private to communal. Dependence on wild-land resources is most likely to ensure their long-term conservation	Examples of sustainable use have involved highly regulated and managed ecosystems; this does not mimic most ICDP sites. Generally, we do not know the ingredients for sustainability: species biology, social issues and incentives, marketing or economics
Incentives	
Appropriate sets of incentives can be readily defined and will influence people to conserve biodiversity	There is enormous complexity in defining different sets of incentives and disincentives that work for the range of stakeholders in most areas of high biodiversity
Management	
Management should be devolved to local control whenever possible	In areas undergoing rapid social change, traditional management systems are often overwhelmed, eroded or non-existent. The potential for devolution may be non-existent or management for biodiversity conservation may not be a local objective
Technology	
Technical and organizational solutions exist to improve resource management and production activities in areas with great biodiversity	Identification, transfer and adoption of new technologies in fragile ecosystems is highly complex and needs to consider a range of site-specific factors, from social stratification to continued operation and maintenance (O&M) to cost factors and organizational capacity to promote change
Poverty mitigation and development	
Rural poverty mitigation and development strategies will lead to conservation and maintain biodiversity	There is little evidence that reducing poverty or promoting local development diminishes pressure on resources. In certain situations, increased income may lead to increased consumption and greater pressure on biodiversity
Social	
Local people are cooperative and live in harmony with one another and with nature	Conflict over resources and competing uses and user groups is rampant, particularly in areas undergoing rapid social change[5]

wildlife through community services, trophy-hunting and tourism illustrates the complexity of defining incentives (Gibson and Marks, 1995).[6] This project identified four difficulties in defining the appropriate incentives: (i) individuals place different values on greater income and development projects; (ii) generational choices and incentives are different, creating conflict; (iii) free-rider problems exist (rational individuals choose to receive hunting benefits while simultaneously enjoying the community projects); and (iv) the unrealistic expectation that all individuals are willing to trade direct access to wildlife resources for access to community facilities and infrastructure, which they may or may not need, want or use.

These constraints highlight some difficulties in structuring incentives, even in a project that is widely regarded as a success story. First, defining which type of incentive or disincentive is the 'right' one, and for which groups of actors, is difficult. Few incentives work well for all stakeholders. Should the target be people who are not threatening biodiversity, or people who are? How do you target certain groups and not get a 'free-rider' problem? When incentives or disincentives have been used successfully, they have been used at sites where there has been relatively little pressure on biodiversity and where pressure has been local. Incentives also work best when there are strong local systems of social control and sanction, but these are rapidly eroding in much of the world.[7] At sites under substantial pressure and with multiple sources of threat, it is necessary to define different incentives and disincentives for each of the different sets of actors causing the different threats. Field experience shows that identifying both the incentives and disincentives for a wide variety of actors can be an overwhelming task (Margoluis and Salafsky, 1998; Stedman-Edwards, 1998).

The second assumption highlighted here is that of linkages, or how to establish a strong and direct linkage between incentives and the conservation objective. Economic incentives and desired conservation outcomes must be clearly linked in the eyes of participants; otherwise the incentives will not work. Many threats are externally generated but acted out locally. For example, even when incentives can be designed for and with local people, corrupt government officials can often counter the incentives, especially in the short term and in the absence of strong political will. While incentives and projects are directed toward small-scale users causing biodiversity loss through poverty-induced environmental degradation, problems and policies elsewhere may propel local actors and their destructive actions. Thirdly, what constitutes the appropriate incentive is likely to change over time. As noted by Reardon and Vosti (1995, p. 48), the causal links between poverty and environment 'change over time, and causal directions may reverse in unexpected ways'. Therefore, even what serves as an incentive at one point in time may not function in the same way at another point in time. As areas change and pressure on biodiversity intensifies, certain incentives will be of greater or lesser value, requiring careful attention and frequent modification, and perhaps even reversal, of incentives. Finally, economic incentives may be overridden by exogenous factors, such as ethical, religious and cultural practices. For example, Aché children in Paraguay are named after a game animal prepared by the mother while pregnant. Despite rapid acculturation, the Aché have explicitly maintained the naming ceremony. Fearing that the pool of names is declining as wildlife declines, they support certain restrictions on hunting, except when these restrictions interfere with naming a child (Hill and Tikuarangi, 1996). In such cases, economic incentives are irrelevant.

Clearly, economic incentives need to be defined and used appropriately within the project context, but there is tremendous variation in the factors that induce people to use and manage biodiversity resources wisely. Defining the appropriate mix of incentives and penalties is challenging and needs to be determined on a site-specific basis, taking into account the scale of the area and the range of actors in the specific setting.

Closely linked to economic incentives, including direct payments of cash, fees, rewards, compensation, grants, subsidies, credit and employment, is the assumption that poverty mitigation will lead to improved biodiversity conservation. For example, it is commonly assumed that poor households will switch from illegal, unsustainable and difficult activities, such as poaching, to legal activities if such activities generate revenue at least equal to illegal gains. The poverty-mitigation approach assumes that poor households have a fixed income need and, if projects can provide an intervention to help households meet this need, destructive practices will end. Yet the poor typically want to do better economically. Rational behaviour for them means continuing illegal activities in the absence of strong deterrents or the presence of a risk they perceive as too great. What risk is 'too great' will vary by site, but creating such risks for local actors can hardly be viewed positively. What has often happened in projects worldwide is that increased income, in the absence of social controls and project linkages, has allowed people to use resources more rapidly: guns are substituted for traditional hunting methods, chainsaws make forest clearing easier and nets make fishing more profitable.

This brief review of the challenges involved with just two of the flawed assumptions used as the basis for ICDP actions provides a window into why ICDPs are so hard to implement. Other chapters in this volume highlight the challenges inherent in the assumptions about technology, in defining appropriate technological packages and in promoting their transfer and adoption. The set of assumptions underlying ICDP design are not flawed in all cases at all sites; rather, these assumptions should not be quickly accepted without a thorough knowledge and understanding of the dynamics existing at a given site. Project design is particularly challenging when all seven of these assumptions form the basis for implementation. While there is a learning curve present in the design of such projects, it is still evident that many are still trying too hard to do too much in the wrong places. Despite these difficulties encountered in trying to conserve biodiversity in park-based approaches, recent analyses give room for cautious optimism, provided that there is a major reorientation in what parks are expected to do and depending on what happens on lands outside protected areas. These issues are summarized in the next section.

Parks in Peril: the Broader Context of Threats

This section reviews selected findings from a recent publication, *Parks in Peril: People, Politics, and Protected Areas* (Brandon *et al.*, 1998) (hereafter referred to as *Parks in Peril*), which analysed nine protected areas that are part of the Nature Conservancy's Parks in Peril (PiP) programme and the activities around those sites. The PiP programme is the largest single programme to support parks in Latin America, and perhaps in the entire tropical world, including over 60 parks in 18 countries, covering an area of over 30 Mha. PiP is administered by the Nature Conservancy through its partner organizations in each country, largely with financial support from USAID. Faced with the task of understanding and responding to the 'outward' aspects of park protection, the Conservancy and its partner organizations in Latin America and the Caribbean decided to undertake a broader analysis of the ecological, social and political issues faced by the parks in their portfolio. Some of the book's general conclusions, highlighted and summarized here, serve as a starting point for thinking about ways in which agricultural intensification, along with other development activities, should be undertaken where viewed from the perspective of biodiversity conservation.

- *Parks must be seen as a cornerstone for biodiversity conservation.* While the mission of parks was originally intended to emphasize protection, parks have increasingly become open to a variety of human uses (IUCN *et al.*, 1980). Even when lands are officially protected, it is vital to realize that biodiversity conservation is often not an integral management objective for these lands. The International Union for World Conservation (IUCN) lists six categories of protected areas, but only one of these explicitly describes management for biodiversity conservation as a primary objective. Instead, many parks have biodiversity conservation and sustainable development as dual management objectives. Also, in many countries, other land-use designations, such as forest reserves, comprise a significant share of the protected lands. For example, while Costa Rica is often lauded for protecting 27% of its lands, at least 15% is legally open to consumptive human uses.[8] Within Latin America and the Caribbean, protected areas in the least restrictive uses occupy twice the land area of more restricted categories. Whether any of these protected areas contribute to biodiversity conservation, and, if so, to what component of biodiversity (genetic, species, ecosystems), is largely dependent on their ownership and management (Redford and Richter, 2000). As lands outside protected areas are increasingly utilized and developed, achieving conservation goals will mean defining what the core management objective is at each site and ensuring that not all protected lands are opened to human use.
- *Improved park management and biodiversity conservation should not be equated with sustainable development.* Many of the lands containing the greatest biodiversity – and often where parks have been established – are among the most remote and marginal regions, with low productive potential and poor access to markets. Findings from *Parks in Peril* and other studies (e.g. Sanjayan *et al.*, 1997) demonstrate that park management activities and the surge of funds for biodiversity conservation have brought with them substantial amounts of development money. This money is often spent on ill-conceived efforts, trying to help 'stabilize' and 'settle' migrant

populations and promote 'sustainable' development in or adjacent to park boundaries. Most efforts at promoting agricultural development within ICDPs are typically small-scale and non-intensive, often emphasizing household consumption at the expense of revenue generation. Attempts to promote agriculture on inappropriate lands are a component of many projects and should be stopped. The reality is that, in many of these settings, productivity will remain low and access to markets scarce, virtually guaranteeing that lives and livelihoods will be imperilled and wildland and wildlife resources threatened. While such development projects may sound 'pro-people', their implementation is likely to condemn the poor to lives of poverty under the guises of 'decentralization' and 'sustainable use'.[9] In addition, these projects are failing in their core objective, which is to conserve biodiversity.

- *Activities within protected areas must include a realistic appraisal of the social context, local livelihood dependence and ecological conditions. Management objectives must be set accordingly.* To date, there has been little rigorous analytical work on what management objectives are appropriate and in what context across an array of protected areas. Rigorous analytical criteria must, at a minimum, include ecological significance; level and extent of threats; livelihood needs and uses; a good understanding of the existing social context and levels of change; and the root causes of biodiversity loss (Margoluis and Salafsky, 1998; Stedman-Edwards, 1998). The type of project and its scope can be readily determined by these factors (Brandon, 1998). For example, in densely populated and socially complex areas, such as Sierra de las Minas Biosphere Reserve in Guatemala, combining local consultation with technical information has produced a system of zoning which, while not completely accepted, has been relatively well understood. For a biosphere to be successful over time, zoning must be based on both conservation objectives and livelihood needs and must be accepted and enforced by local communities (see Lehnhoff and Nuñez, 1998). Especially within populated zones in biosphere reserves, agricultural intensification should be viewed as one essential component of meeting local livelihood needs.
- *Social change is the critical determinant of project scope and scale, time horizon, budget and probable success.* The hardest places for projects to succeed are in those situations that are undergoing a rapid process of social change. Long-term approaches or small-scale solutions are likely to have little impact in areas undergoing rapid regional changes. At sites undergoing rapid change, there is a need for a more complicated process of management. But, most importantly, efforts in all regards must be both more intensive (in terms of technical and financial resources) and more extensive (in terms of the geographical area and the time required to stabilize parks). Actions not only must address the localized concerns around the park, but must begin to influence the processes that are driving regional change. These influences are often external to regions and rooted in government policies or programmes, such as road construction and consequent changes

in land uses (Stedman-Edwards, 1998). Addressing changes will usually require immense amounts of money and political will.

- *Stereotypes are fatal to new solutions.* The *Parks in Peril* study identified numerous examples where the 'best practices' challenged conventional wisdom. For example, one commonly held view is that it is best to locate buffer areas outside park boundaries. While this is certainly true in ecological terms, what is best in ecological terms may not be the best in social terms. If a park is sufficiently large to meet the habitat requirements of the species within, the best boundaries may indeed be intensive land uses (allowing for wildlife corridors) outside the park. Such boundaries are clear, dramatic and obvious. They may reduce wildlife–people conflicts and they are automatically enforced by the residents of these areas. The case study in Belize, citing intensive Mennonite farmers with lands abutting the park, stated: 'They purchase, delineate and defend their large farms from trespassers. A hunter wishing to poach a peccary in northern Rio Bravo Conservation Management Area must first cross miles of private Mennonite farms, a difficult prospect' (Wallace and Naughton-Treves, 1998, p. 225). Ironically, few ICDPs have sought to promote agricultural intensification adjacent to parks or in strategically located areas away from parks, based on the perception that this may, in fact, be better in social, economic and ecological terms than traditional agricultural practices throughout a park.

These lessons offer a brief summary of selected findings of *Parks in Peril*. Along with other conclusions of the report, they highlight the need for new and creative ways of thinking about park management. Specifically, the book makes the case that 'protected areas are extremely important for the protection of biodiversity, yet requiring them to carry the entire burden for biodiversity conservation is a recipe for ecological and social failure' (Brandon *et al*., 1998, p. 2). What is increasingly evident is that, while well-managed systems of protected areas are essential for maintaining biodiversity, greater attention needs to be given to innovative thinking about these areas and their place in the broader landscape.

Agricultural Intensification and Parks: Future Directions

There are a number of preliminary assertions that can be made regarding where and how to best link agricultural intensification and park management. This section addresses some future directions in thinking about these linkages, briefly focusing on the role of intensification: (i) as related to parks and protected areas specifically; (ii) within the context of ecoregional planning; and (iii) within the broader policy context.

Intensification and protected areas

Two key challenges for conservationists are to change the expectations that parks are supposed to be the cornerstone of sustainable development activities and to refocus their attention and actions on biodiversity conservation. There has been a tremendous polarization among those interested primarily in biodiversity conservation and those principally concerned with human welfare. The latter group have unrealistically expected conservation projects and parks to 'cure structural problems such as poverty, unequal land distribution and resource allocation, corruption, economic injustice, and market failure' (Brandon, 1998, p. 418). But this is exactly what park advocates are implicitly expecting parks to do,[10] when the solutions to these problems lie elsewhere. These structural problems need solutions, but it is unreasonable to expect parks to provide the solutions for whole regions. Parks will be crippled until there is clarity over their mission and greater attention is given to dealing with the root causes of biodiversity loss as generated elsewhere. Until there is strong and strategic analysis, followed by actions, to solve the underlying policies that lead to poor resource management and biodiversity loss, efforts to safeguard parks, unfortunately, will remain of paramount importance.

Greater creativity will be needed in the future, particularly in areas where the potential for sustainable livelihoods is only possible at low densities and cycles of migration and destruction are high. Such creativity will mean making compromises and becoming innovative about how and where to target efforts. Agricultural intensification can be one strategic element of project activities of parks, both in stable areas and in areas of rapid social change. In the former, ICDP approaches are appropriate, since there is adequate time for them to be effective. In this context, agricultural intensification efforts may provide a useful way to stabilize local land uses.[11] Areas in the midst of rapid social change may benefit from intensification as well. For example, granting land tenure to residents who surround a park and assisting them in intensifying agriculture would probably turn them into *de facto* guards to stop migrants from claiming residents' lands or infringing on parklands. In such cases, intensifying agriculture around a park would be of vital importance in stabilizing land uses, increasing local incomes, absorbing surplus labour and limiting migration. Such approaches may work particularly well when watersheds are involved, since the linkages between conservation needs and development benefits and the desires of different stakeholder groups and users can often be readily defined.

Intensification and ecoregional planning

Within the conservation community, there has been an evolution in thinking about how and where to prioritize investments on biodiversity outside protected areas. In the late 1980s, the key approach to biodiversity conservation was based on 'hot spots', or identifying areas of species richness either by dealing with regions within countries or by dealing with biodiversity protection within particular countries. The emphasis at the national level, then, focused on identifying countries that possessed the greatest biodiversity, and these became known as the 'megadiversity' countries (Myers, 1988; Mittermeier and Werner, 1990). These approaches have been overlaid with the concept of ecoregions –

identifying all of the 'ecosystem and habitat types in order to conserve the distinct communities of organisms they contain' (Dinerstein *et al.*, 1995, p. 2). In contrast to trying to save biodiversity in the countries with the greatest variety of organisms – the 'hot spots' or 'megadiversity' countries – the ecoregional approach ignores geopolitical boundaries and focuses on representative areas with distinct natural communities in blocks that are sufficiently large to deal with natural disasters (e.g. fire) and long-term environmental degradation (e.g. changes stemming from global warming) (Dinerstein *et al.*, 1995).

While the ecoregional approach represents a more balanced attempt to save 'some of everything', the 'hot spots' approach attempts to prioritize investment and concentrate on the areas with the greatest biodiversity. Although the former is 'fairer' in ecological terms, it is far more daunting when viewed from political, social and economic perspectives. Just as biodiversity is scale-dependent, so too are the forces that shape how biodiversity is used, managed or destroyed. Biodiversity loss is not uniform – it is characterized by different regional patterns and historical trends (Fairhead and Leach, 1995; Rudel and Roper, 1996, 1997; Stedman-Edwards, 1998) – and, at the regional level, it is greatly influenced by a set of forces above (e.g. national) and below (e.g. community or local). From a policy perspective, one element of the hot spot or megadiversity approach that makes it more attractive than the ecoregional approaches is that, in megadiversity countries, the policies leading to the biodiversity loss and the reforms needed to improve asset distribution, poverty, land use and biodiversity protection can be readily identified. Also, ecological weight can be given to prioritizing national-level policy interventions. Implementing these interventions, however, would rely on national and international political will.

Conservation at an ecoregional scale will only happen if there is a clear understanding of the factors shaping biodiversity loss within an ecoregional context. These can be identified and probably modelled to inform conservation planning. The ecoregional emphasis cuts across geopolitical boundaries; by its very nature and scale, it is less focused on the 'big policy picture' and is more clearly tied to regional-level concerns. This requires addressing social, political, and economic forces leading to biodiversity loss at the ecoregional level. It also requires understanding and addressing the poverty–environmental linkages in different localities and realizing that long-term approaches or small-scale solutions will do little to stop biodiversity loss in areas undergoing rapid regional changes. Realistic assumptions must underlie the design of regional projects, with the phasing in of appropriate responses, given a realistic time frame and financial support appropriate to the scale of the area.

Addressing issues at the ecoregional scale may mean that conservationists will have to become advocates for the construction of 'growth poles', areas close enough to parks for cities to act as magnets and yet far enough away to ensure that resources are not depleted for urban markets (Brandon and Wells, 1992; Rudel and Roper, 1996). In some areas, it may mean large regional projects designed to 'pull' the marginal and landless poor out of frontier areas and promote intensive agricultural development. In other areas, where the potential for agricultural intensification or rural industrialization is poor, it may mean coming up with other alternatives to provide stable and secure livelihoods. Ecoregional problem-solving will also require broader conceptual thinking about tradeoffs within regional levels. For example, there is little potential for improving agriculture in and around the Calakmul Biosphere Reserve on the Yucatán Peninsula, an area that is receiving high numbers of refugees stemming from the poverty and political instability rampant in Chiapas, Mexico (Stedman-Edwards, 1998). Creative solutions are needed to protect the reserve. As distasteful as it may seem to some to promote more beach resorts along this peninsula, conservationists may be forced to explore such options. For such areas, one can imagine all the features desired: training, employment and higher wages for local people; more 'green' features, which offer a better quality of life for residents; quality social services; and lower density and improved site designs with reduced adverse environmental impacts.

The bottom line is that, in some ecoregions, the best solution may be to plan for and draw large numbers of people from one part of the ecoregion (e.g. park) to another (e.g. town). The tremendous amount of information and planning it will take to make ecoregional planning successful on a worldwide scale should not be underestimated. Effective actions will have to address the social and political forces generated at levels above and below the ecoregional level. There are tradeoffs inherent in what can happen and where – triage within ecoregions may be necessary. In some cases, it may be possible to save whole ecoregions; in others, the most realistic approach may be saving representative remnants of habitats in parks. But, unless ecological goals are quickly and effectively overlaid with social reality, the result will be little more than inefficiently squandering resources.

Intensification, biodiversity loss and the policy context

The linkages between policy and agricultural intensification are addressed in greater detail elsewhere in this volume. But it is vital to reiterate the message that there is an integral tie between biodiversity conservation and agricultural intensification, both inside parks and elsewhere in the landscape. Land distribution and use outside parks is likely to be the most essential determinant of pressure on parks and on the level of biodiversity conservation outside parks. The overall policy context – 'getting the policies right' – is the swiftest and most direct way to influence the links between poverty, land use and biodiversity loss. Other work confirms that:

> the sustainability of natural resource use is influenced by population pressure, but this exercises a much less critical impact than the overall policy framework. . . various agricultural and other policies whose effect is to constrain the poor's access to land encourages environmental degradation.
>
> (Heath and Binswanger, 1996).

Closely related to the policy challenges is a significant change in the nature of rural poverty, which is increasingly tied to landlessness and the insufficiency of land that is available. Approximately 30 million people in developing countries are landless and 138 million are near-landless (WRI, 1996). Of the land available, an estimated 38% of the 1.5 billion acres of cropland worldwide are degraded, and productivity losses from soil degradation may be greater than 20% in some Asian and Middle Eastern countries (WRI, 1996). This increased marginality is both a cause and an effect of rural poverty and is closely linked to rural environmental degradation and biodiversity loss.

Addressing biodiversity loss must also occur at national levels, with efforts to effectively deal with the linkages between growth and poverty, and assets and resources and their distribution. For example, a key determinant of poverty and aggregate growth is the initial distribution of assets, both physical and human (Birdsall and Londoño, 1997). Not surprisingly, large-scale agriculture and logging operations often benefit those who are already better off, because they are able to invest in land and new technologies. One result is that about 80% of deforestation in the Amazon is tied to large-scale ranchers, rather than to peasant farmers (Rudel and Roper, 1996, 1997). Increasing land values, fuelled by land scarcity and speculation, often displace smallholders and the near landless. Investments intended to alleviate poverty among the landless poor may be captured by the non-poor, who are able to take advantage of opportunities presented by the construction of roads, which stimulate new commerce (FAO, 1993d). This pattern was evident during the Green Revolution and may be recurring in the context of globalization and liberalization, as those who are better off can more readily capture benefits, leading to increased inequality. The cycle reproduces itself when the poor, who have sold their land or been pushed off it, relocate to marginal areas, often forested, and begin an unsustainable process of slash-and-burn agriculture, leading to deforestation. This latter cycle has been termed an 'immiseration' model of deforestation, pushed by peasant farmers and shifting cultivators, in contrast to the 'frontier' model (Rudel and Roper, 1996, 1997).

The development community has often shied away from such analysis and from recommending policy reforms addressing these issues over the last two decades, partly as a result of past failures in these areas and, more substantially, from timidity. Yet research and recommendations to address 'property rights, land reform, access of the poor to legal systems and credit, and fair competition' (Birdsall and Londoño, 1997), which would formerly have been found exclusively within political economy and political ecology, are increasingly proposed by neoclassical economists. But, despite rhetoric surrounding these issues, there has been little action. A notable absence of an aggressive push for such reforms can be found within the mainstay lending policies of development institutions such as the World Bank, the world's largest environmental agency.[12] Appropriately organized efforts to promote agricultural intensification are vital as a means of improving production on poor and degraded lands and increasing productivity on underutilized lands. However, such efforts will only be beneficial to biodiversity conservation – rather than leading to increased biodiversity loss – if they are accompanied by policies that begin to address the root causes of rural poverty.

Notes

The views expressed herein do not necessarily represent those of Conservation International.

1 The term 'parks' is used interchangeably with the more cumbersome 'protected areas'.

2 Key factors that influenced performance were: (i) baseline data collection and a good understanding of the ecosystems, threats and socio-economic context; (ii) involvement of local people in all phases of project design and implementation in an active capacity; (iii) collaboration among governments, donors and executing agencies and a willingness to undertake 'innovative' management structures; (iv) an ability to balance the enforcement and regulatory components of the project with development objectives and incentives; (v) an ability to influence the broader policy environment, which affects projects; (vi) long-term commitment of financial and technical support; and (vii) enforcement of park regulations.

3 Other projects address policy reforms, research, park infrastructure needs, ecological monitoring, training, etc.

4 See Brandon and Margoluis (1996) for a review of how assumptions in one area can lead to flaws in project design.

5 See Radford (1990), Agrawal and Gibson (1999) and Leach *et al.* (1999).

6 Note that in terms of developing incentives and direct linkages, the African context where there are the 'big five' species that attract tourists is a much 'easier' context than developing linkages and tourism in tropical moist forests, where there may be little to see (see Brandon, 1996).

7 Note that there are examples of projects pushing participation and democratization, while lamenting the diminished power of traditional authority structures for social control over resource use.

8 These percentages are estimates and are based on national designations of protection and use. They change if protected areas are added or deleted or if categories of protection are changed (WCMC, 1992, 1996).

9 I am not suggesting that traditional populations be moved from their lands. In fact, traditional populations should be accorded land rights and provided with the assistance they need to be capable managers – everything from global positioning system (GPS) use, boundary demarcation, timber management and harvesting to pricing policies and investment management (Mansour and Redford, 1996; Brandon, 1997). I discuss the many areas of the world, particularly in Latin America but also elsewhere, where peasants are pushed on to the worst land for agriculture (see, for example, Heath and Binswanger, 1996; Rudel and Roper, 1996; case studies for WWF's *Root Causes of Biodiversity Loss Project* (WWF, 1997); Stedman-Edwards, 1998). In areas where the social context is sufficiently stable, decision making and authority can be effectively decentralized; however, we must also recognize that such efforts will rarely lead to management for the sake of biodiversity conservation – they may produce greater vegetation or tree cover, but biodiversity in such systems is a correlate to management regimes, rather than an end in itself.

10 The bar for what parks should achieve was set at an impossible height in 1989. Parks were expected to move beyond meeting local development needs and become engines of regional growth, or 'critical elements of regionally envisioned harmonious landscapes' (McNeely, 1989, pp.156–157). At the 1992 World Parks Congress, the expectations of what parks should accomplish were ratcheted up to say that parks should provide tangible benefits at local, national and international levels (Barzetti, 1993).

11 Of course, requisite attention must be given to ensure that such intensification is consistent with local production systems and needs and does not lead to migration.

12 For a review of the World Bank's role versus a more potentially proactive role for the World Bank, see Brandon (1995).

23 Balancing Development and Environmental Goals through Community-based Natural Resource Management

NORMAN UPHOFF

Cornell International Institute for Food, Agriculture and Development, Cornell University, Ithaca, New York, USA

Framing the Problem

For at least a decade now, 'sustainable development' has been both a leading criterion and an objective for public and private decision-making (WCED, 1987b). Importance and some urgency are attached to finding effective and efficient ways whereby desires for increased production of economic goods and services can be satisfied while at the same time preserving the resource base upon which economic activities, and indeed all human life, depend – productive soil, clean and abundant water, pure air and diverse biological resources.

In the agricultural sector, world food requirements are likely to double within the next 30–40 years. Intensification of production easily comes into conflict with environmental conservation. Fortunately, the demand for food will not be continuously ratcheted up by a rapidly growing global population, as feared until recently. With a declining growth rate, world population is more likely to increase by half than to double, as was projected when global growth was over 2% annually. But there will still be much increased demand for food that comes from hoped-for rises in incomes. And one should never forget that there are currently very large unmet food needs to be satisfied in the world. About 800 million persons currently lack sufficient food, and this number will not go down without substantial increases in supply and lower prices for food (Conway, 1997). Doubling world food production will tax our ingenuity as well as our resources.

Intensification of agriculture is needed in part because the arable land base is limited and, in many areas, it is contracting, due to soil erosion and degradation and to the expansion of urban centres. The most suitable land for production is already being cultivated. Arable land per capita will decline by at least one-third in the next 30–40 years, while the water available for agricultural production will diminish by at least this much in per capita terms. So ways must be found

(eds D.R. Lee and C.B. Barrett)

to produce more food from the available land and water supplies used in conjunction with available labour, capital and biological resources.

More-favoured areas

Raising production in the better-endowed areas, where the Green Revolution has been most successful, must be a major part of the intensification effort. Unfortunately, increases in food production using high-yielding varieties (HYVs) of rice, wheat and maize are slowing and even stagnating (Conway, 1997). The continuing productivity of the better-endowed areas can no longer be taken for granted. In large areas of South Asia and China, where about 22.5 million ha are planted with HYVs, yields are no longer rising and water-tables for irrigation are falling, putting the food supply for almost 2 billion (10^9) people at risk. Agricultural methods and technologies that capitalize on the potentials of biology, in terms of both biotechnology and the principles of agroecology, will be essential for achieving sustainable yield improvements.

Less-favoured areas

An important question is how much agriculture in areas that have less favourable soil and climatic conditions, more difficult topography and less infrastructure can be intensified to halt the encroachment of shifting cultivation on forested and other protected areas, where biodiversity is endangered by extensive agricultural practices. Between 200 and 300 million people depend on shifting cultivation for all or a large part of their sustenance, and a considerably larger proportion of land area than of agricultural population is involved in extensive production systems. In most areas where extensive agriculture is practised, soils, topography and climate are not favourable for very high yields. These could probably be increased with agroecologically sound practices, but most technologies currently being promoted are not well suited to the resource endowments of these areas. Finding alternatives to slash-and-burn agriculture is thus a parallel focus for efforts to promote agricultural intensification. Happily, there is evidence, such as that of Hazell and Fan (Chapter 9 of this volume), that investments in currently less well-endowed areas can have high returns.

Agronomics

Biophysical and technical factors are often considered to be the starting-point of efforts to intensify agricultural production. Economic and social initiatives cannot succeed if they run contrary to agronomic possibilities. Madagascar is a good example of a country where intensification is urgently needed to raise production and incomes in order to avoid further encroachment upon its rich and diverse biological resources. The Cornell International Institute for

Food, Agriculture and Development (CIIFAD) has experience there in trying to improve agricultural production, focused on preserving valuable rainforest ecosystems.

Fortunately, some encouraging opportunities for increasing yields of the staple crop, rice, under lowland, irrigated conditions have recently appeared. Yields are being doubled, and more, without the requirement of external inputs. A system of rice intensification gives these results by radically altering plant, soil, water and nutrient management practices (de Laulaníe, 1993; CIIFAD, 1997, 1998; Uphoff, 2000a). Returns to land and labour are increased several-fold for farmers using this system. Substantial improvements in upland agriculture have also been observed on the forest margins from simple agro-forestry innovations that stabilize and raise production through better soil management and polycropping in place of slash-and-burn farming. A combination of laying out contours, planting leguminous shrubs and using compost, mulches and green manures can conserve and enrich the soil, raise yields and generate year-round income (CIIFAD, 1995, 1996, 1997). That there are some good agronomic possibilities for intensification that are compatible with and supportive of environmental conservation means that economic, social and institutional constraints are more likely to be determinant.

Economics

The economic costs and benefits of intensification – the main focus of this volume – are, of course, critical. In the area around Ranomafana in Madagascar and also in the Great Afram Plains in Ghana, where CIIFAD is also involved in integrated conservation and development efforts, farmers need to be assured of sufficient and steady returns from their labour to justify production practices that are environmentally benign. Given the interconnected ways that commodity and credit markets operate in both countries, farmers receive very low farm-gate prices at harvest time, when they must urgently sell much of their crop to pay off accumulated debts. Given the high interest rates charged by private traders and moneylenders, it is not economic for farmers to withhold their produce for later sale. Within a few months, they need to purchase these same commodities for household consumption, for 50 or 100% (or more) than they received when selling the same foodstuffs not long before. Such food deficits and financial needs add to the agricultural pressures that are placed on unexploited forest areas (Barrett, 1997a).

These price fluctuations can be explained in terms of supply-and-demand interaction; supply is greater than demand at harvest time, and demand becomes greater as household supplies dwindle. But a system of poverty and debt drives this system, which links commodities and credit, and the latter system (commodities and credits) perpetuates the first (poverty and debt). Farmers are obliged to sell their products to middlemen at prices shielded from competition, due to buyer coordination. The need for loans arises, to a large extent, from the low prices that farmers receive and from the much higher prices they must pay subsequently when buying food to feed their families.

Price differences during the year also reflect a lack of storage facilities in villages and the associated risks of holding grain stocks. To address this problem, in both Ghana and Madagascar, CIIFAD has helped communities to establish their own granaries, with revolving funds, from which they can get paid for their produce at harvest time. At that time, they receive the (low) market price for their rice (or maize or groundnuts) and retain the right to buy it back later at a lower price than prevails in the market, reimbursing charges for interest, storage and handling to the community fund. This is a popular scheme, although not without its pitfalls.

This experimentation is mentioned with the caveat that a buy-back system has yet to be devised that is fully efficient and sustainable, without dangers of loss or failure. More experience and adaptations are needed. Technical agricultural advances, such as those described above, need to be accompanied by institutional development that can help rural households in vulnerable environments to retain more of the value-added that they produce. Otherwise, they will be pressed to exploit the natural resources around them more aggressively. Many of the economic difficulties of farmers in Madagascar and Ghana result from the institutional contexts in which they live, not just from the limitations of their natural environment. While balancing development and conservation goals requires more than resource assessments and more than farm management analyses, we need to assess the economics of both existing and new systems of production whenever changes are contemplated.

Social and institutional considerations

This chapter considers some of the non-economic aspects of balancing environmental and development objectives, focusing on social and institutional issues and alternatives. There is growing interest in the potentials of community-based natural resource management (CBNRM). As this subject can become very broad, the focus here is on conservation and development issues with regard to vulnerable ecosystems, where protected-area management and integrated conservation and development projects (ICDPs) are relevant. Somewhat different issues arise in more-favoured areas, although the principles of participatory development and the institutional alternatives to be modified and meshed for the sake of balancing conservation and development objectives are similar.

Institutional Options for Natural Resource Management

More than 30 years ago, Garrett Hardin (1968) maintained that individual incentives will normally defeat efforts to sustain the natural resource base of economic activities like raising cattle on commonly held rangelands or fishing in publicly accessible bodies of water. This argument could easily be extended to the extraction of forest resources or groundwater and even to agriculture, where soils can be exhausted by overuse, reducing benefits for future generations. The 'tragedy of the commons', according to Hardin's analysis, will arise wherever individuals bear only part of the cost of a benefit that would accrue to them from exploiting a natural resource that is not state-regulated or privately owned. He concluded that the exploitation of natural resources should not be left to communities in some kind of common-property regime, because the incentives under such a system are too strong for non-sustainable exploitation. Either state institutions will have to govern access to natural resources – rationing them equitably and authorizing extraction at rates and in places that will not exhaust them – or all resources have to be placed under a private-property regime, which links private benefits from exploiting resources with the costs of their diminution.

Hardin's analysis delineated three institutional alternatives for resource management – public, private and community – with the latter considered ineffective for maintaining environmental resources. In the 30+ years since this analysis was advanced, its conclusion has been contested from different perspectives, although even its critics agree that it has usefully raised some institutional and policy questions that need attention (Feeny *et al.*, 1990).

- A first and most fundamental objection was that Hardin equated 'common property' with open access, which is quite a different kind of property regime, involving no control over access to resources, not even social controls (Berkes, 1989; Bromley and Feeny, 1992). Hardin's conclusions were directed to unregulated tenure systems, rather than to the management and maintenance of common-property resources.
- Secondly, in empirical terms, many community systems of management and regulation have, in fact, been able to limit and sustain resource use over many years (e.g. Netting, 1976; Alexander, 1982; Gilles and Jamtgaard, 1982; Sandford, 1983). So community management is feasible even if there are many ways in which it can break down.
- Thirdly, the assertion that 'rational' behaviour will undermine common-property regimes is logically flawed (Kimber, 1981). Hardin's argument assumed, similarly to the 'prisoner's dilemma' situation so often analysed in game theory, that there would be no communication among people. It also assumed no continuing social relationships that would enable people to fashion a system of management with rules, limits and sanctions, which produce a collective optimum in place of individual optimization efforts which exhaust resources (Ostrom, 1986). If people are indeed rational, even in the absence of communication, they can see that pursuing unchecked individual self-interest will be self-defeating, which itself creates incentives to devise effective management regimes (Kimber, 1981; Ostrom, 1990).

In recent years, there has been an upswing of interest in devolution or the sharing of responsibilities for natural resource management (NRM) with local residents (e.g. Western and Wright, 1994); these approaches have been termed 'community NRM' and 'community-based NRM'. It is important to distinguish between the two, because the first is too limiting in its solutions to the brokering of conservation and development aims.

- With community NRM, communities, clans or other local social organizations manage resources autonomously, according to traditions, interests and values that are given weight within local systems of culture and social relationships.
- Community-based NRM is more complex, depending primarily, but not exclusively, upon local management. Communities *per se* are not the only focus of management responsibility. Groups that are smaller than communities may have responsibility within a broader framework of decision-making, and localities (sets of communities having economic and/or social connections) may coordinate plans and supervise activities within larger domains, such as a landscape or a watershed. Ecosystems are generally larger than and cross-cut the boundaries of any particular village domain. By including responsibilities below and above the community level, CBNRM provides for more opportunities and scope of management than communities alone.

CBNRM involves a variety of stakeholders outside the locality who participate in decisions and enforcement of NRM along with members/residents of communities, groups and localities. This approach supports collaborative, multiparty systems for managing natural resources, where communities are the fulcrum for decision-making and implementation, but their members/residents are not the only actors (Uphoff, 1998a). Indeed, the pendulum of enthusiasm can swing too far in favour of 'localism', ignoring contradictions between local decision-making and the aims of resource conservation (Herring, 1998).

A recent and already influential volume entitled *Last Stand* (Kramer *et al.*, 1997) has argued for the more rigorous maintenance of protected areas for the sake of preserving biodiversity. It concludes that the beneficiaries of such preservation efforts extend far beyond the locality, and that this broader set of persons should be expected (required) to share in the costs of preservation since they derive benefit from this. Imposing all of these costs on those persons and communities within or near endangered ecosystems is untenable and unsustainable. This line of argument in support of protected areas should not be taken, however, as an argument against community involvement in NRM. Rather, it argues against simple and unconditional devolution of responsibility to communities that may have a direct interest in preserving soil, forest or water resources but less interest in maintaining biodiversity. By themselves, they may lack the coherence and power to enforce conservation decisions, even if their will is strong, against dissident members or against external encroachers, who may have government support or complicit acquiescence.

Many conservation biologists and their allies believe that more effort needs to be made to preserve biodiversity as valuable in itself, not compromising this objective for the sake of economic development, which they believe has occurred with ICDPs (Kramer *et al.*, 1997; on ICDPs, see Wells and Brandon, 1992; Brandon, 1997). They contend that balancing conservation and development objectives is not a reasonable policy goal where biodiversity is endangered. Preservation, they think, must be given priority where there would be irreversible losses of species and ecosystems, especially when their possible value in economic or ecological terms is not fully known. This concern, they believe, justifies absolute protection and addressing economic production and income needs elsewhere, where ecological disruption would be less.

Arguments in favour of more effective protection of vulnerable ecosystems can, however, just as well imply that communities should become more, rather than less, involved in responsibility for NRM. Real community participation in most ICDPs thus far has been fairly minimal, so it is hard to make a case that it is local participation and devolution of responsibility that has undermined ICDP efforts to preserve certain endangered ecosystems and species. It is true that the ICDPs that have been criticized in the literature have been expensive and slow to show progress. But, given the magnitude of what is expected of them, most have been underfunded or expected to reverse in a few years complex relationships and patterns of behaviour that have operated for a long time. Many have been rather bureaucratic undertakings, operating in a top-down manner, regarding local residents more as adversaries to be deterred or as beneficiaries to be bought off than as partners to be engaged in constructing and carrying out regimes of sustainable resource use and protection. The failings of bureaucracies, whether governmental or non-governmental, should not be displaced upon communities, which amounts to blaming victims rather than culprits.[1]

Encroachments on endangered ecosystems designated for protection usually reflect some conjunction of weak government agencies and the private profit-seeking forces of individual, self-interested economic rationality. Thus, the two institutional pillars that Hardin looked to for resource preservation as an alternative to community management are more likely to be sources of environmental debilitation than of salvation. No consideration of institutional options in the abstract can produce any consistent *a priori* preference for one given channel for NRM, however, because the effectiveness of public-sector, private-sector and community institutions is not intrinsic. Rather, it is contingent on various factors, such as prevailing cultural values, which reinforce some kinds of institutions and diminish others, the exercise of leadership through one institutional channel or another and respective capacities to enforce decisions.

One can identify likely strengths and weaknesses of various categories of institutions engaged in NRM, but these reflect tendencies or probabilities, not certainties. Such considerations are listed in Table 23.1, with non-governmental organizations (NGOs) considered separately as a fourth set of actors. NGOs are best regarded as a distinct, not-for-profit part of the private sector (Uphoff, 1993, 1996c). Universities or research institutions can be a further discrete set of actors, with different strengths and weaknesses for NRM (Uphoff, 1998b).

Government actors in NRM have the advantages of operating with authority and the capacity to mobilize large amounts of resources. Their limitations – bureaucratic constraints and pace and politicization – are mirrored in a positive way by private-sector actors, who are free from such strictures but are limited in terms of where they will work and for what objectives, not preserving the environment or benefiting the poor as such. Community actors exhibit some of the strengths and weaknesses of both, being particularly suited to flexible and innovative initiatives and having the advantage of prospectively being in the vicinity of the managed resources for years to come.

Table 23.1. Common strengths and weaknesses of alternative institutions for natural resource management (NRM) (adapted from Uphoff, 1998a).

	Common strengths	Common weaknesses
Government institutions	• Can invoke the authority of the state to enforce decisions • Can have personnel specialized and trained for NRM tasks • Power to levy taxes and prohibit certain behaviours within framework of policy	• Often limited budget, staff and facilities for effective action • Can be influenced by economic interests concerned more with extraction than with conservation • Prone to partisan or sectional actions lacking legitimacy
Private enterprises	• Can reduce costs to both government and communities if NRM is conducted successfully • Greater efficiency in use of capital and labour resources • Likely to be innovative in devising new approaches to NRM	• No built-in incentives for taking inter- or intragenerational welfare into account • No interest in NRM where there is no profit to be gained • Both the environment and the poor can lose out to interests of narrow set of beneficiaries
Community organizations	• More flexibility than government organizations • Most access to local residents' knowledge of resources • Stake in the productivity and sustainability of resources	• Management capacity is subject to fluctuation over time • Persons with special interests can dominate/veto decisions and action • Local conflicts can interfere with NRM
Non-governmental organizations (NGOs)	• Often have high degree of commitment to conservation • Often able to provide or access relevant expertise • Access to financial resources government/communities cannot muster • Can operate pragmatically, without much bureaucracy	• Financial resource base seldom steady or secure, so cannot assure long-term management • Can be constrained by factionalism or partisanship; internal conflicts • Can be object of jealousy and even obstruction from government • Can be paternalistic and unresponsive to local needs and interests
Universities and research institutions	• Able to bring knowledge to bear in solving problems • Outside power structure, able to work at grass roots like NGOs	• Limited management capacity • May not be accepted as 'neutral' • Likely to have less sustained institutional commitment

Efforts to preserve endangered natural resources have gone through several phases, starting with the designation of protected areas. This strategy made little effort to relate or optimize conservation and development objectives, regarding the latter as inimical to the former. The best way to support conservation was thought to be the prohibition of all economic activities within, or sometimes even around, vulnerable ecosystems. A second phase was the inception of the ICDPs referred to above, where, as a tradeoff or quid pro quo for placing certain protected areas beyond the pale of human activities, development activities were supported in the peripheral or buffer zones around designated ecosystems (Buck and Uphoff, 1997; Ferraro and Kramer, 1997).

Balancing the respective objectives of conservation and development within ICDPs was a matter of investing in enough of the latter to relieve pressure on the ecosystem to be preserved. The infrastructure, technical assistance and services provided could be viewed variously as compensation, reward or bribe for local people not having access to the protected area. Unfortunately, the implication of this is that preserving the area is in the interest more of outsiders than of local people, making continuing inducements necessary to maintain the area intact.

A third approach has involved local residents in assessing the status, vulnerabilities and potentials of natural resources together with government, NGO, private-sector or university actors. This was expected to lead to management agreements that serve the interests and needs of all parties in some optimizing fashion, with the long-term preservation of valued natural resources as a 'bottom-line' constraint. Community-based plans are developed for demarcated zones of responsibility, possibly using participatory rural appraisal (PRA) techniques, and supported by mapping facilities, such as geographical information systems (GIS), which are becoming increasingly common and widely available. When conflicts between or within communities arise in the course of identifying and evaluating resources, assessing the sustainability of resource uses, clarifying rights of access and devising techniques of conflict resolution can be accomplished, with NGO, university or government agents as facilitators.

In CBNRM, communities' interests in preserving forests, watersheds or other landscapes, as well as vulnerable flora and fauna, are clarified and reinforced through discussion and consultation. As stated already, there is no simple delegation or devolution of authority over natural resources to communities. Rather, agreements are arrived at within and between communities, to which outside organizations are privy and associated. Local responsibility, knowledge and initiative provide the main operational basis for NRM and natural resource protection, with some combination of government, NGO, private-sector and/or university agents as partners.

A limitation in this process is the willingness, as well as the ability, of communities to become partners in efforts to conserve natural resources in their area. The image outsiders commonly have of local residents in upland, forested, rangeland or coastal areas is of resource users unmindful of the consequences of their extractive actions or driven by extreme poverty to mine whatever productive value they can get from available resources. If communities are ignorant or desperate, not much can be done by voluntary measures to protect vulnerable ecosystems.

It should not be assumed, according to populist predilections, that this image of rural people as persons who are multiply marginalized and consigned to the margins of vulnerable ecosystems is totally wrong. Certainly, forces of poverty can impel certain destructive practices, and degradation can be incremental enough for its causes, if not its effects, to go unnoticed. But one can find in communities around the world a growing consciousness of the deterioration of forest, water, soil and biological resources, leading to local willingness to help reverse downward trends. This makes CBNRM more feasible than before.

Voices from Communities

CIIFAD is involved with government, NGO and university partners in a number of countries to advance the knowledge base and human and institutional capabilities for sustainable agricultural and rural development. Together with these partners, it engages with communities to identify or devise, evaluate and diffuse technologies, incentives, institutional arrangements and ideas that can improve rural households' livelihoods and security in ways compatible with preserving natural environments.

Some examples of local interest and initiative for sustaining natural resources indicate that, whatever may have been thought and done about natural resources in the past, there is now considerable willingness to engage in collaborative efforts to promote conservation at the same time as development goals are being pursued in qualified ways. Below are reported some conversations with residents in a number of communities visited over the past 5 years in a variety of settings where conservation and development objectives have come into conflict, and where efforts were being made to resolve the conflict with support for conservation outcomes. These cases are reasonably representative of the thinking of rural communities where there has been some process of explicitly and openly assessing current natural resource status, trends and options. In the absence of challenges and deliberations, however, past practices and thinking are likely to persist.

Dominican Republic

In October 1994, CIIFAD organized an international workshop on integrated watershed analysis and management, with a field trip in the Nizao watershed, located west of Santo Domingo. This watershed feeds a series of four dams, which provide hydroelectricity and water supplies for the capital city, as well as irrigation water for farms downstream.

Unfortunately, the dams have been silting up more rapidly than expected, endangering the huge investments made. Shifting cultivation and deforestation have been blamed for this problem, and several thousand farm households in the upper reaches of the watershed had been forcibly relocated, with more planned for removal by the government, even though there was no evidence that their activities were primarily responsible for the silt accumulation.[2]

A government ban on all tree felling had discouraged tree planting by farmers in the watershed, and official reforestation efforts were meagre and unsuccessful. To promote planting, the government promised that farmers could cut down, once grown, any trees that they planted. Free seedlings were offered, although there was no clear process for documenting who had planted certain trees, weakening the promise in farmers' eyes.

The La Esperanza coffee cooperative, with 800 members in the watershed, had been resisting government relocation plans before workshop participants arrived to study the biophysical and socioeconomic dynamics of the reservoir. The cooperative's leaders had, we learned, come to their own conclusion that deforestation was contributing to soil erosion, if not siltation. The cooperative had enacted a rule that, when members planted new trees, which was encouraged, they should agree never to cut down more than two out of five. Moreover, they agreed that, when they did this, they would replace the two trees immediately with new seedlings (Uphoff, 1994). Thus, farmers were prepared to enforce upon themselves more restrictive rules regarding tree cutting than the government provided.

Indonesia

In 1995, in the village of Sesaot on the island of Lombok, conflict had arisen with the Forestry Department over reclassification of a nearby forested area as 'protected forest'. The government wanted to ensure a continuing supply of water for irrigation systems in the valley downstream from where the village was located. With the facilitation of an Indonesian NGO, villagers were now cooperating with the Department. They had established a committee called Partnership for Forest Protection, which was patrolling the forest boundaries and reporting any illicit extraction of timber that they could observe (including some by government personnel).

During the course of a year, agreement had been reached on establishing a 12 ha pilot project for community forestry within the protected forest boundary on an area already degraded. This was planted with durian, rambutan, jackfruit and other trees and was subsequently judged the most successful reforestation effort in the province. Farmers recognized the adverse consequences of cutting down the forest and were willing to help preserve it, organizing themselves to manage their part of the forest in ways that contributed to both economic income and environmental conservation (Fisher, 1999).

Sri Lanka

In March 1996, we visited the village of Dothulagala in the Nilwala watershed in the southern part of this country as part of an assessment of a US Agency for International Development (USAID)-funded project for Shared Control of Natural Resources (SCOR), designed and managed by the International Irrigation Management Institute. This project was introducing participatory watershed management, based on previous successful experience with participatory irrigation water management, using institutional organizers as catalysts to mobilize collective action and community efforts (Wijayaratna, 1994, 1997; Uphoff, 1996b).

Dothulagala had already set up its own NGO, with the local Buddhist priest as patron, to protect and restore the environment. Villagers thought that the deforestation of hillsides was reducing their water-supply. One commented that streams were now drying up as quickly after 1 week without rain as previously after 1 month without rain. Vigilante action had been taken to burn down at night the huts of persons who were cutting down trees up on the hillsides, but this was seen as a short-term solution, not viable for the long run (Uphoff, 1996a).

The hillsides belonged to a tea plantation, owned by the state but now managed by a private company, which had no financial interest in protecting the forest. Through SCOR project committees, responsibility for the forested hillsides was assigned by the company to the Forest Department, which in turn enlisted the assistance of the local government, which deputized members of a village forest protection committee to patrol the forest and stop or report any encroachment. Villagers, who had initiated a process of establishing protection, were determined that the remaining forest should be preserved intact and even restored through local efforts.

Madagascar

In September 1997, with Cornell and Malagasy colleagues, I visited the village of Riambondrona, almost an hour's walk from the main road, to respond to a request for assistance in changing its farming system from slash-and-burn to more settled cultivation. In this part of Madagascar, the use of fire to clear forest and capture nutrients for field crops is part of the culture, not just of the farming system.

Using PRA mapping methods, villagers identified the natural resources and utilization activities in all directions from Riambondrona. They discussed resource problems and trends, as well as activities that could counter the problems enumerated. It was recognized that continuing to practise shifting cultivation was likely to reduce the water-supply on which their all-important rice crop depended. The community developed its own action plan, including setting aside 2 ha of prime lowland to test and demonstrate the intensification of rice and other production.

Other villages in the area around Ranomafana National Park made similar requests for assistance, once they saw that there were some good alternatives for lowland and upland farming being introduced in the 30 communities situated closest to the Park boundaries. It took several years to demonstrate effective agricultural technologies and establish rapport with communities previously dealt with peremptorily by government agents, but acceptance is now accelerating.

Ghana

In March 1998, the village of Domi in the Greater Afram Plains in the centre of the country was visited. This area is under strong environmental pressure, as population has grown and the climate has become drier, something attributed by villagers to the reduction in forest cover through slash-and-burn agriculture and charcoal making. The previous summer, a Ghanaian student had worked with Domi and two other villages to introduce community-based land-use planning and management. Through a diagnostic process, consensus was built up on the need for actions to protect the environment and on a plan of action.

The chief of the village banned farming around the sources of the six streams that were identified as originating within his jurisdiction. There were also plans to re-establish tree cover around them. Villagers acknowledged that, due to the loss of forest, some species of plants were disappearing, some of which had value for medicinal purposes. They agreed to cooperate with university researchers to evaluate and protect these plants.

Recent conflicts with the Department of Wildlife over expansion of the Kogyae Strict Nature Reserve, which put Domi within the expanded boundaries, were being resolved amicably. The Department agreed that simply expelling villagers, which it legally had the power to do, would not provide long-term protection. A new system of community protection was being devised, where all burning would be banned in a buffer zone around the reserve and villagers would undertake new conservation farming practices to protect and improve the soil as well as the larger set of environmental resources (Uphoff, 1998b).

While these reports are not based on any scientific sampling, they are informed by more intensive interaction with villagers than is common with survey methodologies. The reports are consistent in revealing an increasing awareness among rural people that measures must be taken to preserve the natural environments on which they depend for livelihoods and cultural values, or these will be lost. Where we have done systematic surveys – for example, in the peripheral zone around Los Haitises National Park in the Dominican Republic (Lizardo, 1996) – we have found villagers both concerned about environmental deterioration and willing to accept limitations upon their access to and use of natural resources, provided that the government does not simply impose restrictions without consultation and with no accommodation for livelihood needs.

There is no guarantee that rural people will cooperate with each other and with outsiders in making plans to conserve natural resources and in implementing these. Much depends on whether the communities have reasonably intact and effective local roles for decision making, resource management, communication, coordination and conflict resolution, all of which contribute to what is increasingly understood as 'social capital' (Uphoff, 2000b). The degree of dependence of economic sustenance and survival upon the resource(s) in question is also important, as is the availability of other economic options. However, if the first is great and the second is limited, this can be translated into a shared desire to regulate and preserve resource use if there are the kind of roles listed above, either indigenous or introduced with outside encouragement. These considerations need to be elaborated.

Motivations for Community Management

For communities to undertake systematic NRM, their members need to be both willing and able to undertake such responsibility. The two dispositions are not really independent of one another, since, where capability is limited, willingness is diminished, if only because collective action is less likely to produce effective results. Factors that impede local abilities to regulate resource use include: acute resource scarcity, such as that caused by rapid population growth; rapid change in community composition or norms; changes in policy and governance that unsettle established understandings and practices; and acute internal conflict and competition within communities. The kind of willingness reported above from communities in Latin America, Asia and Africa will not automatically lead to successful CBNRM.

One factor that affects both willingness and ability could be called a sense of community, of common interest, identity and fate. Where a strong sense of individualism prevails – with people attaching little, if any, value to others' welfare – it is harder to get collective action to conserve natural resources. What is difficult to predict is how people will respond to resource scarcity. When natural resources are limited relative to demand, this can elicit either conflict or cooperation. Conflict can break out based on zero-sum intentions to divide up limited resources; negative-sum consequences can occur whenever 'the pie is made smaller', due to competitive actions that seek to achieve benefits at others' expense. Conversely, cooperation that has zero-sum motivation, seeking to share limited resources equitably and efficiently, may lead to positive-sum results, increasing 'the size of the pie'. When renewable (flow) resources are involved rather than non-renewable (stock) resources, quite different social dynamics can accompany natural processes.[3]

In most of the situations where we have worked, concern for conservation is most clearly motivated by the need to sustain water-supplies for agriculture and domestic purposes, although these are linked with forest and soil resources and less directly with preservation of biological resources. The reduction or vulnerability of water-supply can be a very compelling reason for changing behaviour regarding natural resource use and for eliciting collective action. One can argue that community assumption of responsibility under such conditions reflects self-interest, even narrow interest in terms of individual or family benefit. But discussions with villagers suggest that their motivation is not so simple. The following discussion is based on observations and conversations with villagers, rather than on a systematic theory, although it is informed by a model for accounting for people's motivation that leads to resource-conserving versus resource-degrading behaviour (Uphoff and Langholz, 1998).

A very important influence on community considerations is the impact of resource losses in the present on the prospects that future generations will be able to maintain a satisfactory way of life in rural areas. This is a very strong motivation for rural people, who want their sons and daughters and their grandchildren to have the option of a fulfilling existence in the countryside. Partly this reflects an attachment to traditional practices and beliefs, but it also represents an understandably negative assessment of urban lifestyle alternatives, associated with crowding, crime, drugs and other abominations from a rural viewpoint. The idea that steps must be taken now to preserve a healthy environment that can support the next generation is very powerful for many people living in poorly endowed areas and on the margins of protected areas.

This valuation cannot be well accounted for by – or integrated into – an evaluation of choices that relies on imputed interest rates to discount the value of future benefit streams. Such calculations make anything more than 10–15 years in the future worth little or nothing, with the assumption that all important values can be denominated in common monetary terms. Quality-of-life considerations and the welfare of succeeding generations do not fit satisfactorily into this logical framework. Incomplete quantification gives unjustifiable emphasis to those elements that can be measured. Also, goods of a more public than private nature, which include many environmental benefits, lose out in evaluations that denominate everything in monetary terms. Much useful analysis and important insights have resulted from such reductionism, but this is better suited to treatments of industrial enterprise, for example, than natural resource conservation.

It is true that one sees little explicit concern in rural communities with biodiversity *per se*.[4] There is, however, considerable understanding that ecosystems and their preservation are important. Disturbances in the water cycle are most noticed, but the health of natural systems, which include flora and fauna together with soil and water resources, also registers in people's consciousness, at least if heightened through discussion. Just as environmental awareness has been growing in countries with more education and wealth over the past 25 years, so are rural people in poorer countries becoming more attuned to shifts in the environment. They may have shrugged off warnings and worries a decade ago, but they have been exposed to some of the same evidence and reports that have moved people in richer countries to take their environment's condition seriously. Indeed, residents in poorer countries who do not have much of a margin of financial resources to fall back on seem to attend more scrupulously to what is happening in their surroundings. Shifts in their environment affect much more than comfort and marginal changes in income, as livelihoods and even survival are at stake.

Conservationists may prefer to be concerned with biodiversity-driving programmes aimed at strengthening both natural resource preservation and economic development, but probably what is most persuasive for resource users in the foreseeable future will be maintaining complexes of soil, forest and water resources, with biological resources a secondary consideration. With enough discussion and planning, these objectives can often be made complementary and positive-sum, so that the limited support for biodiversity protection narrowly defined becomes part of and contributes to broader systems of natural resource management, which conserve genetic endowments that have been created over millennia.

Summary and Conclusions

Any consideration of relationships between conservation and development has to consider tradeoffs and complementarities, but it should not be assumed that the former will always dominate, with outcomes necessarily cast in zero-sum terms. Complementary outcomes are possible in those situations which build on positive-sum relationships. In the agronomic area, we have seen technical innovations that can raise agricultural production in ways that are compatible with maintaining soil fertility and which reduce the need to expand cultivated area at the expense of remaining forest areas. This makes economic and agronomic considerations complementary and supportive of natural resource conservation.

Similarly, when considering the bundle of natural resources of concern – soil, forest, water and biological resources – we find that they are, for the most part, complementary. Losing any of these – particularly within the complex of soil, forest and water resources – diminishes the others; when enhancing the quantity and quality of any one of these, the other two benefit. It is fortunate that there is a strong symbiotic relationship between biota and the other three, because, although rural people often perceive little stake in preserving biodiversity, the preservation of soil, forest and water systems supports biodiversity conservation. To the extent that livelihoods depend on conserving watersheds and maintaining the integrity of the hydrological cycle, the flora and fauna, which are codependent on this cycle, can 'free-ride' on this protection.

Not all economic interests are compatible with environmental conservation. There are many short-term economic gains that can be reaped at the expense of ecosystems and their constituent elements. But, when a time dimension is added to the calculations of net value, unless heavily discounted, the social value favours some optimization between exploitation and conservation. If there is privatization of resource control, it may be attractive to extract short-term benefits by depleting the natural assets' long-term potentials. However, if there is community-based management, to return to our discussion above, these calculations are not purely individual. Negative externalities for others in the current generation who are living nearby and negative effects on the prospects of future generations are factored into such considerations. Tradeoffs between immediate individual benefits, on the one hand, and others' costs (especially intergenerationally), on the other, get weighed in a process of community-based decision-making.

To have a systematic balancing of tradeoffs between economic and environmental values, both in the present and in the future, there need to be institutions that carry out the functions noted above: decision-making, resource mobilization and management, communication and coordination, and conflict resolution. These can be performed formally or informally. At local levels, the latter mode is common and often more effective than the former would be, because consensus and social sanctions are invoked.

This analysis could be interpreted as giving grounds for much optimism about the future of environmental conservation, but that would be reading too much into it. The message from the reports from the field above and from the accompanying analysis is essentially that pessimistic prognoses, based on assumptions that there must always be tradeoffs, are premature. There remain positive-sum opportunities; not everything has to be cast in zero-sum terms. There are appropriate agronomic practices that can produce economic net benefits that support intensified agricultural practices, which favour, rather than ravage, the environment.

Rural people are increasingly appreciating the need to revise their farming systems to accommodate the long-term sustainability of associated ecosystems. While conserving biodiversity may not be a priority for many communities, maintaining ecosystems that conserve soil, forest and water resources is easily understood as a widespread need, and fortuitously this is what endangered flora and fauna also require. Reaching agreements that utilize natural resources in sustainable ways and that enforce them will, of course, depend on the operation and effectiveness of local institutions, rather than just on appeals to individual rationality.

The effectiveness of action at group, community and locality levels depends, it must be added, on more than just local customs, precedents, values and human resources. There is an important role for government, private-sector, NGO and university/research institutions as complements and as supports for decision-making and follow-up at local levels. It is necessary to consider what will be economically attractive courses of action for resource users and stakeholders in natural resources of all types, and to figure out what balance of incentives will favour resource-conserving over resource-degrading behaviour (Uphoff and Langholz, 1998). But with such incentives we need to see what kinds of community interests and sanctions can be mobilized as conducive to achieving that same result, with government and other outside sanctions reinforcing such directions of activity. While tradeoffs are always present, identifying complementarities will produce the best prospects for finding ways of sustaining ecosystems, not privileging economic needs, but rather seeking to serve them in ways compatible with environmental vitality.

Notes

1 Vulnerable ecosystems are as (or more) often damaged by timber merchants, mechanized fishing vessels or traders who offer large payments for endangered species as by local residents, who are themselves excluded or manipulated by 'outsiders'. It may be objected that community management and protection in such situations will be inadequate, because local authorities do not have enough power to enforce regimes of protection. But then the alternative of central-government responsibility for conservation is likely to be similarly flawed, being seldom any stronger at local levels.

2 Subsequent research indicated that the main source of the siltation was from slopes nearer to the dams, which had been disturbed by construction work for the dams themselves and associated roads (Nagle, 1997).

3 This was demonstrated with the distribution of scarce water-supply in the Gal Oya irrigation scheme in Sri Lanka, where a regime of cooperation was established, more quickly than anyone expected, during severe water shortfall when the reservoir for the country's largest irrigation system was only 25% full at the start of the planting season (Uphoff, 1996b). A system of cooperation was able to reduce seepage and conveyance losses through voluntary cleaning of distribution channels and sharing water between head-enders and tail-enders, even with more deprived farmers on other channels. This had the consequence of, in effect, doubling the supply of water, a quintessentially scarce resource. When dealing with natural resources, at least those that are renewable, positive-sum solutions are possible. For a detailed and independent evaluation of the Gal Oya case, see Amarasinghe *et al.* (1998).

4 Reflecting on her observations of indigenous communities during fieldwork in western India, Baviskar comments: 'While reverence for nature is evident in the myths

and many ceremonies which attempt to secure nature's cooperation, that ideology does not translate into a conservationist ethic or a set of ecologically sustainable practices' (1995, p. 173).

24 Assessing Tradeoffs and Synergies among Agricultural Intensification, Economic Development and Environmental Goals: Conclusions and Implications for Policy

DAVID R. LEE,[1] CHRISTOPHER B. BARRETT,[1] PETER HAZELL[2] AND DOUGLAS SOUTHGATE[3]

[1]Department of Applied Economics and Management, Cornell University, Ithaca, New York, USA; [2]Environment and Production Technology Division, International Food Policy Research Institute (IFPRI), Washington, DC, USA; [3]Department of Agricultural Economics, Ohio State University, Columbus, Ohio, USA

Introduction

The underlying issue addressed by this volume is whether and to what extent three sets of goals – agricultural intensification and increased food production, economic growth and poverty alleviation, and environmental sustainability – are generally synergistic and complementary, or whether attaining one or more of these objectives typically requires that tradeoffs be made with respect to one or more of the others. The preceding chapters offer both conceptual and empirical evidence that agricultural intensification is a necessary but not sufficient condition for environmentally sustainable economic progress under widely varying circumstances in the developing world. When agricultural productivity growth does not increase in step with human demand for food and other farm products, smallholders are generally pushed to expand cultivation and grazing into environmentally fragile margins, where yields are low, poverty remains acute and environmental degradation can be severe. Intensification strategies are thus the appropriate cornerstone of efforts to enhance food security while mitigating environmental losses. However, even successful intensification efforts are not always environmentally benign. One of the most important challenges facing those seeking to enhance both environmental and development goals is to identify how agricultural intensification in poor countries can be made consistent

(eds D.R. Lee and C.B. Barrett)

with the pursuit of environmental objectives in more places and at shorter as well as longer time horizons.

In the ongoing debate over sustainable development – particularly in the wake of such developments as the 1982 World Parks Congress, the 1987 report of the World Commission on Environment and Development (the Brundtland Commission Report), the 1992 United Nations (Rio) Conference on Environment and Development and the World Bank's *World Development Report 1992* on the environment – the complementarity of these objectives has often been accepted on faith rather than demonstrated. As reviewed in Chapter 1, there have been numerous reasons for the oftentimes uncritical acceptance of the 'complementarity' viewpoint. The success of technology-led 'Green Revolution'- type intensification in parts of Asia and Latin America led to the popular view that, by increasing food production on existing lands, agricultural settlement and production pressures on forests, wetlands and other fragile areas might be reduced, thereby 'saving' land resources. The historical record of the 'agricultural transformation' (Timmer, 1989) and the extensive evidence on farm–non-farm growth linkages (Rosegrant and Hazell, 2000) have also emphasized agriculture's diverse contributions to economic growth: providing inexpensive wage goods and raw commodities to maintain low input costs for other sectors; absorbing surplus rural labour; generating savings for investment; and providing a market for non-agricultural goods and services.

As many countries have made the transition from a dependence on agriculture and other primary commodities to more urbanized economies based on manufacturing and services, enjoying reduced poverty and increased demand for environmental amenities along the way, it has become tempting to overgeneralize these long-run synergies as inevitably characterizing the growth and development process. Early evidence on the environmental Kuznets curve (EKC), for example, claimed positive 'global' synergies between increased per capita income growth and a variety of environmental indicators. These findings were sometimes based on overly simplistic generalizations about the observed correlation between income growth and the increased demand for environmental amenities in many Western industrialized countries. Complementarities between environmental conservation and economic development objectives were also asserted in the context of integrated conservation and development projects (ICDPs), which were widely, although sometimes uncritically, promoted by conservation organizations and development non-governmental organizations (NGOs) during the 1980s and 1990s, in seeking simultaneously to protect parks while enhancing living standards in surrounding areas (McNeely, Chapter 21; Brandon, Chapter 22).

The limitations of these different arguments have become increasingly apparent over time. More recent work on the EKC has shown that many environmental indicators do not unambiguously improve with increased incomes. Some, such as household waste production and energy consumption, generally increase with income growth. High-income countries have long wrestled with growth-related problems, including air and water pollution, waste disposal and siting issues and, more recently, greenhouse-gas emissions and global warming. However, lower-income regions, such as India's Punjab (Pingali and Rosegrant,

Chapter 20) and the Andean highlands (Crissman *et al.*, Chapter 8) often exhibit analogous signs of the high environmental costs of intensified agriculture. These include depletion of water resources, salinization and micronutrient depletion in soils, chemical contamination of groundwater (with associated health impacts) and reduced genetic diversity, both above- and below-ground. In industrialized countries as well, the negative environmental externalities associated with intensified agriculture have become major policy issues. Finally, despite substantial investments by donors, the ICDP approach to conservation and development thus far counts relatively few successes and the untested, implicit assumptions underpinning the ICDP model have become increasingly more apparent (see discussion in Brandon, Chapter 22).

In short, the common assumption of inherent complementarity between agricultural intensification, economic development and environmental goals does not appear to stand up well to empirical scrutiny. Whatever synergies may exist across space and in the long run can be complicated by many factors at the micro level and in the short run. In these cases, tradeoffs may be more common than synergies (Vosti and Reardon, 1997). For instance, even on simply theoretical grounds, the common assumption that intensified agriculture will necessarily reduce pressures for agricultural extensification and deforestation is demonstrably incorrect (Pagiola and Holden, Chapter 5; Angelsen and Kaimowitz, Chapter 6). Outcomes depend fundamentally on assumptions about the type of technological change (labour- versus capital-intensive) that occurs, the subsector in which it occurs (intensive versus extensive sector) and the possibility of imperfections in input (particularly labour) or product markets, among other factors. Even parsimonious theoretical models clearly indicate that the question of whether there are synergies or tradeoffs among agricultural intensification, poverty alleviation and environmental goals is fundamentally an empirical one.

And the empirical record is a mixed one. There do exist many important examples demonstrating the potential synergies between pursuit of household food security, economic growth and environmental sustainability objectives through agricultural intensification. Some promising cases of intensification strategies representing 'win–win' (or, on occasion, 'win–win–win') outcomes emerge from the 'best-bet' technologies analysed by the three case studies from the Consultative Group on International Agricultural Research's (CGIAR) Alternatives to Slash-and-Burn (ASB) Programme (Gockowski *et al.*, Chapter 11; Tomich *et al.*, Chapter 12; Vosti *et al.*, Chapter 13). These technologies – typically based on diversified farming systems oriented toward high-value and/or agroforestry products – demonstrate that food security, income generation and biodiversity conservation goals can indeed be accomplished jointly. The question is less whether synergies exist than how best to achieve them through project or policy interventions that stimulate a 'virtuous circle' of soil nutrient improvement, higher agricultural productivity, improved rural livelihoods and reduced human pressure on renewable natural resources.

Yet, in other cases, clear tradeoffs are evident in achieving economic and environmental goals. The potato–pasture system of the Andean highlands provides an excellent example (Crissman *et al.*, Chapter 8), where increased chemical use may conflict with environmental and health objectives. The same is true in the irrigated rice and rice–wheat systems of Asia, where water and chemical use patterns associated with intensification have caused significant environmental degradation (Pingali and Rosegrant, Chapter 20). In addition, there are typically tradeoffs among goals pursued in different ecosystems within a country or region, as in the case of India (Hazell and Fan, Chapter 9). The scarcity of resources available with which to address the myriad problems facing developing countries confronts decision-makers with difficult tradeoffs at all turns.

A central conclusion of this volume is that, under most circumstances, agricultural intensification is necessary but not sufficient to achieving food security, poverty alleviation and environmental goals. In most rural areas, agriculture remains the backbone of the rural economy, the major source of employment and the dominant land use. It is difficult or impossible to conceive of solutions being reached to underlying food production, income and environmental problems in the absence of sustainable agricultural intensification strategies.

At the same time, the empirical record shows that agricultural intensification strategies are often not enough. The record is replete with examples where agricultural intensification, in and of itself, has failed to generate broad-based development outcomes. Various factors regularly appear in the stories of intensification strategies that have failed to bring about their declared objectives: concentrated land and resource ownership; inappropriate tenure regimes for land and water rights; rural factor and product market failures, especially in labour and financial markets; policy-induced distortions in the allocation of inputs such as water or chemical fertilizers; failure to address rural employment needs; and insufficient political will to address structural problems seriously, especially, perhaps, in marginal lands. Moreover, since the overriding objective of rural households is to secure a sustainable livelihood (Vosti and Reardon, 1997), agricultural development may represent only one of a variety of strategies they may employ, along with off-farm employment, non-farm activities and seasonal migration. This is the case, for example, with intensified horticultural production in parts of Latin America and Asia close to cities and alternative employment opportunities, and is increasingly true of many parts of rural Africa as well (Reardon, 1997; Barrett and Reardon, 2000).

Modelling and Measurement Issues

The evidence presented in the preceding chapters underscores the fact that the associations among agricultural, demographic, economic and environmental variables are complex and often context-specific. Even in a single small country, there may appear multiple pathways of agricultural development, with decidedly different impacts on outcome variables of interest (Pender *et al.*, Chapter 10). Making sense of why one observes complementarities among food security, poverty alleviation and environmental sustainability objectives in some settings, or as reflected in some indicators but not in others, remains a significant challenge for both researchers and practitioners. Achieving a 'Doubly Green

Revolution' (Conway, Chapter 2) will depend, among other things, on the improved measurement of multiple biophysical, demographic and economic variables and at multiple levels of analysis. A key challenge for researchers is the development of sound models and analytical approaches that can explain the observed interactions among biophysical, demographic and economic phenomena and can predict likely outcomes of alternative courses of action with some accuracy.

The challenges presented by these tasks stem from three fundamental aspects. First, the multidimensional nature of explaining agricultural intensification and associated analytical outcomes means that models can quickly become intractable. Substantial progress has been made in integrating biophysical and economic models; see, for example, Ruben *et al.* (Chapter 7), Crissman *et al.* (Chapter 8), and Schipper *et al.* (Chapter 14). These models represent some of the most promising avenues for future research exploring intensification-related problems and tradeoffs. Yet the many necessary simplifying assumptions reduce the generalizability of the results of any one modelling exercise. Secondly, questions of agricultural intensification and environmental sustainability are intrinsically dynamic and subject to uncertainty, which adds further complication. Questions of non-linear dynamics, threshold effects and risk preferences can feature prominently, although there are as yet no well-received methods for accommodating these important aspects of the problem. Thirdly, whether they live in favoured or marginal environments, many of the world's poorest people live in areas commonly characterized by relatively weak states, markets and formal institutions. Often, the core problem is less one of correcting a single failed component of a system, than the simultaneous failure of various potential loci of authority for rectifying poverty, food security and environmental problems.

Even if scientists could offer universally accepted models capturing the underlying relationships well, the identification and measurement of the concepts of interest need further attention. For example, period or area average rates of growth in agricultural output, yields per unit of cultivated area and output per agricultural worker may be poor indicators of trends in underlying food security when market access and productive assets are inequitably distributed (Conway, Chapter 2). Similarly, measures of biodiversity often differ substantially, depending on whether they measure on-farm biodiversity, that in parks and protected areas, above-ground versus below-ground biodiversity or that of fauna or flora.

It is not only the site or level of analysis of the variable that matters, but also the identification of proxy variables to represent more complex concepts, a key matter, particularly when societal valuations of these variables are ranked in ambiguous ways. The ASB case studies (Chapters 11–13), for example, demonstrate that carbon sequestration objectives can be addressed not only through the conservation of natural tropical forests – the typical indicator used for carbon sequestration capacity – but also through increases in the production and productivity of tree-crop perennials (cacao, coffee, rubber). Yet the latter indicator is not consistent with other biodiversity conservation objectives – the protection of native habitat for endangered mammals, for example. Even if tropical forest cover were always the 'right' measure, available large-scale data are commonly based on small samples, questionable extrapolations and obsolete information (Rudel and Roper, 1997; Kaimowitz and Angelsen, 1998). The multiplicity of indicators (imperfectly) related to economic, agronomic, population and environmental variables of interest makes it essential that researchers', practitioners' and policy-makers' efforts specifically identify their objectives and how those objectives are to be assessed – by measures of time, space and scale – particularly given that tradeoffs in achieving competing objectives are commonly involved.

Just as nascent efforts to improve integrated biophysical and economic modelling are likely to enhance our understanding of what makes agricultural, economic and environmental goals complementary in a given place and point in time, so too do relatively recent improvements in data collection methods and analysis show real promise. Longitudinal data gleaned from satellite imagery are becoming more available. Field information is increasingly being georeferenced to facilitate the storage and analysis of multivariate data in geographical information systems. Furthermore, participatory research methods are evolving that can identify local issues and indicators rapidly and reliably. All of these should enable increasing precision and relevance of research incorporating economic, social and environmental objectives.

Conditioning Factors and Complementary Development Outcomes

What factors, then, can be shown to influence the environmental and poverty outcomes associated with agricultural intensification? The preceding chapters show how various combinations of the conditioning factors that follow often prove critical to simultaneously achieving agricultural intensification, poverty reduction and environmental protection goals.

- *Population pressure* (whether measured by rural population density or other measures) is commonly an important determinant of intensification strategies, for example, in Central America (Pender *et al.*, Chapter 10) and in India's semi-arid tropics (Kerr *et al.*, Chapter 16). However, the simple population growth–technological change–food production relationship posited by Boserup (1965) does not hold universally. Other economic, market and political factors can override or dominate the role of population growth in driving rural land-use changes (Bilsborrow and Carr, Chapter 3). Moreover, in some settings, such as the humid forests of Central Africa, waiting for increased population density to induce intensification is probably a recipe for ecological disaster (Gockowski *et al.*, Chapter 11).
- Type of *technological change* (labour-intensive versus labour-saving, for example) and associated rates of adoption exert critical influences on the outcomes of intensification efforts. Mechanization can displace agricultural labourers, some of whom increase the ranks of the rural poor and landless, while others become colonists, degrading fragile forest margins. However, land improvements that increase on-farm labour productivity on existing

plots (e.g. bunds, ridges, irrigation, agroforestry) can reduce extensification pressures, even though farmers' economic requirements for adoption may not always correspond with potential agronomic benefits (Neill and Lee, 2000; see also Reardon *et al.*, Chapter 19).

- *Agroecological conditions* are, of course, a key factor determining the inherent potential for achieving productivity growth as a result of intensification efforts. Moreover, these conditions also help determine the underlying comparative advantage of different regions for agricultural versus non-agricultural land uses (as well as type of agricultural use) and thereby heavily influence spatial land-use patterns (Sanchez *et al.*, Chapter 17; Staal *et al.*, Chapter 18).
- The *development pathway*, as Pender *et al.* (Chapter 10) show, can create significant differences in resource exploitation strategies and sustainability outcomes among rural smallholders, based on a unique confluence of different agroecological characteristics, cropping patterns, population densities and access to infrastructure and markets. The dynamics, synergies and tradeoffs that exist within these pathways can make them difficult to 'scale up' to broader regions (see discussion below).
- *Labour-market conditions* are important in influencing seasonality in labour demand, the (non-)existence of off-farm agricultural and non-agricultural employment opportunities and the household's demand for leisure. These significantly affect household incentives to invest in technologies and practices that make agricultural intensification more sustainable.
- *Roads, infrastructure and market access* are often critical factors in improving the income-generation prospects of farmers by increasing access to inputs and product markets, reducing the risk associated with investments in sustainable production technologies and generally creating more profitable alternatives to traditional agriculture – all with the potential of reducing pressures for extensification (Reardon *et al.*, Chapter 19). On the other hand, in newly opened 'frontier' areas subject to colonization pressures, roads and improved infrastructure may represent a 'double-edged sword' by simultaneously facilitating access to hitherto unsettled lands, making environmental degradation much more likely, especially where neither local institutions nor the central government have the capability to enforce property rights and other laws.
- *Policy changes*, macroeconomic and sectoral, play a major role in affecting the behaviour of individual decision-makers and thus the likelihood of achieving joint social objectives through reinforcing (or offsetting) the incentives presented by other technical and economic factors.
- *Land-tenure* and *property-rights regimes* can complement and reinforce market-based reforms in assuring rights to land and other productive assets, creating private and community incentives for sustainable resource use and thus achieving results from practices as varied as increased fertilizer use, conservation technologies and tree planting (Otsuka and Place, Chapter 15).
- *Local institutions and participation* improve the design and increase the legitimacy of intensification strategies, by ensuring that the articulation and

results of these strategies are broad-based. In many cases, ensuring local participation and community involvement in programme design and implementation is not only desirable, but essential to project or programme success, given that local people are those whose decisions, first and foremost, affect local resource use (Kerr *et al.*, Chapter 16; McNeely, Chapter 21; Uphoff, Chapter 23).

These (and other) conditioning factors jointly determine whether a given intensification strategy is likely to prove successful or not in simultaneously accomplishing multiple objectives.

The enormous diversity in the forms that these factors assume in specific circumstances makes generalization difficult. However, it appears that there is an increased likelihood of achieving complementary production, economic growth and environmental goals where:

- population density is high enough to stimulate intensification and technological change, but not so high that environmental resources (e.g. soils and forests) are substantially degraded;
- alternative off-farm and non-farm employment opportunities are available to soak up available labour and to complement on-farm income sources;
- technology is tailored to local conditions, generally implying labour-intensive technologies that absorb labour while also increasing its productivity;
- cropping systems are diversified, providing alternative income sources, reducing both price and yield risk and generating favourable biodiversity outcomes, particularly in rain-fed areas (important exceptions exist in some locations where food security concerns are of such overriding importance that monoculture crop production will predominate);
- physical and institutional infrastructure (e.g. roads, communications, grades and standards, contract law enforcement) facilitate market access and thus tend to improve the profitability of intensification strategies – although in areas with significant unsettled marginal lands, road and infrastructural development may be counterproductive in stimulating spontaneous settlement, logging and other extractive activities;
- property rights are secure and rural households are more likely to enjoy the results of investments they make;
- public institutions and infrastructure are effective and reinforce market-oriented growth;
- local organizations are effective in facilitating collective action and complementing the services provided by formal government institutions;
- prudent macroeconomic policies are pursued, in that the central government permits no more than modest deviations from equilibrium exchange rates and fiscal and external budget balances, while exercising monetary restraint in order to maintain stable aggregate price and employment patterns;
- development is participatory, with development strategies not imposed 'from above', but through taking into account local people's needs and priorities.

Characteristics of Sustainable Systems

The breadth of empirical studies evaluated in this volume is necessarily limited and is primarily focused on crop–tree systems and rain-fed areas, rather than major cereal- and livestock-producing areas. With this limitation in mind, several specific agriculture- and agroforestry-based systems identified in these studies appear to represent 'best bets' for addressing joint production, economic and environmental goals under distinct agroecological and socioeconomic conditions. These include, among others, the 'horticulture pathway' found in Central America (Pender *et al.*, Chapter 10); the intensive cocoa and mixed fruit system in Cameroon (Gockowski *et al.*, Chapter 11); the community-based managed forest and rubber agroforests of Indonesia (Tomich *et al.*, Chapter 12); the managed forest and coffee/*bandarra* system in Brazil (Vosti *et al.*, Chapter 13); the diversified tropical product system of Costa Rica (Schipper *et al.*, Chapter 14); and the mixed crop–livestock systems of sub-Saharan Africa (Staal *et al.*, Chapter 18).

While these systems differ in many respects, they have several important elements in common. First, they are generally mixed or diversified systems, relying not simply on crop monocultures. Monocultural systems (e.g. rice) are critical for addressing global food needs and – notwithstanding the sustainability concern raised here (Pingali and Rosegrant, Chapter 20) – have significant potential for becoming sustainable if environmental constraints are addressed. However, diversified systems, by including multiple crops (and livestock) products, simultaneously address both agroecological and economic constraints. These include technical risk factors (e.g. those stemming from rainfall and other climatic factors, pest infestations and such), economic risk, stemming from price fluctuations and marketing constraints, and seasonality in labour demands.

Secondly, these systems often incorporate perennial crops or agroforestry, not simply annual crops. These elements provide significant economic benefits in the form of reduced risk exposure, more flexible labour requirements, more favourable long-term price trends and the economic advantages associated with higher-valued products. They also confer substantial agronomic benefits, such as climatic risk reduction, improved soil quality and moisture retention.

Thirdly, these systems are also frequently characterized by significant environmental benefits, including increasing on-farm biodiversity, carbon sequestration, soil and water purification and retention functions. While it is almost tautological to say that agricultural species diversification contributes to biodiversity conservation, what is noteworthy is the range of environmental benefits that appear to be causally related to diversified livelihood strategies. For example, adoption of tree crops and agroforestry systems tends to reduce sediment displacement and transport, as well as stream-flow variability, often to the advantage of downstream populations. To the extent that these functions, taken together, enhance biodiversity conservation, they enhance the resilience of agroecosystems in confronting environmental and market shocks (see Perrings, Chapter 4).

It is important to note that growing human needs for crop and livestock products around the world will continue to be met, in many cases, through domestic and international markets, in which the diversified crop–tree systems mentioned often play, at best, only a minor role. Supplying these markets will continue to rely on the production from what are often large-scale and crop monoculture-based farms, where the multiple agronomic, economic and environmental benefits identified above will be difficult to realize (Pingali and Rosegrant, Chapter 20). The ultimate beneficiaries of more efficient global food markets and lower food prices are, of course, the billions of net food consumers, including hundreds of millions of urban and rural poor. Yet, even in these environments, there is considerable potential for increasing food production in ways that are environmentally more sustainable. Sanchez *et al.* (Chapter 17) and Staal *et al.* (Chapter 18) identify a host of farmer practices and policies that stand to increase the productivity of the mixed crop–livestock systems common in Africa. Similarly, Pingali and Rosegrant (Chapter 20) specify various technology- and policy-based initiatives that could help resolve many of the sustainability concerns currently facing the highly productive rice and rice–wheat systems of the Indo-Gangetic plains.

Replication and Limitations

In considering the generalizability of these and other systems and the extent to which they may create the conditions for and meet criteria of sustainable agricultural intensification, several other cross-cutting themes merit brief discussion. To begin, site specificity is important. 'Geography matters' in determining: the key underlying role of agroecological conditions; the loci of increased rural population densities, which drive land-use changes; the importance of transportation and marketing costs; the spatial distribution of biodiversity, cultivated area and livestock; the proximity of off-farm employment alternatives; and in influencing the distribution of underlying productive assets and wealth. There are countless examples; the natural regeneration of agroforests, for example, is typically dependent on the particular mosaic of systems and species present (Tomich *et al.*, Chapter 12). Given the particularity of agroclimatic and economic conditions at any given site, it is not surprising that the 'best-bet' systems identified above, while 'optimal' under specific agroecological and economic conditions, may be difficult (and expensive) to replicate across a wide set of biophysical and economic conditions. This suggests that the effort to identify appropriate sustainable systems in specific locations is a complex one, not only due to the number of circumstances that must be addressed, but because site specificity and locational attributes may make it difficult to 'borrow' solutions from elsewhere.

These factors help to account for the increased interest in 'ecoregional' or 'landscape' approaches to development, which integrate conservation and economic objectives. An ecoregional framework has become increasingly common, not only as an integrating approach for agronomic and economic modelling (Ruben *et al.*, Chapter 7), but also in terms of the role that 'agroecological zoning' might play in permitting rural development strategies to take advantage of agroecosystem heterogeneity (Vosti and Reardon, 1997). A 'landscape' or ecoregional framework is also being viewed as a feasible alternative to traditional ICDP strategies, in that these approaches are seen as potentially better able to accommodate distinct human activities and land uses across space (Brandon, Chapter 22). But the enormity of the task confronting researchers and development practitioners seeking to identify promising technological options for particular agroecosystems is still present. For example, sustainable intensification in low-potential areas typically requires different strategies from those in high-potential areas, so countries with considerable geographical heterogeneity, such as India, Kenya or Mexico, must design and pursue a portfolio of strategies if they are to simultaneously advance agricultural, economic and environmental objectives throughout the nation. In many places, this may overtax institutional capacity, particularly at a time when devolved institutional management is increasingly the rule.

A second, and related, cross-cutting issue is the challenge created by scaling up from farm to watershed or market levels (Fresco, 1995). Human capital, financing and institutional constraints, which may be minimal in a small development project, may be critical in a larger-scale one. Local institutions that 'work' in a limited setting may be difficult to scale up to a larger one. In product markets, the 'fallacy of composition' problem may be critical when moving from the farm to the regional or national market level. Once multiple farmers adopt promising cropping alternatives, increased aggregate supply drives prices down and, if demand for the product is price-inelastic, farmers may end up worse off, following the logic of immiserizing growth. Regardless of whether price falls so much as to make farmers worse off in long-run equilibrium, the benefits of adoption accrue disproportionately to early adopters of new crops, so the economic and social effects of intensification depend very much on adoption patterns.

Scaling problems related to policy specificity and policy appropriateness may imply that it would be most effective to 'scale down' policy to the regional or local level to take maximum advantage of unique local conditions and thus support distinct 'development pathways' (Pender *et al.*, Chapter 10). However, as with institutions, scaling down or fine-tuning national or sectoral-level policies to make them locally appropriate is a tall order under any circumstances. This is particularly so in an era of reduced central-government interventions across the board and widespread market liberalization, which increasingly leaves private decision-makers, rather than public-sector policy-makers, as the major arbiters of resource-use decisions. This reinforces the importance, stressed by Perrings (Chapter 4), of designing and using economic instruments, wherever possible, to stimulate socially desired resource use decisions through introducing and redesigning systems of economic incentives.

A third concern, noted in many of the chapters in this volume, is that associated with policy distortions and market failures, which regularly exacerbate the problems associated with achieving multiple goals. Inappropriate (or non-existent) water-pricing policies in intensively irrigated agricultural systems have contributed to high levels of food production, but also provide incentives for inefficient water use, which, in turn, exacerbates agroecosystem sustainability problems, such as soil salinity build-up, waterlogging and other changes in soil physical characteristics. In forest margin areas, currency overvaluation and poorly integrated markets limit the economic incentives facing farmers, while, simultaneously, insecure property rights constrain investments in furthering the productive capacity of the land. As a result, the sustainable management of both intensified agriculture and forests is discouraged. Marketing reforms, such as the termination of African parastatals' crop-marketing monopolies, may have unintended adverse consequences in situations where those distortions formerly helped resolve underlying market failures (e.g. for seasonal credit) (Reardon *et al.*, Chapter 19). The problems caused by the unintended effects of policy changes or the lack of appropriate policies and institutions can be severe.

Finally, it is true, almost by definition, that in trying to address multiple goals – the essence, after all, of the 'synergy' argument – the likelihood of encountering obstacles that prevent the attainment of those goals is increased. This is one of the central dilemmas of applied development projects that address multiple goals (e.g. ICDPs), for example. For this reason, sustainably intensifying agriculture – as assessed by the multiple criteria of enhanced food security and household incomes, decreased poverty and enhanced environmental outcomes – makes the focus on institutional development, on local participation and on the resolution of social conflict that much more important.

Policies for Sustainable Agricultural Intensification

Perhaps it is inevitable that a book such as this one accentuates caveats, extenuating circumstances, and the like. After all, the very nature of scholarly enquiry is to subject broad generalizations to close scrutiny. Accordingly, the researchers who have contributed to this volume emphasize that, while agricultural intensification often yields direct environmental benefits and is a necessary condition for sustainable economic progress in rural areas, it is not an environmental panacea. None the less, there is widespread support for the vigorous pursuit of environmentally sustainable approaches to agricultural intensification, having the joint goals identified throughout this volume. Bilsborrow and Carr (Chapter 3) remind us that determining where exactly various populations stand in their respective demographic transitions is a considerable challenge, which implies that future growth paths are difficult to predict. Regardless, it is clear that human numbers, and therefore demand for agricultural commodities, will continue to increase for the foreseeable future in Africa, Asia and Latin America. In the absence of improved strategies for intensification, the pressure to pursue unsustainable practices on existing croplands and to convert marginal land and tropical forests into cropland and pasture will be all but irresistible.

Conway (Chapter 2) argues that truly sustainable food security can only be achieved by a second 'Doubly Green' Revolution, which, in addition to being technologically productive and economically viable, is equitable, sustainable and consistent with long-term desired environmental outcomes. This is an ambitious but attainable goal. As numerous chapters in this volume make clear, progress toward these ends is made immeasurably more difficult in the absence of an appropriate policy environment. As important as they may be in specific localities, micro- and farm-level interventions, even if successful, are unlikely to solve the critical food security, economic development and environmental outcomes identified in this volume at a scale large enough to achieve major impact and in an acceptable time frame. A proper policy environment is indispensable in ensuring that agricultural intensification strategies are likely to work at the desired spatial and temporal levels. The success in 'scaling up' and replicating local projects and initiatives is also likely to be heavily dependent on a supportive policy environment. Unlike unidimensional market-based reforms, which cannot be expected to solve the multiple problems associated with sustainable intensification strategies, the policy components that are most likely to lead to these complementary outcomes are diverse and include the following:

- Policies that promote improved technologies and management practices, both for monocultural systems and for mixed diversified farming systems, and which are appropriate to and tailored to local conditions. This not only includes investments in improved and appropriate technologies, but systems that are characterized by labour intensities that aid, rather than worsen, local labour market and resource-use patterns.
- Infrastructure policies that promote market access, input provision and crop diversification by farmers, but which do not lead to indiscriminate extensification on the agricultural frontier.
- Policies that promote farm diversification into agroforestry and other high-value products, agroindustrialization and increased non-farm employment – all of which have the ability to provide alternative employment and income-generation opportunities to complement traditional income sources from crops and livestock.
- Tenurial policies that promote enhanced incentives for farm household investments in land-use conservation practices, perennial crops, tree planting and other investments that afford joint long-run economic and environmental pay-offs, typically, although not exclusively, through individualized land ownership.
- Market-based macroeconomic and sectoral reforms, which provide the overall structure of market and price incentives, critical to determining economic feasibility.
- Local institutional development and participatory approaches that involve communities and households in the design and implementation of development projects.

Technologies and farmers' practices, if poorly designed or inappropriately used, can increase production, but in ways that degrade natural resources. New technologies have often been developed with a narrow focus on short-term profitability to farmers and without due consideration of their longer-term sustainability. The development of powerful pesticides and herbicides, for example, has reduced costs and improved yields, but has often had negative effects on the environment and on long-term yields. Other technologies may spread from the agroclimatic zones or farming systems for which they were developed to other, less suitable areas, where they may degrade resources. Many national agricultural research and extension systems have yet to integrate environmental concerns successfully into their agenda. Too little consideration is given to the sustainability features of recommended technologies and practices and to the technology and management problems of more fragile areas, where resource degradation is considerable. National and international research institutions must continue to strengthen and broaden their efforts to address these obstacles.

Although past patterns of agricultural growth have sometimes led to negative environmental effects and have failed to solve rural poverty and food insecurity – even as they have contributed to national food needs and export earnings – this is not an inevitable outcome of agricultural growth. Rather, it reflects inappropriate economic incentives for managing modern inputs in intensive farming systems, insufficient investment in many heavily populated underdeveloped areas, inadequate social and poverty concerns and political systems that are often biased against rural people. Appropriate government policies and investments, improved human and institutional development and a broadened agricultural research agenda are all important means of redressing these past problems. With these changes and investments in place, there is no reason why agricultural development cannot be much more effective in simultaneously contributing to food production, income growth, poverty alleviation and environmental sustainability goals.

References

Abrol, I.P. (1987) Salinity and food production in the Indian sub-continent. In: Jordan, W.R. (ed.) *Water and Water Policy in World Food Supplies*. Texas A&M University Press, College Station, Texas, pp. 109–113.

Adger, W.N. and Brown, K. (1994) *Land Use and the Causes of Global Warming*. John Wiley & Sons, New York.

Agne, S. (1996) *Economic Analysis of Crop Protection Policy in Costa Rica*. Pesticide Policy Project Publication Series No. 4, Institute of Horticultural Economics, University of Hannover , Germany.

Agrawal, A. and Gibson, C.C. (1999) Enchantment and disenchantment: the role of community in natural resource conservation. *World Development* 27, 629–649.

Ahmed, J., Falcon, W. and Timmer, P. (1989) *Fertilizer Policy for the 1990s*. Discussion Paper No. 293, Harvard Institute for International Development, Cambridge, Massachusetts.

Ahmed, N. (1985) Fertilizer efficiency and crop yields in Pakistan. *Phosphorus in Agriculture* 89, 17–32.

Alberts, J. (1977) *Migración hacia áreas metropolitanas de América Latina: un estudio comparativo*. Centro Latinoamericano de Demografía, Santiago, Chile.

Aldy, J.E., Hrubovcak, J. and Vasavada, U. (1998) The role of technology in sustaining agriculture and the environment. *Ecological Economics* 26(1), 81–96.

Alexander, P. (1982) *Sri Lankan Fishermen: Rural Capitalism and Peasant Society*. Australian National University Monograph on South Asia No. 7, Australian National University, Canberra.

Alexandratos, N. (ed.) (1995) *World Agriculture: Towards 2010, a FAO Study*. John Wiley & Sons, Chichester, UK.

Alfaro, R., Bouma, J., Fresco, L.O., Jansen, D.M., Kroonenberg, S.B., van Leeuwen, A.C.J., Schipper, R.A., Sevenhuysen, R.J., Stoorvogel, J.J. and Watson, V. (1994) Sustainable land use planning in Costa Rica: a methodological case study on farm and regional levels. In: Fresco, L.O., Stroosnijder, L., Bouma, J. and van Keulen, H. (eds) *The Future of the Land: Mobilising and Integrating Knowledge for Land Use Options*. John Wiley & Sons, Chichester, UK, pp. 183–202.

Alfsen, K.H., Bye, T., Glomsrod, S. and Wiig, H. (1997) Soil degradation and economic development in Ghana. *Environment and Development Economics* 2, 119–143.

al Hassan, R., Kydd, R. and Warner, M. (1996) *Review of Critical Issues for Natural Resources Policy in Ghana's Northern Region.* Briefing Paper 4, Project on the Dynamics of Smallholder Agriculture in the Guinea Savannah Zone of West Africa, with particular reference to Northern Region Ghana, Wye College, Ashford, Kent, UK.

Ali, M. (1996) Quantifying the socio-economic determinants of sustainable crop production: an application to wheat cultivation in the Tarai of Nepal. *Agricultural Economics* 14, 45–60.

Aliadi, A. and Djatmiko, W.A. (1998) *Economic Analysis of NTFP Extraction in Rantau-Pandan, Province of Jambi.* Southeast Asia Policy Research Working Paper, International Centre for Research in Agroforestry (ICRAF), Bogor, Indonesia.

Alston, J.M., Norton, G.W. and Pardey, P.G. (1995) *Science Under Scarcity. Principles and Practice for Agricultural Research Evaluation and Priority Setting.* Cornell University Press, Ithaca, New York.

Altamirano, T. and Hirabayashi, L.R. (1997) (eds) *Migrants, Regional Identities and Latin American Cities.* American Anthropological Association, Arlington, Virginia.

Altieri, M., Benito, C., Gomez-Lobo, A., Faeth, P., Tomic, T. and Valenzuela, J. (1993) Wheat farming in central Chile. In: Faeth, P. (ed.) *Agricultural Policy and Sustainability: Case Studies from India, Chile, the Philippines and the United States.* World Resources Institute (WRI), Washington, DC, pp. 15–37.

Amarasinghe, U.A., Saktivadivel, R. and Murray-Rust, H. (1998) *Impact Assessment of Rehabilitation Intervention in the Gal Oya Left Bank.* Research Report No. 18, International Irrigation Management Institute, Colombo.

Amend, S. and Amend, T. (eds) (1995) *National Parks Without People? The South American Experience.* The World Conservation Union (IUCN), Gland, Switzerland and Parques Nacionales y Conservación Ambiental, Quito, Ecuador.

Anderson, D. (1987) *The Economics of Afforestation: a Case Study in Africa.* World Bank Occasional Paper No. 1, Johns Hopkins University Press, Baltimore, Maryland.

Anderson, J.R. (1985) Assessing the impact of farming systems research: framework and problems. *Agricultural Administration* 20, 225–235.

Anderson, J.R. (ed.) (1994) *Agricultural Technology: Policy Issues for the International Community.* CAB International, Wallingford, UK.

Anderson, K. and Blackhurst, R. (1992) Trade, the environment and public policy. In: Anderson, K. and Blackhurst, R. (eds) *The Greening of World Trade Issues.* Harvester Wheatsheaf, Hemel Hempstead, UK, pp. 3–22.

Angelsen, A. (1995) Shifting cultivation and 'deforestation': a study from Indonesia. *World Development* 23(10), 1713–1729.

Angelsen, A. (1996) *Deforestation: Population or Market Driven? Different Approaches in Modelling Agricultural Expansion.* Chr. Michelsen Institute, Fantoft, Norway.

Angelsen, A. (1999) Agricultural expansion and deforestation: modelling the impact of population, market forces and property rights. *Journal of Development Economics* 58(1), 185–218.

Antle, J.M. and Heidebrink, G. (1995) Environment and development: theory and international evidence. *Economic Development and Cultural Change* 43(3), 603–625.

Antle, J.M. and McGuckin, T. (1993) Technological innovation, agricultural productivity, and environmental quality. In: Carlson, G.A., Zilberman, D. and Miranowski, J.A. (eds) *Agricultural and Environmental Resource Economics.* Oxford University Press, Oxford, pp. 175–220.

Antle, J.M., Cole, D.C. and Crissman, C.C. (1998) Further evidence on pesticides, productivity, and farmer health: potato production in Ecuador. *Agricultural Economics* 2(18), 199–208.

Antle, J.M., Stoorvogel, J.J., Crissman, C.C. and Bowen, W. (1999) *TRADEOFF: a Decision Support System for Policy Decision-makers*, Version 2.1.

Antle, J.M., Stoorvogel, J., Crissman, C.C. and Bowen, W. (2000) *The Tradeoff Model: a Policy Decision Support System for Agriculture, Quantitative Approaches in Systems Analysis.* AB-DLO and C.T. de Wit Graduate School for Production Ecology, Wageningen, the Netherlands.

Arifin, B. and Hudoyo, A. (1998) *Does Shifting Cultivation Really Cause Deforestation? Economic Analysis of Shifting Cultivation and Five-year Bush Fallow in Lampung Province*. Southeast Asia Policy Research Working Paper, International Centre for Research in Agroforestry (ICRAF), Bogor, Indonesia.

Arrow, K., Bolin, B., Costanza, R., Dasgupta, P., Folke, C., Holling, C.S., Jansson, B.-O., Levin, S., Mäler, K.-G., Perrings, C. and Pimentel, D. (1995) Economic growth, carrying capacity, and the environment. *Science* 268(5210), 520–521.

Arrow, K., Cropper, M.L., Eads, G.C., Hahn, R.W., Lave, L.B., Noll, R.G., Portney, P.R., Russel, M., Schmallensee, R., Smith, V.K. and Stavins, R.N. (1996) Is there a role for benefit–cost analysis in environmental, health and safety regulation? *Science* 272, 221–222.

Ash, H.B. (1941) *Lucius Junius Moderatus Columella on Agriculture*, Vol. 1. Loeb Classical Library, Harvard University Press, Cambridge, Massachusetts, and William Heinemann, London.

Askari, H. and Cummings, J.T. (1976) *Agricultural Supply Response: a Survey of the Econometric Evidence*. Praeger Publishers, New York.

Assoumou, J. (1977) *L'economie du cacao*. Delarge, Paris.

Avila, M. (ed.) (1994) *Alternatives to Slash-and-burn in South America: Report of Research-site Selection in Acre and Rondônia States of Amazon Region, Brazil.* Conducted from 31 August to 15 September 1992. International Centre for Research in Agroforestry, Nairobi, Kenya.

Awudu, A. and Hazell, P. (1998) The impact of policy reforms and trade liberalization on fragile lands in Sub-Saharan Africa. Paper presented at IFPRI/DSE/NARO/EC-sponsored International Conference on Strategies for Poverty Alleviation and Sustainable Resource Management in the Fragile Lands of Sub-Saharan Africa, Entebbe, Uganda, May 1998.

Bade, J., Hengsdijk, H., Kruseman, G., Ruben, R. and Roebeling, P. (1997) *Farm Household Modelling System for the Analysis of Sustainable Land Use and Food Security: Theoretical and Mathematical Description*. DLV Report No. 7, AB-DLO, Wageningen, The Netherlands.

Badiane, O. and Delgado, C.L. (1995) *A 2020 Vision for Food, Agriculture and the Environment in sub-Saharan Africa*. Food, Agriculture and the Environment Discussion Paper 4, International Food Policy Research Institute (IFPRI), Washington, DC.

Baker, D. and Dvorak, K. (1993) Resource management practices in Cameroon. Mimeo, International Institute of Tropical Agriculture (IITA) Humid Forest Ecoregional Centre, Yaoundé, Cameroon.

Baland, J.-M. and Platteau, J.-P. (1996) *Halting Degradation of Natural Resources. Is There a Role for Rural Communities?* Clarendon Press, Oxford.

Bandy, D., Garrity, D.P. and Sánchez, P.A. (1993) The worldwide problem of slash and burn agriculture. *Agroforestry Today* 5(3), 2–6.

Barber, C.V., Afiff, S. and Purnomo, A. (1995) *Tiger by the Tail? Reorienting Biodiversity Conservation and Development in Indonesia.* World Resources Institute (WRI), Washington, DC, and the Pelangi Institute, Jakarta.

Barbier, B. (1994) Modélisation agronomique et économique de la durabilité d'un système agraire villageois. PhD thesis, Centre de Coopération Internationale en Recherche Agronomique pour le Développement (CIRAD), Montpellier, France.

Barbier, B. (1998) Induced innovation and land degradation: results from a bioeconomic model of a village in West Africa. *Agricultural Economics* 19(1–2), 15–25.

Barbier, B. and Bergeron, G. (1999) Impact of policy interventions on land management in Honduras: results of a bio-economic model. *Agricultural Systems* 60, 1–16.

Barbier, E.B. (1990) The farm-level economics of soil conservation: the uplands of Java. *Land Economics* 66, 199–211.

Barbier, E.B. (1992) Rehabilitating gum arabic systems in Sudan: economic and environmental implications. *Environmental and Resource Economics* 2, 341–358.

Barbier, E.B. (1997) Introduction to the environmental Kuznets curve special issue. *Environment and Development Economics* 2(4), 369–382.

Barbier, E.B. and Burgess, J.C. (1996) Economic analysis of deforestation in Mexico. *Environment and Development Economics* 1(2), 203–239.

Barlow, C. (1986) *Oil Palm as a Smallholder Crop.* Occasional Paper No. 21, Palm Oil Research Institute of Malaysia.

Barlow, C., Jayasuriya, S.K. and Tan, C.S. (1994) *The World Rubber Industry.* Routledge, London.

Barraclough, S.L. and Ghimire, K.B. (1995) *Forests and Livelihoods: the Social Dynamics of Deforestation in Developing Countries.* St Martin's Press, New York.

Barrett, C.B. (1995) Madagascar: an empirical test of the market relaxation–state compression hypothesis. *Development Policy Review* 13(4), 391–406.

Barrett, C.B. (1996) On price risk and the inverse farm size–productivity relationship. *Journal of Development Economics* 51(2), 193–215.

Barrett, C.B. (1997a) Food marketing liberalization and trader entry: evidence from Madagascar. *World Development* 25(5), 763–777.

Barrett, C.B. (1997b) Liberalization and food price distributions: arch-m evidence from Madagascar. *Food Policy* 22(2), 155–173.

Barrett, C.B. (1998) Immiserized growth in liberalized agriculture. *World Development* 26(5), 743–753.

Barrett, C.B. (1999) Stochastic food prices and slash-and-burn agriculture. *Environment and Development Economics* 4(2), 161–176.

Barrett, C.B. and Arcese, P. (1995) Are integrated development–conservation projects (ICDPs) sustainable? On the conservation of large mammals in sub-Saharan Africa. *World Development* 23(7), 1073–1084.

Barrett, C.B. and Carter, M.R. (1999) Microeconomically coherent agricultural policy reform in Africa. In: Paulson, J. (ed.) *African Economies in Transition*, Vol. 2: *The Reform Experiences.* Macmillan, London.

Barrett, C.B. and Reardon, T. (2000) *Asset, Activity and Income Diversification Among African Agriculturalists: Some Practical Issues.* Project report to the BASIS Collaborative Research Support Program, Ithaca, New York.

Barrett, S. (1991) Optimal soil conservation and the reform of agricultural pricing policies. *Journal of Development Economics* 36, 167–187.

Barrett, V., Lassiter, G., Wilcock, D., Baker, D. and Crawford, E. (1982) *Animal Traction in Eastern Upper Volta: a Technical, Economic and Institutional Analysis.* Michigan State University International Development Paper No. 4, Department of Agricultural Economics, Michigan State University, East Lansing, Michigan.

Barzetti V. (ed.) (1993) *Parks and Progress: Protected Areas and Economic Development in Latin America and the Caribbean.* IUCN – The World Conservation Union, Washington, DC.

Bates, R. (1981) *Markets and States in Tropical Agriculture: the Political Basis of Agricultural Policies.* University of California Press, Los Angeles.

Bautista, R.M. (1990) Development strategies, foreign trade regimes, and agricultural incentives in Asia. *Journal of Asian Economics* 1, 115–134.

Baviskar, A. (1995) *In the Belly of the River: Tribal Conflicts over Development in the Narmada Valley.* Oxford University Press, Delhi.

Beckerman, W. (1992) Economic growth and the environment: whose growth? whose environment? *World Development* 20(4), 481–496.

Behnke, R.H., Sconnes, I. and Kerven, C. (eds) (1993) *Range Ecology at Disequilibrium: New Models of Natural Variability and Pastoral Adaptation in African Savannas.* Overseas Development Institute, International Institute for Environment and Development, Commonwealth Secretariat, London.

Bekunda, M.A., Bationo, A. and Ssali, H. (1997) Soil fertility management in Africa: a review of selected research trials. In: Buresh, R.J., Sanchez, P.A. and Calhoun, F. (eds) *Replenishing Soil Fertility in Africa.* Soil Science Society of America Special Publication 51, Soil Science Society of America, Madison, Wisconsin, pp. 63–80.

Bell, C., Hazell, P. and Slade, R. (1982) *Project Evaluation in Regional Perspective: a Study of an Irrigation Project in Northwest Malaysia.* Johns Hopkins University Press, Baltimore, Maryland.

Bender, W. and Smith, M. (1997) *Population Bulletin: Population, Food, and Nutrition.* Population Reference Bureau, Washington, DC.

Berkes, F. (ed.) (1989) *Common Property Resources: Ecology and Community-Based Sustainable Development.* Belhaven Press, London.

Besley, T. (1995) Property rights and investment incentives. *Journal of Political Economy* 103(5), 903–937.

Bilsborrow, R.E. (1987) Population pressures and agricultural development in developing countries: a conceptual framework and recent evidence. *World Development* 15, 183–203.

Bilsborrow, R.E. (1997) Overview and assessment of the state of knowledge. In: Bilsborrow, R.E. (ed.) *Migration, Urbanization, and Development: New Directions and Issues.* UN Population Fund, New York.

Bilsborrow, R.E. and Geores, M. (1992) *Rural Population Dynamics and Agricultural Development: Issues and Consequences Observed in Latin America.* Cornell International Institute for Food, Agriculture and Development, Ithaca, New York.

Bilsborrow, R.E. and Geores, M. (1994) Population change and agricultural intensification in developing countries. In: Arizpe, L., Stone, M.P. and Major, D.C. (eds) *Population and Environment: Rethinking the Debate.* Westview Press, Oxford.

Binswanger, H.P. and McIntire, J. (1987) Behavioral and material determinants of production relations in land abundant tropical agriculture. *Economic Development and Cultural Change* 36(1), 73–99.

Binswanger, H. and Ruttan, V. (eds) (1978) *Induced Innovation: Technology, Institutions, and Development.* Johns Hopkins University Press, Baltimore, Maryland.

Binswanger, H., Mundlak, Y., Yang, M.C. and Bowers, A. (1987) On the determinants of cross-country aggregate agricultural supply. *Journal of Econometrics* 36, 111–131.

Binswanger, H., Khandker, S. and Rosenzweig, M. (1989) *How Infrastructure and Financial Institutions Affect Agricultural Output and Investment in India.* Working Paper No. 163, World Bank, Washington, DC.

Biodiversity Support Program (1996) *Biodiversity Conservation Network, 1996 Annual Report: Stories from the Field and Lessons Learned.* Biodiversity Support Program, Washington, DC.

Birckhead, J., de Lacy, T. and Smith, L. (eds) (1992) *Aboriginal Involvement in Parks and Protected Areas.* Australian Institute of Aboriginal and Torres Strait Islander Studies, Canberra.

Birdsall, N. and Londoño, J.L. (1997) Asset inequality matters: an assessment of the World Bank's approach to poverty reduction. *AEA Papers and Proceedings* 87(2), 32–37.

Blackburn, H. (1998) Livestock production, the environment and mixed farming systems. In: Nell, A.J. (ed.) *Livestock and the Environment International Conference. Proceedings of International Conference on Livestock and the Environment, Wageningen, 16–20 June 1997.* International Agricultural Centre, Wageningen, the Netherlands, pp. 114–123.

Blaikie, P. (1987) *The Political Economy of Soil Degradation in Developing Countries.* Longman Scientific, London.

Bojanic, A. and Echeverria, R.G. (1990) *Retornos a la inversión en investigación agrícola en Bolivia: el caso de la soya.* Staff Notes No. 90–94, International Service for National Agricultural Research (ISNAR), The Hague, the Netherlands.

Bond, M.E. (1983) Agricultural responses to prices in sub-Saharan countries. *IMF Staff Papers* 30(4), 703–726.

Bonneuil, C. and Kleiche, M. (1993) *Du jardin d'essais colonial à la station expérimentale 1880–1930: Eléments pour une histoire du CIRAD.* Centre de Coopération Internationale en Recherche Agronomique pour le Développement (CIRAD), Paris.

Borlaug, N.E. (1987) *Making Institutions Work – a Scientist's Viewpoint.* Texas A&M University Press, College Station.

Borlaug, N.E. and Dowswell, C.R. (1994) Feeding a human population that increasingly crowds a fragile planet. In: *Transactions of the 15th World Congress of Soil Science, Acapulco, Mexico.* International Soil Science Society, Chapingo, Mexico.

Borrini-Feyerabend, G. (ed.) (1997) *Beyond Fences: Seeking Social Sustainability in Conservation.* World Conservation Union (IUCN), Gland, Switzerland.

Bosc, P.M. and Hanak-Freud, E. (1993) *Economic and Technical Factors Influencing Regional Research Priorities in the Humid and Sub-humid Zones of West and Central Africa.* Centre de Doopération Internationale en Recherche Agronomique pour le Développement (CIRAD), Paris.

Boserup, E. (1965) *The Conditions of Agricultural Growth: the Economics of Agrarian Change Under Population Pressure.* Aldine, Chicago.

Boserup, E. (1981) *Population and Technological Change: a Study of Long-Term Trends.* University of Chicago Press, Chicago.

Bosworth, D., Dawkins, P. and Stromback, T. (1996) *The Economics of the Labor Market.* Longman, Harlow, UK.

Bouma, J., Kuyvenhoven, A., Bouman, B.A.M., Luyten, J.C. and Zandstra, H.G. (eds) (1995) *Eco-regional Approaches for Sustainable Land Use and Food Production.* Kluwer Academic Publishers, Dordrecht, the Netherlands.

Bouman, B.A.M., Nieuwenhuyse, A. and Hengsdijk, H. (1998a) *PASTOR: a Technical Coefficient Generator for Pasture and Livestock Systems in the Humid Tropics, Version 2.0. A User's Guide.* Quantitative Approaches in Systems Analysis No. 18, AB-DLO-PE, Wageningen, the Netherlands.

Bouman, B.A.M., Nieuwenhuyse, A., Hengsdijk, H. and Jansen, H.G.P. (1998b) An integrated methodology for sustainable land use exploration using GIS. In: *Proceedings of the First International Conference on Geospatial Information in*

Agriculture and Forestry, Lake Buena Vista, Florida, 1–3 June 1998, Vol. II. ERIM International, Ann Arbor, Michigan, pp. 230–237.

Bouman, B.A.M., Schipper, R.A., Nieuwenhuyse, A., Hengsdijk, H. and Jansen, H.G.P. (1998c) Quantifying economic and environmental trade-offs in land use exploration at the regional level: a case study for the Northern Atlantic Zone of Costa Rica. *Ecological Modeling* 114, 95–109.

Bouman, B.A.M., Jansen, H.G.P., Schipper, R.A., Hengsdijk, H. and Nieuwenhuyse, A. (eds) (1999a) *Tools for Land Use Analysis at Different Scales With Case Studies for Costa Rica*. Kluwer Academic Publishers, Dordrecht, the Netherlands.

Bouman, B.A.M., Jansen, H.G.P., Schipper, R.A., Nieuwenhuyse, A., Hengsdijk, H. and Bouma, J. (1999b) A methodology for integrated biophysical and economic land use analysis at different scales. *Agriculture, Ecosystems and Environment* 75(1–2), 55–73.

Brandon, K. (1995) The World Bank and the environment. In: *Bretton Woods: Looking to the Future*. Bretton Woods Commission, Washington, DC, pp. C133–C142.

Brandon, K. (1996) *Ecotourism and Conservation: a Review of Key Issues*. Environment Department Paper No. 33, World Bank, Washington, DC.

Brandon, K. (1997) Policy and practical considerations in land-use strategies for biodiversity conservation. In: Kramer, R., von Schaik, C. and Johnson, J. (eds) *Last Stand: Protected Areas and the Defense of Tropical Biodiversity*. Oxford University Press, New York, pp. 90–114.

Brandon, K. (1998) Perils to parks: the social context of threats. In: Brandon, K., Redford, K. and Sanderson, S. (eds) *Parks in Peril: People, Politics, and Protected Areas*. Island Press, Covelo, California.

Brandon, K. and Margoluis, R. (1996) Structuring ecotourism success: framework for analysis. In: Malek-Zadeh, E. (ed.) *The Ecotourism Equation: Measuring the Impacts*. Yale University School of Forestry and Environmental Studies, New Haven, Connecticut.

Brandon, K. and Wells, M. (1992) Planning for people and parks: design dilemmas. *World Development* 20(4), 557–570.

Brandon, K., Redford, K. and Sanderson, S. (eds) (1998) *Parks in Peril: People, Politics, and Protected Areas*. Island Press, Covelo, California.

Breman, H. (1990) Integrating crops and livestock in southern Mali: rural development or environmental degradation? In: Rabbinge, R. (ed.) *Theoretical Production Ecology: Reflections and Prospects*. Pudoc Simulation Monograph No. 34, Wageningen, the Netherlands, pp. 277–294.

Breman, H. and Sissoko, K. (eds) (1998) *Intensification agricole au Sahel*. Karthala, Paris.

Bromley, D.W. and Feeny, D. (eds) (1992) *Making the Commons Work: Theory, Practice and Policy*. Institute of Contemporary Studies Press, San Francisco.

Brons, J., Diarra, S., Dembélé, J., Bagayoka, S. and Djouara, H. (1994) *Diversité de gestion de l'exploitation agricole: étude sur les facteurs d'intensification agricole au Mali-Sud*. IER-DRSPR Document 94/33, IER-DRSPR, Sikasso.

Brooke, A., Kendrick, D. and Meerhaus, A. (1992) *GAMS Release 2.25: a User's Guide*. Scientific Press, San Francisco.

Brown, L. and Haddad, L. (1994) *Agricultural Growth as the Key to Poverty Alleviation*. 2020 Brief No. 7, International Food Policy Research Institute (IFPRI), Washington, DC.

Brown, L.R. (1997) Facing the prospect of food scarcity. In: Brown, L.R. (ed.) *State of the World 1997*. W.W. Norton, New York, pp. 23–41.

Buck, L. and Uphoff, N. (1997) *Déjà vu* all over again: dissatisfaction with integrated conservation and development projects. Unpublished paper, Cornell International Institute for Food, Agriculture and Development, Ithaca, New York.

Budidarsono, S. (1998) *Profitability Assessment of Transmigration Land Use System in Dryland Peneplain Zone of Lampung: Continuous Annual Food Crop Farming System, Degraded to* Imperata cylindrica *Grassland.* International Centre for Research in Agroforestry (ICRAF), Bogor, Indonesia.

Bulte, E. (1997) Essays in economics of renewable resources. PhD thesis, Wageningen Agricultural University, Wageningen, the Netherlands.

Bulte, E. and van Soest, D. (1999) A note on soil depth, failing markets and agricultural pricing. *World Development* 58, 245–254.

Bultena, G.L., Duffy, M.D., Jungst, S.E., Kanwar, R., Menzel, B.W., Misra, M.K., Singh, P.T., Thompson, J.R., van der Valk, A.W. and Wilham, R.L. (1996) In: Srivastava, J.P., Smith, N.J.H. and Forno, D.A. (eds) *Effects of Agricultural Development on Biodiversity: Lessons from Iowa, in Biodiversity and Agricultural Intensification: Partners for Development and Conservation.* World Bank, Washington, DC, pp. 80–94.

Bumb, B.L. and Baanante, C.A. (1996) *The Role of Fertilizer in Sustaining Food Security and Protecting the Environment to 2020.* Food, Agriculture and the Environment Discussion Paper 17, International Food Policy Research Institute (IFPRI), Washington, DC.

Bunch, R. (1988) Guinope integrated development program. In: Conroy, C. and Litvinoff, M. (eds) *The Greening of Aid: Sustainable Livelihoods in Practice.* Earthscan Publications, London, pp. 40–44.

Buol, S.W. (1995) Sustainability of soil use. *Annual Review of Ecological Systems* 26, 25–44.

Buresh, R.J., Sanchez, P.A. and Calhoun, F.J. (eds) (1997a) *Replenishing Soil Fertility in Africa.* SSSA Special Publication 51, Soil Science Society of America, Madison, Wisconsin.

Buresh, R.J., Smithson, P.C. and Hellums, D.T. (1997b) Building soil phosphorus capital in Africa. In: Buresh, R.J., Sanchez, P.A. and Calhoun, F. (eds) *Replenishing Soil Fertility in Africa.* SSSA Special Publication 51, Soil Science Society of America, Madison, Wisconsin, pp. 111–151.

Byerlee, D. (1980) Rural labor markets in West Africa with emphasis on the semi-arid tropics. In: *Proceedings of the International Workshop on Socioeconomic Constraints to Development of Semi-arid Tropical Agriculture, 19–23 February 1979, Patancheru, India.* International Crops Research Institute for the Semi-arid Tropics (ICRISAT), Pantancheru, Andhra Pradesh, India.

Byerlee, D. (1990) *Technical Change, Productivity, and Sustainability in Irrigated Cropping Systems of South Asia: Emerging Issues in the Post-Green Revolution Era.* CIMMYT Economics Working Paper 90/06, CIMMYT, Mexico, DF.

Byerlee, D. and Moya, P. (1993) *Impacts of International Wheat Breeding Research in the Developing World, 1966–90.* International Wheat and Maize Improvement Center (CIMMYT), Mexico, DF.

Byiringiro, F. and Reardon, T. (1996) Farm productivity in Rwanda: effects of farm size, erosion, and soil conservation investments. *Agricultural Economics* 15(2), 127–136.

CAMASE (Concerted Action for the Development and Testing of Quantitative Methods for Research on Agricultural Systems and the Environment) (1998) URL: http://www.bib.wau.nl/camase

Carpentier, C.L., Vosti, S.A. and Witcover, J. (2000) *BrasilBEM: a Household–Farm Bioeconomic Model for the Western Brazilian Amazonian Forest Margin*. Environment and Production Technology Division Discussion Paper, International Food Policy Research Institute (IFPRI), Washington, DC.

Casper, R. and Landsmann, J. (eds) (1992) *The Biosafety Results of Field Tests of Genetically Modified Plants and Microorganisms*. Biologische Bundensansalt für Landund Fortwirtschaft, Braunschweig, Germany.

Cassells, D.S., Bonell, M., Hamilton, L.S. and Gimour, D.A. (1987) The protective role of tropical forests: a state-of-knowledge review. In: Vergara, N.T. and Briones, N.D. (eds) *Agroforestry in the Humid Tropics*. East-West Center, Honolulu, Hawaii.

Cassman, K.G. and Pingali, P.L. (1993) Extrapolating trends from long-term experiments to farmers fields: the case of irrigated rice systems in Asia. In: *Proceedings of the Working Conference on Measuring Sustainability Using Long-Term Experiments, Rothamsted Experimental Station, 28–30 April 1993*. Agricultural Science Division of the Rockefeller Foundation, New York.

Cassman, K.G., De Datta, S.K., Olk, D.C., Alcantara, J., Samson, M., Descalsota, J. and Dizon, M. (1994) Yield decline and nitrogen balance in long-term experiments on continuous, irrigated rice systems in the tropics. *Advances in Soil Science*, Special Issue.

CAST (Council for Agricultural Science and Technology) (1999) *Animal Agriculture and Global Food Supply*. Ames, Iowa, 92 pp.

Celis, R. (1989) Factor substitution in linear programming: a methodology and an application to cane alcohol substitution in Costa Rica. PhD thesis, University of New Mexico, Albuquerque, New Mexico.

CGIAR (Consultative Group for International Agricultural Research) (1993) *The Eco-regional Approach to Research in the CGIAR*. Report of the TAC/Centre Directors Working Group, Technical Advisory Committee Secretariat, FAO, Rome.

CGIAR (Consultative Group for International Agricultural Research) (1995) Report of the task force on eco-regional approaches to research. In: *CGIAR Mid-term Meeting, Nairobi, Kenya, 22–26 May 1995*.

CGIAR (Consultative Group on International Agricultural Research) (1997) *CGIAR: Consultative Group on International Agricultural Research*. CGIAR Secretariat, Washington, DC.

CGIAR/TAC (Consultative Group on International Agricultural Research/Technical Advisory Committee) (1988) *Sustainable Agricultural Production: Implications for International Agricultural Research*. CGIAR meeting at FAO, May 1988, *Rome*.

CGIAR/TAC (Consultative Group on International Agricultural Research/Technical Advisory Committee) (1998) *CGIAR Study on Marginal Lands*. CGIAR, Washington, DC.

Chakraborty, M. (1994) An analysis of the causes of deforestation in India. In: Brown, K. and Pearce, D. (eds) *The Causes of Tropical Deforestation: the Economic and Statistical Analysis of Factors Giving Rise to the Loss of Tropical Forests*. University College, London Press, London, pp. 226–239.

Chambers, R. (1988) *Managing Canal Irrigation: Practical Analysis from South Asia*. Press Syndicate of the University of Cambridge, Cambridge, UK.

Chambers, R. (1997) *Whose Reality Counts? Putting the Last First*. Intermediate Technology Publications, London.

Chen, N., Valente, P. and Zlotnic, H. (1998) What do we know about recent trends in urbanization? In: Bilsborrow, R. (ed.) *Migration, Urbanization and Development:*

New Directions and Issues. United Nations Population Fund, Norwell, Massachusetts, and Kluwer Academic Publishers, Dordrecht, the Netherlands, pp. 59–88.

Chomitz, K.M. and Kumari, K. (1996) *The Domestic Benefits of Tropical Forests: a Critical Review*. Working Paper No. GEC 96-19, Centre for Social and Economic Research on the Global Environment (CSERGE), University of East Anglia, Norwich, UK.

Church, P. and Brandon, K. (1995) *Strategic Approaches to Stemming the Loss of Biological Diversity*. Center for Development Information and Evaluation, US Agency for International Development, Washington, DC.

CIIFAD (Cornell International Institute for Food, Agriculture and Development) (1995, 1996, 1997, 1998) *Annual Reports, 1994–95, 1995–96, 1996–97, 1997–98*. Cornell International Institute for Food, Agriculture and Development, Ithaca, New York.

CIMMYT (International Wheat and Maize Improvement Center) (1996) *1995–96 CIMMYT World Wheat Facts and Trends: Global Trends in Wheat Diversity and International Flows of Wheat Genetic Diversity*. International Wheat and Maize Improvement Center (CIMMYT), Mexico, DF.

Clay, D.C. and Reardon, T. (1997) Population and sustainability: understanding population, environment and development linkages. In: D'Souza, G. and Gebremedhin, G. (eds) *Sustainability in Agricultural and Rural Development*. Dartmouth Publishing, Aldershot, UK.

Clay, D.C., Byiringiro, F., Kangasniemi, J., Reardon, T., Sibomana, B., Uwamariya, L. and Tardif-Douglin, D. (1995) *Promoting Food Security in Rwanda through Sustainable Agricultural Productivity: Meeting the Challenges of Population Pressure, Land Degradation, and Poverty*. Michigan State University International Development Paper No. 17, Department of Agricultural Economics, Michigan State University, East Lansing, Michigan.

Clay, D.C., Reardon, T. and Kangasniemi, J. (1998) Sustainable intensification in the highland tropics: Rwandan farmers' investments in land conservation and soil fertility. *Economic Development and Cultural Change* 46(2), 351–378.

Cleaver, K.M. (1993) *A Strategy to Develop Agriculture in Sub-Saharan Africa and a Focus for the World Bank*. World Bank Technical Paper No. 203, Africa Technical Department Series, World Bank, Washington, DC.

Cleaver, K.M. and Schreiber, G.A. (1994) *Reversing the Spiral; the Population, Agriculture and Environment Nexus in Sub-Saharan Africa*. World Bank, Washington, DC.

Clegg, M.T., Giddings, L.V., Lewis, C.S. and Barton, J.H. (eds) (1993) *Report of the International Consultation on Rice Biosafety in Southeast Asia, 1–2 September 1992*. Technical Paper, Biotechnology Series No. 1, World Bank, Washington, DC.

Collier, P. and Lal, D. (1984) Why poor people get rich: Kenya 1960–79. *World Development* 12(10), 1007–1018.

Conservation International (1993) *Plan Annuel de Travail. Projet de Conservation et Développement de la Réserve Naturelle Intégrale No. 3 de Zahamena*. Conservation International, Antananarivo, Madagascar.

Conway, G.R. (1993) Sustainable agriculture: the trade-offs with productivity, stability and equitability. In: Barbier E.B. (ed.) *Economics and Ecology: New Frontiers and Sustainable Development*. Chapman & Hall, London.

Conway, G.R. (1997) *The Doubly Green Revolution: Food for All in the 21st Century*. Penguin Books, London.

Conway, G.R. and Barbier, E. (1990) *After the Green Revolution: Sustainable Agriculture for Development*. Earthscan Publications, London.

Conway, G.R. and Pretty, J.N. (1991) *Unwelcome Harvest: Agriculture and Pollution*. Earthscan Publications, London.

Conway, G.R., Lele, U., Peacock, J. and Pineiro, M. (1994) *Sustainable Agriculture for a Food Secure World*. Consultative Group on International Agricultural Research, Washington, DC, and Swedish Agency for Research Co-operation with Developing Countries, Stockholm.

Cornwall, A., Gujit, I. and Welbourn, A. (1994) Acknowledging process: methodological challenges for agricultural research and extension. In: Scoones, I. and Thompson, J. (eds) *Beyond Farmer First: Rural People's Knowledge, Agricultural Research and Extension Practice*. Intermediate Technology Publications, London, pp. 98–117.

Costanza, R., d'Arge, R., De Groot, R., Grasso, M., Hannon, B., Limburg, K., Naeem, S., O'Neill, R.V., Paruelo, J., Raskin, R.G., Sutton, P. and van den Belt, M. (1997) The value of the world's ecosystem services and natural capital. *Nature* 387, 253–260.

Coxhead, I. and Shively, G. (1995) *Measuring the Environmental Impacts of Economic Change: the Case of Land Degradation in Philippine Agriculture*. Agricultural Economics Staff Paper Series No. 384, University of Wisconsin, Madison, Wisconsin.

Craven, I. and de Fretes, Y. (1987) *Arfak Mountains Nature Conservation Area. Management Plan 1988–1992*. World Wildlife Fund (WWF), Bogor, Indonesia.

Crissman, C.C., Cole, D.C. and Carpio, F. (1994) Pesticide use and farm worker health in Ecuadorian potato production. *American Journal of Agricultural Economics* 76(3), 593–597.

Crissman, C., Antle, J. and Capalbo, S. (1997) *Economic, Environmental, and Health Trade-Offs in Agriculture: Pesticides and the Sustainability of Andean Potato Production*. Kluwer Academic Publishers, Dordrecht, the Netherlands.

Crissman, C.C., Antle, J.M. and Capalbo, S.M. (eds) (1998) *Economic, Environmental, and Health Tradeoffs in Agriculture: Pesticides and the Sustainability of Andean Potato Production*. Kluwer Academic Publishers, Boston.

Cropper, M. and Griffiths, C. (1994) The interaction of population growth and environmental quality. *American Economic Review* 84(2), 250–254.

Cunningham, A.B. and Mbenkum, F.T. (1993) *Sustainability of Harvesting* Prunus africana *Bark in Cameroon: a Medicinal Plant in the International Trade*. People and Plants Working Paper 2, UNESCO, Paris.

Daily, G. (ed.) (1997) *Nature's Services: Societal Dependence on Natural Systems*. Island Press, Washington, DC.

Dalton, T.J. (1996) Soil degradation and technical change in southern Mali. PhD thesis, Purdue University, West Lafayette, Indiana.

Dang, H. (1991) *Human Conflict in Conservation: Protected Areas, the Indian Experience*. Development Alternatives, New Delhi, India.

Dasgupta, P. (1993) *An Inquiry into Wellbeing and Destitution*. Clarendon Press, Oxford.

Dasgupta, P. (1995a) The population problem: theory and evidence. *Journal of Economic Literature* 33, 1897–1902.

Dasgupta, P.S. (1995b) Population, poverty, and the local environment. *Scientific American*, February, 40–45.

Dasgupta, P. (1996) The economics of the environment. *Environment and Development Economics* 1, 387–421.

David, C.C. (1990) *Determinants of Rice Price Protection in Asia*. Social Science Division Paper No. 90–22, International Rice Research Institute (IRRI), Los Baños, Laguna, Philippines.

David, C.C. and Otsuka, K. (1994) *Modern Rice Technology and Income Distribution in Asia*. Lynne Rienner Publishers, Boulder, Colorado.

Davis, K. (1963) The theory of change and response in modern demographic history. *Population Index* 29, 345–366.

de Almeida, O.T. and Uhl, C. (1995) Developing a quantitative framework for sustainable resource-use planning in the Brazilian Amazon. *World Development* 23(10), 1745–1764.

De Datta, S.K. (1981) *Principles and Practices of Rice Production*. Wiley-Interscience Publication, John Wiley & Sons, New York.

De Datta, S.K., Gomez, K.A. and Descalsota, J.P. (1988) Changes in yield response to major nutrients and in soil fertility under intensive rice cropping. *Soil Science* 146, 350–358.

De Graaff, J. (1996) The price of soil erosion: an economic evaluation of soil conservation and watershed management. PhD thesis, Wageningen Agricultural University, Wageningen, The Netherlands.

de Haan, C. (1997) Balancing livestock and environment; the study framework. In: Nell, A.J. (ed.) *Livestock and the Environment International Conference. Proceedings of International Conference on Livestock and the Environment, Wageningen, 16–20 June 1997*. International Agricultural Centre, Wageningen, the Netherlands, pp. 34–40.

de Haan, C., Steinfeld, H. and Blackburn, H. (1997) *Livestock and the Environment: Finding a Balance*. Wrenmedia, Suffolk, UK.

Deininger, K. and Minten, B. (1996) *Poverty, Policies, and Deforestation: the Case of Mexico*. Research Project on Social and Environmental Consequences of Growth-Oriented Policies, Working Paper No. 5, World Bank Policy Research Department, Washington, DC.

de Janvry, A., Fafchamps, M. and Sadoulet, E. (1991) Peasant household behaviour with missing markets: some paradoxes explained. *Economic Journal* 101, 1400–1417.

de Laulaníe, H. (1993) Le système de riziculture intensive malgache. *Tropicultura (Brussels)* 11(3), 110–114.

Delgado, C. and Pinstrup-Andersen, P. (1993) Agricultural productivity in the Third World: patterns and strategic issues. Keynote address. In: *1993 AAEA/IFPRI Pre-conference Workshop, Post Green Revolution Agricultural Development Strategies in the Third World: What Next?* Orlando, Florida.

Delgado, C., Hopkins, J. and Kelly, V. (1994) Agricultural growth linkages in sub-Saharan Africa: a synthesis. In: *Proceedings of Agricultural Growth Linkages in Sub-Saharan Africa Workshop*. International Food Policy Research Institute (IFPRI), Washington, DC.

Delgado, C., Rosegrant, M., Steinfeld, H., Ehui, S. and Courbois, C. (1999) *Livestock to 2020: the Next Food Revolution*. Food, Agriculture, and the Environment Discussion Paper 28, International Food Policy Research Institute (IFPRI), Washington, DC.

DE/MAP (Direcçao de Economia/Ministério de Agricultura e Pescas) and DNER/MAP (Direcçao Nacional de Extensao Rural/Ministério de Agricultura e Pescas) (1998) Is agricultural intensification profitable for Mozambican smallholders? An appraisal of the inputs subsector and the 1996/97 DNER/SG2000 Program. Paper presented at the IFA Regional Conference for Africa, Maputo, Mozambique, 8–12 June 1998.

Dembélé, N.N. and Savadogo, K. (1996) The need to link soil fertility management to input/output market development in West Africa: key issues. Paper presented at the International Fertilizer Development Centre Seminar, Lomé, Togo, 19–22 November.

Dent, J.B. and Thornton, P.K. (1988) The role of biological simulation models in farming systems research. *Agricultural Administration and Extension* 29, 111–122.

Desai, G.M. (1986) Fertilizer use in India: the next stage in policy. *Indian Journal of Agricultural Economics* 41, 248–270.

Desai, G.M. (1988) Development of fertilizer markets in South Asia: a comparative overview. Mimeo, International Food Policy Research Institute (IFPRI), Washington, DC.

Desai, G.M. and Gandhi, V. (1989) Phosphorus for sustainable agriculture growth in Asia – an assessment of alternative sources and management. In: *Proceedings of the Symposium on Phosphorus Requirements for Sustainable Agriculture in Asia and Pacific Region, Los Baños, Laguna, Philippines, 6–10 March 1989.* International Rice Research Institute (IRRI), Los Baños, Laguna, Philippines.

DeShazo, R.P. and DeShazo, J.R. (1995) An economic model of smallholder deforestation: a consideration of the shadow value of land on the frontier. In: Sandbukt, Ø. (ed.) *Management of Tropical Forests: Towards an Integrated Perspective.* Centre for Development and the Environment, Oslo, pp. 153–165.

de Wit, C.T. (1992) Resource use efficiency in agriculture. *Agricultural Systems* 40, 125–151.

de Wit, C.T., van Keulen, H., Seligman, N.G. and Spharim, I. (1988) Application of interactive multiple goal programming techniques for analysis and planning of regional agricultural development. *Agricultural Systems* 26, 211–230.

Deybe, D. (1994) Vers un agriculture durable. PhD thesis, Centre de Coopération Internationale en Recherche Agronomique pour le Développement (CIRAD), Montpellier, France.

Deybe, D. and Robilliard, A.S. (1996) *MATA: an Integrated Micro, Meso and Macroeconomic Model for Decision Making in Agriculture: Some Applications for Burkina Faso.* Note et Documents URPA No. 57, Centre de Coopération Internationale en Recherche Agronomique pour le Développement (CIRAD), Paris.

DGEC (Dirección General de Estadística y Censos) (1987) *Censo de población 1984*, Vols 1 and 2. Ministerio de Economía, Industria y Comercio, San José, Costa Rica.

DGEC (Dirección General de Estadística y Censos) (1997a) *Costa Rica: calculo de población por provincia, cantón y distrito al 1ª de Julio de 1996.* Ministerio de Economía, Industria y Comercio, San José, Costa Rica.

DGEC (Dirección General de Estadística y Censos) (1997b) *Encuesta de hogares de proposito multiples modulo de empleo julio de 1996.* Ministerio de Economía, Industria y Comercio, San José, Costa Rica.

Diagana, B., Kelly, V. and Kébé, M. (1996) *L'Offre agricole suite à la dévaluation: pourquoi une réponse si faible au Sénégal?* Policy Brief, Institut Sénégalais de Recherches Agricoles, Institut du Sahel, Kaolack, and Michigan State University, East Lansing, Michigan.

Diegues, A.C. (1992) *The Social Dynamics of Deforestation in the Brazilian Amazon: an Overview.* Discussion Paper 36, United Nations Research Institute for Social Development (UNRISD), Geneva.

Dillon, J.L. and Virmani, S.S. (1985) The farming systems approach. In: Muchow, R.C. (ed.) *Agro-research for Semi-arid Tropics.* Northwest Australia University Press, Brisbane, Australia, pp. 507–532.

Dinerstein, E., Olson, D.M., Graham, D.J., Webster, A.L., Primm, S.A., Bookbinder, M.P. and Ledec, G. (1995) *A Conservation Assessment of the Terrestrial Ecoregions of Latin America and the Caribbean*. World Bank, Washington, DC.

Dioné, J. (1989) Informing food security policy in Mali: interactions between technology, institutions, and market reforms. PhD dissertation, Michigan State University, East Lansing, Michigan.

Dogra, B. (1986) The Indian experience with large dams. In: Goldsmith, E. and Hildyard, N. (eds) *The Social and Environmental Effects of Large Dams*, Vol. 2. Wadebridge Ecological Centre, London, pp. 201–208.

Dommen, A.J. (1988) *Innovation in African Agriculture*. Westview Press, Boulder, Colorado.

Donovan, W.G. (1996) The role of inputs and marketing systems in the modernization of agriculture. In: Breth, S.A. (ed.) *Achieving Greater Impact from Research Investments in Africa*. Sasakawa Africa Association, Mexico City, pp. 178–194.

Dreze, J. and Srinivasan, P. (1996) Poverty in India: regional estimates, 1987–8. Mimeo, Delhi School of Economics and Indira Gandhi Institute of Development Research.

D'Souza, G.E. and Gebremedhin, T.G. (eds) (1998) *Sustainability in Agricultural and Rural Development*. Dartmouth Publishing, Aldershot, UK.

Duloy, J.H. and Norton, R.D. (1973) CHAC: a programming model of Mexican agriculture. In: Goreux, L.M. and Manne, A.S. (eds) *Multi-level Planning: Case Studies in Mexico*. North-Holland, Amsterdam, pp. 291–337.

Duloy, J.H. and Norton, R.D. (1983) The CHAC demand structures. In: Norton, R.D., and Solís, L. (eds) *The Book of CHAC: Programming Studies for Mexican Agriculture*. Johns Hopkins University Press, Baltimore, Maryland, pp. 58–82.

Duxbury, I. (1995) The significance of greenhouse gas emissions from soils of tropical agroecosystems. In: Lal, R., Kimble, J., Levine, E. and Stewart, B.A. (eds) *Soil Management and Greenhouse Effects: Advances in Soil Science*. CRC Press, Boca Raton, Florida.

Dyson, T. (1996) *Population and Food: Global Trends and Future Prospects*. Routledge, London.

Edwards, C.A. and Wali, M.K. (1993) The overall impact of agriculture in developed and developing countries on the global environment: selected topics and conclusions. *Agriculture, Ecosystems and Environment* 46(1–4), ix–xii.

Ehui, S. and Spencer, D. (1993) Measuring the sustainability and economic viability of tropical farming systems: a model from sub-Saharan Africa. *Agricultural Economics* 9, 279–296.

Eicher, C.K. (1995) Zimbabwe's maize-based green revolution: preconditions for replication. *World Development* 23(5), 805–818.

Eicher, C. and Baker, D. (1982) *Research on Agricultural Development in Sub-Saharan Africa: a Critical Survey*. MSU International Development Paper 1, Michigan State University Press, East Lansing, Michigan.

Fairhead, J. and Leach, M. (1995) False forest history, complicit social analysis: rethinking some West African environmental narratives. *World Development* 23(6), 1023–1035.

Faminow, M.D. and Vosti, S.A. (1998) Livestock–deforestation links: policy issues in the western Brazilian Amazon. In: Nell, A.J. (ed.) *Livestock and the Environment International Conference, Wageningen, The Netherlands*. World Bank, New York, Food and Agriculture Organization of the United Nations, Rome, and the International Agricultural Centre.

Faminow, M.D., de Sá, C.P. and Oliveira, S. (1997a) Development of an investment model for the smallholder cattle sector in the Western Amazon: I. Preliminaries. Mimeo, International Food Policy Research Institute (IFPRI), Washington, DC.

Faminow, M.D., Oliveira, S. and de Sá, C.P. (1997b) Conserving forest through improved cattle production technologies. Policy brief mimeo, International Food Policy Research Institute (IFPRI), Washington, DC.

Faminow, M., Dahl, C., Vosti, S., Witcover, J., Carpentier, C. and Oliveira, S. (1999) Smallholder risk, cattle, and deforestation in the Western Brazilian Amazon. *World Animal Review* 93, 16–23.

Fan, S. and Hazell, P. (1999) *Are Returns to Public Investment Lower in Less-favored Rural Areas? An Empirical Analysis of India.* Environment and Production Technology Division Discussion Paper No. 43, International Food Policy Research Institute (IFPRI), Washington, DC.

Fan, S., Hazell, P. and Thorat, S. (1998) *Government Spending, Growth and Poverty: an Analysis of Interlinkages in Rural India.* Environment and Production Technology Division Discussion Paper No. 33, International Food Policy Research Institute (IFPRI), Washington, DC.

FAO (Food and Agriculture Organization of the United Nations) (1984) *Production Yearbook.* FAO, Rome.

FAO (Food and Agriculture Organization of the United Nations) (1992) *World Food Supplies and Prevalence of Chronic Undernutrition in Developing Regions as Assessed in 1992.* Document ESS/MISC/1992, FAO, Rome.

FAO (Food and Agriculture Organization of the United Nations) (1993a) *Agriculture: Towards 2010.* FAO, Rome.

FAO (Food and Agriculture Organization of the United Nations) (1993b) *Guidelines for Land Use Planning.* FAO Development Series 1, FAO, Rome.

FAO (Food and Agriculture Organization of the United Nations) (1993c) *Production Yearbook.* FAO, Rome.

FAO (Food and Agriculture Organization of the United Nations) (1993d) *Strategies for Sustainable Agriculture and Rural Development: the Role of Agriculture, Forestry, and Fisheries.* FAO, Rome.

FAO (Food and Agriculture Organization of the United Nations) (1995) *Production Yearbook.* FAO, Rome.

FAO (Food and Agriculture Organization of the United Nations) (1997a) *Production Yearbook.* FAO, Rome.

FAO (Food and Agriculture Organization of the United Nations) (1997b) *State of the World's Forests.* Words and Publications, Oxford, UK.

FAO (Food and Agricultural Organization of the United Nations) (1998) *FAO Expert Consultation on Policies for Animal Production and Natural Resources Mangement.* FAO Report AGA-802, Brasilia, Brazil.

FAOSTAT (1997) FAO on-line database, http://apps.fao.org, 1997 reported data.

FAOSTAT TS: SOFA '99, available with FAO (1999) *The State of Food and Agriculture.* FAO, Rome.

Farrington, J., Turton, C. and James, A.J. (1999) (eds) *Participatory Watershed Development: Challenges for the Twenty-first Century.* Oxford University Press, New Delhi.

Feder, G., Just, R.E. and Zilberman, D. (1985) Adoption of agricultural innovations in developing countries: a survey. *Economic Development and Cultural Change* 33(2), 255–298.

Feeny, D., Berkes, F., McCay, B. and Acheson, J.M. (1990) The tragedy of the commons: twenty-two years later. *Human Ecology* 18(1), 1–19.

Ferraro, P.J. and Kramer, R.A. (1997) Compensation and economic incentives: reducing pressure on protected areas. In: Kramer, R., von Schaik, C. and Johnson, J. (eds) *Last Stand: Protected Areas and the Defense of Tropical Biodiversity*. Oxford University Press, New York, pp. 187–211.

Ferraro, P.J. with Tshombe, R., Mwinyihali, R. and Hart, J.A. (1997) *Projets intégrés de conservation et de développement: un cadre pour promouvoir la conservation et la gestion des ressources naturelles*. Wildlife Conservation Society Working Paper No. 6, Wildlife Conservation Society, Bronx, New York.

Findlay, R. and Lundahl, M. (1994) Natural resources, 'vent-for-surplus,' and the staples theory. In: Meier, G.M. (ed.) *From Classical to Development Economics*. St Martin Press, London, pp. 68–93.

Fisher, L. (1999) Beyond the *berugaq*: conflict, policy and decision-making for forest and conservation management in Nusa Tenggara, Indonesia. PhD thesis, Department of Natural Resources, Cornell University, Ithaca, New York.

Forster, N. and Stanfield, D. (1993) *Tenure Regimes and Forest Management Case Studies in Latin America*. Land Tenure Center, University of Wisconsin, Madison.

Foster, A.D., Rosenzweig, M.R. and Behrman, J.R. (1997) Population growth, income growth, and deforestation: management of village common land in India. Mimeo, Brown University, Providence, Rhode Island, and University of Pennsylvania, Philadelphia, Pennsylvania.

Foster, A.D., Behrma, J.R. and Rosenzweig, M.R. (1999) *Population Growth, Technical Change, and Forest Degradation*. Department of Economics Working Paper, Brown University, Providence, Rhode Island.

Franzel, S., Jaenicke, H. and Janssen, W. (1996) *Choosing the Right Trees: Setting Priorities for Multipurpose Tree Improvement*. ISNAR Research Report No. 8, International Service for National Agricultural Research (ISNAR), The Hague.

Frederiksen, H.D., Berkoff, J. and Barber, W. (1993) *Water Resources Management in Asia*, Vol. 1, *Main Report*. World Bank Technical Paper No. 212, World Bank, Washington, DC.

Freeman, H.A., Roe, T.L. and Smith, J. (1997) The effects of policies on farming households' decisions: economic evidence and environmental implications in northern Nigeria. *Environment and Development Economics* 2, 307–321.

Fresco, L.O. (1995) Agro-ecological knowledge at different scales. In: Bouma, J., Kuyvenhoven, A., Bouman, B.A.M., Luyten, J.C. and Zandstra, H.G. (eds) *Eco-regional Approaches for Sustainable Land Use and Food Production*. Kluwer Academic Publishers, Dordrecht, The Netherlands, pp. 133–141.

Fresco, L.O. and Steinfeld, H. (1997) A food security perspective to livestock and the environment. In: Nell, A.J. (ed.) *Livestock and the Environment International Conference. Proceedings of International Conference on Livestock and the Environment, Wageningen, 16–20 June 1997*. International Agricultural Centre, Wageningen, the Netherlands, pp. 5–20.

Fresco, L.O., Huizing, H., van Keulen, H., Luning, H.A. and Schipper, R.A. (1992) *Land Evaluation and Farming Systems Analysis for Land Use Planning*. FAO Working Document, Food and Agriculture Organization of the United Nations (FAO), Rome.

Freud, C., Freud, E.H., Richard, J. and Thènevin, P. (1997) *L'Arachide au Senegal: Un moteur en panne*. Karthala–CIRAD (Centre de Coopération Internationale en Recherche Agronomique pour le Développement), Paris.

Friedlander, D. (1969) Demographic responses and population change. *Demography* 6, 359–381.

Fujisaka, J.S., Harrington, L. and Hobbs, P. (1994) Rice–wheat in South Asia: systems and long-term priorities established through diagnostic research. *Agricultural Systems* 46, 169–187.

Fujisaka, S., Bell, W., Thomas, N., Hurtado, L. and Crawford, E. (1996) Slash-and-burn agriculture, conversion to pasture, and deforestation in two Brazilian Amazon colonies. *Agriculture Ecosystems and Environment* 1015, 1–16.

Gachengo, C.N. (1996) Phosphorus release and availability on addition of organic materials to phosphorus fixing soils. MSc thesis, Moi University, Eldoret, Kenya.

Gallagher, K.D., Kenmore, P.E. and Sogawa, K. (1994) Judicial use of insecticides deters planthopper outbreaks and extends the life of resistant varieties in Southeast Asian rice. In: Denno, R.F. and Perfect, T.J. (eds) *Ecology and Management of Planthoppers*. Chapman & Hall, London, pp. 599–614.

Garland Bedoya, E. (1991) *The Social and Economic Causes of Deforestation in the Peruvian Amazon Basin: Natives and Colonists*. Institute for Development Anthropology, Binghamton, New York.

Garrity, D.P. (1995) Improved agroforestry technologies for conservation farming: pathways toward sustainability. In: *Proceedings of the International Workshop on Conservation Farming for Sloping Uplands in Southeast Asia: Challenges, Opportunities and Prospects*. Proceedings No, 14, International Board for Soil Research and Management (IBSRAM), Bangkok, Thailand, pp. 145–168.

Garrity, D.P. (1996) Tree–soil–crop interactions on slopes. In: Ong, C.K. and Huxley, P.A. (eds.) *Tree–Crop Interactions: a Physiological Approach*. CAB International, Wallingford, UK, pp. 299–318.

Garrity, D.P., Soekardi, M., van Noordwijk, M., de la Cruz, R., Pathak, P.S., Gunasena, H.P.M., Van So, N., Huijin, G. and Majid, N.M. (1997) The *Imperata* grasslands of tropical Asia: area, distribution, and typology. *Agroforestry Systems* 36(103), 3–29.

Gaye, M. and Sène, M. (1994) Les fortes densités de semis de l'arachide au Sénégal: motivations paysannes et interprétation agronomique. Mimeo, Institut Sénégalais de Recherches Agricoles, Kaolack, Senegal.

Gazmuri Schleyer, R. and Rosegrant, M.W. (1996) Chilean water policy: the role of water rights, institutions, and markets. *International Journal of Water Resources Development* 12(1), 33–48.

Geurts, J.A.M.M., Jansen, H.G.P. and van Tilburg, A. (1997) *Domestic Demand for Food in Costa Rica: a Double Hurdle Analysis*. CATIE Technical Series No. 287, Centro Agronómico Tropical de Investigación y Enseñanza, Turrialba, Costa Rica.

Gibson, C. and Marks, S.A. (1995) Transforming rural hunters into conservationists: an assessment of community-based wildlife management programs in Africa. *World Development* 23(6), 941–957.

Gilbert, A. (1994) *The Latin American City*. Latin America Bureau, New York. (Distribution in North America by Monthly Review Press.)

Giller, K.E., Cadisch, G., Ehaliotis, C., Adams, E., Sakala, W.D. and Mafongoya, P.L. (1997) Building soil nitrogen capital in Africa. In: Buresh, R.J., Sanchez, P.A. and Calhoun, F. (eds) *Replenishing Soil Fertility in Africa*. SSSA Special Publication 51, Soil Science Society of America, Madison, Wisconsin, pp. 151–192.

Gilles, J. and Jamtgaard, C. (1982) 'Overgrazing' in pastoral areas: the commons reconsidered. *Sociologia Ruralis* 21(2), 129–141.

Gillis, M. (1988) Indonesia: public policies, resource management, and the tropical forest. In: Repetto, R. and Gillis, M. (eds) *Public Policies and the Misuse of Forest Resources*. Cambridge University Press, New York, pp. 43–113.

Gillison, A.N. (ed.) (1999) *Aboveground Biodiversity Assessment Working Group Summary Report 1996–1998. ASB Phase II Final Report.* Centre for International Forestry Research (ICRAF), Bogor, Indonesia.

Gillison, A.N. and Liswanti, N. (eds) (1999) An intensive biodiversity baseline study in Jambi Province, Central Sumatra, Indonesia: preliminary report. In: Gillison, A.N. (compiler) *Above-ground Biodiversity Assessment Working Group Summary Report 1996–1998.* Part C. Alternatives to Slash and Burn Projects. Centre for International Forestry Research (CIFOR), Bogor, Indonesia.

Gladwin, C.H., Buhr, K.L., Goldman, A., Hiebsch, C., Hildebrand, P.E., Kidder, G., Langham, M., Nkedi-Kizza, P. and Williams, D. (1997) Gender and soil fertility in Africa. In: Buresh, R.J., Sanchez, P.A. and Calhoun, F. (eds) *Replenishing Soil Fertility in Africa.* SSSA Special Publication 51, Soil Science Society of America, Madison, Wisconsin, pp. 219–236.

Gockowski, J., Tonye, J., Baker, D., Weise, S., Ndoumbé, M., Tiki-Manga, T. and Fouaguégué. A. (1997) Characterization and diagnosis of farming systems in the ASB forest margins benchmark of southern Cameroon. Mimeo, International Institute of Tropical Agriculture (IITA), Yaoundé, Cameroon.

Godoy, R., O'Neill, K., Groff, S., Kostishack, P., Cubas, A., Demmer, J., McSweeney, K., Overman, J., Wilkie, D., Brokaw, N. and Martinez, M. (1997) Household determinants of deforestation by Amerindians in Honduras. *World Development* 25(6), 977–987.

Goetz, R.U. (1997) Diversification in agricultural production: a dynamic model of optimal cropping to manage soil erosion. *American Journal of Agricultural Economics* 79, 341–356.

GOI (Government of India) (1991) *WARASA: National Watershed Development Project for Rainfed Areas (NWDPRA) Guidelines*, 2nd edn. Ministry of Agriculture, Government of India, New Delhi.

Goklany, I.M. and Sprague, M.W. (1992) *Sustaining Development and Biodiversity: Productivity, Efficiency, and Conservation.* Cato Institute, Washington, DC.

Goldney, D. and Bauer, J. (1998) Integrating conservation and agricultural production: fantasy or imperative? In: Pratley, J. and Robertson, A. (eds) *Agriculture and the Environmental Imperative.* CSIRO Publishing, Collingwood, Australia, pp. 15–54.

Gradwohl, J. and Greenberg, R. (1988) *Saving the Tropical Forests.* Island Press, Washington, DC.

Greeley, M. (1992) *Agricultural Biotechnology, Poverty and Employment: the Policy, Context and Research Priorities.* Paper for the Technology and Employment Branch of the International Labour Organization, Institute for Development Studies, University of Sussex, Brighton, UK.

Greene, W.H. (1990) *Econometric Analysis.* Macmillan Publishing, New York.

Greenland, D.J. and Kowal, J.M.L. (1960) Nutrient content of a moist tropical forest of Ghana. *Plant and Soil* 12, 841–844.

Grigg, D. (1993) *The World Food Problem,* 2nd edn. Blackwell, Oxford.

Grossman, G.M. and Krueger, A.B. (1993) Environmental impacts of the North American Free Trade Agreement. In: Garber, P. (ed.) *The US–Mexico Free Trade Agreement.* MIT Press, Cambridge, Massachusetts.

Guillaumont, P. (1992) Adjustment policy and agricultural development. In: Peters, G.H. and Stanton, B.F. (eds) *Sustainable Agricultural Development: The Role of International Cooperation. Proceedings of the 21st International Conference of Agricultural Economists.* Dartmouth Publishing, Brookfield, Vermont.

Guyer, J. (1984) *Family and Farm in Southern Cameroon.* African Research Studies No. 15, African Studies Center, Boston University, Boston, Massachusetts.

Hadi, P.U. and Budhi, G.S. (1998) *Analysis of the Economic Efficiency and Comparative Advantage of the Sumatran Small-holder Rubber Using 'PAM' Method.* Southeast Asia Policy Research Working Paper, International Centre for Research in Agroforestry (ICRAF), Bogor, Indonesia.

Hardin, G. (1968) The tragedy of the commons. *Science* 168, 1243–1248.

Harrington, L.W. (1994) Indicators and adoption: economic issues in research on the sustainability of agriculture. Mimeo, International Maize and Wheat Improvement Center (CIMMYT), Mexico.

Harrington, L.W., Morris, M., Hobbs, P.R., Pal Singh, V., Sharma, H.C., Singh, R.P., Chaudhary, M.K. and Dhiman, S.D. (eds) (1992) *Wheat and Rice in Karnal and Kurukshetra Districts, Haryana, India – Practices, Problems and an Agenda for Action. Report from an Exploratory Survey, 22–29 March 1992.* Haryana Agricultural University (HAU), Indian Council for Agricultural Research (ICAR), International Maize and Wheat Improvement Center (CIMMYT), International Rice Research Institute (IRRI), Los Baños, Laguna, Philippines.

Hayami, Y. (1994) Peasant and plantation in Asia. In: Meier, G.M. (ed.) *From Classical Economics to Development Economics.* St Martin's Press, New York.

Hayami, Y. and Otsuka, K. (1993) *The Economics of Contract Choice: an Agrarian Perspective.* Clarendon Press, Oxford.

Hayami, Y. and Ruttan, V.W. (eds) (1985) *Agricultural Development: an International Perspective*, 2nd edn. Johns Hopkins University Press, Baltimore, Maryland.

Hazell, P. (1987) Economic policy for diversification. In: Davis, T.J. and Schirmer, I.A. (eds) *Sustainability Issues in Agricultural Development: Proceedings of the Seventh Agriculture Sector Symposium.* World Bank, Washington, DC.

Hazell, P. (1992) The appropriate role of agricultural insurance in developing countries. *Journal of International Development* 4(6), 567–581.

Hazell, P. and Garrett, J. (1996) *Reducing Poverty and Protecting the Environment: the Overlooked Potential of Less-favored Lands.* 2020 Brief 39, International Food Policy Research Institute (IFPRI), Washington, DC.

Hazell, P. and Hojjati, B. (1995) *Farm–Nonfarm Growth Linkages in Zambia.* EPTD Discussion Paper No. 8, International Food Policy Research Institute (IFPRI), Washington, DC.

Hazell, P. and Norton, R.D. (1986) *Mathematical Programming for Economic Analysis in Agriculture.* Macmillan Publishing, New York.

Hazell, P. and Ramasamy, C. (1991) *The Green Revolution Reconsidered: the Impact of High-yielding Rice Varieties in South India.* Johns Hopkins University Press, Baltimore, Maryland.

Hazell, P. and Reardon, T. (1998) Interactions among the rural nonfarm economy, poverty, and the environment in resource-poor areas. Paper presented at an IFPRI/ World Bank Workshop on Strategies for Stimulating Growth of the Rural Nonfarm Economy in Developing Countries, Airlie Conference Center, Warrenton, Virginia, May 1998.

Hazell, P., Pomareda, C. and Valdes, A. (eds) (1986) *Crop Insurance for Agricultural Development: Issues and Experience.* Johns Hopkins University Press, Baltimore, Maryland.

Heath, J. and Binswanger, H. (1996) Natural resource degradation effects of poverty and population growth are largely policy-induced: the case of Colombia. *Environment and Development Economics* 1, 65–83.

Hebinck, P. and Ruben, R. (1997) Rural diversity upside down: heterogeneity and policy development. In: *Proceedings 14th International Symposium on Sustainable Farming Systems*, November 1996, Colombo, Sri Lanka.

Hecht, S. and Cockburn, A. (1990) *The Fate of the Forest.* Harper Collins, New York.

Heerink, N.B.M. and Ruben, R. (1996) Economic approaches for the evaluation of low external input agriculture. *Tijdschrift voor Sociaalwetenschappelijk Onderzoek in de Landbouw* 11, 294–308.

Heilig, G.K. (1999) Population, food demand, and land use in China. In: Williams, M.A. (ed.) *Options: China's Changing Land.* International Institute for Applied Systems Analysis (IIASA), Vienna.

Heisey, P.W. and Mwangi, W. (1997) Fertilizer use and maize production in sub-Saharan Africa. In: Byerlee, D. and Eicher, C.K. (eds) *Africa's Emerging Maize Revolution.* Lynne Reinner, Boulder, Colorado.

Heisey, P.W., Smale, M., Byerlee, D. and Sonza, E. (1997) Wheat rusts and the costs of genetic diversity in the Punjab of Pakistan. *American Journal of Agricultural Economics* 79, 726–737.

Hengsdijk. H., Quack, W., Bakker, E.J. and Ketelaars, J.J.M.H. (1996) *A Land Use System and Technology Generator for the Koutiala Region in South Mali.* DLV Report No. 5, AB-DLO/Wageningen Agricultural University, Wageningen, the Netherlands.

Hengsdijk, H., Nieuwenhuyse, A. and Bouman, B.A.M. (1998) *LUCTOR 2.0. Land Use Crop Technical Coefficient Generator: a Model to Quantify Crop Systems in the Atlantic Zone of Costa Rica.* Quantitative Approaches in Systems Analysis No. 17, AB-DLO-PE, Wageningen, the Netherlands.

Heong, K.L., Aquino, G.B. and Barrion, A.T. (1992) Population dynamics of plant and leafhoppers and their natural enemies in rice ecosystems in the Philippines. *Crop Protection* 11, 371–379.

Herring, R.J. (1998) Celebrating the local: scale and orthodoxy in political ecology. Working paper for workshop on Knowledge and Authority in Nature, Cornell University, Ithaca, New York, 18 April 1998.

Heywood, V. (ed.) (1995) *Global Biodiversity Assessment.* Cambridge University Press, Cambridge.

Hill, K. and Tikuarangi, T. (1996) The Mbaracayu Reserve and the Aché of Paraguay. In: Redford, K. and Mansour, J. (eds) *Traditional Peoples and Biodiversity Conservation in Large Tropical Landscapes.* Nature Conservancy, Rosslyn, Virginia.

Hinchcliffe, F., Thompson, J., Pretty, J., Guijt, I. and Shah, P. (eds) (1999) *Fertile Ground: the Impacts of Participatory Watershed Management.* Intermediate Technology Publications, London.

Hobbs, P. and Morris. M. (1996) *Meeting South Asia's Future Food Requirements from Rice–Wheat Cropping Systems: Priority Issues Facing Researchers in the Post-Green Revolution Era.* Natural Resource Group Working Paper 96-01, International Maize and Wheat Improvement Center (CIMMYT), Mexico, DF.

Hobbs, P.R., Harrington, L.W., Adhikary, C., Giri, G.S., Upadhyay, S.R. and Adhikary, B. (1996) *Wheat and Rice in the Nepal Tarai: Farm Resources and Production Practices in Rupandehi District.* No. AECO95A, Annual Report R95A 53, International Maize and Wheat Improvement Center (CIMMYT), Mexico, DF.

Hoekstra, D. and Corbett, J.D. (1995) *Sustainable Agricultural Growth for the Highlands of East and Central Africa: Prospects to 2020.* International Food Policy Research Institute (IFPRI), Washington, DC.

Hoekstra, S.J. (1995) Field research on transportation costs, farm-gate and farmer market prices in the Atlantic Zone of Costa Rica. MSc thesis, REPOSA, Guápiles, Costa Rica.

Hoff, K., Braverman, A. and Stiglitz, J.E. (1993) *The Economics of Rural Organization. Theory, Practice and Policy*. Oxford University Press, New York.

Holden, S.T. (1993a) Peasant household modeling: farming systems evolution and sustainability in Northern Zambia. *Agricultural Economics* 9, 241–267.

Holden, S.T. (1993b) The potential of agroforestry in the high rainfall areas of Northern Zambia: a peasant programming model approach. *Agroforestry Systems* 24, 39–55.

Holden, S.T. (1997) Adjustment policies, peasant household resource allocation and deforestation in Northern Zambia: an overview and some policy conclusions. *Forum for Development Studies* (1), 117–134.

Holden, S.T. and Pagiola, S. (forthcoming) *Intensification vs Extensification: an Analysis of Farm Household Choices*. World Bank, Washington, DC.

Holden, S.T. and Shanmugarathan, N. (1994) *Structural Adjustment, Production Subsidies and Sustainable Land Use*. Discussion Paper No. D-11/1994, Department of Economics and Social Sciences, Agricultural University of Norway, Ås, Norway.

Holden, S.T., Shiferaw, B. and Wik, M. (1998) Poverty, market imperfections, and time preferences: of relevance for environmental policy? *Environment and Development Economics* 3, 83–104.

Holling, C.S. (1973) Resilience and stability of ecological systems. *Annual Review of Ecology and Systematics* 4, 1–23.

Holling, C.S. (1986) Resilience of ecosystems: local surprise and global change. In: Clark, W.C. and Munn, R.E. (eds) *Sustainable Development of the Biosphere*. Cambridge University Press, Cambridge.

Holling, C.S. (1992) Cross-scale morphology, geometry and dynamics of ecosystems. *Ecological Monographs* 62, 447–502.

Holling, C.S., Schindler, D.W., Walker, B.W. and Roughgarden, J. (1995) Biodiversity in the functioning of ecosystems: an ecological synthesis. In: Perrings, C.A., Midler, K.-G., Folke, C., Holling, C.S. and Jansson, B.-O. (eds) *Biodiversity Loss: Economic and Ecological Issues*. Cambridge University Press, Cambridge, pp. 44–83.

Holmgren, P., Masakha, E.J. and Sjoholm, H. (1994) Not all African land is degraded: a recent survey of trees on farms in Kenya reveals rapidly increasing forest resources. *Ambio* 23, 390–395.

Homer-Dixon, T.F., Boutwell, J.H. and Rathfens, G.W. (1993) Environmental change and violent conflict. *Scientific American* 268(2), 16–23.

Hooper, D.U. and Vitousek, P.M. (1997) The effects of plant composition and diversity on ecosystem processes. *Science* 277, 1302–1305.

Houghton, J.T., Meira Filho, L.G., Callander, B.A., Harris, N., Kattenberg, A. and Maskell, K. (1996) *Climate Change 1995: The Science of Climate Change. Contribution of the Working Group I to the Second Assessment Report of the Intergovernmental Panel on Climate Change*. Cambridge University Press, Cambridge.

Houghton, R.A. (1994) Land-use change and tropical forests. *BioScience* 44, 305–331.

Howard, J. and Mungoma, C. (1997) Zambia's stop-and-go revolution: the impact of policies and organizations on the development and spread of maize technology. In: Byerlee, D. and Eicher, C.K. (eds) *Africa's Emerging Maize Revolution*. Lynne Reinner, Boulder, Colorado.

Hoyt, E. (1992) *Conserving the Wild Relatives of Crops*. WWF, IUCN and IBPGR, Gland, Switzerland, and Rome, Italy.

Hu, D., Rerdy, R. and Pagoulatos, A. (1997) Dynamic optimal management of wind erosive rangelands. *American Journal of Agricultural Economics* 79, 327–340.

Humphries, S. (1996) Milk cows, migrants, and land markets: unravelling the complexities of forest to pasture conversion in northern Honduras. Mimeo, Department of Sociology and Anthropology, University of Guelph, Ontario.

Ibrahim, M.N.M, Staal, S.J., Thorpe, W. and Tanner, J. (eds) (1999) *Appraisal of the Sri Lanka Sairy Sub-sector*, Vol. 1, *Synthesis of Main Results*, Vol. 2, *Main Report*. Ministry of Livestock Development and Estate Infrastructure, University of Peradeniya and the International Livestock Research Institute (ILRI), Colombo, Sri Lanka.

IBSNAT (International Benchmark Sites Network for Agrotechnology Transfer) (1990) *Documentation for IBSNAT Crop Model Input and Output Files. Version 1.1: For the Decision Support System for Agrotechnology Transfer (DSSAT V. 2.1)*. Technical Report 5, University of Hawaii, Honolulu, Hawaii.

ICRAF (International Centre for Research in Agroforestry) (1996) *Annual Report for 1995*. International Centre for Research in Agroforestry, Nairobi, Kenya.

ICRAF (International Centre for Research in Agroforestry) (1998) *Annual Report for 1997*. International Centre for Research in Agroforestry, Nairobi, Kenya.

IFPRI (International Food Policy Research Institute) (1996) *Feeding the World, Preventing Poverty and Protecting the Earth: a 2020 Vision*. Discussion Paper, International Food Policy Research Institute (IFPRI), Washington, DC.

IGPB Terrestrial Carbon Working Group (1998) The terrestrial carbon cycle: implications for the Kyoto Protocol. *Science* 280, 1393–1394.

Indian Journal of Agricultural Economics (1991) Subject 1: watershed development, 46 (July–September).

IRRI (International Rice Research Institute) (1997) *World Rice Statistics*. International Rice Research Institute, Manila, Philippines.

IUCN (World Conservation Union) (1998) *1997 United Nations List of National Parks and Protected Areas*. IUCN, Gland, Switzerland.

IUCN (World Conservation Union), UNEP (United Nations Environment Programme) and WWF (World Wildlife Fund) (1980) *World Conservation Strategy: Living Resource Conservation for Sustainable Development*. IUCN, UNEP and WWF, Gland, Switzerland.

IUCN (World Conservation Union), UNEP (United Nations Environment Programme) and WWF (World Wildlife Fund) (1991) *Caring for the Earth: a Strategy for Sustainable Living*. IUCN, UNEP and WWF, Gland, Switzerland.

Izac, A.-M.N. (1997) Ecological economics of investing in natural resource capital in Africa. In: Buresh, R.J., Sanchez, P.A. and Calhoun, F. (eds) *Replenishing Soil Fertility in Africa*. SSSA Special Publication 51, Soil Science Society of America, Madison, Wisconsin, pp. 237–251.

Jaenicke, H., Franzel, S. and Boland, D.J. (1996) Towards a method to set priorities among species for tree improvement research: a case study from West Africa. *Journal of Tropical Forest Science* 7, 490–506.

Jana, S. and Cooke, S. (1996) Biodiversity and the World Bank's agricultural portfolio. In: Srivastava, J.P, Smith, N.J.H. and Forno, D.A. (eds) *Biodiversity and Agricultural Intensification: Partners for Development and Conservation*. World Bank. Washington, DC, pp. 107–120.

Jansen, D.M., Stoorvogel, J.J. and Jansen, H.G.P. (1997) A quantitative tool for land-use analysis. In: Teng, P.S., Kropff, M.J., ten Berge, H.F.M., Dent, J.B., Lansigan, F.P. and van Laar, H.H. (eds) *Applications of Systems Approaches at the Farm and Regional Levels*, Vol. 1. Kluwer Academic Publishers, Dordrecht, the Netherlands, pp. 399–411.

Jansen, H.G.P. and Stoorvogel, J.J. (1998) Quantification of aggregation bias in regional agricultural land use models: application to Guácimo county, Costa Rica. *Agricultural Systems* 58, 417–439.

Jansen, H.G.P. and van Tilburg, A. (with J. Belt and S. Hoekstra) (1996) *Agricultural Marketing in the Atlantic Zone of Costa Rica: a Production, Consumption and Trade Study of Agricultural Commodities Produced by Small and Medium-scale Farmers.* CATIE Serie Técnica, Informe Técnico No. 271, Centro Agronómico Tropical de Investigación y Enseñanza, Turrialba, Costa Rica.

Jansen, H.G.P., Uytewaal, E. and Stoorvogel, J.J. (1998) Health externalities and pesticide use in the Atlantic zone of Costa Rica: an economic evaluation. Paper presented at International Conference on Pesticide Use in Developing Countries: Impact on Environment and Health, San José, Costa Rica, 23–27 February 1998.

Jansen, J.C.M. and de Wit, J. (1996) *Environmental Impact Assessment of Livestock Production in Mixed Systems in the (Sub) Humid Zones.* Consultancy Report, International Agriculture Centre, Wageningen, the Netherlands.

Jodha, N.S. (1997) Management of common property resources in selected dry areas of India. In: Kerr, J., Marothia, D., Singh, K., Ramasamy, C. and Bentley, W. (eds) *Natural Resource Economics: Theory and Application in India.* Oxford and IBH, New Delhi.

Jones, D.W. and O'Neill, R.V. (1992) Endogenous environmental degradation and land conservation: agricultural land use in a large region. *Ecological Economics* 6, 79–101.

Jones, D.W. and O'Neill, R.V. (1993) Rural land use, urban unemployment, and deforestation: the influence of development policies with a non-tradable good. *Resources and Energy Economics* 15, 353–370.

Jones, D.W. and O'Neill, R.V. (1994) Development policies, rural land use, and tropical deforestation. *Regional Science and Urban Economics* 24, 753–771.

Jones, D.W., Dale, V.H., Beaucamp, J.J., Pedlowski, M.A. and O'Neill, R.V. (1995) Farming in Rondonia. *Resource and Energy Economics* 17, 155–188.

Jones, J.W., Thornton, P.K. and Hansen, J.W. (1997) Opportunities for systems approaches at the farm scale. In: Teng, P.S., Kropff, M.J., ten Berge, H.M.F., Dent, J.B., Lansigan, F.P. and van Laar, H.H. (eds) *Applications of Systems Approaches at the Farm and Regional Levels*, Vol. 1. Kluwer Academic Publishers with IRRI, Dordrecht, pp. 1–18.

Kaimowitz, D. and Angelsen, A. (1998) *Economic Models of Tropical Deforestation: a Review.* Centre for International Forestry Research (CIFOR), Bogor, Indonesia.

Kaimowitz, D., Thiele, G. and Pacheco, P. (1999) The effects of structural adjustment on deforestation and forest degradation in lowland Bolivia. *World Development* 27(3), 505–520.

Kangasniemi, J. (1998) People and bananas on steep slopes: agricultural intensification under demographic pressure and environmental degradation in Rwanda. PhD dissertation, Michigan State University, East Lansing, Michigan.

Kante, B., Reardon, T. and Keita, A. (1995) *Dévaluation du franc CFA: analyse typologique des mesures d'accompagnement.* Working Paper No. DT 95-11, PRISAS, Institut du Sahel, Mali.

Kaosard, M., Patmasiriwat, D., DeShazo, J.R. and Panayotou, T. (1994) *Green Finance: a Case Study of Khao Yai.* Thailand Development Research Institute Foundation, Bangkok, Thailand, and Harvard Institute for International Development, Boston, Massachusetts.

Kapkiyai, J. (1996) Dynamics of soil organic carbon, nitrogen and microbial biomass in a long-term experiment as affected by inorganic and organic fertilizers. MSc thesis, Department of Soil Science, University of Nairobi, Nairobi, Kenya.

Karson, M.J. (1982) *Multivariate Statistical Methods.* Iowa State University Press, Ames, Iowa.

Katila, M. (1995) Modeling deforestation in Thailand. Paper presented at the IUFRO World Congress, Tampere, Finland.

Katz, R.W. and Brown, R.G. (1992) Extreme events in a changing climatic: variability is more important than averages. *Climatic Change* 21, 289–302.

Kaufmann, D. and O'Connell, S.A. (1991) The macroeconomics of the unofficial foreign exchange market in Tanzania. In: Chhibber, A. and Fischer, S. (eds) *Economic Reform in Sub-Saharan Africa*. World Bank, Washington, DC.

Kelly, V. (1988) Factors affecting the demand for fertilizer in Senegal's peanut basin. PhD dissertation, Michigan State University, East Lansing, Michigan.

Kelly, V., Reardon, T., Fall, A.A., Diagana, B. and McNeilly, L. (1993) *Final Report for the IFPRI/ISRA Study of Consumption and Supply Impacts of Agricultural Price Policies in the Peanut Basin and Senegal Oriental*. International Food Policy Research Institute (IFPRI), Washington, DC.

Kelly, V., Diagana, B., Reardon, T., Gaye, M. and Crawford, E. (1996) *Cash Crop and Foodgrain Productivity in Senegal: Historical View, New Survey Evidence, and Policy Implications*. Michigan State University International Development Paper No. 20, Michigan State University, East Lansing, Michigan.

Kemf, E. (ed.) (1993) *Law of the Mother: Protecting Indigenous People in Protected Areas*. Sierra Club, San Francisco.

Kenmore, P. (1991a) Getting policies right, keeping policies right: Indonesia's integrated pest management policy, production and environment. Paper presented at the Asia Region and Private Enterprise Environment and Agriculture Officers' Conference, Sri Lanka.

Kenmore, P. (1991b) *How Rice Farmers Clean Up the Environment, Conserve Biodiversity, Raise More Food, Make Higher Profits: Indonesia's IPM – a Model for Asia*. Food and Agriculture Organization, Manila.

Kerr, J. (1996) *Sustainable Development of Rainfed Agriculture in India*. Environment and Production Technology Division Discussion Paper No. 20, International Food Policy Research Institute (IFPRI), Washington, DC.

Kerr, J. and Sanghi, N.K. (1992) *Indigenous Soil and Water Conservation in India's Semi-arid Tropics*. Gatekeeper Series No. 34, Sustainable Agriculture Programme, International Institute of Environment and Development, London.

Kerr, J., Hazell, P. and Jha, D. (1998a) *Sustainable Development of Rainfed Agriculture in India: a Synthesis of Findings of the World Bank–ICAR Project on Sustainable Rainfed Agriculture Research and Development*. International Food Policy Research Institute (IFPRI), Washington, DC.

Kerr, J., Pangare, G., Pangare, V.L., George, P.J. and Kolavalli, S. (1998b) *The Role of Watershed Projects in Developing Rainfed Agriculture in India*. Report submitted to Indian Council of Agricultural Research and World Bank, International Food Policy Research Institute, Washington, DC.

Kerr, J., Pangare, G., Pangare, V.L. and George, P.J. (1999) Effects of watershed project subsidies on soil and water conservation investments in India's semi-arid tropics. In: Sanders, D., Juszar, P., Sombatpanit, S. and Enters, T. (eds) *Incentives in Soil Conservation: From Theory to Practice*. Science Publishers, Enfield, New Hampshire.

Kerr, J., Pangare, G. and Pangare, V.L. (2001) *The Role of Watershed Projects in Developing Rainfed Agriculture in India*. Research Report, International Food Policy Research Institute (IFPRI), Washington, DC (in press).

Kesseba, A.M. (1995) Strategies for sustainable agriculture in sub-Saharan Africa: some issues and options. In: Kang, B.T., Osiname, A.O. and Larbi, A. (eds) *Alley Farming*

Research and Development: Proceedings of the International Conference on Alley Farming. International Institute of Tropical Agriculture, Ibadan, Nigeria.

Kiepe, P. (1995) *No Runoff, No Soil Loss: Soil and Water Conservation in Hedgerow Barrier Systems*. Discussion Paper, Wageningen Agricultural University, Wageningen, the Netherlands.

Kiepe, P. and Rao, M.R. (1994) Management of agroforestry for the conservation and utilization of land and water resources. *Outlook in Agriculture* 23(1), 17–25.

Kijima, Y., Sakurai, T. and Otsuka, K. (2000) *Iriaichi*: collective vs. individualized management of community forests in post-war Japan. *Economic Development and Cultural Change* (in press).

Kim, T.-K., Hayes, D.J. and Hlam, A. (1992) Technology adoption under price uncertainty. *Journal of Development Economics* 38(1), 245–253.

Kimber, R. (1981) Collective action and the fallacy of the liberal fallacy. *World Politics* 33(2), 178–196.

Kiss, A. (ed.) (1990) *Living with Wildlife: Wildlife Resource Management with Local Participation in Africa*. World Bank, Washington, DC.

Knickel, K.W. (1994) MSSB – a concept and microcomputer application for agro-environmental land use analysis and planning. In: Fresco, L.O., Stroosnijder, L., Bouma, J. and van Keulen, H. (eds) *The Future of the Land: Mobilizing and Integrating Knowledge for Land Use Options*. John Wiley & Sons, Chichester, UK, pp. 337–344.

Kolbert, E. (1986) For wildlife the times are very good. *New York Times*, 23 November, Sec. E, 9.

Koop, G. and Tole, L. (1999) Is there an environmental Kuznets curve for deforestation? *Journal of Development Economics* 58(1), 231–244.

Kothari, A., Pande, P., Singh, S. and Variava, D. (1989) *Management of National Parks and Sanctuaries in India: a Status Report*. Environmental Studies Division, Indian Institute of Public Administration, New Delhi, India.

Kothari, A., Singh, N. and Suri, S. (eds) (1996) *People and Protected Areas: Towards Participatory Conservation in India*. Sage Publications, New Delhi, India.

Kothari, A., Vania, F., Das, P., Christopher, K. and Jha, S. (eds) (1997) *Building Bridges for Conservation: Towards Joint Management of Protected Areas in India*. Indian Institute of Public Administration, New Delhi, India.

Kotto-Same, J., Woomer, P.L., Moukam, A. and Zapfack, L. (1997) Carbon dynamics in slash-and-burn agriculture and land use alternatives of the humid forest zone in Cameroon. *Agriculture, Ecosystems and Environment* 65, 245–256.

Kramer, R.A., Sharma, N. and Munasinghe, M. (1995) *Valuing Tropical Forests: Methodology and Case Study of Madagascar*. Environment Paper No. 13, World Bank, Washington, DC.

Kramer, R.A., von Schaik, C. and Johnson, J. (eds) (1997) *Last Stand: Protected Areas and the Defense of Tropical Biodiversity*. Oxford University Press, New York.

Krautkraemer, J.A. (1994) Population growth, soil fertility, and agricultural intensification. *Journal of Development Economics* 44, 403–428.

Krueger, A.O., Schiff, M. and Valdés, A. (1988) Agricultural incentives in developing countries: measuring the effect of sectoral and economywide policies. *World Bank Economic Review* 2, 255–271.

Kruseman, G. and Bade, J. (1998) Agrarian policies for sustainable land use: bio-economic modelling to assess the effectiveness of policy instruments. *Agricultural Systems* 58(3), 465–481.

Kruseman, G. and Ruben, R. (1998) Transaction costs and incentives for sustainable land use: a meta modelling approach for Southern Mali. Paper presented at the European

Association of Agricultural Economics Seminar on Agricultural Markets Beyond Liberalization, September 1998, Wageningen, the Netherlands.

Kruseman, G., Ruben, R., Hengsdijk, H. and van Ittersum, M. (1995) Farm household modelling for estimating the effectiveness of price instruments in land use policy. *Netherlands Journal of Agricultural Science* 43, 111–123.

Kruseman, G., Bade, J., Ruben, R. and Hengsdijk, H. (1997) *Farm Household Modelling System: Theoretical and Mathematical Description*. DLV Report No. 7, AB-DLO/ Wageningen Agricultural University, Wageningen, the Netherlands.

Kundu, D.K. and Ladha, J.K. (1995) Efficient management of soil and biologically fixed N2 in intensively cultivated ricefields. *Soil Biology and Biochemistry* 27(4/5), 431–439.

Kutcher, G.P. and Scandizzo, P.L. (1981) *The Agricultural Economy of Northeast Brazil*. Johns Hopkins University Press, Baltimore, Maryland.

Kuyvenhoven, A., Ruben, R. and Kruseman, G. (1995) Options for sustainable agricultural systems and policy instruments to reach them. In: Bouma, J., Kuyvenhoven, A., Bouman, B.A.M., Luyten, J. and Zandstra, H. (eds) *Eco-regional Approaches for Sustainable Land Use and Food Production. Proceedings of a Symposium on Eco-regional Approaches in Agricultural Research, 12–16 December 1994, ISNAR, The Hague*. Kluwer Academic Publishers, Dordrecht, the Netherlands, pp. 187–212.

Kuyvenhoven, A., Ruben, R. and Kruseman, G. (1998) Technology, market policies and institutional reform for sustainable land use in Southern Mali. *Agricultural Economics* 19, 53–62.

Kwesiga, F.R., Franzel, S., Place, F., Phiri, D. and Simwanza, C.P. (1998) *Sesbania sesban* improved fallows in eastern Zambia: their inception, development and farmer enthusiasm. In: Buresh, R.J. and Cooper, P.J. (eds) *The Science and Practice of Short-term Improved Fallows*. Kluwer Academic Publishers, Dordrecht, the Netherlands.

Ladipo, D.O., Fondoun, J.M. and Nganga, N. (1996) Domestication of bush mango (*Irvingia* sp.): some exploitable intraspecific variations in West Africa. In: Leakey, R.R.B., Temu, A.B., Melnyk, M. and Vantomme, P. (eds) *Domestication and Commercialization of Non-timber Forest Products in Agroforestry Systems*. Non-wood Forest Products 9, FAO, Rome, pp. 193–205.

LaFrance, J.T. (1992) Do increased commodity prices lead to more or less soil degradation? *Australian Journal of Agricultural Economics* 36(1), 57–82.

Lal, R., Kimble, J., Levine, E. and Whitman, C. (1995) World soils and greenhouse effect: an overview. In: Lal, R., Kimble, J., Levine, E. and Stewart, B.A. (eds) *Soils and Global Change*. Lewis Publishers, Boca Raton, Florida, pp. 1–7.

Lampietti, J. and Dixon, J.A. (1995) *To See the Forest for the Trees: a Guide to Non-timber Forest Benefits*. Environment Department Paper No. 13, World Bank, Washington, DC.

Larson, P., Freudenberger, M. and Wyckoff-Baird, B. (1998) *WWF Integrated Conservation and Development Projects: Ten Lessons from the Field 1985–1996*. World Wildlife Fund, Washington, DC.

Leach, M., Mearns, R. and Scoones, I. (1999) Environmental entitlements: dynamics and institutions in community-based natural resource management. *World Development* 27, 225–247.

Leakey, R.R.B. (1996) Definition of agroforestry revisited. *Agroforestry Today* 8(1), 5–7.

Leakey, R.R.B. and Izac, A.-M.N. (1996) Linkages between domestication and commercialization of non-timber forest products: implications for agroforestry. In: Leakey, R.R.B., Temu, A.B., Melnyk, M. and Vantomme, P. (eds) *Domestication and Commercialization of Non-timber Forest Products in Agroforestry Systems.* Non-wood Forest Products 9, FAO, Rome, pp. 1–7.

Leakey, R.R.B. and Jaenicke, H. (1995) The domestication of indigenous fruit trees opportunities and challenges for agroforestry. In: Leakey, R.R.B., Temu, A.B. and Melnyk, M. (eds) *Proceedings of the 4th International BIO-REFOR Workshop.* FAO, Rome, pp. 1–8.

Leakey, R.R.B., Temu, A.B., Melnyk, M. and Vantomme, P. (eds) (1996) *Domestication and Commercialization of Non-timber Forest Products in Agroforestry Systems.* Non-wood Forest Products 9, FAO, Rome.

Lehnhoff, A. and Nuñez, O. (1998) Sierra de las Minas Biosphere Reserve, Guatemala. In: Brandon, K., Redford, K. and Sanderson, S. (eds) *Parks in Peril: People, Politics, and Protected Areas.* Island Press, Covelo, California.

Lekasi, J.K., Tanner, J.C., Kimani, S.K. and Harris, P.J.C. (1998) *Manure Management in the Kenya Highlands: Practices and Potential.* Emmerson Press for Natural Resources Systems Programme, Department of International Development (DFID), UK, and the Henry Doubleday Research Association, Kenilworth, UK, 35 pp.

Lele, U. (1975) *The Design of Rural Development: Lessons from Africa.* Johns Hopkins University Press, Baltimore, Maryland.

Lele, U. (1989) *Agricultural Growth, Domestic Policies, the External Environment and Assistance to Africa: Lessons of a Quarter Century.* Managing Agricultural Development in Africa Discussion Paper 1, World Bank, Washington, DC.

Lele, U. and Stone, S. (1989) *Population Pressure, the Environment, and Agricultural Intensification in Sub-Saharan Africa: Variations on the Boserup Hypothesis.* Managing Agricultural Development in Africa (MADIA) Discussion Paper No. 4, World Bank, Washington, DC.

Leonard, H.J. (1987) *Natural Resource and Economic Development in Central America: an Environmental Perspective.* Transaction Books, New Brunswick.

Leonard, H.J. (1989) Overview: environment and the poor. In: Leonard, H.J. (ed.) *Environment and the Poor: Development Strategies for a Common Agenda.* US – Third World Policy Perspectives No. 11, Overseas Development Council, Washington, DC.

Lewandrowski, J., Tobey, J. and Cook, Z. (1997) The interface between agricultural assistance and the environment: chemical fertilizer consumption and area expansion. *Land Economics* 73(3), 404–427.

Lewis, C. (1996) *Managing Conflicts in Protected Areas.* World Conservation Union (IUCN), Gland, Switzerland.

Linnemann, H. (1979) *MOIRA: Model of International Relations in Agriculture.* North Holland Publishing, Amsterdam.

Lipton, M. (1991) Market relaxation and agricultural development. In: Colclough, C. and Manor, J. (eds) *States or Markets? Neo-liberalism and the Development Policy Debate.* Clarendon Press, Oxford.

Little, P.D. and Watts, M. (1994) *Living Under Contract: Contract Farming and Agrarian Transformation in Sub-Saharan Africa.* University of Wisconsin Press, Madison, Wisconsin.

Lizardo, J. (1996) *Population and Environment Survey in Los Haitises National Park.* Report by IEPD/PROFAMILIA for UNDP/GEF and CIIFAD in collaboration with Pedro Henriquez Ureña University, Santo Domingo, Dominican Republic.

Lopes, A.S. (1980) Micronutrients in soils of the tropics as constraints to food production. In: *Priorities for Alleviating Soil-related Constraints to Food Production in the Tropics.* International Rice Research Institute, Los Baños, Laguna, Philippines, and the New York State College of Agriculture and Life Sciences, Cornell University, Ithaca, New York.

Lopez, R. (1998) Where development can or cannot go: the role of poverty–environment linkages. In: *Annual World Bank Conference on Development Economics 1997*. World Bank, Washington, DC.

Losch, B., Fusillier, J.L. and Dupraz, P. (1990) *Stratégies des production en zone cafière et cacoayer du Cameroun: quelles adaptions à la crise?* Centre de Coopération Internationale en Recherche Agronomique pour le Développement (CIRAD), Montpellier.

Low, A. (1986) *Agricultural Development in Southern Africa: Farm-household Economics and the Food Crisis*. James Currey, London.

Low, A. (1993) The low input, sustainable agriculture (LISA) prescription: a bitter pill for farm households in Southern Africa. *Project Appraisal* 8(2), 97–101.

Lutz, E., Pagiola, S. and Reiche, C. (eds) (1994) *Economic and Institutional Analysis of Soil Conservation Projects in Central America and the Caribbean*. World Bank Environment Paper No. 8, World Bank, Washington, DC.

Lynam, J.K. and Herdt, R.W. (1989) Sense and sustainability: sustainability as an objective in international agricultural research. *Agricultural Economics* 3, 381–398.

MacCleery, D.W. (1993) *American Forests: a History of a Resiliency and Recovery*. US Forest Service (Department of Agriculture), in cooperation with The Forest History Society, Durham, North Carolina.

McConnell, K.E. (1983) An economic model of soil erosion. *American Journal of Agricultural Economics* 65, 83–89.

McCown, R.L., Cox, P.G., Keating, B.A., Hammer, G.L., Carberry, P.S., Probert, M.E. and Freebairn, D.M. (1994) The development of strategies for improving agricultural systems and land use management. In: Goldsworthy, P.R. and Penning de Vries, F.W.T. (eds) *Opportunities, Use, and Transfer of Systems Research Methods in Agriculture to Developing Countries*. Kluwer Academic Publishers, Dordrecht, pp. 81–96.

Machfudh and Endom, W. (1998) *Economic Analysis of Large Scale Logging*. Southeast Asia Policy Research Working Paper, International Centre for Research in Agroforestry (ICRAF), Bogor, Indonesia.

McIntire, J., Bourzat, D. and Pingali, P. (1992) *Crop–Livestock Interaction in Sub-Saharan Africa*. World Bank, Washington, DC, 246 pp.

McKean, M.A. (1992) Management of traditional common lands (*Iriaichi*) in Japan. In: Bromley, D.W. (ed.) *Making the Commons Work: Theory, Practice, and Policy*. ICS Press, San Francisco.

MacKinnon, K. (1997) The ecological foundations of biodiversity protection. In: Kramer, R., van Schiak, C. and Johnson, J. (eds) *Last Stand, Protected Areas and the Defence of Tropical Biodiversity*. Oxford University Press, New York, pp. 36–63.

McLeod, R. (1997) Postscript to the survey of recent developments: on causes and cures for the Rupiah crisis. *Bulletin of Indonesian Economic Studies* 33(3), 35–52.

McNaughton, S.J. (1985) Ecology of a grazing ecosystem: the Serengeti. *Ecological Monographs* 55, 259–294.

McNeely, J.A. (1975) *Wildlife in the Lower Mekong Basin*. Mekong Committee/ESCAP, Bangkok.

McNeely, J.A. (1987) How dams and wildlife can coexist: natural habitats, agriculture, and major water resource development projects in tropical Asia. *Conservation Biology* 1, 228–238.

McNeely, J.A. (1988) *Economics and Biological Diversity: Developing and Using Economic Incentives to Conserve Biological Resources.* IUCN (International Union for Conservation of Nature and Natural Resources), Gland, Switzerland.

McNeely, J.A. (1989) Protected areas and human ecology: how national parks can contribute to sustaining societies to the twenty-first century. In: Western, D. and Pearl, M. (eds) *Conservation for the Twenty-first Century.* Oxford University Press, Oxford.

McNeely, J.A. and Guruswamy, L. (1998) Conclusion: how to save the biodiversity of planet earth. In: Guruswamy, L. and McNeely, J.A. (eds) *Protection of Global Diversity: Converging Strategies.* Duke University Press, Durham, North Carolina, pp. 372–387.

Malthus, T.R. (1798, 1803) *On Population (First Essay on Population,* 1798, and *Second Essay on Population,* 1803). Reprint, Modern Library for Random House, New York.

Mamingi, N. (1997) The impact of prices and macroeconomic policies on agricultural supply: a synthesis of available results. *Agricultural Economics* 16, 17–34.

Mann, C. (1998) *Higher Yield for All Smallholders through 'Best Bet' Technology: the Surest Way to Restart Economic Growth in Malawi.* Soil Fertility Network Research Results Working Paper No. 3, CIMMYT, Harare, Zimbabwe.

Mansour, J. and Redford, K. (eds) (1996) *Traditional Peoples and Biodiversity Conservation in Large Tropical Landscapes.* Nature Conservancy, Rosslyn, Virginia.

Manyong, V.M., Smith, J., Weber, G.K., Jagtap, S.S. and Oyewole, B. (1996) *Macrocharacterization of Agricultural Systems in Central Africa: an Overview.* Resource and Crop Management Research Monograph No. 22, International Institute of Tropical Agriculture (IITA), Ibadan, Nigeria.

Maredia, M. and Howard, J. (1997) *Facilitating Seed Sector Transformation in Africa: Key Findings from the Literature.* Policy Synthesis No. 33, Department of Agricultural Economics, Michigan State University, East Lansing, Michigan.

Maredia, M.K. and Byerlee, D. (eds) (1999) *The Global Wheat Improvement System: Prospects for Enhancing Efficiency in the Presence of Spillovers.* CIMMYT Research Report No. 5, International Wheat and Maize Improvement Center (CIMMYT), Mexico, DF.

Margoluis, R. and Salafsky, N. (1998) *Measures of Success: a Systematic Approach to Designing, Managing, and Monitoring Community-oriented Conservation Projects.* Island Press, Covelo, California.

Mariko, D., Chohin-Kuper, A. and Kelly, V. (1999) *La Filière riz à l'Office du Niger au Mali: Une nouvelle dynamique depuis la dévaluation du FCFA.* INSAH/IER Working Paper, INSAH/IER, Bamako, Mali.

Maroko, J., Buresh, R.J. and Smithson, P.C. (1999) Soil phosphorus fractions in unfertilized fallow–maize systems on two tropical soils. *Soil Science Society of America Journal* 63(2), 320–326.

Maryani, R. and Irawanti, S. (1998) *Economic Analysis of Land Use System for Large Scale Plantations of Oil Palm and Industrial Timber Estate.* Southeast Asia Policy Research Working Paper, International Centre for Research in Agroforestry (ICRAF), Bogor, Indonesia.

Mather, A.S. (1992) The forest transition. *Area* 24(4), 367–379.

Mattos, M.M. and Uhl, C. (1994) Economic and ecological perspectives on ranching in the eastern Amazon. *World Development* 22(2), 145–158.

Mausolff, C. and Farber, S. (1995) An economic analysis of ecological agricultural technologies among peasant farmers in Honduras. *Ecological Economics* 12, 237–248.

May, R.M. (1993) Resisting resistance. *Nature* 361, 593–594.

Meadows, D.H., Meadows, D.L. and Randers, J. (1992) *Beyond the Limits: Confronting Global Collapse, Envisioning a Sustainable Future*. Chelsea Green Publishing Company, Post Mills, Vermont.

Meertens, H.C.C., Fresco, L.O. and Stoop, W.A. (1996) Farming systems dynamics: impact of increasing population density and the availability of land resources on changes in agricultural systems: the case of Sukumaland, Tanzania. *Agriculture, Ecosystems, and Environment* 56(3), 203–215.

Meinzen-Dick, R., Mendoza, M.S., Sadoulet, L., Abiad-Shields, G. and Subramanian, A. (1994) Sustainable water user associations: lessons from a literature review. Paper presented at the World Bank Water Resources Seminar, Lansdowne, Virginia, 13–15 December 1994.

Melack, J.M. and MacIntyre, S. (1992) Phosphorus concentrations, supply and limitation in tropical African rivers and lakes. In: Tiessen, H. and Frossard, E. (eds) *Phosphorus Cycles in Terrestrial and Aquatic Ecosystems. Proceedings Regional Workshop 4: Africa*. SCOPE/UNEP, University of Saskatchewan, Saskatoon, Canada, pp. 1–10.

Mendez, J.C. (1995) Détermination des objectifs des petit producteurs de café du Guatemala: une application du paradigme décisionnel multicritère. In: Reyniers, F.N. and Benoît-Cattin, M. (1995) *Couplage de modèles en agriculture*. Collection Colloques, Centre de Coopération Internationale en Recherche Agronomique pour le Développement (CIRAD), Montpellier, France, pp. 40–46.

Michon, G. and de Foresta, H. (1995) The Indonesian agroforest model: forest resource management and biodiversity conservation. In: Halladay, P. and Gilmour, D.A. (eds) *Conserving Biodiversity Outside Protected Areas: the Role of Traditional Agro-ecosystems*. World Conservation Union (IUCN), Gland, Switzerland, pp. 90–106.

Migot-Adholla, S., Hazell, P., Blarel, B. and Place, F. (1991) Indigenous land rights systems in sub-Saharan Africa: a constraint on productivity? *World Bank Economic Review* 5(1), 155–175.

MINAGRI (Cameroon Ministry of Agriculture) (1987) *1984 Agricultural Census–Traditional Sector (Results of the Center and South Provinces)*, Vols 2.E and 2.F. MINAGRI, Yaoundé, Cameroon.

Ministry of Forest and Soil Conservation (Nepal) (1994) *Sedimentation Survey of Kulekhani Reservoir, December 1993*. Research and Soil Conservation Sections, Ministry of Forest and Soil Conservation, Kathmandu, Nepal.

Mishra, P.K. (1996) *Agricultural Risk, Insurance and Income: a Study of the Impact and Design of India's Comprehensive Crop Insurance Scheme*. Avebury Publishing, Aldershot, UK.

Mitchell, D.O. and Ingco, M.D. (1995) Global and regional food demand and supply prospects. In: Islam, N. (ed.) *Population and Food in the Early Twenty-first Century: Meeting Future Food Demands of an Increasing Population*. International Food Policy Research Institute (IFPRI), Washington, DC, pp. 49–60.

Mittermeier, R. and Werner, T. (1990) Wealth of plants and animals unites 'mega-diversity' countries. *Tropics* 4, 4–5.

Mohammed, T.A. (1990) *Peri-urban Cattle Agropastoralists in the Derived Savannahs of Oyo State, South West Nigeria*. Research Report, International Livestock Centre for Africa (ILCA), Ibadan, Nigeria.

Monke, E. and Pearson, S.R. (1989) *The Policy Analysis Matrix for Agricultural Development*. Cornell University Press, Ithaca, New York.

Monke, E. and Pearson, S. (1991) Introduction. In: Pearson, S., Falcon, W., Heytens, P., Monke, E. and Naylor, R. (eds) *Rice in Indonesia*. Cornell University Press, Ithaca, New York.

Moody, K. (1994) Postplanting weed control in direct-seeded rice. In: Ali, A.H. (ed.) *Proceedings of the National Workshop on Direct Seeding Practices and Productivity*. Malaysian Agricultural Research and Development Institute (MARDI), Penang, Malaysia, pp. 249–265.

Mooney, H.A. and Bernardi, G. (eds) (1990) *Introduction of Genetically-Modified Organisms into the Environment, Scientific Committee on Problems of the Environment*. John Wiley & Sons, Chichester, UK.

Moormann, F.R. and van Breemen, N. (1978) *Rice: Soil, Water, Land*. International Rice Research Institute (IRRI), Los Baños, Laguna, Philippines.

Morris, C.T. and Adelman, I. (1988) *Comparative Patterns of Economic Development*. Johns Hopkins University Press, Baltimore, Maryland.

Mosher, W.T. (1980) The theory of change and response: an application to Puerto Rico, 1940 to 1970. *Population Studies* 34, 45–58.

Mudimu, G. (1996) An analysis of incentives for adopting soil and moisture conservation tillage practices by smallholder farmers in Zimbabwe: a case study of Kandeya communal land. Mimeo, Department of Agricultural Economics and Extension, University of Zimbabwe, Harare, Zimbabwe.

Munasinghe, M. and Cruz, W. (1995) *Economywide Policies and the Environment*. World Bank Environment Paper 10, World Bank, Washington, DC.

Muñoz Braz, E., de Magalhães de Oliveira, S.J., Carvalho dos Santos, J., Gonçalves Cordeiro, D., Franke, I.L., de Araújo Gomes, T.C., Borges de Araujo, H.J., Correia da Costa, R.S., Mendes, Â.M., Rodrigues, V., Barros, E., Vosti, S., Witcover, J., Carpentier, C., Valentim, J.F., Huang, S.P., de Souza Moreira, F.P., de Sá, C.P., do Amaral, E., Rossi, L.M., Feigl, B., d'Oliveira, M.V.N. and Carneiro, J. (2000) *Relatório do Projeto Alternativas para a Agricultura de Derruba e Queima – ASB/Brasil – Fase II*. Alternatives to Slash and Burn Project, Embrapa, Acre and Rondônia, Brazil (in press).

Murphree, M.W. (1994) The role of institutions in community-based conservation. In: Western, D. and Wright, R.M. (eds) *Natural Connections: Perspectives in Community-based Conservation*. Island Press, Washington, DC, pp. 403–427.

Murray, G.F. (1994) Technoeconomic, organizational, and ideational factors as determinants of soil conservation in the Dominican Republic. Environment Paper No. 8. In: Lutz, E., Pagiola, S. and Reiche, C. (eds) *Economic and Institutional Analyses of Soil Conservation Projects in Central America*. World Bank, Washington, DC.

Murwira, K.H., Swift, M.J. and Frost, P.G.H. (1995) Manure as a key resource in sustainable agriculture. In: Powell, J.M., Fernández-Rivera, S., Williams, T.O. and Renard, C. (eds) *Livestock and Sustainable Nutrient Cycling in Mixed Farming Systems of Sub-Saharan Africa*, Vol. II *Technical Papers*. International Livestock Centre for Africa (ILCA), Addis Ababa, Ethiopia, pp. 1–48.

Mustafa, U. (1991) Economic impact of land degradation (salt affected and waterlogged soils) on rice production in Pakistan's Punjab. PhD dissertation, College of Economics and Management, University of the Philippines at Los Baños (UPLB), College, Laguna, Philippines.

Myers, N. (1988) Threatened biotas: hotspots in tropical forests. *Environmentalist* 8, 178–208.

Myint, H. (1965) *The Economics of the Developing Countries*. F.A. Praeger, New York.

Naeem, S., Thompson, L.J., Lawler, S.P., Lawton, J.H. and Woodfin, R.M. (1995) Empirical evidence that declining species diversity may alter the performance of terrestrial ecosystems. *Philosophical Transactions of the Royal Society of London Series B* 347, 249–262.

Nagle, G. (1997) The use of cesium-137 in the study of hillslope erosion in a tropical watershed. PhD dissertation, Department of Natural Resources, Cornell University, Ithaca, New York.

NAS (National Academy of Sciences) (1989) *Irrigation Induced Water Quality Problems – What Can Be Learned from the San Joaquin Valley Experience.* National Academy Press, Washington, DC.

Nations, J.D. (1992) Terrestrial impacts in Mexico and Central America. In: Downing, T.E., Hecht, S.B., Pearson, H.A. and Garcia-Downing, C. (eds) *Development or Destruction: the Conversion of Tropical Forest to Pasture in Latin America.* Westview Press, Boulder, Colorado.

NBSS and LUP (National Bureau of Soil Survey and Land Use Planning) (1992) *Agro-ecological Regions of India*, 2nd edn. NBSS and LUP Publication No. 24, National Bureau of Soil Survey and Land Use Planning, Nagpur.

Neidecker-Gonzales, O. and Scherr, S.J. (eds) (1997) *Desarrollo agrícola, sostenibilidad y alivio de la pobreza en América Latina. El papel de las regiones de laderas. Memoria de la conferencia celebrada del 4 al 8 de diciembre, 1995 en Tegucigalpa, Honduras.* IFPRI, IICA, SRN and DSE, Washington, DC.

Neill, S.P. and Lee, D.R. (2000) Explaining the adoption and disadoption of sustainable agriculture: the case of cover crops in Northern Honduras. *Economic Development and Cultural Change* (in press).

Netting, R., Stone, P. and Stone, G. (1989) Koyfar cash-cropping: choice and change in indigenous agricultural development. *Human Ecology* 17, 299–319.

Netting, R.M. (1976) What Alpine peasants have in common: observations on communal tenure in a Swiss village. *Human Ecology* 4(1), 135–146.

Newbery, D.M.G. and Stiglitz, J.E. (1981) *The Theory of Commodity Price Stabilization: a Study in the Economics of Risk.* Clarendon Press, Oxford.

Newmark, T. (1998) Carbon sequestration and cocoa production: financing sustainable development by trading carbon emission credits. Paper presented at the First International Sustainable Cocoa Workshop, 29 March–3 April, 1998, Panama City, Panama.

Niang, A., Gathumbi, S. and Amadalo, B. (1996) The potential of short-duration improved fallow for crop productivity enhancement in the highlands of western Kenya. In: Mugah, J.O. (ed.) *People and Institutional Participation in Agroforestry for Sustainable Development.* Kenya Forestry Research Institute, Muguga, Kenya, pp. 218–230.

Niang, A., Recke, H., Place, F., Kiome, R., Nandwa, S. and Nyamai, D. (1997) *Soil Fertility Recapitalization and Replenishment Project in Western Kenya.* ICRAF, KARI and KEFRI, Nairobi, Kenya.

Niang, A.I., Styger, E., Gahamany, A., Hoekstra, D. and Coe, R. (1998) Fodder quality improvement through contour planting of legume–shrub/grass mixtures in croplands of the Rwanda highlands. *Agroforestry Systems* 39, 263–274.

Nieuwenhuyse, A. (1996) *Soils, Geology, and Soil-related Sustainability Aspects of the Perhumid Tropical Limon Basin, Costa Rica.* Serie Técnica, Informe Técnico No. 272, Centro Agronómico Tropical de Investigación y Enseñanza, Turrialba, Costa Rica.

Njoroge, M. and Rao, M.R. (1994) Barrier hedgerow intercropping for soil and water conservation in sloping lands. In: *8th International Soil Conservation Conference 2.* Indian Association of Soil and Water Conservationists, Dehra Dun, India, pp. 1427–1441.

Norbu, S.L. (1985) Management issues in Nepal's national parks. In: McNeely, J.A. and Thorsell, J.W. (eds) *People and Protected Areas in the Hindukush–Himalaya.*

International Centre for Integrated Mountain Research, Kathmandu, Nepal, pp. 123–126.

NRC (National Research Council) (1993) Sustainable land use options. In: *Sustainable Agriculture and the Environment in the Humid Tropics*. National Academy Press, Washington, DC.

Nyasimi, M., Niang, A., Amadalo, B. and Obonyo, E. (1997) *Using the Wild Sunflower,* Tithonia, *in Kenya*. ICRAF, KARI and KEFRI, Nairobi, Kenya (English and Kiswahili versions).

Nziguheba, G., Palm, C.A., Buresh, R.J. and Smithson, P.J. (1998) Soil phosphorus fractions and adsorption as affected by organic and inorganic sources. *Plant and Soil* 198, 159–168.

Oates, J.F. (1999) *Myth and Reality in the Rain Forest: How Conservation Strategies are Failing in West Africa*. University of California Press, Berkeley, California.

OECD (Organization for Economic Cooperation and Development) (1994) *The Use of Environmental Indicators for Agricultural Policy Analysis*. OECD Joint Working Party of the Committee for Agriculture and the Environment Committee, Paris.

Omore, A., Staal, S.J. and Thorpe, W. (eds) (1998) *The Tanzanian Dairy Sub-sector: a Rapid Appraisal*, Vol. 1, *Executive Summary*, Vol. 2, *Targeting Dairy Development*, Vol. 3, *Main Report*. MoAC/Sokoine/ILRI Collaborative Research Report, Ministry of Agriculture and Cooperatives, Tanzania, Sokoine University of Agriculture, Morogoro, Tanzania, and the International Livestock Research Institute, Nairobi, Kenya.

Ostrom, E. (1986) *How Inexorable is the Tragedy of the Commons? Institutional Arrangements for Changing the Structure of Social Dilemmas*. Distinguished Research Lecture, Indiana University, Bloomington, Indiana, 3 April 1986.

Ostrom, E. (1990) *Governing the Commons: the Evolution of Institutions for Collective Action*. Cambridge University Press, Cambridge.

Ostrom, E. (1992) *Crafting Institutions for Self-governing Irrigation Systems*. Institute for Contemporary Studies, San Francisco, California.

Ostrom, E. (1994) *Neither Market nor State: Governance of Common-pool Resources in the Twenty-first Century*. Lecture Series No. 2, IFPRI, Washington, DC.

Ostrom, E. (1998) Scales, policentricity, and incentives: designing complexity to govern complexity. In: Guruswamy, L. and McNeely, J.A. (eds) *Protection of Global Diversity: Converging Strategies*. Duke University Press, Durham, North Carolina, pp. 145–163.

Otsuka, K. and Place, F. (1998) Land tenure and the management of land and trees: issues and analytical framework. Paper presented at Workshop on Land Tenure and the Management of Land and Trees: Community and Household Case Studies from Asia and Africa, Tokyo Metropolitan University, 1–3 July 1998, Tokyo.

Otsuka, K. and Tachibana, T. (2000) Evolution and consequences of community forest management in the Hill Region of Nepal. In: Hayami, Y. and Aoki, M. (eds) *Community and Market in Economic Development*. Clarendon Press, Oxford.

Otsuka, K., Aidoo, J.B. and Quisumbing, A. (1998a) Causes and consequences of changing land tenure institutions in western Ghana. Paper presented at Workshop on Land Tenure and the Management of Land and Trees: Community and Household Case Studies from Asia and Africa, Tokyo Metropolitan University, 1–3 July 1998.

Otsuka, K., Suyanto, S. and Tomich, T. (1998b) Does land tenure insecurity discourage tree planting? Evolution of customary land tenure and agroforestry management in Sumatra. Paper presented at Workshop on Land Tenure and the Management of

Land and Trees: Community and Household Case Studies from Asia and Africa, Tokyo Metropolitan University, 1–3 July 1998, Tokyo.

Otsuka, K., Quisumbing, A., Payongayong, E. and Aidoo, J.B. (1999) Land tenure and the management of land and trees: the case of customary land areas of Ghana. Mimeo, International Food Policy Research Institute (IFPRI), Washington, DC.

Owens, T. and Hoddinott, J. (1998) Investing in development or investing in relief: quantifying the poverty trade-offs using Zimbabwe household panel data. Mimeo, International Food Policy Research Institute (IFPRI), Washington, DC.

Ozorio de Almeida, A.L. and Campari, J.S. (1995) *Sustainable Settlement in the Brazilian Amazon*. Oxford University Press, Oxford, UK.

Pagiola, S. (1995) The effect of subsistence requirements on sustainable land use practices. Paper presented at the Annual Meetings of the American Agricultural Economics Association, Indianapolis, 6–9 August 1995.

Pagiola, S. (1996) Price policy and returns to soil conservation in semi-arid Kenya. *Environmental and Resource Economics* 8, 255–271.

Pagiola, S., Kellenberg, J., Vidaeus, L. and Srivastava, J. (1997) *Mainstreaming Biodiversity in Agricultural Development: Toward Good Practice*. Environment Paper Number 15, World Bank, Washington, DC.

Palanisami, K. (1994) Cost of water in different uses in the Bhavani basin. Paper presented at the Policy Advisory Committee Meeting on the Sharing of Common Water Resources in Bhavani Basin, Madras Institute of Development Studies, Madras, India.

Palm, C.A. (1995) Contribution of agroforestry trees to nutrient requirements of intercropped plants. *Agroforestry Systems* 30, 105–124.

Palm, C.A., Swift, M.J. and Woomer, P.L. (1996) Soil biological dynamics in slash-and-burn agriculture. *Agriculture, Ecosystems and Environment* 58, 61–74.

Palm, C.A., Myers, R.J.K. and Nandwa, S.M. (1997) Combined use of organic and inorganic nutrient sources for soil fertility maintenance and replenishment. In: Buresh, R.J., Sanchez, P.A. and Calhoun, F. (eds) *Replenishing Soil Fertility in Africa*. SSSA Special Publication 51, Soil Science Society of America, Madison, Wisconsin, pp. 193–218.

Palm, C.A., Woomer, P.L., Alegre, J., Castilla, C., Cordeiro, D., Hairiah, K., Kotto-Same, J., Moukam, A., Njomgang, R., Ricse, A., Rodrigues, V. and van Noordwijk, M. (1999) Carbon sequestration and trace gas emissions in slash and burn and alternative land uses in the Tropics. *Alternatives to Slash and Burn Phase II Final Report*. TSBF, Nairobi, Kenya.

Palm, C.A. Woomer, P.L., Alegre, J.C., Arevalo, L., Castilla, C.E., Cordeiro, D.G., Feigl, B., Hariah, K., Kotto-Same, J., Mendes, A., Moukam, A., Murdiyarso, D., Njomgang, R., Parton, W.J., Ricse, A., Rodrigues, V., Sitompul, S.M. and van Noordwijk, M. (2000) Carbon sequestration and trace gas emissions in slash and burn and alternative land uses in the humid tropics. In: Sanchez, P.A., Ericksen, P.J. and Juo, A.S.R. (eds) *Alternatives to Slash and Burn Agriculture: a Global Synthesis*. American Society of Agronomy Special Publication.

Panayotou, T. (1993) *Green Markets: the Economics of Sustainable Development*. ICEG Sector Studies No. 7, ICS Press, San Francisco.

Panayotou, T. (1995) Environmental degradation at different stages of economic development. In: Ahmed, I. and Doeleman, J.A. (eds) *Beyond Rio: the Environmental Crisis and Sustainable Livelihoods in the Third World*. St Martin's Press, New York, pp. 13–36.

Panayotou, T. and Ashton, P. (1992) *Not By Timber Alone: Economics and Ecology for Sustaining Tropical Forests*. Island Press, Washington, DC.

Panayotou, T. and Sungsuwan, S. (1994) An econometric analysis of the causes of tropical deforestation: the case of northeast Thailand. In: Brown, K. and Pearce, D. (eds) *The Causes of Tropical Deforestation: the Economic and Statistical Analysis of Factors Giving Rise to the Loss of Tropical Forests*. University College London Press, London, pp. 192–210

Pardey, P.G., Roseboom, J. and Beintema, N.M. (1995) *Investments in African Agricultural Research*. EPTD Discussion Paper No. 14, International Food Policy Research Institute (IFPRI), Washington, DC.

Parker, I.M., Simberloff, D., Lonsdale, W.M., Goodell, K., Wonham, M., Williamson, M., von Holle, B., Moyle, P.B., Byers, J.E. and Goldwasser, L. (1999) Impact: toward a framework for understanding the ecological effects of invaders. *Biological Invasions* 1, 3–19.

Parry, M. (1990) *Climate Change and World Agriculture*. Earthscan Publications, London.

Parsons, J.J. (1972) The diffusion of African pasture in the American tropics. *Journal of Range Management* 25, 12–17.

Pearce, D.W. (1999a) Economics and biodiversity conservation in the developing world. *Environment and Development Economics* 4(2), 230–233.

Pearce, D.W. (1999b) *Economics and Environment: Essays on Ecological Economics and Sustainable Development*. Edward Elgar, Cheltenham, UK.

Pearce, D.W. and Moran, D. (1994) *The Economic Value of Biodiversity*. Earthscan, London.

Pearce, D.W. and Warford, J. (1993) *World Without End*. Oxford University Press, Oxford.

Pender, J.L. (1996) Discount rates and credit markets: theory and evidence from rural India. *Journal of Development Economics* 50, 157–196.

Pender, J. (1998) Population growth, agricultural intensification, induced innovation and natural resource sustainability: an application of neoclassical growth theory. *Agricultural Economics* 19, 99–112.

Pender, J. and Scherr, S.J. (1997) *Community Survey of Hillside Development Patterns in the Central Region of Honduras*. Mimeo, Environment and Production Technology Division, International Food Policy Research Institute (IFPRI), Washington, DC.

Pender, J. and Scherr, S.J. (1999) *Organizatonal Development and Natural Resource Management: Evidence from Central Honduras*. Environment and Production Technology Division Discussion Paper No. 49, International Food Policy Research Institute (IFPRI), Washington, DC.

Pender, J., Place, F. and Ehui, S. (1999a) *Strategies for Sustainable Agricultural Development in the East African Highlands*. Environment and Production Technology Division Discussion Paper No. 41, International Food Policy Research Institute (IFPRI), Washington, DC.

Pender, J., Scherr, S. and Durón, G. (1999b) *Pathways of Development in the Hillsides of Honduras: Causes and Implications for Agricultural Production, Poverty, and Sustainable Resource Use*. Environment and Production Technology Division Discussion Paper No. 45, International Food Policy Research Institute (IFPRI), Washington, DC.

Perrings, C. (1989) An optimal path to extinction? Poverty and resource degradation in the open agrarian economy. *Journal of Development Economics* 30, 1–24.

Perrings, C. (1995) Biodiversity conservation as insurance. In: Swanson, T. (ed.) *Economics and Ecology of Biodiversity Decline*. Cambridge University Press, Cambridge, pp. 69–77.

Perrings, C. and Ansuategi, A. (2000) Environmental sustainability and economic growth. *Journal of Economic Studies* 27(1/2), 19–54.

Perrings, C. and Stern, D.I. (2000) Modelling loss of resilience in agroecosystems: rangelands in Botswana. *Environmental and Resource Economics*, 16(2), 185–210.

Perrings, C., Mäler, K.-G., Folke, C., Holling, C.S. and Jansson, B.-O. (eds) (1994) *Biodiversity Conservation: Problems and Policies*. Kluwer Academic Press, Dordrecht.

Perrings, C., Mäler, K.-G., Folke, C., Holling, C.S. and Jansson, B.-O. (eds) (1995) *Biological Diversity: Economic and Ecological Issues*. Cambridge University Press, Cambridge.

Peters, C., Gentry, A.H. and Mendelsohn, R.O. (1989) Valuation of an Amazonian rain forest. *Nature* 339, 655–656.

Petersen, W. (1972) *Readings in Population*. Macmillan, New York.

Pichón, F.J. (1997a) Colonist land-allocation decisions, land use, and deforestation in the Ecuadorian Amazon frontier. *Economic Development and Cultural Change* 44, 707–744.

Pichón, F.J. (1997b) Settler households and land-use patterns in the Amazon frontier: farm-level evidence from Ecuador. *World Development* 25, 67–91.

Pichón, F.J. and Bilsborrow, R.E. (1999) Land-use systems, deforestation, and demographic factors in the humid tropics: farm-level evidence from Ecuador. In: Bilsborrow, R.E. and Hogan, D. (eds) *Population and Deforestation in the Humid Tropics*. International Union for the Scientific Study of Population, Liège, Belgium.

Pieri, C.M.G. (1989) *Fertility of Soils: a Future for Farming in the West African Savannah*. Springer-Verlag, Berlin.

Pieri, C.M.G. (1998) Soil fertility improvement: key connection between sustainable land management and rural well being. In: *16th World Congress of Soil Science*, Montpellier, France.

Pimentel, D. (1995) Environmental and economic costs of soil erosion and conservation benefits. *Science* 267, 24.

Pimentel, D., Stachow, U., Takacs, H., Brubaker, W., Dumas, A.R., Meaney, J.J. O'Neil, J.A.S., Onsi, D.E. and Corzilius, D.B. (1992) Conserving biological diversity in agricultural–forestry systems. *Bioscience* 42(5), 354–362.

Pimentel, D., Harvey, C., Resosudarmo, P., Sinclair, K., Kurz, D., McNair, M., Crist, S., Shpritz, L., Fitton, L., Saffouri, R. and Blair, R. (1995) Environmental and economic costs of soil erosion and conservation benefits. *Science* 267, 1117–1123.

Pingali, P. and Binswanger, H. (1984) *Population Density and Farming Systems: the Changing Locus of Innovation and Technical Change*. Working Paper No. ARV24, World Bank, Washington, DC.

Pingali, P.L. and Gerpacio, R.V. (1997) *Towards Reduced Pesticide Use for Cereal Crops in Asia*. Economics Working Paper No. 97-04, International Wheat and Maize Improvement Center (CIMMYT), Mexico, DF.

Pingali, P.L. and Heisey, P. (1996) Cereal crop productivity in developing countries: past trends and future prospects. In: *Conference Proceedings of Global Agricultural Science Policy for the Twenty-First Century, 26–28 August 1996, Melbourne, Australia*. Victoria Department of Natural Resources and Environment, Melbourne, Australia, pp. 51–94.

Pingali, P.L. and Roger, P.A. (eds) (1995) *Impact of Pesticides on Farmer Health and the Rice Environment*. Kluwer Academic Publishers, Norwell, Massachusetts, and the International Rice Research Institute (IRRI), Los Baños, Laguna, Philippines.

Pingali, P.L. and Rosegrant, M. (1994) *Confronting the Environmental Consequences of the Green Revolution in Asia*. Environment and Production Technology Division

Discussion Paper No. 2, International Food Policy Research Institute (IFPRI), Washington, DC.

Pingali, P.L., Hossain, M. and Gerpacio, R.V. (1997) *Asian Rice Bowls: the Returning Crisis?* International Rice Research Institute (IRRI), Los Baños, Laguna, Philippines, and CAB International, Wallingford, UK.

Pinstrup-Andersen, P. (1994) *World Food Trends and Future Food Security.* Food Policy Report, International Food Policy Research Institute (IFPRI), Washington, DC.

Pinstrup-Andersen, P. and Hazell, P. (1985) The impact of the Green Revolution and prospects for the future. *Food Reviews International* 1(1), 1–25.

Pinstrup-Andersen, P., Pandya-Lorch, R. and Rosegrant, M.W. (1997) *The World Food Situation: Recent Developments, Emerging Issues, and Long-term Prospects.* 2020 Vision Food Policy Report, International Food Policy Research Institute (IFPRI), Washington, DC.

Place, F. and Hazell, P. (1993) Productivity effects of indigenous land tenure in sub-Saharan Africa. *American Journal of Agricultural Economics* 75(1), 10–19.

Place, F. and Otsuka, K. (1998a) Land tenure systems and agricultural productivity in Uganda. Paper presented at Workshop on Land Tenure and the Management of Land and Trees: Community and Household Case Studies from Asia and Africa, Tokyo Metropolitan University, 1–3 July 1998, Tokyo.

Place, F. and Otsuka, K. (1998b) Tenure, agricultural investment, and productivity in the customary tenure sector of Malawi. Paper presented at Workshop on Land Tenure and the Management of Land and Trees: Community and Household Case Studies from Asia and Africa, Tokyo Metropolitan University, 1–3 July 1998, Tokyo.

Place, F. and Otsuka, K. (2000a) Population, land tenure, and natural resource management: the case of customary land tenure area in Malawi. *Journal of Environmental Economics and Management* (in press).

Place, F. and Otsuka, K. (2000b) Population pressure, land tenure, and tree resource management in Uganda. *Land Economics* (in press).

Poffenberger, M. (ed.) (1990) *Keepers of the Forest: Land Management Alternatives in Southeast Asia.* Kumarian Press, West Hartford, Connecticut.

Ponnamperuma, F.N. (1974) Micronutrient limitations in acid tropical rice soils. In: Bornemisza, E. and Alvarado, A. (eds) *Soil Management in Tropical America.* Soil Science Department, North Carolina State University, Raleigh, North Carolina, pp. 330–347.

Postel, S. (1989) *Water for Agriculture: Facing the Limits.* Worldwatch Paper 93, Worldwatch Institute, Washington, DC.

Postel, S. (1993) Water and agriculture. In: Gleick, P.H. (ed.) *Water in Crisis: a Guide to the World's Fresh Water Resources.* Oxford University Press, New York.

Prescott-Allen, R. and Prescott-Allen, C. (1984) Park your genes: protected areas as *in situ* genebanks for the maintenance of wild genetic resources. In: McNeely, J.A. and Miller, K.R. (eds) *National Parks, Conservation and Development: the Role of Protected Areas in Sustaining Society.* Smithsonian Institution Press, Washington, DC, pp. 634–638.

Prescott-Allen, R. and Prescott-Allen, C. (1986) *The First Resource: Wild Species in the North American Economy.* Yale University Press, New Haven, Connecticut.

Probert, M.E., Okalebo, J.R. and Jones, R.K. (1995) The use of manure on smallholders' farms in semi-arid eastern Kenya. *Experimental Agriculture* 31, 371–381.

Putterman, L. (1995) Economic reform and smallholder agriculture in Tanzania: a discussion of recent market liberalization, road rehabilitation, and technology dissemination efforts. *World Development* 23(2), 311–326.

Queen Elizabeth House (1996) World cereals markets: a review of the main model. Mimeo, Food Studies Group, Queen Elizabeth House, Oxford.

Quiñones, M.A., Borlaug, N.E. and Dowswell, C.R. (1997) A fertilizer-based green revolution for Africa. In: Buresh, R.J., Sanchez, P.A. and Calhoun, F. (eds) *Replenishing Soil Fertility in Africa*. Soil Science Society of America Special Publication 51, Soil Science Society of America, Madison, Wisconsin, pp. 81–96.

Qureshi, J.N. (1987) The cumulative effects of N–P fertilizers, manure and crop residues on maize grain yields, leaf nutrient contents and some chemical properties at Kabete. In: *CIMMYT National Maize Agronomy Workshop*. Kenya Agricultural Research Institute, Nairobi, Kenya.

Qureshi, J.N. (1991) The cumulative effects of N–P fertilizers manure and crop residues on maize and bean yields and some soil chemical properties at Kabete. In: *Recent Advances in KARI's Research Programmes*. Kenya Agricultural Research Institute, Nairobi, Kenya, pp. 160–167.

Rabbinge, R. (1990) *Theoretical Production Ecology: Reflections and Prospects*. Pudoc Simulation Monograph No. 34, Wageningen, the Netherlands.

Rabbinge, R. and van Ittersum, M.K. (1994) Tension between aggregation levels. In: Fresco, L.O., Stroosnijder, L., Bouma, J. and van Keulen, H. (eds) *The Future of the Land: Mobilizing and Integrating Knowledge for Land Use Options*. John Wiley & Sons, Chichester, pp. 31–40.

Rabbinge, R., Leffelaar, P.A. and van Latesteijn, H.C. (1994) The role of systems analysis as an instrument in policy making and resource management. In: Goldsworthy, P.R. and Penning de Vries, F.W.T (eds) *Opportunities, Use, and Transfer of Systems Research Methods in Agriculture to Developing Countries*. Kluwer Academic Publishers, Dordrecht, pp. 67–79.

Raloff, J. (1997) Can grain yields keep pace? Stiff challenges face today's breeders and tomorrow's farmers. *Science News* 152(7), 104–105.

Randhawa, N.S. (n.d.) Some concerns for future of Punjab agriculture. Mimeo, New Delhi.

Rao, M.R., Niang, A., Kwesiga, F., Duguma, B., Franzel, S., Jama, B. and Buresh, R. (1998) Soil fertility replenishment in sub-Saharan Africa: new techniques and the spread of their use on farms. *Agroforestry Today* 10(2), 3–8.

Rasmussen, L.N. and Meinzen-Dick, R. (1995) *Local Organizations for Natural Resource Management: Lessons from Theoretical and Empirical Literature*. Environment and Production Technology Division Discussion Paper No. 11, International Food Policy Research Institute (IFPRI), Washington, DC.

Reardon, T. (1995) Sustainability issues for agricultural research strategies in the semi-arid tropics: focus on the Sahel. *Agricultural Systems* 48(3), 345–360.

Reardon, T. (1997) Using evidence of household income diversification to inform study of the rural nonfarm labor market in Africa. *World Development* 25(5), 735–748.

Reardon, T. and Islam, N. (1989) Issues of sustainability in agricultural research in Africa. In: *Proceedings of the Symposium on the Sustainability of Production Systems in Sub-Saharan Africa, NORAGRIC, Agricultural University of Norway, Ås, Norway, 4–7 September 1989*. Occasional Paper Series C, NORAGRIC, Ås, Norway.

Reardon, T. and Vosti, S. (1992) Issues in the analysis of the effects of policy on conservation and productivity at the household level in developing countries. *Quarterly Journal of International Agriculture* 31(4), 380–396.

Reardon, T. and Vosti, S. (1995) Links between rural poverty and the environment in developing countries: asset categories and investment poverty. *World Development* 23(9), 1495–1506.

Reardon, T. and Vosti, S.A. (1997a) Policy analysis of conservation investments: extensions of traditional technology adoption research. In: Vosti, S.A. and Reardon, T. (eds) *Sustainability, Growth, and Poverty Alleviation: a Policy and Agroecological Perspective*. Johns Hopkins University Press, Baltimore, Maryland.

Reardon, T., and Vosti, S.A. (1997b) Poverty–environment links in rural areas of developing countries. In: Vosti, S.A. and Reardon, T. (eds) *Sustainability, Growth, and Poverty Alleviation: a Policy and Agroecological Perspective*. Johns Hopkins University Press, Baltimore, Maryland.

Reardon, T., Hopkins, J. and Kelly, V. (1992) Structural adjustment impacts on real incomes and demand patterns of urban and rural households in the Sahel. Paper presented at the 29th Seminar of the European Association of Agricultural Economics, Food and Agricultural Policies Under Structural Adjustment, Hohenheim, Germany, 21–25 September 1992.

Reardon, T., Crawford, E. and Kelly, V. (1994) Links between nonfarm income and farm investment in African households: adding the capital market perspective. *American Journal of Agricultural Economics* 76(5), 1172–1176.

Reardon, T., Crawford, E. and Kelly, V. (1995) *Promoting Farm Investment for Sustainable Intensification of African Agriculture*. International Development Paper No. 18, Michigan State University, East Lansing, Michigan.

Reardon, T., Kelly, V., Crawford, E., Diagana, B., Dione, J., Savadogo, K. and Boughton, D. (1997) Promoting sustainable intensification and productivity growth in Sahel agriculture after macroeconomic policy reform. *Food Policy* 22(4), 317–328.

Redford, K.H. (1990) The ecologically noble savage. *Orion* 9, 24–29.

Redford, K.H. and Richter, B. (2000) Conservation of biodiversity in a world of use. *Conservation Biology* (in press).

Reid, W.V., Barnes, J.N. and Blackwelder, B. (1988) *Bankrolling Successes: a Portfolio of Sustainable Development Projects*. Environmental Policy Institute and National Wildlife Federation, Washington, DC.

Reijntjes, C., Haverkrot, B. and Waters-Bayer, A. (1992) *Farming for the Future: An Introduction to Low-external-input and Sustainable Agriculture*. Macmillan Press, London.

Reil, C., Scoones, I. and Toulmin, C. (1996) *Sustaining the Soil: Indigenous Soil and Water Conservation in Africa*. Earthscan Publications, London.

Reilly, J. (1996) Agriculture in a changing climate: impacts and adaptation. In: Watson, R.T., Zinyowera, M.C. and Moss, R.H. (eds) *Climate Change 1995: Impacts, Adaptations and Mitigation of Climate Change: Scientific–Technical Analyses*. Contribution of Working Group II to the Second Assessment Report of the Intergovernmental Panel on Climate Change. Cambridge University Press, Cambridge, pp. 427–467.

Reyniers, F.N. and Benoît-Cattin, M. (1995) *Couplage de modèles en agriculture*. Collection colloques, Centre de Coopération Internationale en Recherche Agronomique pour le Développement (CIRAD), Montpellier, France.

Ricardo, D. (1887) *Letters of David Ricardo to Thomas Malthus, 1810–1823*. Clarendon Press, London.

Richards, M. (1997) *Missing a Moving Target? Colonist Technology Development on the Agricultural Frontier*. Overseas Development Institute, London.

Rissler, J. and Mellon, M. (1996) *The Ecological Risks of Engineered Crops*. MIT Press, Cambridge, Massachusetts.

Roberts, P.C., Rust, R.H. and Larson, W.E. (1995) Site-specific management for agricultural systems. In: *Proceedings of Soil-specific Management for Agricultural Systems*

2nd International Conference. Department of Soil Science and Minnesota Extension Service, University of Minnesota, Minneapolis, Minnesota.

Robison, D.M. and McKean, S.J. (eds) (1992) *Shifting Cultivation and Alternatives: an Annotated Bibliography, 1972–1989*. CAB International, in association with CIAT, Wallingford, UK.

Rogers, P., Lydon, P., Seckler, D. and Pitman, G.T.K. (1994) *Water and Development in Bangladesh: a Retrospective on the Flood Action Plan*. Synthesis report prepared for the Bureau of Asia and the Near East of the US Agency for International Development by the Irrigation Support Project for Asia and the Near East, Irrigation Support Project for Asia and the Near East (ISPAN), Arlington, Virginia.

Rola, A.C. and Pingali, P.L. (1993) *Pesticides, Rice Productivity, and Farmers' Health – an Economic Assessment*. International Rice Research Institute (IRRI), Los Baños, Laguna, Philippines and World Resources Institute (WRI), Washington, DC.

Romero, C. (1993) A research on Andalusian farmers' objectives: methodological aspects and policy implications. Paper presented at the VII European Agricultural Association of Agricultural Economics (AEEA) Congress, Stresa, Italy.

Rosegrant, M.W. (1995) Water transfers in California: potentials and constraints. *Water International* 20, 72–78.

Rosegrant, M.W. (1997) *Water Resources in the Twenty-First Century: Challenges and Implications for Action*. Food, Agriculture, and the Environment Discussion Paper No. 20, International Food Policy Research Institute (IFPRI), Washington, DC.

Rosegrant, M.W. and Hazell, P.B.R. (2000) *Transforming the Rural Asian Economy: the Unfinished Revolution*. Oxford University Press, Hong Kong.

Rosegrant, M.W. and Pingali, P.L. (1994) Policy and technology for rice productivity growth in Asia. *Journal of International Development* 6(6), 665–688.

Rosegrant, M.W. and Ringler, C. (1997) World food markets into the 21st century: environmental and resource constraints and policies. *Australian Journal of Agricultural and Resource Economics* 41(3), 401–428.

Rosegrant, M.W., Agcaoili-Sombilla, M. and Perez, N.D. (1995) *Global Food Projections to 2020: Implications for Investment*. Food, Agriculture and the Environment Discussion Paper 5, International Food Policy Research Institute (IFPRI), Washington, DC.

Rosenzweig, C. and Parry, M.L. (1994) Potential impact of climate change on world food supply. *Nature* 367, 133–138.

Rothman, D.S. (1998) Environmental Kuznets curves – real progress or passing the buck? A case for consumption-based approaches. *Ecological Economics* 25(2), 177–194.

Ruben, R., Kruseman, G., Hengsdijk, H. and Kuyvenhoven, A. (1996) The impact of agrarian policies on sustainable land use. In: Teng, P.S., Kropff, M.J., ten Berge, H.F.M., Dent, J.B., Lansigan, F.P. and van Laar, H.H. (eds) *Applications of Systems Approaches at the Farm and Regional Level, Proceedings SAAD-2 Symposium*. Kluwer Academic Publishers, Dordrecht, pp. 65–82.

Ruben, R. Heerink, N.B.M. and Mol, E. (1998a) Economic evaluation of low and high external input agriculture: a production function analysis for Malang, Indonesia. In: Wossink, G.A.A., van Kooten, G.C. and Peters, G.H. (eds) *Economics of Agrochemicals*. Dartmouth Publishing, Aldershot, UK, pp. 94–106.

Ruben, R., Moll, H.A.J. and Kuyvenhoven, A. (1998b) Integrating agricultural research and policy analysis: analytical framework and policy applications for bio-economic modelling. *Agricultural Systems* 58(3), 331–349.

Rudel, T.K. (1983) Roads, speculators, and colonization in the Ecuadorian Amazon. *Human Ecology* 11(4), 385–403.

Rudel, T.K. and Richards, S. (1990) Urbanization, roads, and rural population change in the Ecuadorian Andes. *Studies in Comparative International Development* 25, 73–89.

Rudel, T.K. and Roper, J. (1996) Regional patterns and historical trends in tropical deforestation, 1976–1990: a qualitative comparative analysis. *Ambio* 25(3), 160–166.

Rudel, T.K. and Roper, J. (1997) The paths to rain forest destruction: crossnational patterns of tropical deforestation, 1975–90. *World Development* 25(1), 53–65.

Ruf, F. (1995) *Booms et crises du cacao: les vertiges de l'or brun*. Editions Karthala, Paris.

Ruf, F. (1998) Cocoa: from deforestation to reforestation. Paper presented at the First International Sustainable Cocoa Workshop, 29 March–3 April 1998, Panama City, Panama.

Ruitenbeek, H.J. (1989) *Economic Analysis of Issues and Projects Relating to the Establishment of the Proposed Cross River National Park (Oban Division) and Support Zone*. World Wide Fund for Nature, London.

Rukuni, M. (1996) *A Framework for Crafting Demand-driven National Agricultural Research Institutes in Southern Africa*. Department of Agricultural Economics Staff Paper No. 96–76, Michigan State University, East Lansing, Michigan.

Runge, C.F. (1985) The innovation of rules and the structure of incentives in open access resources. *American Journal of Agricultural Economics* 67, 368–372.

Rusike, J., Reardon, T., Howard, J. and Kelly, V. (1997) *Developing Cereal-based Demand for Fertilizer among Smallholders in Southern Africa: Lessons Learned and Implications for Other African Regions*. Policy Synthesis No. 30, Food Security Project, Michigan State University, East Lansing, Michigan.

Ruthenberg, H. (1980) *Farming Systems in the Tropics*. Clarendon Press, Oxford.

Ruttan, V. (1990) Models of agricultural development. In: Eicher, C. and Staatz, J. (eds) *Agricultural Development in the Third World*, 2nd edn. Johns Hopkins University Press, Baltimore, Maryland.

Sachs, J. (1997) The limits of convergence of nature, nurture and growth. *The Economist*, 14 June 1997, 19–22.

Sader, S.A., Reining, C., Sever, T., Smoot, J.C. and Richards, M. (1997) Human migration and agricultural expansion: a threat to the Maya tropical forests. *Journal of Forestry* 95, 27–30.

Sadoulet, S. and de Janvry, A. (1995) *Quantitative Development Policy Analysis*. Johns Hopkins University Press, Baltimore, Maryland.

Sahn, D. (ed.) (1989) *Seasonal Variability in Third World Agriculture: the Consequences for Food Security*. Johns Hopkins University Press, Baltimore, Maryland.

Sajjapongse, A. and Syers, K. (1995) Tangible outcomes and impacts from the ASIALAND management of sloping lands network. In: *Proceeding of the International Workshop on Conservation Farming for Sloping Uplands in Southeast Asia: Challenges, Opportunities and Prospects*. Proceedings No. 14, IBSRAM, Bangkok, Thailand, pp. 3–14.

Sakurai, T., Rayamajhi, S., Pokharel, R. and Otsuka, K. (1998) Private, collective, and centralized community management: a comparative study of timber forest and plantation management in inner Tarai of Nepal. Paper presented at Workshop on Land Tenure and the Management of Land and Trees: Community and Household Case Studies from Asia and Africa, Tokyo Metropolitan University, 1–3 July 1998.

Salehi-Isfahani, D. (1988) Technology and preferences in the Boserup model of agricultural growth. *Journal of Development Economics* 28(2), 175–191.

Samad, M., Merrey, D., Vermillion, D., Fuchs-Casrsch, M., Mohtadullah, K. and Lenton, R. (1992) Irrigation management strategies for improving the performance of irrigated agriculture. *Outlook on Agriculture* 21(4), 279–286.

Sambrook, R.A., Pigozzi, B.W. and Thomas, R.N. (1999) Population pressure, deforestation, and land degradation: a case study from the Dominican Republic. *Professional Geographer* 51, 25–39.

Sanabria, H. (1993) *The Coca Boom and Rural Social Change in Bolivia*. University of Michigan Press, Ann Arbor.

Sanchez, P.A. (1976) *Properties and Management of Soils in the Tropics*. John Wiley & Sons, New York.

Sanchez, P.A. (1995) Science in agroforestry. *Agroforestry Systems* 30, 5–55.

Sanchez, P.A. and Buol, S.W. (1975) Soils of the tropics and the world food crisis. *Science* 18, 598–603.

Sanchez, P.A. and Cochrane, T.T. (1980) Soil constraints in relation to major farming systems of tropical America. In: Metz, J.F. and Brady, N.C. (eds) *Soil-related Constraints to Food Production in the Tropics*. International Rice Research Institute, Cornell University, Ithaca, New York.

Sanchez, P.A. and Leakey, R.R.B. (1997) Land-use transformation in Africa: three determinants for balancing food security with natural resource utilization. *European Journal of Agronomy* 7, 15–23.

Sanchez P.A. and Palm, C.A. (1996) Nutrient cycling and agroforestry in Africa. *Unasylva* 185(47), 24–28.

Sanchez, P.A., Bandy, D.E., Villachica, J.H. and Nicholaides, J.J. (1982) Amazon basin soils: management for continuous crop production. *Science* 216(4548), 821–827.

Sanchez, P.A., Palm, C.A. and Smyth, T.J. (1990) Approaches to mitigate tropical deforestation by sustainable soil management practices. In: Scharpenseel, H.W., Schomaker, M. and Ayoub, A. (eds) *Soils on a Warmer Earth: Effects of Expected Climate Change on Soil Processes, with Emphasis on the Tropics and Sub-tropics. International Workshop on Effects of Expected Climate Change on Soil Processes in the Tropics and Sub-tropics, Nairobi, Kenya*. Elsevier Science, New York, pp. 211–220.

Sanchez, P.A., Izac, A.-M.N., Valencia, I. and Pieri, C. (1996) Soil fertility replenishment in Africa: a concept note. In: Breth, S.A. (ed.) *Achieving Greater Impact from Research Investments in Africa*. Sasakawa Africa Association, Mexico City, pp. 200–207.

Sanchez, P.A., Buresh, R.J. and Leakey, R.R.B. (1997a) Trees, soils and food security. *Philosophical Transactions of the Royal Society of London (Series B)* 353, 949–961.

Sanchez, P.A., Shepherd, K.D., Soule, M.J., Place, F.M., Buresh, R.J., Izac, A.-M.N., Mokwunye, A.U., Kwesiga, F.R., Ndiritu, K.D. and Woomer, P.L. (1997b) Soils fertility replenishment in Africa: an investment in natural resource capital. In: Buresh, R.J., Sanchez, P.A. and Calhoun, F. (eds) *Replenishing Soil Fertility in Africa*. Soil Science Society of America Special Publication 51, Soil Science Society of America, Madison, Wisconsin, pp. 1–46.

Sandbrook, R. (1992) From Stockholm to Rio. In: Quarrie, J. (ed.) *Earth Summit 1992*. Regency Press, London.

Sanders, J.H., Shapiro, B.I. and Ramaswamy, S. (1996) *The Economics of Agricultural Technology in Semi-arid Sub-Saharan Africa*. Johns Hopkins University Press, Baltimore, Maryland.

Sanderson, S.E. and Redford, K.H. (1997) Biodiversity politics and the contest for ownership of the world's biota. In: Kramer, R., von Schaik, C. and Johnson, J. (eds) *Last Stand: Protected Areas and the Defence of Tropical Biodiversity*. Oxford University Press, Oxford.

Sandford, S. (1983) *Management of Pastoral Development in the Third World*. John Wiley & Sons, London.

Sanjayan, M.A., Shen, S. and Jansen, M. (1997) *Experiences with Integrated Conservation Development Projects in Asia.* World Bank Technical Paper No. 388, World Bank, Washington, DC.

Sansoucy, R. (1995) Livestock – a force for food security and sustainable development. *World Animal Review* 84/85, 5–17.

Sargent, M., Lichte, J., Matlon, P. and Bloom, R. (1981) *An Assessment of Animal Traction in Francophone West Africa.* African Rural Economy Working Paper No. 34, Department of Agricultural Economics, Michigan State University, East Lansing, Michigan.

Savadogo, K. (1997) *Politiques nationales, commerce agricole régional et sécurité alimentaire en Afrique de l'Ouest: les enjeux de la concertation.* A report to CILSS/FERAP, Ouagadougou, Burkina Faso.

Savadogo, K., Reardon, T. and Pietola, K. (1998) Adoption of improved land-use technologies to increase food security in Burkina Faso: relating animal traction, productivity, and nonfarm income. *Agricultural Systems* 58(3), 441–484.

Sayer, J. (1991) *Rainforest Buffer Zones: Guidelines for Protected Area Managers.* World Conservation Union (IUCN), Gland, Switzerland.

Scherr, S.J. and Hazell, P. (1994) *Sustainable Agricultural Development Strategies in Fragile Lands.* Environment and Production Technology Division, Discussion Paper No. 1, International Food Policy Research Institute (IFPRI), Washington, DC.

Scherr, S.J., Bergeron, G., Pender, J. and Barbier, B. (1996) *Policies for Sustainable Development in Fragile Lands: Methodology Overview.* Fragile Lands Program, Environment and Production Technology Division Discussion Paper, International Food Policy Research Institute (IFPRI), Washington, DC.

Schiff, M. and Valdés, A. (1992) *The Political Economy of Agricultural Pricing Policy.* Johns Hopkins University Press, Baltimore, Maryland.

Schindler, D.W. (1990) Experimental perturbations of whole lakes as tests of hypotheses concerning ecosystem structure and function. In: Proceedings of 1987 Crafoord Symposium. *Oikos* 57, 25–41.

Schipper, R.A. (1996) Farming in a fragile future: economics of land use with applications in the Atlantic Zone of Costa Rica. PhD thesis, Wageningen Agricultural University, Wageningen, the Netherlands.

Schipper, R.A., Jansen, H.G.P., Stoorvogel, J.J. and Jansen, D.M. (1995) Evaluating policies for sustainable land use: a sub-regional model with farm types for Costa Rica. In: Bouma, J., Kuyvenhoven, A., Bouman, B.A.M., Luyten, J. and Zandstra, H. (eds) *Eco-regional Approaches for Sustainable Land Use and Food Production. Proceedings of a Symposium on Eco-regional Approaches in Agricultural Research, 12–16 December 1994, ISNAR, The Hague.* Kluwer Academic Publishers, Dordrecht, the Netherlands, pp. 377–395.

Schipper, R.A., Jansen, H.G.P., Bouman, B.A.M., Hengsdijk, H., Nieuwenhuyse, A. and Sáenz, F. (1998) *Evaluation of Development Policies using Integrated Bio-economic Land Use Models: Applications to Costa Rica.* Research Program on Sustainability in Agriculture, Phase 2, Report No. 136, Centro Agronómico Tropical de Investigación y Enseñanza, Turrialba, Costa Rica.

Schipper, R.A., Jansen, H.G.P., Bouman, B.A.M., Hengsdijk, H. and Nieuwenhuyse, A. (2000) Integrated biophysical and socio-economic land use analysis at the regional level. In: Bouman, B.A.M., Jansen, H.G.P., Schipper, R.A., Hengsdijk, H. and Nieuwenhuyse, A. (eds) *Tools for Land Use Analysis at Different Scales, with Case Studies for Costa Rica.* Kluwer Academic Publishers, Dordrecht, the Netherlands (in press).

Schmink, M. (1994) *The Socioeconomic Matrix of Deforestation*. Population and Environment, Hacienda Cocoyoc, Morelos, Mexico.

Schultz, T.W. (1964) *Transforming Traditional Agriculture*. Yale University Press, New Haven, Connecticut.

Sedjo, R.A. (1995) Forests: conflicting signals. In: Bailey, R. (ed.) *The True State of the Planet*. Free Press, New York, pp. 177–210.

Semgalawe, Z.M. (1998) Soil conservation and agricultural sustainability: the case of north-eastern mountain slopes, Tanzania. PhD dissertation, Wageningen University, Wageningen, the Netherlands.

Sen, A. (1982) *Poverty and Famines: an Essay on Entitlement and Deprivation*. Oxford University Press, New York.

Sène, A. (1998) *Impact de la dévaluation du FCFA sur les producteurs d'arachide au Sénégal*. Kaolack, Institut Sénégalais de Recherches Agricoles (INSAH/PRISAS), Bamako, Mali.

SEPSA (Secretaría Ejecutiva de Planificación Sectoral Agropecuaria) (1997) *Políticas del sector agropecuario (revisión y ajuste)*. Ministry of Agriculture and Livestock (MAG), San José, Costa Rica.

Seré, C. and Steinfeld, H. (1996) *World Livestock Production Systems: Current Status, Issues and Trends*. FAO Animal Production and Health Paper, FAO, Rome.

Seroa da Motta, R. (1997) The economics of biodiversity in Brazil: the case of forest conversion. In: *OECD, Investing in Biological Diversity: Proceedings of the OECD International Conference on Incentive Measures for the Conservation and the Sustainable Use of Biological Diversity, Cairns, 25–28 March 1996*. Organization for Economic Cooperation and Development (OECD), Paris.

Serrao, E.A. and Toledo, J.M. (1990) The search for sustainability in Amazonian pastures. In: Anderson, A.B. (ed.) *Alternatives to Deforestation: Steps Toward the Sustainable Use of the Amazon Rain Forest*. Columbia University Press, New York, pp. 195–214.

Shepherd, K.D. and Soule, M.J. (1998) Soil fertility management in west Kenya: dynamic simulation of productivity, profitability and sustainability at different resource endowment levels. *Agriculture, Ecosystems and Environment* 71, 131–145.

Shively, G. and Martinez, E. (1999) Can agricultural intensification stop deforestation? Irrigation, employment, and room for cautious optimism in southern Palawan, the Philippines. Paper presented at the CIFOR Workshop on Technological Change in Agriculture and Deforestation, CATIE, Turrialba, Costa Rica, 11–13 March 1999.

Shroeder, P.E., Dixon, R.K. and Winjun, J.K. (1993) Forest management and agroforestry to sequester and conserve atmospheric carbon dioxide. *Unasylva* 173(44), 52–60.

Sidhu, D.S. and Byerlee, D. (1991) Technical change and wheat productivity in post-green revolution Punjab. *Economic and Political Weekly* 26(52), A159–A166.

Silva, J. (1987) Responses of savannas to stress and disturbance: species dynamics. In: Walker, B.H. (ed.) *Determinants of Tropical Savannas*. IRL Press, Oxford, pp. 141–156.

Simon, J. (1977) *The Economics of Population Growth*. Princeton University Press, Princeton, New Jersey.

Simon, J. (1981) *The Ultimate Resource*. Princeton University Press, Princeton, New Jersey.

Simons, A.J. (1996) ICRAF's strategy for domestication of non-wood tree products. In: B., Leakey, R.R.B., Temu, A.B., Melnyk, M. and Vantomme, P. (eds) *Domestication and Commercialization of Non-timber Forest Products in Agroforestry Systems*. Non-wood Forest Products 9, FAO, Rome, pp. 8–22.

Simons, A.J., Dawson, I.K., Duguma, B. and Tchoundjeu, Z. (1998) Passing problems: prostate and prunus. *HerbalGram* 43, 49–53.

Singh, I., Squire, L. and Strauss, J. (1986) *Agricultural Household Models: Extensions, Applications, and Policy*. Johns Hopkins University Press, Baltimore, Maryland.

Slingenbergh, J. and Wint, W. (1997) Livestock geography and land use. In: Nell, A.J. (ed.) *Livestock and the Environment International Conference. Proceedings of International Conference on Livestock and the Environment, Wageningen, 16–20 June 1997*. International Agricultural Centre, Wageningen, the Netherlands, pp. 198–209.

Smaling, E.M.A. (1993a) An agroecological framework for integrated nutrient management with special reference to Kenya. PhD thesis, Wageningen Agricultural University, Wageningen, the Netherlands.

Smaling, E.M.A. (1993b) Soil nutrient depletion in sub-Saharan Africa. In: van Reuler, H. and Prins, W.H. (eds) *The Role of Plant Nutrients for Sustainable Food Crop Production in Sub-Saharan Africa*. Dutch Association of Fertilizer Producers, Leidschendam, the Netherlands.

Small, L.E. and Carruthers, I. (1991) *Farmer-financed Irrigation: the Economics of Reform*. Published in association with the International Irrigation Management Institute, Cambridge University Press, Cambridge, UK.

Smith, J.W., Naazie, A., Largi, A., Agyemang, K. and Tarawali, S. (1997) Integrated crop–livestock systems in sub-Saharan Africa: an option or an imperative? *Outlook on Agriculture* 26(4), 237–260.

Smithson, P.J. and Sanchez, P.A. (1999) *Marginal Soils and Options for African Agricultural Development*. Japan International Research Centre for Agricultural Sciences (JIRCAS), Tsukuba, Japan.

Sneath, D. (1998) State policy and pasture degradation in Inner Asia. *Science* 281, 1147–1148.

Soares, O. (1997) *Rondônia Produz*. Senado Federal PTB, Brasilia DF, Brazil.

SODECAO (1995) *Bilan du processus de prise en charge par les groupements de producteurs de la commercialisation du cacao*. Direction de Programmes Agricoles, Yaoundé, Cameroon.

Somda, Z.C. (1995) Feed factors affecting nutrient excretion by ruminants and the fate of nutrients when applied to soil. In: Powell, J.M., Fernandez-Rivera, S., Williams, T.O. and Renard, C. (eds) *Livestock and Sustainable Nutrient Cycling in Mixed Farming Systems of sub-Saharan Africa*. Vol. II: *Technical Papers. Proceedings of an International Conference held in Addis Ababa, Ethiopia, 22–26 November 1993*. International Livestock Centre for Africa (ILCA), Addis Ababa, Ethiopia.

Southgate, D. (1990) The causes of land degradation along spontaneously expanding agricultural frontier in the Third World. *Land Economics* 66, 93–101.

Southgate, D. (1994) Tropical deforestation and agricultural development in Latin America. In: Brown, K. and Pearce, D. (eds) *The Causes of Tropical Deforestation: the Economic and Statistical Analysis of Factors Giving Rise to the Loss of Tropical Forests*. University College London Press, London.

Southgate, D. (1998) *Tropical Forest Conservation: an Economic Assessment of the Alternatives in Latin America*. Oxford University Press, Oxford.

Southgate, D., Bravo-Ureta, B. and Whitaker, M. (1999) Public policy, agricultural development, and the fate of forests in South America: Ecuador's experience versus Chile's. In: *CIFOR Workshop on Technological Change and Deforestation, Turrialba, Costa Rica, 11–13 March 1999*. Centre for International Forestry Research (CIFOR), Bogor, Indonesia.

Spencer, D.S.C. (1985) A research strategy to develop appropriate technologies for small farm development in sub-Saharan. In: *Appropriate Technology in Semi Arid West Africa*. Purdue University Press, West Lafayette, Indiana.

Spencer, D.S.C. and Badiane, O. (1995) Agriculture and economic recovery in African countries. In: Peters, G.H. and Hedley, D.D. (eds) *Agricultural Competitiveness: Market Forces and Policy Choice – Proceedings of the Twenty-Second International Conference of Agricultural Economists*, Hants, England. Dartmouth Publishing, Dartmouth, UK.

Spencer, J.E. (1966) *Shifting Cultivation in Southeast Asia*. University of California Press, Berkeley, California.

Sperling, L. and Scheidegger, U. (1995) *Participatory Selection of Beans in Rwanda: Results, Methods and Institutional Issues*. Gatekeeper Series No. 51, International Institute for Environment and Development, London.

Spiertz, J., van Keulen, H. and van der Pouw, B. (1994) Land use options and environmental goals. In: Fresco, L.O., Stroosnijder, L., Bouma, J. and van Keulen, H. (eds) *The Future of the Land: Mobilizing and Integrating Knowledge for Land Use Options*. John Wiley & Sons, Chichester, UK, pp. 57–64.

Staal, S.J., Chege, L., Kenyanjui, M., Kimari, A., Lukuyu, B., Njubi, D., Owango, M., Tanner, J., Thorpe, W. and Wambugu, M. (1997) *Characterisation of Dairy Systems Supplying the Nairobi Milk Market: a Pilot Survey in Kiambu District for the Identification of Target Groups of Producers*. KARI/MoA/ILRI Collaborative Research Project Report, Ministry of Agriculture, Kenya Agricultural Research Institute, and the International Livestock Research Institute, Nairobi, Kenya.

Staal, S.J., Kenyanjui, M., Owango, M., Muriuki, H., Lukuyu, B., Musembi, F., Bwana, O., Gichungu, V., Omore, A., Njubi, D., Baltenweck, I., Njoroge, L. and Thorpe, W. (1999) *Dairy Systems Characterisation of the Nairobi Milk Shed: Application of Spatial and Household Analysis*. KARI/MoA/ILRI Collaborative Research Project Report, Ministry of Agriculture, Kenya Agricultural Research Institute, and the International Livestock Research Institute, Nairobi, Kenya.

Stedman-Edwards, P. (1998) *Root Causes of Biodiversity Loss: an Analytical Approach*. World Wide Fund for Nature, Macroeconomics for Sustainable Development Program, Washington, DC.

Steenhuijsen Piters, B. (1995) Diversity of fields and farmers. PhD thesis, Wageningen Agricultural University, Wageningen, the Netherlands.

Steinfeld, H., de Haan, C. and Blackburn, H. (1997) *Livestock and the Environment – Issues and Options*. Wrenmedia, Suffolk, UK.

Stern, C. (1976) *Las migraciones rural–urbanas*. Centro de Estudios Sociológicos, Colegio de México, Mexico.

Stern, D.I. (1998) Progress on the environmental Kuznets curve? *Environment and Development Economics* 3(2), 173–196.

Stern, D.I., Common, M.S. and Barbier, E.B. (1996) Economic growth and environmental degradation: the environmental Kuznets curve and sustainable development. *World Development* 24(7), 1151–1160.

Stewart, D. (1994) *After the Trees: Living on the Transamazonian Highway*. University of Texas Press, Austin, Texas.

Stomph, T.J., Fresco, L.O. and van Keulen, H. (1995) Land use systems evaluation: concepts and methodology. *Agricultural Systems* 44, 243–255.

Stone, B. (1986) Chinese fertilizer application in the 1980s and 1990s: issues of growth, balances, allocation, efficiency and response. In: *China's Economy Looks Toward the Year 2000*, Vol. 1. Congress of the United States, Washington, DC, pp. 453–493.

Stone, R. (1992) Researchers score victory over pesticides – and pests – in Asia. *Science* 256, 5057.

Stone, R.D. (1991) *Wildlands and Human Needs: Reports from the Field.* World Wildlife Fund (WWF), Washington, DC.

Stonich, M. (1989) The dynamics of social processes and environmental destruction: a Central American case study. *Population and Development Review* 15, 269–297.

Stoorvogel, J.J., Schipper, R.A. and Jansen, D.M. (1995) USTED: a methodology for quantitative analysis of land use scenarios. *Netherlands Journal of Agricultural Science* 43, 5–18.

Suri, V. and Chapman, D. (1998) Economic growth, trade and energy: implications for the environmental Kuznets curve. *Ecological Economics* 25(2), 195–208.

Suyanto, S. and Otsuka, K. (1998) From deforestation to development of agroforests in customary land tenure areas of Sumatra. Mimeo, Tokyo Metropolitan University, Tokyo.

Suyanto, S., Tomich, T. and Otsuka, K. (1998a) Land tenure and farm management efficiency: the case of paddy and cinnamon production in customary land areas of Sumatra. 1998. Paper presented at Workshop on Land Tenure and the Management of Land and Trees: Community and Household Case Studies from Asia and Africa, Tokyo Metropolitan University, 1–3 July 1998, Tokyo.

Suyanto, S., Tomich, T. and Otsuka, K. (1998b) Land tenure and farm management efficiency: the case of smallholder rubber production in customary land areas of Sumatra. Mimeo, Tokyo Metropolitan University, Tokyo.

Svendsen, M. and Changming, L. (1990) Innovations in irrigation management and development in Hunnan Province. *Irrigation and Drainage Systems* 4, 195–214.

Swanson, T. (1997) *Global Action for Biodiversity: an International Framework for Implementing the Convention on Biological Diversity*. Earthscan, London.

Swift, M. (ed.) (1998) *Belowground Biodiversity Assessment: Working Group Summary Report.* ASB Phase II Final Report, Tropical Soil Biology and Fertility Programme (TSBF), Nairobi, Kenya.

Tachibana, T., Nguyen, T.M. and Otsuka, K. (1998) From deforestation to reforestation through tenure reforms: the case of the northern Hill Region of Vietnam. Paper presented at Workshop on Land Tenure and the Management of Land and Trees: Community and Household Case Studies from Asia and Africa, Tokyo Metropolitan University, 1–3 July 1998, Tokyo.

Tandon, H.L.S. (1987) *Phosphorus Research and Agricultural Production in In*dia. Fertilizer Development and Consumption Organization, New Delhi, India.

Taylor, J.E. and Adelman, I. (1996) *Village Economies: the Design, Estimation, and Use of Villagewide Economic Models.* Cambridge University Press, Cambridge.

Teitelbaum, M.S. (1975) Relevance of demographic transition theory for developing countries. *Science* 188, 420–425.

Templeton, S. and Scherr, S.J. (1999) Effects of demographic and related microeconomic change on land quality in hills and mountains of developing countries. *World Development* 27(6), 903–918.

Thapa, G., Koirala, G. and Otsuka, K. (1998) Population pressure, natural resource management, and agricultural intensification: the case of hill villages in Nepal, 1982 and 1997. Paper presented at Workshop on Land Tenure and the Management of Land and Trees: Community and Household Case Studies from Asia and Africa, Tokyo Metropolitan University, 1–3 July 1998, Tokyo.

Thirtle, C.G. (1985) Accounting for increasing land–labour ratios in developed country agriculture. *Journal of Agricultural Economics* 36(1), 1–14.

Tiffen, M. (1995) Population density, economic growth and societies in transition: Boserup reconsidered in a Kenyan case-study. *Development and Change* 26, 31–65.

Tiffen, M. and Mortimore, M. (1994) *Land Resource Management in Machakos District, Kenya 1930–1990.* World Bank Environment Paper 5, World Bank, Washington, DC.

Tiffen, M., Mortimore, M. and Gichuki, F. (1994) *More People, Less Erosion: Environmental Recovery in Kenya.* John Wiley & Sons, New York.

Tilman, D. and Downing, J.A. (1994) Biodiversity and stability in grasslands. *Nature* 367, 363–365.

Tilman, D., Knops, J. and Wedin, D. (1997) The influence of functional diversity and composition on ecosystem processes. *Science* 277, 1300–1302.

Timmer, P. (1989) The agricultural transformation. In: Chenery, H.B. and Srinivasan, T.N. (eds) *Handbook of Development Economics*, Vol. 1. Elsevier Science, Amsterdam.

Timmer, P., Falcon, W. and Pearson, S. (1983) *Food Policy Analysis.* Johns Hopkins University Press, Baltimore, Maryland.

Tinker, P.B., Ingram, J.S.I. and Struwe, S. (1994) Effects of slash-and-burn agriculture and deforestation on climate change. In: Sanchez, P.A. and van Houeten, H. (eds) *Alternatives to Slash-and-burn Agriculture. Symposium ID-6, 15th International Soil Science Congress, Acapulco, Mexico and Nairobi, Kenya.* International Centre for Research in Agroforestry (ICRAF), Paris, and International Society of Soil Science, Vienna, Austria.

Tinker, P.B., Ingram, J.S.I. and Struwe, S. (1996) Effects of slash-and-burn agriculture and deforestation on climate change. *Agriculture, Ecosystems and Environment* 58(1), 13–22.

Tolba, M.K., El-Kholy, O.A., El-Hinnawi, E., Holdgate, M.W., McMichael, D.F. and Munn, R.E. (eds) (1992) *The World Environment 1972–1992: Two Decades of Challenge.* United Nations Environment Programme, Chapman & Hall, London.

Tomich, T.P. (1991) Smallholder rubber development in Indonesia. In: Perkins, D.H. and Roemer, M. (eds) *Reforming Economic Systems in Developing Countries.* Harvard University Press, Cambridge, Massachusetts, pp. 249–270.

Tomich, T.P. and Mawardi, M.S. (1995) Evolution of palm oil trade policy in Indonesia, 1978–1991. *Elaeis: Journal of the Palm Oil Research Institute of Malaysia* 7(1), 87–102.

Tomich, T.P. and van Noordwijk, M. (1995) What drives deforestation in Sumatra? Paper presented at the ASB Indonesia and ICRAF Southeast Asia Regional Symposium on Montane Mainland Southeast Asia in Transition, 13–16 November 1995, Chiang Mai, Thailand.

Tomich, T.P., Kilby, P. and Johnston, B.F. (1995) *Transforming Agrarian Economies. Opportunities Seized, Opportunities Missed.* Cornell University Press, Ithaca, New York.

Tomich, T.P., Kuusipalo, J., Menz, K. and Byron, N. (1997) *Imperata* economics and policy. *Agroforestry Systems* 36(1–3), 233–261.

Tomich, T.P., van Noordwijk, M., Budidarsono, S., Gillison, A., Kusumanto, T., Murdiyarso, D., Stolle, F. and Fagi, A.M. (eds) (1998a) *Alternatives to Slash-and-burn in Indonesia: Summary Report and Synthesis of Phase II.* ASB-Indonesia Report No. 8, ASB-Indonesia and International Centre for Research in Agroforestry (ICRAF) Southeast Asia, Bogor, Indonesia.

Tomich, T.P., van Noordwijk, M., Vosti, S.A. and Witcover, J. (1998b) Agricultural development with rainforest conservation methods for seeking best bet alternatives to slash-and-burn, with applications to Brazil and Indonesia. *Agricultural Economics* 19, 159–174.

Tourte, R., Pocthier, G., Ramond, C., Monnier, J., Nicou, R., Poulain, J.F., Hamon, R. and Charreau, C. (1971) Thèmes légers – thèmes lourds – systèmes intensifs: vies différentes ouvertes au développement agricole du Sénégal. *Agronomie Tropicale* 26, 632–645.

Tribe, D. (1994) *Feeding and Greening the World: the Role of International Agricultural Research*. CAB International, Wallingford, UK.

Tripp, R., Anandajayasekeram, P., Byerlee, D. and Harrington, L. (1990) Farming systems research revisited. In: Eicher, C.K. and Staatz, J.M. (eds) *Agricultural Development in the Third World*. Johns Hopkins University Press, Baltimore, Maryland, pp. 384–399.

Turner, B.L. (ed.) (1990) *The Earth as Transformed by Human Action*. Cambridge University Press, Cambridge, UK.

Turner, B.L., Hyden, G. and Kates, R. (1993) *Population Growth and Agricultural Change in Africa*. University Press of Florida, Gainesville, Florida.

Turner, B.L., II and Ali, A.M.S. (1996) Induced intensification: agricultural change in Bangladesh with implications for Malthus and Boserup. *Proceedings of the National Academy of Sciences USA* 93, 14984–14991.

Turner, B.L., II, Hanham, R.Q. and Portararo, A.V. (1977) Population pressure and agricultural intensity. *Annals of the Association of American Geographers* 67(3), 384–396.

Turral, H. (1995) *Devolution of Management in Public Irrigation Systems: cost shedding, empowerment, and performance. A review*. Working Paper 80, Overseas Development Institute, Regent's College, London, UK.

Udry, C. (1990) Credit markets in northern Nigeria: credit as insurance in a rural economy. *World Bank Economic Review* 4, 251–270.

UN (United Nations) (1997) *World Population Prospects: the 1996 Revision*. Population Divison, Department of Economic and Social Affairs of the United Nations Secretariat, New York.

UN (United Nations) (1998) *World Population Projections to 2150*. Population Division, Department of Economic and Social Affairs, United Nations, New York.

UN (United Nations) (1999) *World Population Prospects: the 1998 Revision*. Population Divison, Department of Economic and Social Affairs of the United Nations Secretariat, New York.

UNEP (United Nations Environment Programme) (1988) *People, Parks and Wildlife: Guidelines for Public Participation in Wildlife Conservation*. UNEP, Nairobi, Kenya.

UNEP (United Nations Environment Programme) (1993) *Convention on Biological Diversity*. United Nations Environment Programme, New York.

UNESCO (United Nations Educational, Scientific and Cultural Organization) (1987) *A Practical Guide to MAB (Man and the Biosphere)*. UNESCO, Paris.

UNICEF (United Nations Children's Fund) (1990) *Strategy for Improved Nutrition of Children and Women in Developing Countries*. UNICEF, New York.

United Nations Administrative Committee on Coordination/Sub-Committee on Nutrition (1992) *Second Report on the World Nutrition Situation*, Vol. 1: *Global and Regional Results*. United Nations, Geneva.

Upadhyaya, H.K. and Otsuka, K. (1998) Community forest management in the Hill Region of Nepal: rules and practices in firewood collection. Paper presented at Workshop on Land Tenure and the Management of Land and Trees: Community and Household Case Studies from Asia and Africa, Tokyo Metropolitan University, 1–3 July 1998, Tokyo.

Uphoff, N. (1986) *Local Institutional Development: an Analytical Sourcebook with Cases*. Kumarian Press, West Hartford, Connecticut.

Uphoff, N. (1993) Grassroots organizations and NGOs in rural development: opportunities with diminishing states and expanding markets. *World Development* 21(4), 607–622.

Uphoff, N. (1994) Report on Integrated Watershed Analysis and Management Workshop, Dominican Republic practicum, Nizao Watershed, 8–15 October 1994. Unpublished report, Cornell International Institute for Food, Agriculture and Development, Ithaca, New York.

Uphoff, N. (1996a) *Highlights from Field Visits to Nilwala and Hurulawewa Watersheds*. Report for International Irrigation Management Institute, Colombo, Sri Lanka.

Uphoff, N. (1996b) *Learning from Gal Oya: Possibilities for Participatory Development and Post-Newtonian Social Science*. Intermediate Technology Publications, London.

Uphoff, N. (1996c) Why NGOs are not a third sector: a sectoral analysis with some thoughts on accountability, sustainability and evaluation. In: Edwards, M. and Hulme, D. (eds) *Beyond the Magic Bullet: NGO Performance in the Post-Cold War World*. Kumarian Press, West Hartford, Connecticut, pp. 23–39.

Uphoff, N. (1998a) Community-based natural resource management: connecting the micro and macro, and people with their environment. Keynote paper for International Conference on Community-Based Natural Resource Management, Economic Development Institute, World Bank, Washington, DC, 10–14 May 1998.

Uphoff, N. (1998b) Report on visit to Domi village No. 1, Greater Afram Plains, Ghana, March 15, 1998. Unpublished report, Cornell International Institute for Food, Agriculture and Development, Ithaca, New York.

Uphoff, N. (2000a) Agroecological implications of the system of rice intensification (SRI) in Madagascar. *Environment, Development and Sustainability* 1(3).

Uphoff, N. (2000b) Understanding social capital: learning from the experience and analysis of participation. In: Serageldin, I. and Dasgupta, P. (eds) *Social Capital: a Multifaceted Perspective*. World Bank, Washington, DC, pp. 215–249.

Uphoff, N. and Langholz, J. (1998) Incentives for avoiding the tragedy of the commons. *Environmental Conservation* 25(3), 251–261.

USAID (US Agency for International Development) (1994) *Biodiversity Conservation and Sustainable Use: USAID Project Profiles*. Environment and Natural Resources Information Center (ENRIC), Datex, Arlington, Virginia.

USDA (US Department of Agriculture) (1994) *Agricultural Resources and Environment Indicators*. Agricultural Handbook No. 705, USDA, Economic Research Service, Natural Resources and Environment Division, Washington, DC.

Utting, P. (1993) *Trees, People, and Power: Social Dimensions of Deforestation and Forest Protection in Central America*. UNRISD Discussion Paper, UNRISD, London.

Vaidyanathan, A. (1991) *Integrated Watershed Development: Some Major Issues*. Founder's Day Lecture, Society for Promotion of Wasteland Development, New Delhi.

Valenzuela de Pisano, I. (1996) *Agricultura y Bosque en Guatemala*. UNRISD, WWF, Universidad Rafael Landivar, Guatemala City, Guatemala.

Vandergeest, P. (1996) Property rights in protected areas: obstacles to community involvement as a solution in Thailand. *Environmental Conservation* 23(3), 259–268.

Vandermissen, A. and Symoens, C. (1997) *The Multifactoral Contribution of Livestock to Development in Asia.* Animal Production and Health, European Commission, Brussels.

van der Pol, F. (1992) *Soil Mining: the Unseen Contributor to Farm Income in Southern Mali.* Bulletin No. 325, Royal Tropical Institute, Amsterdam.

van der Valk, A.L. (1998) Elasticities ¡Claro que sí!: an analysis of the domestic Costa Rican food demand. MSc thesis, Department of Marketing and Marketing Research, Wageningen Agricultural University, Wageningen, the Netherlands.

van Keulen, H. and Breman, H. (1990) Agricultural development in the Western African Sahelian region: a cure against land hunger? *Agriculture, Ecosystems and Environment* 32, 177–197.

van Kooten, K., Weisenel, M.P. and Chinthammit, F. (1990) Valuing trade-offs between net return and stewardship practices: the case of soil conservation in Saskatchewan. *American Journal of Agricultural Economics* 73, 104–113.

van Noordwijk, M., Tomich, T.P., Winahyu, R., Murdiyarso, D., Suyanto, S., Partoharjono, S. and Fagi, A.M. (eds) (1995) *Alternatives to Slash-and-burn in Indonesia: Summary Report of Phase 1.* ASB-Indonesia Report No. 4, ASB-Indonesia Consortium and International Centre for Research in Agroforestry (ICRAF), Bogor, Indonesia.

van Noordwijk, M., Tomich, T.P., de Foresta, H. and Michon, G. (1997) To segregate or to integrate? The question of balance between production and biodiversity conservation in complex agroforestry systems. *Agroforestry Today* 9(1), 6–9.

Van Pelt, M.J.F. (1993) *Ecological Sustainability and Project Appraisal.* Aldershot, Avebury, UK.

Van Rheenen, T. (1995) Farm household level optimal resource allocation: an explorative study in the limestone area of East Java. PhD thesis, Wageningen Agricultural University, Wageningen, the Netherlands.

van Staveren, J.M. (1980) Stages in regional planning. In: van Staveren, J.M. and van Dusseldorp, D. (eds) *Framework for Regional Planning in Developing Countries.* ILRI, Wageningen, the Netherlands, pp. 50–61.

van Zyl, J., Binswanger, H. and Thirtle, C. (1995) *The Relationship between Farm Size and Efficiency in South African Agriculture.* Policy Research Working Paper 1548, World Bank, Agriculture and Natural Resources Department, Washington, DC.

Vavilov, M.I. (1997) *Five Continents.* International Plant Genetic Resources Institute, Rome, Italy.

Veeneklaas, F.R. (1990) Dovetailing technical and economic analysis. PhD thesis, Erasmus University, Rotterdam, the Netherlands.

Veeneklaas, F.R., Van Keulen, H., Cissé, S., Gosseye, P. and van Duivenbooden, N. (1994) Competing for limited resources: options for land use in the Fifth Region of Mali. In: Fresco, L.O., Stroosnijder, L., Bouma, J. and van Keulen, H. (eds) *The Future of the Land: Mobilizing and Integrating Knowledge for Land Use Options.* John Wiley & Sons, Chichester, pp. 227–247.

Vermillion, D.L. (ed.) (1996) *The Privatization and Self-Management of Irrigation, Final Report.* Submitted to the Deutsche Geellschaft für Technische Zusammenarbeit (GTZ), by the International Irrigation Management Institute, Colombo, Sri Lanka.

Viyas, V.S. (1983) Asian agriculture: achievements and challenges. *Asian Development Review* 1(1), 27–44.

Vosti, S.A. and Reardon, T. (eds) (1997) *Sustainability, Growth, and Poverty Alleviation: a Policy and Agroecological Perspective.* Johns Hopkins University Press, Baltimore, Maryland.

Vosti, S.A. and Witcover, J. (1996) Slash-and-burn agriculture: household perspectives. *Agriculture, Ecosystems and Environment* 58, 23–38.

Vosti, S.A., Witcover, J. and Carpentier, C.L. (1998a) *Convenio de cooperación técnica regional no reembolsable, No. ATN/SF-4827-RG, Programa 1994 de tecnología agropecuaria regional en América Latina y El Caribe (Arresting Deforestation and Resource Degradation in the Forest Margins of the Humid Tropics: Policy, Technology, and Institutional Options for Western Brazil).* Inter-American Development Bank (IDB), Washington, DC.

Vosti, S.A., Witcover, J., Gockowski, J., Tomich, T., Carpentier, C.L., Faminow, M., Oliveira, S. and Diaw, C. (1998b) Socioeconomic issues linked to best bet land use systems. Mimeo, Environment and Production Technology Division, International Food Policy Research Institute (IFPRI), Washington, DC.

Vosti, S.A., Witcover, J., Oliveira, S. and Faminow, M. (1998c) Policy issues in agroforestry: technology adoption and regional integration in the western Brazilian Amazon. *Agroforestry Systems* 38, 195–222.

Vosti, S.A., Carpentier, C.L., Witcover, J. and Valentim, J. (2000a) Intensified small-scale cattle production systems in the western Brazilian Amazon: will they be adopted and can they save the forest? In: Angelsen, A. and Kaimowitz, D. (eds) *Agricultural Technologies and Tropical Deforestation.* CAB International, Wallingford, UK.

Vosti, S.A., Witcover, J., Gockowski, J., Tomich, T.P., Carpentier, C.L., Faminow, M., Oliveira, S. and Diaw, C. (2000b) *Socioeconomic Issues Linked to Best Bet Land Use Systems.* Embrapa Technical Report.

Waage, J.K. (1991) Biodiversity as a resource for biological control. In: Hawksworth, D.L. (ed.) *The Biodiversity of Micro-organisms and Invertebrates: Its Role in Sustainable Agriculture.* CAB International, Wallingford, UK, pp. 149–163.

Wade, R. (1988) *Village Republics: Economic Conditions for Collective Action in South India.* Cambridge University Press, Cambridge.

Wagenet. R.J. and Hutson, J.L. (1989) *Leaching Estimation and Chemistry Model: a Process-based Model of Water and Solute Movement, Transformations, Plant Uptake and Chemical Reactions in the Unsaturated Zone.* Water Resources Institute, Cornell University, Ithaca, New York.

Waggoner, P.E. (1994) *How Much Land can Ten Billion People Spare for Nature? Task Force Report.* Council for Agricultural Science and Technology No. 121, Report 0194–4088, Council for Agricultural Science and Technology, Ames, Iowa.

Waggoner, P.E., Ausubel, J.H. and Wernick, I.K. (1996) Lightening the tread of population on the land: American examples. *Population and Development Review* 22(3), 531–545.

Walker, B.H. (1988) Autecology, synecology, climate and livestock as agents of rangelands dynamics. *Australian Range Journal* 10, 69–75.

Walker, B.H. (1993) Rangeland ecology: understanding and managing change. *Ambio* 22(2–3), 80–87.

Walker, B.H. and Noy-Meir I. (1982) Aspects of the stability and resilience of savanna ecosystems. In: Huntley, B.J. and Walker, B.H. (eds) *Ecology of Tropical Savannas.* Springer, Berlin, pp. 577–590.

Walker, T. and Ryan, J. (1990) *Village and Household Economies in India's Semi-Arid Tropics.* Johns Hopkins University Press, Baltimore, Maryland.

Wallace, A. and Naughton-Treves, L. (1998) Belize: Rio Bravo Conservation and Management Area. In: Brandon, K., Redford, K. and Sanderson, S. (eds) *Parks in Peril: People, Politics, and Protected Areas.* Island Press, Covelo, California.

Wallis, J.A.N. (1997) *Intensified Systems of Farming in the Tropics and Subtropics.* World Bank Discussion Paper No. 364, World Bank, Washington, DC.

WCED (World Commission on Environment and Development) (1987a) *Food 2000: Global Policies for Sustainable Agriculture*. Zed Books, London.

WCED (World Commission on Environment and Development) (1987b) *Our Common Future*. Oxford University Press, New York.

WCMC (World Conservation Monitoring Center) (1992) *Global Biodiversity: Status of the Earth's Living Resources*. Chapman & Hall, London.

WCMC (World Conservation Monitoring Center) (1996) *Tropical Moist Forests and Protected Areas*. WCMC Digital Files (CD ROM), London.

Weber, J. (1977) *Structures agraires et évolution de milieux ruraux: le cas de la region cacaoyère du Centre-Sud Cameroon*. ORSTOM, Paris.

Weight, D. and Kelly, V. (1998) *Restoring Soil Fertility in Sub-Saharan Africa: Technical and Economic Issues*. Policy Synthesis No. 37, Food Security II Project, Department of Agricultural Economics, Michigan State University, East Lansing, Michigan.

Weischet, W. and Caviedes, C.N. (1993) *The Persisting Ecological Constraints of Tropical Agriculture*. Longman Scientific and Technical Series, Longman Group UK, Essex, UK.

Weise, S. (ed.) (1998) *Agronomic Sustainability Working Group Summary Report*. ASB Phase II Final Report, International Institute for Tropical Agriculture, Yaoundé, Cameroon.

Weise, S.F. (compiler) (2000a) *Agronomic Sustainability: ASB Phase II Working Group Report on Methodology*. IITA, Yaoundé, Cameroon.

Weise, S.F. (compiler) (2000b) *Agronomic Sustainability: ASB Phase II Working Group Report: a First Attempt at Cross-site Comparisons*. IITA, Yaoundé, Cameroon.

Wells, M.P. (1997) *Economic Perspectives on Nature Tourism, Conservation and Development*. World Bank Environment Department Papers No. 55, World Bank, Washington, DC.

Wells, M.P. and Brandon, K. (1992) *People and Parks: Linking Protected Area Management with Local Communities*. World Bank, World Wildlife Fund, and US Agency for International Development, Washington, DC.

West, P.C. and Brechin, S.R. (eds) (1991) *Resident Peoples and National Parks: Social Dilemmas and Strategies in International Conservation*. University of Arizona Press, Tucson, Arizona.

Western, D. and Pearl, M. (eds) (1989) *Conservation for the Twenty-first Century*. Oxford University Press, Oxford.

Western, D. and Wright, R.M. (eds) (1994) *Natural Connections: Perspectives in Community-based Conservation*. Island Press, Washington, DC.

Westphal, E. (1981) *L'Agriculture autochtone au Cameroun: les techniques culturales, les sequences de culture, les plantes alimentaires et leur consommation*. Miscellaneous Papers 20, Landbouwhogeschool, Wageningen, the Netherlands.

Wiersum, K.F. (1986) The effect of intensification of shifting cultivation in Africa on stabilizing land-use and forest conservation. *Netherlands Journal of Agricultural Science* 34(4), 485–488.

Wijayaratna, C.M. (1994) Shared control of natural resources (SCOR): an integrated watershed approach to optimize production and protection. *Sri Lanka Journal of Agricultural Economics* 2, 1.

Wijayaratna, C.M. (1997) Shared management of watershed resources. In: Lawrence, J. (ed.) *Partners in the Harvest*. Australian Centre for International Agricultural Research, Canberra, pp. 37–45.

Williams, R.G. (1986) *Export Agriculture and the Crisis in Central America*. University of North Carolina Press, Chapel Hill.

Williamson, M. (1996) *Biological Invasions*. Chapman & Hall, London.

Wilson, E.O. (1984) *Biophilia: the Human Bond with Other Species.* Harvard University Press, Cambridge, Massachusetts.

Wilson, E.O. (1993) Biophilia and the conservation ethic. In: Kellert, S.T. and Wilson, E.O. (eds) *The Biophilia Hypothesis.* Island Press, Washington, DC, and Shearwater Books, Covelo, California.

Winrock International Institute (1992) *Assessment of Animal Agriculture in Sub-Saharan Africa.* Winrock, Morrilton, Arkansas.

Wint, W. and Bourn, D. (1994) *Anthropogenic and Environmental Correlates of Livestock Distribution in Sub-Saharan Africa.* Research Report, Overseas Development Administration (ODA), London.

Witcover, J., Vosti, S.A., de Almedia Barbosa, F. R., Batista, J., Beatriz, V., Böklin, G., de França, S.B., Castilla, C., Fujisaka, S., Gigueiredo Gallo, S.L., Garcia, J.H., Gomes Cabral, W., Hurtado, L., Leite, A., Nogueira Leite, F.M., Souza, D.A., de Souza Marinho, J.T., Rocha, K. and Rosa Neto, C. (1996) *Alternatives to Slash-and-burn Agriculture (ASB): a Characterization of Brazilian Benchmark Sites of Pedro Peixoto and Theobroma, August/September 1994.* Environment and Production Technology Division MP-8 Working Paper No. 3, International Food Policy Research Institute (IFPRI), Washington, DC.

Wolf, J., Breman, H. and Van Keulen, H. (1991) *Bio-economic Capability of West African Drylands.* CABO-DLO Report 147, Wageningen, the Netherlands.

Wood, H., McDaniel, M. and Warner, K. (eds) (1995) *Community Development and Conservation of Forest Biodiversity Through Community Forestry.* Regional Community Forestry Training Centre, Kasetsart University, Bangkok, Thailand.

Woomer, P.L., Palm, C.A., Qureshi, J.N. and Kotto-Same, J. (1997) Carbon sequestration and organic resource management in African smallholder agriculture. In: Lal, R., Kimble, J.M., Follett, R.A. and Steward, B.A. (eds) *Management of Carbon Sequestration in Soil.* CRC Press, Boca Raton, Florida, pp. 153–173.

World Bank (1988) *Rural Development: World Bank Experience, 1965–86.* Operations Evaluation Department, World Bank, Washington, DC.

World Bank (1990) *World Bank Development Report 1990.* World Bank, Washington, DC.

World Bank (1992) *World Development Report 1992: Development and the Environment.* Oxford University Press, New York.

World Bank (1995a) *Guatemala: Land Tenure and Natural Resources Management.* World Bank, Washington, DC.

World Bank (1995b) *Mainstreaming Biodiversity in Development.* Environment Department Working Paper No. 29, World Bank, Washington, DC.

World Bank (1995c) *Recapitalization of Soil Productivity in Sub-Saharan Africa.* Discussion Paper, World Bank, Washington, DC.

World Bank (1995d) *Towards Environmentally Sustainable Development in Sub-Saharan Africa: a World Bank Agenda.* Report No. 15111-AFR, World Bank, Washington, DC.

World Bank (1996a) *African Development Indicators.* World Bank, Washington, DC.

World Bank (1996b) *From Plan to Market: World Development Report 1996.* Oxford University Press, Oxford.

World Bank (1996c) *Natural Resource Degradation in Sub-Saharan Africa: Restoration of Soil Fertility.* Africa Region, World Bank, Washington, DC.

World Bank (1997) *World Development Report 1997: the State in a Changing World.* Oxford University Press, Oxford.

World Bank (1998a) *Biodiversity in World Bank Projects: a Portfolio Review.* Environment Department Working Paper No. 59, World Bank, Washington, DC.

World Bank (1998b) *World Development Indicators*. World Bank, Washington, DC.

Worrest, R.C. and Grant, L.D. (1989) Effects of ultraviolet-B radiation on terrestrial plants and marine organisms. In: Russell-Jones, R. and Wigley, T. (eds) *Ozone Depletion: Health and Environmental Consequences*. John Wiley & Sons, Chichester, UK.

WRI (World Resources Institute) (1992) *World Resources 1992–1993*. Oxford University Press, Oxford.

WRI (World Resources Institute) (1994) *World Resources 1994–1995*. Oxford University Press, Oxford.

WRI (World Resources Institute) (1996) *World Resources, 1996–1997: a Guide to the Global Environment*. Oxford University Press, New York.

WRI (World Resources Institute) (1997) *World Resources 1996–1997: the Urban Environment*. Oxford University Press, New York.

WRI (World Resources Institute) (1998) *World Resources 1998–1999*. Oxford University Press, New York.

WRR (1992) *Ground for Choices: Four Perspectives for the Rural Areas in the European Community*. Report for the Netherlands Scientific Council for Government Policy, SDU, The Hague.

WWF (World Wide Fund for Nature) (1997) *Root Causes of Biodiversity Loss Project*. Macroeconomics for Sustainable Development Programme, WWF, Washington, DC.

Young, R.A. and Haveman, R.H. (1985) Economic of water resources: a survey. In: Kneese, A.V. and Sweeney, J.L. (eds) *Handbook of Natural Resource and Energy Economics*, Vol. II. Elsevier Science Publishers, Amsterdam.

Zeller, M., Schreider, G., von Braun, J. and Heidhues, G. (1997) *Rural Finance for Food Security for the Poor: Implications for Research and Policy*. Food Policy Review No. 4, International Food Policy Research Institute (IFPRI), Washington, DC.

Zimmerer, K.S. (1993) Soil erosion and labor shortages in the Andes with special reference to Bolivia, 1953–91: implications for 'conservation-with-development'. *World Development* 21, 1659–1675.

Index

Note: page numbers in *italics* refer to figures and tables

Acacia senegal 62–63
Africa
  cotton schemes 372
  economic crisis 41
  food security crisis 325
  governance 326
  infrastructure 326
  labour-enhancing technologies 366–367
  poverty 325
  soil fertility depletion 326–328, *329*, 330–333
aggregate functions in tradeoffs model 143
aggregation 136
agricultural development, broad-based for less-favoured lands 164
agricultural growth
  environmental threats 73
  extensive 58
  India *154–155*, 156–157
    spillover effects 162
  intensive 58
agricultural intensification 3
  Amazon region 249, *250*, 251
  Andes region 135–136
  bioeconomic modelling 131
  Boserupian perspective 4–5
  capital-deficient 367, 368
  capital-led 367, 368, 369
  carbon stocks 203, 258–259
  Central Africa 201–202
  Central African rainforest 213–216
  complementarities 451, 453
  Congo Basin 199–200
  costs of change 71
  decisions 77–81
  definition 356
  deforestation 8–9, 260
  efficiency 79
  endogenous 4–5
  environment 4–5, 11
  environmental costs 453
  environmental indicators 7–8
  failure to achieve goals 454
  farm-household decision-making 75, 82
  ICDPs 10
  institutional barriers to adoption 229–230
  insurance mechanisms 71
  Latin America 49, 50–53
  less-favoured lands 164
  limiting factors 14–15, 54–55
  livestock–people link 354–355
  model intractability 455
  obstacles 262–263
  plant biodiversity 207–208
  policies for sustainable 462–464
  policy barriers to adoption 229–230
  policy-led strategies 5
  population growth 51–52
  principal components analysis 211–213
  productivity decline 15
  resource pricing 71
  strategies 5, 451
  sustainable
    in Mali 126–130
    in sub-Saharan Africa 118
  types 261–262
  working definition 246–249
  see also sustainable agricultural intensification; unsustainable agricultural intensification

agricultural production
complementary economic growth and environmental goals 458
tradeoffs of systems 140
agricultural productivity
incentives 14
population growth 36
agricultural research 403–404
agricultural risk 71
agricultural transformation 452
agriculture
discrimination against in developing countries 79–80
policy bias against 115–116
see also intensive agriculture
agroecological conditions 457, 460–461
agroecological information 125–126
agroecological input–output coefficients 127
agroecological zones 136–137, 460–461
agroecosystems
agricultural intensification 58
biodiversity 70
loss 59–65
genetic diversity 71
price change impact 70
agroforestry 459
carbon stocks 211
deforestation deflection 342–343
intensiveness 297–298
management and communal land tenure 294–298
natural regeneration 460
net present value of agroforestry 237–238
agroindustrial development in Rwanda 377–378
agronomic sustainability
global environmental concerns 232
plot-level 232
Sumatra 226
agronomics 434–435
Alternatives to Slash-and-Burn (ASB)
agricultural intensification promotion 434
Brazil 246
carbon data 202
characterization survey 204–205
economic growth/environment tradeoffs 208–209, *210*, 211
Global Initiative 13–14
global programme 221
matrix for Sumatran forest margins *230*, 231–235
matrix framework for Amazon region 252, *253*
meta land uses 223, *224*
outcomes 453
programme 197–198
research in Indonesia 242
Amazon region
agricultural intensification
context 249, *250*, 251
measures 255–256
agriculture sustainability 49
carbon sequestration 260
institutional requirements for intensification 257–258
labour
availability 251, 260
requirements for intensification 256–257
land-use system evaluation 252, *253*, 254–259
land-use trajectory 247, *248*
markets 262–263
meta land-use practices 252, 259
oil exploitation 50–51
plant biodiversity 260
population 50–51
small-scale agriculture intensification 245–246
ammonia 25
anaemia 18
Andes region 135–136
animal manure 333, 353
shortage in Africa 368
animal waste 350
Asia
food self-sufficiency 384
intensive food systems 383–386
macroeconomic policies 396
productivity slow-down 384–386
rice
monoculture 384–386
yield decline 385
sustainable resource management 384
trade restrictions/tariffs 396
unsustainable management practices 396
wheat productivity 385

Bacillus thuringiensis (*Bt*) gene 30
banana growing in Rwanda 377–378
best-bet technologies 453, 460
biocide active ingredients (BIOA) 275, 277
biocide indicator (BIOI) 275, 277
biocides, taxing 276–278, 282
biodiversity 12
agricultural intensification 207–208
agroecosystems 70
agronomic sustainability 232
community concern 447
conservation 417
goals 453
incentives 68–70
objectives 455–456
protected areas 428
conserving 399
ecological impact 64
ecological services 59
economic incentives 421, 423–424
hot spots 428, 429
ICDPs for implementation of conservation 421

local value 59
loss 57–59, 430–431
agroecosystems 59–65
deforestation 60
desertification 60–61
external costs 59–65
habitat loss 60–61
measures 455
national level issue 431
parks in conservation 425
poverty reduction 431
preservation 438–439
protected areas 402–404
protection 434
with soil fertility replenishment 342–343
savannah rangelands 63
soil fertility depletion in Africa 327–328
value 61–64
bioeconomic land-use models in Costa Rica 267–268
biocide taxing 276–278
downward-sloping demand functions 271–272
economic aspects 271–273
labour use 282
natural forest conservation 278–279, *280*, 282
policy simulations 273–279, *280*, 281
technological innovation 282
technology improvement 275–276
upward-sloping labour supply function 272–273
wage increases 279, *280*, 281, 282
bioeconomic modelling 12–13
aggregate 126–130
aggregation level of analysis 120
agricultural intensification 131
approaches 119–126
biophysical optimization model 126–130, 133
biophysical processes 120
calibration of model 126
characteristics of models 117–118
ecoregional analysis 117–119
ecoregional development 131
empirical assessment 126–130
endogenous prices for non-tradable commodities 126–130
explanatory models 121–122
exploratory models 122–124
extending 131–132
farm management analysis 123
farm-household behaviour 125–126
farm-household modelling procedures 124
farm-household welfare objectives 126–130
farming systems analysis and research 121, 122
fertilizer subsidy impact 127–130, 131
general-equilibrium framework 125
institutional constraints 126
integrated 118–119, 130
land degradation and economic development 120
land evaluation 123
land use alternatives 14
market constraints 126
market-clearing procedures 131
micro–macro aggregation 126
multiple-goal linear programming 123–124
optimal control models 122
policy applications 130–132
predictive models 124–126
production function analysis 121
recursive 126–130
regional development model *128*, 129, 130
social accounting matrices 121–122
structural adjustment programmes in sub-Saharan Africa 118
sustainable development 116
technical coefficient generators 122–123
technology choice 125
validation of model 126
bioeconomic village models 125
biological control 402–403
agents 65
biophysical optimization model 126–130
bioeconomic modelling 133
sustainability indicators 128
biophysical processes
bioeconomic modelling 116, 120
relationship with social processes 119
biotechnology
Doubly Green Revolution 29–30
impact on wild species/traditional crops 65
Brazil
Alternatives to Slash-and-Burn (ASB) 246
deforestation 51
land tenure 106–107
population density 51
soybean technology 106–107
see also Amazon region
Brundtland hypothesis 69, 70 70
Brundtland Report (WCED) 3, 58, 66
Burkina Faso, cotton/maize zones 378
buy-back systems 436

capital
inefficient use 368
markets in Sumatra 240
carbon
fixation 61
organic fertilizers 333
organic input 335
soil fertility depletion in Africa 327–328

carbon dioxide 25
  atmospheric emissions 327–328
  emissions 73
  sink 202
carbon sequestration 61
  agronomic sustainability 232
  Amazon region 260
  objectives 455
  soil fertility replenishment 341–342
carbon stocks
  agricultural intensification 203, 258–259
  agroecosystem range 205, *206*
  agroforestry 211, 214, 236
  Central African rainforest 202–205, *206*, 207
  cocoa growing 203–204, 214
  crop–fallow rotational systems 202–203
  cropping systems 209, *210*
  policy-led extensification 204–205, *206*, 207
  Sumatra 226
  time-averaged 218
Caring for the Earth (IUCN, UNEP, WWF) 3–4, 9, 10
cash crops
  deforestation 103
  production in Kenya 53
cattle production
  Kenya 352–353
  wheat growing area 352
Central African rain-forest
  agricultural intensification 199–200, 201–202, 213–216
  carbon stocks 202–205, *206*, 207
  cropping system 200–201
  environment 197–198
  food production potential 216
  market access 214
  perennial crops 214
  political resource allocation 215
  resource-use intensification 197–198
  sustainable technology systems 197–198
cereal crops
  Asia 383, 384
  intensification-induced productivity decline 385–386
  pests 391–392
  price protection 396
  profitability 396–397
  resistant varieties 392
chainsaws 107
Chayanovian model 91, 96
childbirth 18
children, malnutrition 22
China, rural population density 41–42
Cinderella species 339
cocoa growing 200–201
  agroforests 214
    biodiversity 207–208
  carbon stocks 203–204
  deforestation 103, 105
  genetic reserves 214
cocoa price decline *206*, 207
coffee/*bandarra* system 256–257
collective management of forest 288
commodity prices 69
common property 437
  forest management 292–294
communal land tenure
  agroforestry management 294–298
communal ownership 285–287, 287
  of land 289–290
community
  biodiversity concerns 447
  characteristics and development pathways 172–173
  resource loss impact 446
  sense of 446
  see also natural resource management, community-based
Comprehensive Watershed Development Project (COWDEP) 309
Congo, agricultural extensification 365
conservation
  annual practices in Honduras 178
  measures in Honduras 183, *184*, 185–186
  rural community 405, 406, 407
  traditional measures 404–405
  see also biodiversity, conservation
Consultative Group for International Agricultural Research (CGIAR) 136, 137, 197–198
contour strips 337
Convention on Biological Diversity (UNEP), Article 11 58, 68
cooperation, community 446
Cornell International Institute for Food, Agriculture and Development (CIIFAD) 434–435, 442
Costa Rica
  gross national product (GNP) 279, *280*, 281
  labour supply 272–273
  land-use analysis 268–271
  land-use policies 273–274, 281–282
  product demand 271–272, 281
  SOLUS 268–269, 274
cotton schemes in Africa 372
cotton/maize zones in Burkina Faso and Mali 378
cover crops 105
credit availability
  rural schemes 373
  Sumatra 240
crisis relief in less-favoured lands 152
crop plants/crops
  cover 105
  genetic base 64
  genetic diversity 71
  genetically engineered 30, 403
  native 402, 403–404
  perennial 459

resistance to pests/disease 64–65
wild relatives 64–65, 403
see also cash crops
crop production
development pathways 181, *182*, 183
intensification 181, 183
Latin America 44–46
crop residues
livestock waste nutrient exchange 362
return to soil 336
crop yields
trends 22, *23*, 24–25
UV radiation 26
crop–fallow rotational systems 10
carbon stocks 202–203
cropland management 299
crop–livestock integration development 362, 363
crop–livestock interactions 356–357
industrial livestock production 360
crop-marketing monopolies 462
cropping systems 13–14
Central African rainforests 208–209, *210*, 211
crop-tree systems, diversified 460
cultivated species, genetic diversity 64–65
currency devaluation 373

Decision Support System for Agrotechnology Transfer (DSSAT) 148
decision-support system 138–139
deforestation 3, 7, 14, 285–287
agricultural intensification 8–9, 260
agroforestry effects 342–343
biodiversity loss 60
Brazil 51
carbon dioxide emissions 73
cash crops 103
cocoa growing 103, 105
Ecuador 50–51
endogenous output price model 100–101, 113–114
endogenous wage model 97–100, 111–112
exports 103
global environmental consequences 226
Honduras 179
immiseration 431
Indonesia 222
labour-intensive technologies 95, 99–100
land tenure 289–292
Latin America 44, 47–49, 54, 107
market access 192
market assumptions 96–97
Nepal 291
patterns 66–67
poverty 66–67
resource degradation 35–36
rubber growing 103–104
soil fertility depletion in Africa 328
state ownership 292
subsidies 67
tax incentives 67
technological change 89–91, 101–103
tradeoffs in Sumatra 236–237
degradation slowing with agricultural intensification 248–249
demographic change 39, *40*
desertification, biodiversity loss 60–61
developing countries
development theory 418
discrimination against agriculture 79–80
food imports 21
genetic engineering 30
meat production 347–348
development aid organizations 5
development pathways 13, 171–172, 457
agriculture 179, 181, *182*, 183, *184*, 185–186
characteristics 176, *177*, 178–179
classification 176
community characteristics 172–173
conceptual framework 172–175
conservation measures 183, *184*, 185–186
crop production 181, *182*, 183
determinants 179, *180*
Honduras 174–175, 176, *177*, 178–179
market access 192
market opportunities 173–174
natural resources
conditions *187*, 188, *189*
management 172–173, 179, 181, *182*, 183, *184*, 185–186, 190–191
organic inputs 183, *184*, 185–186
outcomes 178–179, 186, *187*, 188, *189*, 190
policy making 192–193
population growth 173
population pressure 191
poverty *187*, 188, *189*, 190
productivity 186, *187*, 188, 190
public policy 173
technology access 192
technology development 173–174
development strategies 1–2
decentralized 1–2
growth-promoting 1
development trilogy 226
disaggregation 136
discount rate income 67–68
disease 18
cereal crops in Asia 391–392
diversification in less-favoured lands 164
domestic plants, wild relatives 402, 403–404
domestic resource cost ratio (DRC) 209, *210*
Dominican Republic, community-based natural resource management 442–443, 445
Doubly Green Revolution 12, 17–20, 27–29, 455, 462–463
biotechnology 29–30
ecology 30–31
participation 31–32

ecological disruption 63
ecological services
  biodiversity 59
  disruption 63
  protected areas 399–402
  value 400–401
econometric models 20–21
economic change 39, *40*
economic development 3
  complementarities 451, 453
  synergy with environmental conservation 452
economic goals, tradeoffs 453–454
economic growth
  complementary agricultural production and environmental goals 458
  environmental Kuznets curve hypothesis 6
  problems 452–453
economic incentives for biodiversity 421, 423
economic liberalization, environmental effects 68–69
economic model, tradeoffs 149
economic surplus, technology improvement 275–276
economic sustainability 2
economics and agricultural intensification 435–436
ecoregional analysis 136–137, 460–461
  bioeconomic modelling 117–119
  tradeoffs assessment 141
ecoregional development
  bioeconomic modelling 131
  sub-Saharan Africa 118
ecoregional planning in protected areas 428–430
ecosystem
  alteration of structure/function 65
  disruption minimizing 261
  endangered 439
  maintenance 449
  managed and biodiversity loss 57–59
  protection of vulnerable 439
  resilience protection 64
  services for water/watershed protection 401
Ecuador
  deforestation 50–51
  pesticide contamination 143
  potato- and milk-based farming 137–138
effective protection coefficient 209, *210*
employment generation 1
  technology improvement 275–276
employment opportunities in Sumatra 228
endogenous prices for non-tradable commodities 126–130
environment 25–27
  agricultural intensification 4–5, 7–8, 11, 202–205, *206*, 207–209, *210*, 211
  Central African rainforest 197–198
  intensification consequences 386–392
  intensification in Sumatra 236–237, 240–241
  livestock waste 350
  policies for livestock intensification 361
  technology-driven intensification 7–9
  tradeoffs
    with farmers' concerns in Amazon region 258–259
    with intensification in Sumatra 236–237
environmental benefits of sustainable systems 459
environmental conservation synergy with economic development 452
environmental costs of agricultural intensification 453
environmental degradation
  agricultural intensification 7–8, 25
  India 162–163
  poverty 58
environmental goals 454
  complementarities 451, 453
  complementary agricultural production and economic growth 458
  tradeoffs 453–454
environmental impact 1
  livestock intensification 355
  tradeoffs model 149
environmental indicators with technology improvement 275–276
environmental Kuznets curve (EKC) 5–6, 58, 66
  global synergies 452
environmental outcomes of agricultural intensification 456
environmental protection 3
environmental sustainability 1, 2
equitable agricultural growth and regional development 151–152
equity 1
  biases in Sumatra 241
exchange rate overvaluation 396
exports
  deforestation 103
  volume 69
extensification 73
  Congo 365
  decisions 76–77
  farm-household decision-making 75, 76–77, 82
  labour allocation 76–77
  Latin America 54
  policy-led and carbon stocks 204–205, *206*, 207
  unsustainable 378–381
extensive agriculture
  endogenous wage model 98, 99
  technological change 102

family size and food security 28
farm investment and financial markets 373

farm management analysis 123
farm-household behaviour 125–126
farm-household decision-making 12
  extensification 75, 76–77, 82
  intensification decisions 75, 77–81, 82
  labour endowment 74–75
  labour market 75
  land allocation 91
  model 74–75, 85–87
  poverty 80–81
farm-household income 129, 130
farm-household intensification 73–74
farm-household modelling procedures 124
farm-household welfare objectives 126–130
farming systems analysis and research 121, 122
farm-level decision models 142–143
farm–non-farm growth linkages 452
fertility, human
  China 41–42
  Latin America 42
  population pressure response 37, *40*
fertilizers
  chemical 55
  Latin America 49, *50*
  nitrogen 25
  policy 395–396
  price 8
  profitability 376
  provision facilitation 337
  soil fertility depletion in Africa 332
  subsidies 395
  subsidy impact on bioeconomic model 127–130, 131
  sustainable agricultural intensification 371–372
  unbalanced application 390
  use in sub-Saharan Africa 199
finance, stimulation of rural 373
financial institutions, state subsidization 373
financial markets, farm investment 373
financial services 371–373
food
  demand calculation 20–21
  equitable access 12, 17
  imports to developing countries 21, *23*
  intensive systems in Asia 383–386
  needs of poor 28
  self-sufficiency in Asia 384
  supplies 11–12
  unmet needs 433
food gap, hidden 22, *23*
food production
  increase 434
  sustainable 29
  world forecasts 20–21
food security 28, 451, 453, 454
  agricultural intensification 257–258
  crisis in Africa 325
  households 229
  Sumatra 235
forest
  clearance incentives 68
  collective management 288
  common property 287
  communal ownership 285–287
  conservation 278–279, *280*
    benefits 62
  conversion in Sumatra 222
  cost of resource protection 293–294
  global cover 35
  joint ownership 287–287
  land returned in US 104
  Latin America 43, 44, *45*
  low-level extraction 258
  management of natural 282
  margin areas 462
  open access 287–288
  redundancy in system 63
  shifting cultivation halting 434
  state ownership 287
  strategies 261–262
  subsidy 278–279
  see also deforestation; reforestation
forest management 14
  community-based 237–238
Forest Margins Benchmark area 197–198
forest products
  Nepal 292
  non-timber 77
  see also non-timber forest products (NTFP)
forestry, managed 258
fuel shortage in India 162–163
fuelwood trees 340

gene sanctuary creation 403
genetic diversity of crops 71
genetic engineering of crops 30, 403
genetically modified organisms 65
geographical information systems (GIS) 137, 138, 139, 148, 150
  natural resource management 441
Ghana
  agroforestry 297, *298*
  communal land tenure 296
  community-based natural resource management 444–445
  cropland management 299
  land tenure 287, 289, 290, *297*
global warming 25–27
globalization 19–20
governance in Africa 326
grain production, IFPRI model *21*
granaries with revolving funds 436
grazing systems 349–350
  livestock intensification 361
Green Revolution 25
  irrigation 393
  rice 384
  second 29
  success 452
  see also Doubly Green Revolution
greenhouse gases 25, *26*

greenhouse gases *continued*
  livestock production 350
  soil fertility depletion in Africa 327
gross domestic product (GDP) 360
  agricultural 80
gross national product (GNP)
  Costa Rica 279, *280*, 281
groundnut growing in Senegal 380–381
groundwater
  depletion 387–388
  India 162, 163
  ineffective management 388
  intensification consequences 386, 387–388
  management reform 394
  resource management 388
gum arabic tree 62–63

habitat loss 60–61
hardpan formation 391
hedgerow barriers 337
high-potential areas for investment 152
high-value crops 339–340
high-yielding varieties (HYVs)
  food production increase 434
  India 158, 159, 161–162
Honduras
  annual conservation practice 178
  common property resources 178
  conservation measures 183, *184*, 185–186
  crop production 178
  deforestation 179
  development pathways 174–175
  human welfare 190–191
  land use change 176, 178
  livestock 178
  natural resource
    conditions *187*, 188, *189*
    management 173–174, 186
  organic inputs 183, *184*, 185–186
  population density 186, *187*, 188
  poverty 179
  productivity outcome 186, *187*, 188
  technical assistance programmes 192–193
household
  age of head 290
  food security 229
    Sumatra 235
  labour constraints 229
  multiple units 52
human needs 3
human population density
  driving agricultural intensification 356
  livestock intensification 354–355
human welfare in Honduras 190–191
hunger 17–20
  poverty 18–19

immigration, ICDPs 413
immiseration 431
income
  conditioning effect 70
  discount rates 67–68
  generation 453
  growth as environmental indicator synergy 452
  insurance against risk 71
  market failure 58
  preferences of people 67
  remittance 374
  see also wages
India
  agricultural growth *154–155*, 156–157
    spillover effects 162
  environmental degradation 162–163
  fuel shortage 162–163
  gene sanctuary creation 403
  groundwater 162, 163
  HYVs 161–162
  labour productivity *154–155*, 156
  land classification 153
  land productivity *154–155*, 156
  landless buffalo milk producers 362
  marginal returns 161–163
  overgrazing 162–163
  poverty functions 160
  production function estimation 160
  rain-fed areas 161–162
  resource degradation 162–163
  rural infrastructure investment 158–159
  rural poverty 157–158
  soil conservation 163
  soil erosion 162
  total factor productivity *154–155*, 156–157
  water conservation 163
  watershed development in semi-arid tropics 303–307
indigenous species deletion/genetic alteration 65
Indonesia
  community-based natural resource management 443
  deforestation 222
  local communities in protected areas management 408–409
information availability 67
infrastructure 457
  Africa 326
  investment in Mali 376–377
inheritance system 295–296
institutional constraints in bioeconomic modelling 126
institutional development in Central African region 201
institutions 457–458
insurance, agricultural for less-favoured lands 166–167

integrated conservation and development projects (ICDPs) 2, 10, 11, 412–414, 415–416
  agricultural development promotion 427
  approach critique 419–421, *422*, 423–424
  assumptions *422*, 453
  biodiversity conservation implementation 421
  biodiversity loss stopping 417
  community involvement 439
  design 413–414
  desired outcomes 423
  development theory 418
  economic and environmental factor synergies 452
  economic burden reduction 419
  economic development promotion 419
  economic incentives 421, 423–424
  establishment time 420
  exogenous factors 423–424
  flawed assumptions 421
  goals 15
  implementation problems 420, 424
  incentives 421, 423
  investment 420
  limitations 439
  linkage between incentives and conservation objective 421, 423
  natural resource preservation 440–441
  policy-level components 420
  project-level components 420
  social development promotion 419
  standardization lack 420
  strategies 418–419
  underlying philosophy 418
integrated pest management (IPM) 31, 150
  Asia 396
  technologies 147–148
integrated poverty index (IPI) 66
integrated systems modelling, ecoregional research 138
intensification hypothesis 51
intensive agriculture
  endogenous wage model 98, 99
  technological change 102
intensive food systems in Asia 383–386
interest rates in Sumatra 227–228
International Food Policy Research Institute (IFPRI) model 20–21
  forecast 28
International Fund for Agricultural Development (IFAD), integrated poverty index (IPI) 66
International Panel on Climate Change 27
International Potato Center (CIP) 136
International Program on Rice Biotechnology 29–30
interventionism 382
intervillage cooperation in protected areas 407
investment
  capture by non-poor 431
  financial markets 373
  high-potential areas 152
  infrastructure *376–377*
  less-favoured lands 152–153
  marginal impacts 161–162
  model for public 158–159
  sustainable agricultural intensification 381
iron
  deficiency 18
  toxicity 390–391
irrigation
  agricultural intensification 79
  communal systems 291
  desertification 61
  Green Revolution 393
  Latin America 49, *50*
  Near East 41
  policy in Asia 393–395
  salinity 387
  tube-well 388
  water rights 321
Irvingia gabonensis 339

Japan, common property forests 292, 293, *294*
joint ownership of land 287–287

Kenya
  cash crop production 53
  cattle production 352–353
  dairy cattle on maize/coffee/tea farms 353
  livestock intensification 358
  population growth 52–53
Kuznets relations 66
  see also environmental Kuznets curve (EKC)

labour
  agricultural intensification 368–369
  allocation
    extensification 76–77
    intensification 77–81
  availability in Amazon region 251, 260
  community voluntary 309
  endowment in farm household decision-making 74–75
  family 96
  hired 96, 97
  household constraints 229
  intensity and technological change 95
  non-farm work 368–369
  oil-palm agroforests in Sumatra 235
  productivity 39, *40*
    in India *154–155*, 156
  requirement for livestock production industrial systems 359–360
  requirements for agricultural intensification 256–257
  returns in technology systems 209, *210*

labour *continued*
supply in Costa Rica 272–273
sustainable agricultural intensification 373–374
use in bioeconomic land-use models in Costa Rica 281
see also shramdan
labour market
conditions 457
farm household decision-making 75
Sumatra 239–240
technological change 93
labour-enhancing technologies 366–367
labour-intensive technologies 104–106
deforestation 95
labour-saving technology 106–107
land
allocation in farm household decision-making 91
classification in India 153
degradation 4
distribution in protected areas 430–431
evaluation 123
and farming systems analysis (LEFSA) sequence 268–269
intensification and agrarian society evolution 37
policy and tenure security 374
productivity in India *154–155*, 156
values 431
land rights, individual 290, 295, 296–297
land tenure 14, 285–287, 457
Brazil 106–107
communal for agroforestry management 294–298
cropland management 299
deforestation 289–292
on-farm tree planting 343
privatization 288
security 68, 374
study sites 287–289
land–labour ratio 9
landscape approach 460–461
land-sparing 7
land-use
analysis
Costa Rica 268–271
LEFSA 268–269
bioeconomic modelling 14, 116
change and global environmental consequences 226
decisions and technological change 93
individual rights 290, 295, 296–297
intensification and environmental degradation 343
policies in Costa Rica 282–283
population growth 36–38
studies and bioeconomic modelling 119–120
Sumatra 223, *224–225*
technology improvement 275–276
trajectory in Amazon region 247, *248*
trends 12
land-use alternatives
bioeconomic models 14
matrix *230*, 231–232
Sumatra 237–241, 242
land-use systems in Amazon region
evaluation 252, *253*, 254–259
traditional 255
Last Stand 438
Latin America
agricultural intensification 49, 50–53
crop production 44–46
deforestation 44, 47–49, 54, 107
extensification 54
fertility 42
fertilizers 49, *50*
forest 43, 44, *45*
loss 47–49
irrigation 49, *50*
migration 48, 54
pasture 44, 45–47, 48
population density 43, 48–49, 54
rural 42
urbanization 43
water use 49, *50*
legume fallows 334
tree 336
less-favoured lands 151–152
agricultural growth potential 163–164
agricultural insurance 166–167
agricultural intensification 164
high returns from investments 434
migration 164
non-farm diversification 164
organization of farmers 165–166
policy environment 167–168
poverty 152–153
public institutions 168
resource management 164–166
returns to public investments 152–153
risk management 166–167
rural infrastructure investment 167
strategies for development 163–168
women farmers 164
livestock
diversity 63
herders in watershed development, India 321
ruminant 351
livestock production
crop–livestock interactions 356–358
demand increase 348
environment 348–350
feed availability 356
global systems 346–348
grazing systems 346
greenhouse gases 350
industrial systems 346, 348, 350
labour requirement 359–360
industrialization constraints 359–360
mixed farming systems 347–348

Nepal 352
resource costs 348
soil fertility 351–353
specialization 357
waste nutrients 362
livestock production intensification 14, 345, 346, 355–360
environmental effects 349–350, 355
fodder 359
grazing systems 361
increasing 354–355
Kenya 358
labour-abundance 360
land-abundance 360
local patterns 358–359
market integration 362, 363
markets 361–363
measures 356–358
policy 361–363
price changes 359
principal components analysis 358
principles 356–358
regional patterns 358–359
livestock–crop production integration 351
livestock–environment interactions, production intensification 345
local institutional development 2
logging
social cooperation 241
Sumatra 233, 234
low-input sustainable agriculture (LISA) 367–368

macroeconomic policies in Asia 396
macronutrients, soil deficiency 390
Madagascar
community-based natural resource management 444
intensification need 434–435
rice growing 379
rice yields 435
maize
legume fallow fertilization 334
yields 24
Zambia 379
Zimbabwe 379
Malawi
agroforestry 298
common property forests 292, 293
cropland management 299
land tenure 287–288, 289, 290–291
Mali 376–377
cotton/maize zones 378
malnutrition 17–20
children 22
hidden food gap 22
Malthus, Thomas 36
Man and the Biosphere Programme (UNESCO) 3
mango, bush 339
marginal impacts, investment 161–162
marginal returns in India 161–163
market(s)
access 457
Central African rainforest 214
development pathways 192
Amazon region 262–263
assumptions in technological change 96–97
constraints in bioeconomic modelling 126
failure 70–71, 461–462
income 58
gap 22
imperfect 369–370
integration for livestock intensification 362, 363
liberalization 369, 371, 378, 382
livestock intensification 361–363
oil-palm agroforestry 238–240
opportunities in development pathways 173–174
reforms 462
rubber agroforestry 238–240
water 394
rights for irrigation 394
watershed management 401
weak 370
market-clearing procedures 131
matrilineal inheritance, uterine 296
mean neurobehavioural score 145
meat consumption, global 347
meat production in mixed farming systems 347
megadiversity countries 428, 429
meta land-use practices in Amazon region 252, 259
methane 25, 350
Mexico, development to protect park 430
micro–macro aggregation 126
micronutrients, soil deficiencies 390–391
migration
Latin America 48, 54
less-favoured lands 164
see also out-migration
millet yields 24
Ministry of Rural Development (MORD; India) 307, 308, 311
miombo woodland 290–291, 293, 328
mixed farming systems 347–348
molecular markers 30

native ecosystem loss 8
natural resource management
agriculture 179, 181, *182*, 183, *184*, 185–186
collective investment 186
community 437–438
community-based 2, 433–436, 437–438, 441, 442–445
social capital 445
complementary resources 448

natural resource management *continued*
- conditions *187*, 188, *189*
- degradation 463
- development pathways 172–173, 190–191
- economic/environmental value tradeoffs 448
- geographical information systems (GIS) 441
- Honduras 173–174, 186
- incentives 14
- institutional options 436–441
- institutions involved 439, *440*
- local resident assessment 441
- mixed system intensification 361
- motivations for community management 445–447
- participatory rural appraisal (PRA) 441
- preservation of endangered 440–441

natural resources
- deterioration 116
- investment with soil fertility replenishment 338

Near East, irrigation 41

Nepal
- common property forests 292–293, 294
- forest products 292
- hydroelectric dam 401
- land tenure 288, 291
- livestock production 352
- local communities in protected areas management 408

net present value (NPV) 208, 209, 226–227
- establishment costs of agroforestry 237–238
- fertility replenishment *338*

Nigeria, population growth 52

nitrates 25

nitrogen
- livestock excretion 351
- mineral fertilizer 334
- organic input 335
- recycling 353
- soil fertility replenishment 333–336, 338
- soil supply dynamics 389–390
- surplus from livestock production 350
- waterlogged soil 387

nitrous oxide 25

non-governmental organizations (NGOs)
- participatory approach 32
- policy impact 374–375
- watershed development, India 321, 322–323

non-timber forest products (NTFP) 233
- social cooperation 241
- Sumatra 240

nutrients
- capital stocks 328, *329*
- cycling 328, *329*
- depletion process 331
- inputs 328, *329*
- recycling 351, 399
- soil fertility depletion in Africa 328, *329*, 330–332

oil exploitation, Amazon region 50–51

oil-palm
- agroforestry markets 238–240
- Sumatra 234

onion growing, Mali 376–377

open access lands 287–288, 289

open-economy farm model of technological change 94–95, 110

open-economy model 94–95

optimal control models 122

organic farming
- carbon provision 333
- soil fertility depletion in Africa 332–333

organic inputs, Honduras 183, *184*, 185–186

Other Asia region, rural population density 42

Our Common Future (WCED) 6

out-migration 37
- China 41–42
- population growth 52
- rural areas 39, *40*, 41

output increase 248–249

overgrazing, India 162–163

ozone 26

parastatals 372, 374, 462

parks *see* protected areas

Parks in Peril 424–427

Parks in Peril (PiP) programme (Nature Conservancy) 425

participatory development 2, 31–32

participatory rural appraisal (PRA) 32
- natural resource management 441

pasture
- improved in Amazon region 256–257
- Latin America 44, 45–47, 48

People and Parks study 418–420

perfect-markets model of technological change 94–95, 110

Peru
- pesticide contamination 143
- potato- and milk-based farming 137–138

pest build-up 386

pesticides 25
- health simulation 144
- human health risk 146–147
- leaching 143, 144–145
- mean neurobehavioural score 145
- modern cereal varieties 392
- policy 395–396
- production value 146–147
- resistance 403
- subsidies 395–396
- tradeoffs curves 145–148
- use in Honduras 181

pest–predator balance 392

Phalaris minor 392

phosphate, Minjingu rock 336
phosphorus
  loss in Africa 331
  mineral fertilizers 335
  organic input 335–336
  rice–wheat system requirements 390
  soil fertility replenishment 333, 335, 336, 338
physical model of tradeoffs model 149
plant biodiversity in Amazon region 260
plantain crops, Central African rainforests 205
policy
  agricultural intensification 430–431
  analysis 115
    matrix 208–209, *210*, 211, *227*
  appraisal in decision-support systems 125
  bias against agriculture 115–116
  Central African region 201
  changes 457
  complementary outcomes 463
  distortions 461–462
  environment in less-favoured lands 167–168
  extractive use of land 408–409
  fertilizers 395–396
  guidelines for local community involvement in protected area management *410*
  intensification 392–397
    obstacles 263
  land 374
  livestock intensification 361–363
  making for development pathways 192–193
  NGOs 374–375
  pesticides 395–396
  price 396–397
  reforms 370–371, 431
    in Mali 376–377
  resource-use intensification *204–205*, *206*, 207
  scaling problems 461
  sectoral level 370–374
  soil fertility depletion in Africa 330
  sustaining intensified agriculture 397
  trade 396–397
  wages 373–374
  see also public policy
political resource allocation 215
pollination 399
pollution
  global 25–27
  intensification consequences 386
  water 8, 25
polygamy 52
population of Amazon region 50–51
population density
  Brazil 51
  Honduras 186, *187*, 188
  Latin America 43, 48–49, 54
population growth 35, 36
  agricultural intensification 51–52
  agricultural productivity 36
  Amazon region 50–51
  development pathways 173
  Kenya 52–53
  land-use responses 36–38
  multiphasic response 52–53
  Nigeria 52
  out-migration 52
  responses 37–38
  rural 39, *40*
  technological innovation 36–37
population pressures 12, 456
  demographic response 37, *40*
  development pathways 191
potassium
  fertilizers 336
  loss in Africa 331
  organic input 335
  recycling maximization 336
  rice–wheat system requirements 390
potato growing 144
poverty
  Africa 325
  alleviation 454
  asset distribution 431
  biodiversity conservation 431
  deforestation 66–67
  environmental degradation 58
  farm-household decision-making 80–81
  fertility relationship 67
  functions 160
  Honduras 179
  hunger 18–19
  information availability 67
  investment in less-favoured lands 152–153
  less-favoured lands 152–153
  market imperfections 370
  outcomes of agricultural intensification 456
  reduction 1
  rural in India 157–159
  spillover effects of agricultural growth 162
price
  fluctuations 435–436
  policy 396–397
principal components analysis
  commercialization/intensification 219
  livestock intensification 358
  resource-use intensification 211–213
product demand in Costa Rica 271–*272*, 281
production
  analysis 143
  ecology models 123
  function
    analysis 121
    estimation in India 160
  growth in rural infrastructure 158–159
  intensification
    livestock–environment interactions 345

production *continued*
livestock–people link 354–355
raising in better-endowed areas 434
productivity
decline with agricultural intensification 15
development pathways 190
slow-down in Asia 384–386
profitability centrality 376–378
property rights 285–287, 457
assignment 70
less-favoured lands 165–166
Sumatra 241
water management 393–394
protected areas
agricultural intensification 427–431
benefit–cost ratios 411–412
biodiversity 402–404
conservation 425, 428
buffer zones 406–407, 408–409, 413, 427
community involvement in management 408–412
costs to surrounding communities 411–412
development to protect 429–430
ecoregional planning 428–430
ecosystem services 399–402
effective management 405–406
exploitation 410–411
farmers 404–408
genetic resources 403
growth poles 429–430
ICDPs 412–414, 415–416
improved management 425–426
intervillage cooperation 407
land distribution 430–431
local community involvement in management 414–415
major threats 406
management objectives 426
management strategies 415
policy changes for ICDP 414
resource management agencies 407–408
rigorous maintenance 438
shifting cultivation halting 434
social change 426, 428
successful projects 415
sustainable livelihood potential 428
watershed protection 401–402
zoning 426
protein biological value 348
Prunus africana 339
public institutions in less-favoured lands 168
public policy
development pathways 173
less-favoured lands 167–168

rainfall 27
ranch development 67
reforestation 7
community-based natural resource management 443
regional development
equitable agricultural growth 151–152
model *128*, 129, 130
sustainable agricultural growth 151–152
regional economic and agricultural land-use model (REALM) 269, 271
remittance income 374
resource degradation in India 162–163
resource management
less-favoured lands 164–166
sustainable in Asia 384
see also natural resource management
resource-use intensification 197–198
economic growth/environment tradeoffs 208–209, *210*, 211
policy *204–205*, *206*, 207
principal components analysis 211–213
technology targeting 211–213
returns to public investments in less-favoured lands 152–153
rice/rice growing
biotechnology 29–30
Green Revolution 384
high-yielding varieties 104
Madagascar 379
Mali 376–377
monoculture 384–386, 389–391, 459
nitrogen requirement obtaining 389
pests 31
yields 24
decline 385
increase 435
rice–wheat continuous cropping systems 385–386
disease control 391
soil nutrients 389–391
risk
aversion 68, 70, 94
management in less-favoured lands 166–167
roads 457
Rock Coral Canyon Reserve (Arizona, USA) 402
rubber
agroforestry markets 237–240
growing and deforestation 103–104
rubber agroforests
productivity 236–237
smallholders 237
Sumatra 234–235
rural community
conservation 405
framing system revision 449
protected areas 405, 408–412
rural infrastructure investment
India 158–159
less-favoured lands 167
rural population density
China 41–42

Latin America 42
Other Asia region 42
Rwanda, agroindustrial development 377–378

safe handling practices 147–148
salinity/salinization 8, 462
crop yields 24
desertification 61
intensification consequences 386–387
savannah rangelands
biodiversity 63
redundancy in system 63
scale of analysis 136
Sclerocarya birrea 339
Senegal groundnut growing 380–381
Sesbanica sesban 334
settlement schemes in Sumatra 221–222
shramdan 309, 313–316
single-family ownership 295
slash-and-burn agriculture 10
alternatives 13–14
see also Alternatives to Slash-and-Burn (ASB)
smallholders
capital market constraints 240
expansion into environmentally fragile margins 451
household savings 240
marketing restrictions 239
production incentives 229, 233–235
rubber agroforests 237
rubber growing 234–235
socioeconomic criteria 228–229
technological innovation adoption 230
tree-crop systems 236–237
weak markets 370
social accounting matrices 121–122
social capital 445
social cooperation, Sumatra 241
social profitability
cropping systems 209, *210*
Sumatra 227
socioeconomic constraints 330
socioeconomic processes
bioeconomic modelling 116
modelling 125–126
relationship with biophysical processes 119
soil
alkalinization 61
degradation assessment 123
hardpan formation 391
macronutrient deficiency 390
nitrogen supply 389–390
nutrient depletion 326–327
intensification consequences 386
nutrients and Asian cereal crops 389–391
organic matter deficit in sub-Saharan Africa 118
physical characteristic change 391
regeneration and ecosystem services from protected areas 399
toxin accumulation 387, 390–391
soil conservation
India 163
structures 337
watershed development, India 312, 313–314, 316–317, *318–319*
soil erosion 61
conservation structures 337
India 162
soil fertility replenishment 341
sub-Saharan Africa 118
soil fertility
declining 14
livestock 351–353
soil fertility depletion in Africa 326–328, *329*, 330–333
environmental factors 327–328
fertilizer use 332
food security 331
intensification/diversification with high-value products 339–340
nutrient capital stocks 328, *329*
nutrient depletion process 330–332
on-farm effects 331–332
organic farming 332–333
replenishment 333–338
responses 332–333
rural area priority 330
socioeconomic factors 326–327
traditional practice breakdown 330
yield decline 331
soil fertility replenishment 333–338
biodiversity protection 342–343
carbon sequestration 341–342
cost sharing mechanisms 338
environmental implications 340–343
natural resource capital investment 338
soil erosion 341
sorghum yields 24
soybean technology 106–107
species redundancy 63
Sri Lanka
community-based natural resource management 443–444
livestock intensification *357*, 358
state ownership of land 287, 291
deforestation 292
structural adjustment
package 69
programmes in sub-Saharan Africa 118
subsidies
deforestation 67
fertilizers 395
forest 278–279
pesticides 395–396
subsistence model of technological change
labour market 93
land-use decisions 93
risk aversion 94

Sumatra
  agronomic sustainability 226
  Alternatives to Slash-and-Burn (ASB) matrix *230*, 231–235
  barriers to land-use alternatives 237–241
  biodiversity 243
  capital markets 240
  carbon stocks 226
  communal land tenure 295, *296*
  credit availability 240
  employment opportunities 228
  environmental impact of intensification 240–241
  equity biases 241
  forest conversion 222
  global environmental criteria 226
  household food security 235
  interest rates 227–228
  labour markets 239–240
  land tenure 287, 289, 290
  land-use 223, *224–225*
    alternatives 242
    interest groups 222
    systems 227
  logging 233, 234
  non-timber forest products (NTFP) 233, 240
  nucleus estate/smallholder (NES) scheme 234
  oil-palm 234
  policy-makers' criteria 226–228
  population density 221
  property rights 241
  rubber agroforests 234–235
  settlement schemes 221–222
  smallholders' socioeconomic criteria 228–229
  social cooperation 241
  social profitability 227
  technical information availability 240
  transmigration 221–222
supply-and-demand interaction 435
sustainability 1
  Amazon region 49
  indicators for biophysical optimization model 128
  multiple goals 462
  operational 136–137
  tradeoffs 140
sustainable agricultural intensification 365–366
  banana agroindustry in Rwanda 377–378
  capital-led 372–373, 381
  concept 366–369
  cotton/maize zones in Africa 378
  farmers' incentives 371–374
  fertilizers 371–372
  financial services 371–372
    policy 372–373
  incentives 370–371, 381
  labour 373–374
  policy instruments 115–116
  rice/onion double cropping in Mali 376–377
  sectoral-level policies 370–374
  vertical coordination 373
  wages policy 373–374
sustainable development 433
  agrotechnical solutions 116
  complementarity of objectives 452
  policies 2–4
sustainable options for land use (SOLUS) 268–269, 274
sustainable systems
  characteristics 459–460
  technology for Central African rainforest 197–198

Tanzania, dairy production 358–359
tax incentives for deforestation 67
taxation, biocides 276–278, 282
technical assistance programmes in Honduras 192–193
technical coefficient generators 122–123
technological change 456–457
  deforestation 89–91
  elastic/inelastic product demand and labour supply 103–104
  exogenous 90, 98
  extensive sector 102
  farm level 91–97
  income effect 96–97
  intensive sector 102
  labour intensity 95, 101
  labour market 93
  labour-intensive 92, 99, 104–106
  labour-saving 92–93
    technology 106–107
  land-use decisions 93
  macro-level 97–101
    endogenous output price model 100–101, 113–114
    endogenous wage model 97–100, 111–112
  market assumptions 96–97
  modelling 91–93
  perfect-markets model 94–95
  pure yield-increasing 92, 99
  recursive models 94
  risk aversion 94
  subsistence model 93–94, 101–103
  sustainability 463
technological innovation
  adoption 10
  bioeconomic land-use models 282
  induced 5
  land-sparing 9
  population growth 36–37
  women 337
technology
  access for development pathways 192
  choice in bioeconomic modelling 116, 125

development for development pathways 173–174
improvement for bioeconomic land-use models in Costa Rica *275–276*
intensification obstacles 263–264
targeting and resource-use intensification 211–213
technology-driven intensification 7–9
temperature change, global warming 26–27
timber harvesting rates 68
time preference 67
Tithonia diversifolia 335–336
total factor productivity in India *154–155*, 156–157
trade
policy 396–397
restrictions/tariffs in Asia 396
tradeoffs assessment 139, 141–143
tradeoffs curves 145–148
tradeoffs methodology 135
ecoregional approach 137
tradeoffs model 138–143
aggregate functions 143
alternative policies 147–148
decision-support system 138–139
disciplinary integration 141–142
dynamic feedback from environmental conditions to production 149–150
economic model 149
environmental impact 149
GIS linkages 148
human health risk 146–147
integration 148–150
pesticides 143–148
physical model 149
policy analysis 141–142
production value 146–147
purpose 138–139
research chain 139
scales 140–141
technological change 147–148
tradeoffs curves 145–148
units of analysis 140–141
transmigration in Sumatra 221–222
tree crops
fuelwood 340
newly domesticated 339–340
on-farm planting 343
timber 68
tropical forest
carbon dioxide sink 202
conservation value 62
tube-wells 388

Uganda
agroforestry 298
cropland management 299
land tenure 288, 289
ultraviolet radiation 26
undernutrition 17–20
unsustainable agricultural intensification 365, 378–381
capital-deficient 381
concept 366–369
groundnut growing in Senegal 380–381
maize growing in Zambia/Zimbabwe 379
rice growing in Madagascar 379
urbanization 452
Latin America 43
US Agency for International Development 5
user-group management of land 291–292

vegetation loss/alteration 61
vegetative propagation 215
Vietnam land tenure 288, 291
vitamin A deficiency 18

wages
increases in bioeconomic land-use models in Costa Rica *279*, *280*, 281, 282
policy for sustainable agricultural intensification 373–374
water
conservation in India 163
ecosystem services 401
markets 394
policy in Asia 393–395
pricing policies 461–462
provision and ecosystem services from protected areas 399
reuse 394
see also groundwater
water allocation
efficient 394–395
incentive-based 394
water management 82
property rights 393–394
water pollution 8
Green Revolution 25
water rights for irrigation 321
markets 394
secure 393–394
water supply
conservation 446
needs 433–434
water use 461–462
Latin America 49, *50*
waterlogging 8
crop yields 24
intensification consequences 386–387
watershed development, India 303–307
analytical approach 307–309
collaborative projects 323
constraints 305–306
data 309–310
dependent variables 311–312
early projects 306
econometric analysis 310–311, 314–317, *318–320*, 321
equity impact 321–322

watershed development, India *continued*
- explanatory variables 312–314
- landless people 306, 321–322, 324
- livestock herders 321–322
- local participation 306–307
- NGO development 307, 308–309, 311, 321, 322–323
- project category determinants 314–316
- projects 307–309
- qualitative data 321–322
- return to cultivation 312, 314, 317, 320–321
- selection criteria for villages 308–309, 314–316
- *shramdan* 309, 313–316
- soil conservation 305, 312, 313–314, 316–317, *318–319*
- variables 311–314
- villages 314–316
- water conservation 305

watershed management 401

watershed protection
- ecosystem services 401
- protected areas 401–402

weather, extreme conditions 27

weed management 392
- cereal crops in Asia 391–392

weediness 30

wheat
- disease resistance 391–392
- nitrogen requirement obtaining 389
- productivity in Asia 385
- yields 24

wheat-growing area cattle production 352

win–win strategies 13, 150
- intensification 453
- livestock production 352–353
- reduced water use 394

women
- farmers on less-favoured lands 164
- technological innovation 337

World Bank 5

World Development Report (World Bank) 6

world food production 73–74
- models 20–21

yield growth decline 22, *23*, 24–25

Zambia, maize 379

Zimbabwe, maize 379

zinc deficiency 390